Microalgae for Wastewater Treatment and Resource Recovery:

Pathways to a Circular Bioeconomy

Imran Ahmad
https://orcid.org/0000-0001-9774-9111
Durban University of Technology, South Africa

Faizal Bux
Durban University of Technology, South Africa

Nilamjyoti Kalita
https://orcid.org/0009-0007-9115-2980
Murkong Selek College, India

IGI Global
Scientific Publishing
Publishing Tomorrow's Research Today

Vice President of Editorial Melissa Wagner
Director of Acquisitions Mikaela Felty
Director of Book Development Jocelynn Hessler
Production Manager Mike Brehm
Cover Design Jose Rosado

Published in the United States of America by
IGI Global Scientific Publishing
701 East Chocolate Avenue
Hershey, PA, 17033, USA
Tel: 717-533-8845 | Fax: 717-533-7115
Website: https://www.igi-global.com E-mail: cust@igi-global.com

Library of Congress Cataloging-in-Publication Data

LCCN: 2026032924 (CIP Data Pending)
ISBN13: 979-8-2600-3080-6
Isbn13Softcover: 979-8-2600-3081-3
EISBN13: 979-8-2600-3082-0
British Cataloguing in Publication Data
A Cataloguing in Publication record for this book is available from the British Library.

Table of Contents

Detailed Table of Contents

The escalating pressures of water scarcity, nutrient depletion, and unsustainable energy demand have necessitated a complex water-nutrient-energy nexus for creation of a circular bioeconomy. Traditional linear economic models treat wastewater with a cycle of inefficiency and environmental degradation, which are usually energy-intensive with no emphasis on resource recovery. Thus, there is an urgent need for circular systems that valorize wastewater to sustainable bioresources. Various microalgae and cyanobacteria act as biocatalysts for wastewater remediation, carbon (C) sequestration, and biomass production. The chapter brings insight into this interplay by outlining the principles of a circular bioeconomy with respect to wastewater treatments and resource recovery techniques. Further, it highlights the techniques of algal cultivation in wastewater, and deciphers the mechanisms of nutrient and pollutant removal process. It significantly focuses on the downstream processing of the algal biomass, exploring pathways for its conversion into valuable bio-products.

Eutrophication, which is usually caused by the disposal of wastewater containing nutrients such as those present in hydroponic effluent, is known to have a negative effect on the quality and balance of water. Therefore, the optimization of the treatment of wastewater in an economical manner is necessary. Phosphorus and nitrogen are nutrients which can be extracted and recovered from wastewater with the help of bacterial and microalgal consortia. This is in accordance with the concept of the circular bio-economy, which is the production of goods from waste. The concept of the use of microalgae-bacteria consortia in the bioremediation of wastewater will be presented briefly in this chapter. The principle of symbiotic connections states that the use of microalgae-bacteria consortia is better than the use of single species in the bioremediation of wastewater. These consortia collect lipids, proteins, and carbohydrates which can be converted into biofuel and other beneficial products, aside from the removal of contaminants from wastewater.

The increasing volume of emerging contaminants in wastewater poses a major threat to the environment. This study explores the ground-breaking potential of the algal-bacterial interacting system as an eco-friendly and sustainable alternative for water purification and resource recovery. This study explains the basic mechanism of algal-bacterial interactions, including symbiosis, gene transfer, and quorum sensing, by enhancing the system efficiency and resilience. These systems demonstrate high proficiency in eliminating pollutants through biodegradation. The harvested algal-bacterial biomass is utilized as a feedstock for the circular economy by enabling the synthesis of biogas, biodiesel, bioplastic, and biofertilizers. Algal-bacterial interactions face some challenges during wastewater treatment and resource recovery. This study connects the microbiology, economics, and environmental studies for ABS as an integrated solution of transforming wastewater treatment to a resource-generating platform by aligning with global sustainability goals.

Microalgae-bacteria consortia are a new and energy-efficient system of wastewater treatment and recovery of resources. This chapter emphasizes the metabolic interactions that are synergistic between algal photosynthesis and bacterial respiration to promote carbon, nitrogen, and phosphorus cycles. Co cultures involving Chlorclla vulgaris and Pseudomonas putida were studied on a laboratory scale with controlled conditions of light and aeration in order to determine the efficiency in the removal of pollutants. The consortia clearly showed up to a 90 percent reduction in chemical oxygen demand and an improvement on ammonia and phosphate removal over monocultures. Metagenomic and metabolomic studies established a nutritional complementation and symbiotic control of growth. The results show that algal-bacterial systems are a sustainable and innovative approach to valorizing wastewater, making it cheaper in terms of energy and yielding biofertilizers and biomass that can be used to create biofuels.

Emerging contaminants refer to natural and human-made and diverse types of chemicals that are not regularly monitored within environments and are by-products of human activity. The wastewater usually has such pollutants in large amounts with industrial, agro industrial as well as agricultural sources with heavy metals, oils and dangerous compounds being discharged. Microalgae have demonstrated great potential in terms of biological assay to remove such contaminants in wastewater. This review identifies the use of microalgae to remove the new pollutants with a demonstration of the important mechanisms which include absorption and biodegradation. It also talks about the contemporary issues and outlooks of microalgae-based wastewater treatments. The review also highlights the effectiveness of the phycoremediation in enabling sustainable wastewater treatment and limiting the emergent contaminants in accordance with the UN sustainable development goals (SDGs).

K. Kesavardhini, Department of Microbiology, Sacred Heart College, Tirupattur, India
P. Saranraj, Department of Microbiology, Sacred Heart College, Tirupattur, India
S. Dayana, Department of Microbiology, Sacred Heart College, Tirupattur, India
A. Sabina Banu, Department of Microbiology, Sacred Heart College, Tirupattur, India
K. Gayathri, Department of Microbiology, Sacred Heart College, Tirupattur, India
B. Lokeshwari, Department of Microbiology, Sacred Heart College, Tirupattur, India

The aquatic environment and human health are seriously threatened by the increasing amount of Emerging Contaminants (ECs) in sewage. Because of their adaptable metabolic as well as capacity to adapt to the environment, microalgae have been given a lot of focus, as a viable and sustainable solution of eliminating these contaminants. Microalgal systems have extremely broad capabilities of eliminating a variety of ECs, including steroids, insecticides, hygiene items, medications, and heavy metals, with the mechanisms of bioaccumulation, biosorption and biodegradation, as well as photodegradation. Besides the elimination of contaminants, microalgae have been found to help in the recovery of nutrients, sequestration of carbon dioxide and biomass to be used in value-added products. Microalgae-based processes are environmentally friendly and efficient in terms of energy usage compared to the conventional treatment technologies. Nevertheless, there are process optimization issues, high scale deployment and harvesting of biomass.

Eshraq Aljamal, Irbid National University, Jordan

Microalgal adsorption offers a sustainable approach to removing emerging contaminants from wastewater systems. This chapter provides an analytical evaluation of microalgal systems, focusing on key physicochemical factors governing surface interactions, including functional group chemistry, surface charge behavior, and solution-dependent speciation. The discussion integrates equilibrium, kinetic, and thermodynamic modeling with analytical techniques such as Fourier-transform infrared spectroscopy and zeta potential analysis to support mechanistic interpretation. Distinctions between surface adsorption and intracellular bioaccumulation are clarified to improve conceptual accuracy. The chapter also examines the limitations of classical modeling approaches in complex wastewater systems and highlights the influence of dissolved organic matter and competing ions. Finally, microalgal treatment systems are discussed within sustainability frameworks, emphasizing scalability, reproducibility, and environmental integration.

B. Lokeshwari, Department of Microbiology, Sacred Heart College, Tirupattur, India
P. Saranraj, Department of Microbiology, Sacred Heart College, Tirupattur, India
Ashish Kumar Nayak, Prasad V. Potluri Siddhartha Institute of Technology, Vijayawada, India
Alexander Machado Cardoso, Rio de Janeiro State University, Brazil
K. Kesavardhini, Department of Microbiology, Sacred Heart College, Tirupattur, India
K. Gayathri, Department of Microbiology, Sacred Heart College, Tirupattur, India

The rapid industrialization and urbanization cause the release of wastewater that is contaminated with heavy metals, dyes, pharmaceuticals, and pathogenic microorganisms, which poses serious hazards to the environment and human health. Traditional physicochemical treatment processes tend to be expensive, energy-intensive, and generate secondary pollutants, prompting a search for sustainable alternatives. The enzymatic transformation, chelation, and cell wall binding are the major remediation pathways used by fungi in the process of detoxification of heavy metals. Although they have great potential, such challenges as harvesting biomass, metal toxicity levels, scale potential, and economic viability continue to be the main limitations. Microalgae and fungi represent a promising and eco-friendly approach for treating toxic wastewater and bioremediating heavy metals, contributing to the development of circular and sustainable environmental management strategies.

The presence of heavy metals and trace organic micropollutants in the effluent water is still a significant issue, since conventional purification processes are unable to eliminate these contaminants, while advanced techniques are expensive and require substantial amounts of energy. Siliceous unicellular microalgae called diatoms have been shown to be a promising and environmentally-friendly technology to purify wastewater. Their specific structure and metabolic activity allow for effective removal of contaminants via the use of EPS, silica frustule adsorption, and intracellular transformation. The chapter examines the mechanisms of the process and discusses the performance of the system in various types of bioreactors, including biofilm reactors, high-rate algal ponds, and algae-bacteria consortia. Innovative approaches like functionalization of the frustule and increased EPS production are also analyzed.

The dual challenge of escalating carbon emissions and increasing aquatic pollution demands integrated biotechnological solutions capable of addressing both environmental and resource recovery goals. Microalgae have emerged as promising candidates for sustainable carbon sequestration and wastewater treatment. This chapter explores the role of microalgae in bioremediating aquatic pollutants, including heavy metals, pesticides, and industrial effluents, while simultaneously contributing to carbon fixation and biomass production. The integration of microalgal cultivation with wastewater valorization is discussed in the context of circular bioeconomy principles, highlighting opportunities for resource recovery in the form of biofertilizers, biofuels, and bioplastics. Furthermore, the chapter assesses the environmental safety and sustainability of using algal biomass derived from polluted

Preface

Growing pressures on freshwater resources, increasing levels of environmental pollution, and the urgent need for sustainable resource management have accelerated the search for advanced and efficient wastewater treatment technologies. Among the most promising solutions are microalgal systems, which offer the unique ability to simultaneously remediate wastewater, recover valuable nutrients, capture carbon dioxide, and generate biomass that can be transformed into a wide range of value-added products. By converting waste streams into renewable resources, microalgae help redefine wastewater from an environmental liability into a critical component of the circular bioeconomy.

Microalgae for Wastewater Treatment and Resource Recovery: Pathways to a Circular Bioeconomy presents a comprehensive overview of the scientific, technological, and engineering advances driving this rapidly expanding field. Bringing together contributions from researchers worldwide, the book examines the biological principles underlying microalgal wastewater treatment while exploring innovative approaches to nutrient recovery, pollutant degradation, biomass valorization, carbon sequestration, and sustainable bioenergy production.

The volume emphasizes the importance of integrated biological systems, including algal-bacterial consortia, advanced bioremediation strategies, adsorption technologies, heavy-metal removal, and wastewater polishing techniques. It further highlights how recovered microalgal biomass can be converted into biofertilizers, biofuels, bioplastics, and other bio-based products, demonstrating the potential of these technologies to support sustainable agriculture, renewable energy, and environmental restoration. Alongside scientific advancements, the book also considers techno-economic assessments, life-cycle analyses, and commercialization challenges that must be addressed to facilitate large-scale implementation.

By bridging environmental biotechnology, microbiology, chemical engineering, materials science, and sustainability studies, this volume provides an interdisciplinary perspective on wastewater valorization through microalgal technologies. It serves as an essential resource for researchers, engineers, environmental scientists, policymakers, industry professionals, and graduate students seeking practical, innovative solutions to advance circular resource management and sustainable development.

CHAPTER OVERVIEW

Chapter 1: Introduces the concept of wastewater valorization within the framework of the circular bioeconomy. It explores how microalgae and cyanobacteria facilitate wastewater remediation, carbon sequestration, nutrient recovery, and biomass production while discussing cultivation methods and downstream conversion into valuable bio-products.

Chapter 2: Examines the role of microalgal-bacterial consortia in nutrient removal and pollutant biotransformation. The chapter highlights the advantages of symbiotic microbial interactions for wastewater treatment, with a focus on lipid, protein, and carbohydrate recovery for biofuel and other high-value applications.

Chapter 3: Explores advanced algal-bacterial interactions for sustainable wastewater treatment and resource recovery. It discusses mechanisms, including symbiosis, quorum sensing, and gene transfer, that enhance pollutant degradation while enabling the production of biogas, biodiesel, bioplastics, and biofertilizers within a circular economy framework.

Chapter 4: Focuses on synergistic algal-bacterial systems for wastewater bioremediation. Through laboratory-scale studies, the chapter demonstrates how co-cultures improve carbon, nitrogen, and phosphorus cycling, significantly increase pollutant removal efficiency, and generate valuable biomass suitable for bioenergy and agricultural applications.

Chapter 5: Reviews the application of microalgal consortia for removing emerging contaminants from wastewater. It examines mechanisms such as biosorption and biodegradation, discusses the contribution of phycoremediation to sustainable wastewater management, and highlights progress toward achieving the global Sustainable Development Goals.

Chapter 6: Investigates the use of microalgae for eliminating emerging contaminants, including pharmaceuticals, pesticides, heavy metals, and personal care products. The chapter also discusses nutrient recovery, carbon sequestration, biomass production, and the technical challenges associated with large-scale implementation.

Chapter 7: Presents an analytical and mechanistic perspective on microalgal adsorption of emerging pollutants. It examines adsorption chemistry, kinetic and thermodynamic modeling, analytical characterization techniques, and factors influencing treatment efficiency in complex wastewater systems.

Chapter 8: Explores heavy metal remediation using microalgae and fungi as environmentally friendly alternatives to conventional wastewater treatment technologies. The chapter discusses detoxification mechanisms, operational challenges, and the potential of integrated biological systems to support sustainable environmental management.

Chapter 9: Examines the use of diatoms for advanced wastewater polishing and the removal of heavy metals and micropollutants. It highlights contaminant uptake mechanisms, bioreactor configurations, extracellular polymeric substances, and emerging innovations that enhance treatment performance.

Chapter 10: Provides a comprehensive overview of microalgae-based bioremediation for aquatic pollutants while integrating carbon sequestration with wastewater valorization. The chapter explores resource recovery pathways, biomass utilization, ecotoxicological considerations, and the contribution of microalgae toward carbon neutrality and environmental sustainability.

Chapter 11: Focuses on the valorization of microalgal biomass into soil amendments and biofertilizers. It discusses the nutrient composition, agricultural applications, commercialization pathways, and the role of microalgae in promoting climate-resilient, sustainable farming systems.

Chapter 12: Concludes the volume with an evaluation of the techno-economic feasibility and environmental sustainability of microalgal bioenergy systems. The chapter reviews techno-economic analysis (TEA), life-cycle assessment (LCA), commercialization barriers, and key performance indicators for scaling microalgal biofuel production.

The chapters presented in this volume demonstrate that microalgae have evolved from a promising research topic into a versatile platform capable of transforming wastewater treatment into a sustainable system of resource recovery. Through innovative approaches to nutrient removal, contaminant degradation, carbon sequestration, biomass valorization, and renewable energy generation, microalgal technologies offer practical solutions to some of the most pressing environmental challenges of the twenty-first century.

Collectively, the contributions emphasize that successful implementation extends beyond biological performance alone. Advances in microbial ecology, systems engineering, process optimization, analytical modeling, techno-economic evaluation, and environmental policy must work together to overcome current limitations in scalability, operational costs, harvesting technologies, and commercialization. Continued interdisciplinary collaboration will be essential to translating laboratory innovations into economically viable and environmentally sustainable industrial applications.

By integrating scientific discovery with engineering practice and circular economy principles, *Microalgae for Wastewater Treatment and Resource Recovery: Pathways to a Circular Bioeconomy* provides a comprehensive foundation for future research and technological development. The knowledge presented throughout this volume highlights the immense potential of microalgae to redefine wastewater as a renewable resource, contributing to cleaner water, reduced carbon emissions, sustainable agriculture, renewable energy production, and a more resilient global bioeconomy.

Imran Ahmad
Durban University of Technology, South Africa

Faizal Bux
Durban University of Technology, South Africa

Nilamjyoti Kalita
Murkong Selek College, Jonai, India

Chapter 1

Wastewater to Bioresources:

Algal-Mediated Resource Recovery in a Circular Bioeconomy Framework

Puja Saikia
http://orcid.org/0009-0000-9452-0047
Dibrugarh University, India

Priyakshi Hazarika
http://orcid.org/0009-0001-1130-4750
Tezpur University, India

Animesh Saikia
Nagaon University, India

Subhrajyoti Bora
http://orcid.org/0009-0001-6240-0224
Gauhati University, India

Prantik Sharma Baruah
Nagaon University, India

Bhaskar Jyoti Parasar
http://orcid.org/0009-0001-7460-9957
Gauhati University, India

ABSTRACT

The escalating pressures of water scarcity, nutrient depletion, and unsustainable energy demand have necessitated a complex water-nutrient-energy nexus for creation of a circular bioeconomy. Traditional linear economic models treat wastewater with a cycle of inefficiency and environmental degradation, which are usually energy-intensive with no emphasis on resource recovery. Thus, there is an urgent need for circular systems that valorize wastewater to sustainable bioresources. Various microalgae and cyanobacteria act as biocatalysts for wastewater remediation, carbon (C) sequestration, and biomass production. The chapter brings insight into this interplay by outlining the principles of a circular bioeconomy with respect to wastewater treatments and resource recovery techniques. Further, it highlights the techniques of algal cultivation in wastewater, and deciphers the mechanisms of

DOI: 10.4018/979-8-2600-3080-6.ch001

nutrient and pollutant removal process. It significantly focuses on the downstream processing of the algal biomass, exploring pathways for its conversion into valuable bio-products.

1. INTRODUCTION: FROM WASTE TO WEALTH

Rapid population and economic growth have led to a higher demand for resources with rising global scarcity. Water being a resource of its own, serves as a transport medium for other resources as well. According to European Investment Bank, each year about 380 billion m^3 of municipal wastewater is generated globally and it is estimated to increase 24% by 2030 and 51% by 2050. The composition of wastewater varies significantly based on the primary sources contributing to it and the degree of wastewater treatment. Wastewater typically consists of more than 90% water by dry weight, while less than 10% is made up of solids, including approximately 40 to 50% organic matter, 30 to 40% inert substances, 10 to 15% non-degradable materials, and 5 to 8% other components (Kehrein et al., 2020). In addition to these components, wastewater contains numerous microorganisms (up to 109 per millilitre), heavy metals, micro-pollutants, and micronutrients. Thus, wastewater, traditionally considered as an environmental liability, has many opportunities for resource recovery.

Conventional wastewater treatments generally involve three stages namely, primary, secondary and tertiary treatments, which involve a combination of physical, chemical and biological processes (Ranjit et al., 2021). However, it has certain disadvantages like large quantity of sludge generation, high odour generation, high operational cost, requirement of a large area, and difficulties in controlling hazardous materials (Hussain et al., 2021). Conventional biological and industrial systems largely operate on a linear economy model, where raw materials are extracted, transformed into products, and ultimately discarded as waste (Roy et al., 2024). Whereas these processes are effective in removing pollutants from water stream, they lead to irreversible loss of resources and energy (Ghimire et al., 2021). In contrast, the emerging idea of a circular bioeconomy challenges the traditional linear model by viewing waste as a resource and focuses on the ongoing flow of materials, energy, and nutrients through natural processes. In this system, wastewater is viewed as a renewable source of nitrogen (N), phosphorus (P), C, trace elements, and water, which provides dual advantages of reducing environmental pollution and creating valuable products that contribute to sustainable development (Soo & Shon, 2024). The biological systems play a central role in enabling circular economy by leveraging naturally occurring microorganisms to degrade organic pollutants, making it more sustainable than chemical-heavy conventional methods (Priyadarshini et al., 2026).

This reduces reliance on synthetic coagulants and disinfectants, minimizing chemical residues in effluent and sludge (Apollo, 2023). Microbial communities in systems like activated sludge or membrane bioreactors (MBRs) thrive on organic matter, converting it into microbial biomass, CO_2, and water through metabolic processes.

Among biological agents, microalgae have gained considerable attention as ideal platforms for wastewater-based resource recovery. Compared to other techniques, microalgae can grow faster, are more adaptable, Photosynthetic efficiency of microalgae ranged from 10–20% in comparison with 1-2% of most terrestrial plants (Singh & Ahluwalia, 2013). Microalgae-based wastewater treatment is emerging as a sustainable platform for nutrient removal and recovery, as well as biomass production by utilizing the ability of strains like *Chlorella, Chlamydomonas* and *Scenedesmus* to absorb nutrient rich municipal, industrial and agricultural effluents (Da Fontoura et al., 2017; Ma et al., 2016; Nagarajan et al., 2020). In contrast to traditional activated sludge and physico-chemical wastewater treatment, microalgae systems can recover upto 90% of dissolved nutrients, generate low residual N and P levels while cutting the energy demand for the overall process. The resulting algal biomass reinforces the circular bioeconomy idea, which views wastewater as a resource rather than a waste, by acting as a flexible feedstock for biofuels, biogas, biochar, and other bio-based products in integrated biorefineries. The exceptional physiological and genetic adaptation of algal systems to the high nutrient concentrations, hazardous substances and fluctuating physical conditions of wastewater further enhance their effectiveness in the circular bioeconomy (Guedes et al., 2019; Nagarajan et al., 2020).

This chapter adopts a 'waste to wealth' approach by analyzing wastewater as a resource enriched substrate and investigates the useful functions of microalgae as biocatalytic agents in circular bioeconomic systems. It outlines the composition and variability of key wastewater streams, emphasizing their latent resource value and the challenges they pose for conventional treatment systems. It focusses on the biological and engineering foundations of algal-mediated resource recovery and the design of integrated bioremediation and biorefinery processes. Furthermore, the chapter highlights the integration of mechanistic principles underlying pollutant removal, metabolic transformation, and adaptive biological regulation to demonstrate the capacity of microalgae to support sustainable and economically feasible circular bioeconomy systems.

2. CHARACTERIZING WASTEWATER FOR HARNESSING RESOURCES

Successful incorporation of wastewater into circular bioeconomic systems depends on comprehensive understanding of its physicochemical and biochemical characteristics. Therefore, characterization of wastewater is a necessary step to determine the suitability of wastewater in biological resource recovery process, especially in the algal mediated systems. This includes essential wastewater descriptors which are measured using specific parameters that guide treatment system design and assess operational effectiveness (Henze, 2008). Physical characteristics comprises flow rate and its daily fluctuations, temperature, color, odor, and the concentration and distribution of solids, generally measured as total suspended solids (TSS) and volatile suspended solids (VSS), which reflect the amount of particulate matter and its organic component (Winkler & Van Loosdrecht, 2022). While, the chemical analysis involves chemical oxygen demand (COD) and biochemical oxygen demand (BOD) as an indicator of organic compounds, levels of nutrients (total N, ammonia, organic N, total P, ortho-phosphate), major ions and salinity (chloride, sulfate), pH, alkalinity, conductivity, dissolved sulfide and cyanide, along with toxic or persistent substances such as heavy metals (Cr, Cu, Ni, Zn, Pb, Cd, Hg) and xenobiotic organics such as detergents, surfactants, phenols, phthalates, polycyclic aromatic hydrocarbons (PAHs), volatile fatty acids (VFA), and chlorinated solvents (Sonune et al., 2015). The biological characteristics are indicated by the presence of microorganisms such as *Escherichia coli*, total coliforms, *Salmonella*, enteric viruses, protozoan cysts, and helminth eggs, which determine the risk of infection and the necessary disinfection protocols (Makuwa et al., 2020). These parameters can show significant fluctuations on both short-term and seasonal bases as a result of variations in household water consumption, industrial activity timelines, and storm water dilution. Therefore, both their specific values and their typical ratios (such as COD:BOD, COD:N, COD:P, VFA:COD) are utilized to evaluate biodegradability ratio, pollution risks, detect unusual discharges, and determine the appropriate physical, chemical, and biological treatment methods and their sizing (Ali Ch. et al., 2022; Choi et al., 2017). Collectively, these characterization parameters serve as the input variables for a coupled biological and engineering decision matrix, where C:N:P ratios, turbidity, and COD:BOD values directly inform strain selection and cultivation system design.

2.1 Variability in wastewater streams

Waste water comprises the effluents from various sources such as domestic, industrial or agricultural fields. Domestic wastewater, also known as sewage, consists of 1% solids and 99% water. The 1% solids contain dissolved, suspended or colloidal organic and inorganic material. The presence of faeces can lead to highly pathogenic bacteria and viruses that may result in diseases such as cholera, typhoid, tuberculosis, or infectious hepatitis. N, P, carbonates, bicarbonates, chlorides, and sulphates make up the inorganic portion. The organic component consists of 90% proteins and carbohydrates, while only 10% is made up of oils. A small amount includes synthetic organic compounds such as surfactants, phenols, pesticides, and trace elements like sulfur, phosphorus, and iron. The inorganic section of sewage includes metals, their salts, and various substances like nitrates, phosphorus, sulfur, chlorine, and their compounds. Sewage is typically known for its strong odor, dark color, pathogenic nature, with BOD and COD values often measured in the hundreds of mg/l, and pH levels ranging from 6 to 9. BOD:COD ratio for sewage typically more than 2.5 which indicates its low biodegradability and conductivity ranges from 2100- 6400 μS/cm which indicates the poor quality of water (Boutin & Eme, 2016).

Industrial wastewater refers to the liquid waste produced from water utilized during various stages of manufacturing or cleaning processes associated with production. Unlike domestic sewage, their composition depends strongly on the specific products, raw materials, and process steps (e.g., washing, cooking, cooling, extraction, reactions, separations), so even effluents from the same industry can vary in both volume and pollutant load, sometimes mixed with a minor fraction of domestic wastewater from worker facilities (Kato & Kansha, 2024; R. Singh et al., 2025). While, agro-industrial wastewater can be viewed as a specific category of industrial wastewater that encompasses effluents produced from large-scale animal farming, fisheries, slaughterhouses, and seed oil manufacturing. The makeup of these effluents varies based on the particular type of industry and the materials involved in processing(Martinez-Burgos et al., 2021). pH of wastewater generated from these agro- industrial sources ranges typically from 6-10, with very high BOD and COD ranging in thousands of mg/L (Anh et al., 2021; De La Mora-Orozco et al., 2018). The values exceed typical discharge limits necessitating pretreatments like anaerobic digestion.

Table 1. Algal strain selection based on wastewater characteristics and valorization pathways. (Adopted from Nagarajan et al. 2020, Calijuri et al. 2025, Dias et al. 2025)

Sl. No.	Wastewater Type	Key Characteristics	Suitable Algal Strains	Rationale	Lead Valorization Pathway
1	Municipal sewage	Moderate N/P (~100 mg/L NH_4^+, ≈10 mg/L PO_4^{3-}), BOD 200-500 mg/L, variable COD:BOD	*Chlorella vulgaris*, *Scenedesmus obliquus*	Fast growth in moderate nutrient loads; high tolerance to microbial competition	Biogas (via AD) + biofertilizer
2	Agricultural runoff	High nitrate (50-200 mg/L), low P, high turbidity	*Haematococcus pluvialis*, *Chlamydomonas reinhardtii*	Adapted to N-rich, light-limited conditions; ability to switch metabolism	High-value pigments (astaxanthin)
3	Industrial/ agro-industrial effluent	High COD (1,000-10,000+ mg/L), variable pH, potential inhibitors	*Spirulina platensis*, *Chlorella sorokiniana*	High tolerance to extreme pH and organic loads; mixotrophic capacity	Lipids for biodiesel + proteins
4	Hydroponic effluent	Low COD, balanced NPK (N/P ≈ 5-10:1)	*Scenedesmus* sp., *Micractinium* sp.	Efficient nutrient recovery from low-carbon streams; lipid accumulation under stress	PUFA-rich biomass (nutraceuticals)

2.2 Suitability for algal growth

The suitability of wastewater for microalgal cultivation depends on key physicochemical parameters like nutrient levels, inhibitory compounds, light availability, and hydraulic retention time (HRT), which collectively regulate algal growth, biomass yield, and pollutant removal efficiency (Calijuri et al., 2025). N and P are essential macronutrients in wastewater for microalgal metabolism and productivity with optimal N/P ratios around 10:1 to 16:1 for promoting growth, while imbalances in these ratios cause osmotic stress or reduced photosynthesis. Although excessive loading can inhibit nutrient uptake, but wastewater often aligns with algal nutrient requirements (C:N:P ≈50:8:1), minimizing synthetic fertilizer use and enabling up to 99% N removal in tolerant strains (Qian et al., 2024; Ramírez Mérida & Rodríguez Padrón, 2023). Algal enzymes, membranes, and electron transport are disrupted by heavy metals, organics, and pharmaceuticals in wastewater. Microalgae can adsorb metals at rates up to 15 mg/g biomass via bioaccumulation. As some species tolerate contaminants better, strain selection is crucial for enhancing remediation in circular bioeconomy systems (Hasan et al., 2023; Kundu et al., 2024). Another factor effecting algal growth is turbidity from suspended solids which reduces photosynthetically active radiation (PAR), limiting C fixation and shifting metabolism. For nutrient recovery, shallow depths (<0.2 m) optimize irradiation. For example, ideal light intensity of 300 $\mu mol m^{-2} s^{-1}$ increases biomass without causing photochemical stress (González-Camejo et al., 2019). HRT regulates nutrient interaction and biomass retention; where 2-6 days of duration optimizes N/P removal (e.g., higher lipids at shorter HRT, starch at longer HRT), avoiding washout or competition (Iman Shayan et al., 2016).

Therefore, wastewater can be a suitable substrate for the recovery of biological resources owing to the essential physicochemical and nutritional characteristics inherent in it. In this regard, by characterizing wastewater in a resource-oriented approach, the transition from pollution removal to biomass generation and valorization is made possible. Such an approach lays the groundwork for harnessing microalgae as potent circularity agents, whose photosynthetic and metabolic potentials facilitate the transformation of the resources taken up from wastewater into valuable bio-based products.

3. ALGAE AS AGENTS FOR CIRCULARITY: PHOTOSYNTHESIS, REMEDIATION, VALORIZATION

The fast-growing urbanization and industrialization along with the utilisation of agro- and petrochemicals, extensive mining operations, intense farming practices, and inappropriate waste disposal have led to the release of xenobiotic compounds and a rise in the concentration of HMs in the lithosphere and hydrosphere (Mukherjee & Bairwa, 2025). Algae is a potent candidate for environmental pollution remediation (Munir et al., 2025). The widespread distribution of algae, along with their simple body organization and versatile metabolic functions, makes them highly suitable agents for remediation of pollutants, from atmospheric contaminants to wastewater pollutants. Algae can inhabit freshwater, marine, and terrestrial environments and can also thrive in thermophilic and other harsh environments which make them even more suitable for remediation (Swaminaathan et al., 2024), supported by their high-capacity to absorb contaminants and high metabolic activity (Chugh et al., 2022). The metabolism of algae is vital to bioremediation since it breaks down pollutants through nutrient uptake and photosynthesis (Sivakumar et al., 2025). These are controlled by environmental variables such as temperature, light, salinity, and pH. Many microalgal species *Euglena* sp., *Oscillatoria* sp., *Chlorella* sp., and *Scenedesmus* sp., capture the pollutants and accumulate them inside the cell body or detoxify them to less toxic form or immobilized state (Touliabah et al., 2022). It is possible to eliminate over 60% of total contaminants in the contaminated water by using algae as an absorbent. Additionally, algae can be added to contaminated water as part of clean technology for bioremediation. The development stage and binding potential are two of the many variables that affect how well algae remove metals.

It has been observed that over 100,000 different dyes are used by industries, where effluent from the textile or dye sector contains high amounts of heavy metals, phosphates, nitrates, BOD, COD, and dye compounds (Selvaraj et al., 2022). This causes environmental damage even at low concentrations of these hazardous pollutants (Priyadarshini et al., 2026). Therefore, it is of high concern and necessary to treat this wastewater and remove the harmful contaminants to an acceptable limit before eliminating it into the water stream. The physicochemical methods for it include filtration, membrane separation, adsorption, ion exchange, coagulation, and flocculation (Bora et al., 2023).

Algae as a significant biorefinery platforms can play a major role in providing long-term answers to pressing global problems including the energy crisis and climate change (Mahmood et al., 2023). Biomass from algae has the potential to produce a broad range of high-value, economically significant chemicals, new materials, and feedstock for a number of industries, such as food, cosmetics, and feed (Zabochnicka et al., 2022). In addition to helping to reduce greenhouse gas emissions, algal biomass is one of the most effective renewable energy sources for producing bio-oil with rising energy demands (EswaryDevi et al., 2023). Because of its low emissions and sufficient physio-chemical properties and biodegradable nature, algal oil has become a preferred primary substrate for the production of biodiesel. The two-step process of transesterification of algal lipids produce more fatty acid methyl esters than direct transesterification, which forms the basis for algal biodiesel production (Ali et al., 2023). There are many more possibility of employing algal biomass to create a range of bioenergy carriers, such as hydrogen, syngas, methane, charcoal, and bio-oils through anaerobic digestion, pyrolysis, gasification and alcohols from reserved carbohydrate fermentation (Das et al., 2023). Creating new technologies for boosting lipid-rich algal biomass would lead to a long-term, low-cost method of producing biofuel. For that, advance culture condition is one of the basic needs to boost biomass. In this case, photobioreactors can boost cell density, biomass production, and photosynthetic efficiency. Basically, photobioreactors are categorized as closed systems, as they offer a sterile, regulated environment for the growth of photosynthetic algae (Ijaz & Ahmed, 2025). By combining light capture, conveyance, distribution, and improved utilization, algal biomass can enhance creation of medicinal items, food, and cosmetics.

Table 2. Some of the algae used effectively for wastewater treatment and their role in the process.

Sl. No.	Algae Species	Application	Reference
1.	*Chlorella*, *Chlamydomonas* and *Scenedesmus*	Absorb nutrient rich municipal, industrial and agricultural effluents	Da Fontoura et al., 2017; Ma et al., 2016; Nagarajan et al., 2020
2.	*Euglena* sp., *Oscillatoria* sp., and *Scenedesmus* sp.	Capture the pollutants and accumulate them inside the cell body	Touliabah et al., 2022
3.	*Chlorella vulgaris*	Co-digestion with food waste enhanced methane yields and hence, resulted a balanced C/N ratio	Y. Zhang et al., 2019
4.	*Chlorella*, *Scenedesmus* and *Micractinium*	Produces lipid-rich microalgal biomass	Chaudhuri & Balasubramanian, 2025
5.	*Spirulina platensis*	Biomass recoveries of 95% for lipid, 88% for protein, and 78% for polyphenols	Khalid et al., 2025
6.	*Chlorella* sp. G049	Yields 23.78% lipid content with 27.78% PUFAs, which is suitable for both biodiesel and nutraceutical use	Sriket et al., 2025
7.	*Furcellaria lumbricalis*	Source of hydrocolloids such as carrageenans	Tuvikene et al., 2010
8.	*Scenedesmus sp.*	High pigment concentration and lipid content	Lee et al., 2016

4. MECHANISMS OF ALGAL MEDIATED POLLUTANT REMOVAL AND RESOURCE RECOVERY

Phycoremediation is gaining prominence rapidly as compared to the other conventional treatment like chemical treatment, biological treatment such as bacterial and fungal treatment, because of its higher efficacy of polluted removal, cost effectiveness, eco friendliness. The algal outer layer consists of proteins and carbohydrates, while polysaccharides, proteins, peptidoglycan, teichuronic acid, and teichoic acid are essential components of the cell wall of algae (Markai et al., 2003; Schiewer & Wong, 2000). Proteins exhibit amino, carboxyl, amine, phosphate, sulphate, imidazole, N, and oxygen groups of the peptide bond, whereas polysaccharides mostly display carboxyl and amino groups (S. Zhang et al., 2024). Algal cell walls are amphoteric in nature because these functional groups are either protonated or deprotonated depending on the pH, which creates the anionic and cationic sites at the algal external surfaces for ion exchange which leads to active and passive transport (Ayele et al., 2021). The extracellular polymeric substances (EPS) layer present in certain algal cells plays a crucial role in pollutant removal (Huang et al., 2022). These are biopolymer with varying structure and composition with its major component as polysaccharides, proteins, nucleic acid and lipids; a very trace amount of uronic acid, humic substances, and inorganic components (Ranjit et al., 2021). Hydroxyl and phosphoric groups in EPS, in addition to the carboxyl group, play a major role in metal ion sequestration (De Philippis et al., 2011; Kanamarlapudi & Muddada, 2017). Uronic acid and sulphate give the largest negative surface charge for interacting with positively charged metal ions (Wang et al., 2013). One aspect of bioaccumulation is the enzymatic sequestration of metal ions inside the cell (Diep et al., 2018). When the extracellular metal ion gradient exceeds the internal metal ion gradient, the transport of metal ions within the cell begins, and bioaccumulation follows (Arunakumara & Zhang, 2008). Heavy metal ions are transported into the periplasm by porins from the outer membrane. Three different proteins known as secondary carriers, channels, and primary active transporters are responsible for moving metal ions from the periplasm into the cytoplasm. Channels mediate energy-intensive passive transport and are dependent on the metal ion gradient. Secondary carriers are energy-dependent and depend on the motive force of proteins. The primary active transporter, a protein complex, uses ATP hydrolysis to move metal ions. In the cytoplasm, metal binding entities combine with metal ions to create complexes.

The first step in remediation is the movement of pollutants from the wastewater into the algal cells through the cell wall or membranes (Mehta & Gaur, 2005) after that depending on the nature, the pollutants go through a series of processes that can be either neutralisation, transformation, or elimination. Basically, all types of algae undergo more or less similar mechanism or pathways to treat the pollutant such as sedimentation, absorption, adsorption, flocculation, cation, anion replacement, accumulation, oxidation/reduction, complexation. In case of the microalgae, they eliminate xenobiotics from waste water prominently though two mechanisms: (A) a non-active adsorption and absorption process (B) a low concentration metabolism dependant absorption into the cell (Upadhyay et al., 2019). Similarly, cyanobacteria have a remarkable ability to capture pollutant from their environment through the process called biosorption. This can occur through two mechanisms: either by interacting with a slimy substance (EPS) they secrete, or directly through their cell walls, regardless of whether the cell is alive or dead (Patil et al., 2025). So, a wide range of algae utilizes biosorption, biodegradation or bioaccumulation, a brief about all the major mechanisms have given below.

Biodegradation- The natural microbiota has the ability of disintegrating the complex organics compound disseminate from industrial waste into simple less toxic molecules. Similarly, the algae possess various enzymes such as reductase, oxidoreductase, and oxidative enzymes, which can break the bonds in more complex chemical compounds and transform them into simpler, less hazardous chemicals

(Ahmed Bakhit & Elhadi Sulieman, 2026). Depending on the type of compound, a wide range of intermediate and final products could be produced.

Biosorption- Biosorption is a process which eliminate polluted from different source like waste water, water, solution by biological agents. In this physicochemical process, the polluted materials which are eliminated are called sorbate and the biological agents such as bacteria and algae are referred as sorbent (Gahlout et al., 2021; Gouda, S & Taha, 2023). These agents might be organic or inorganic, soluble or insoluble, dry biomass (dead) or wet biomass (alive). Biosorption is considered to be the primary method for removing heavy metals from wastewater that follows several mechanisms such as adsorption, ion exchange, precipitation, surface complexation, and chelation (Bhatt et al., 2022). Considering the cyanobacterial species which cell wall is generally charged with negatively charged ions, making the cell wall a good sorbent for metal ions from waste water and other polluted sites. Moreover, peptidoglycan carboxyl groups provide better adsorption sites for heavy metals. Metal binding to lipopolysaccharides, lipids, exopolysaccharides, and membrane proteins can result in adsorption. Consequently, metal ions can be translocated inside the cells through concentration gradient of metal ions between the external and intracellular environments and transformed into less harmful forms or retained in vacuoles (Van Ho et al., 2002). Depending on the algal biomass involved in the biosorption process it is categorised into two different ways i.e., passive biosorption and active biosorption.

Figure 1. A schematic diagram illustrating the key biological and physicochemical processes employed by an algal cell for the remediation of toxic pollutants. The mechanism shows extracellular adsorption and ion exchange, intracellular bioaccumulation and biotransformation process.

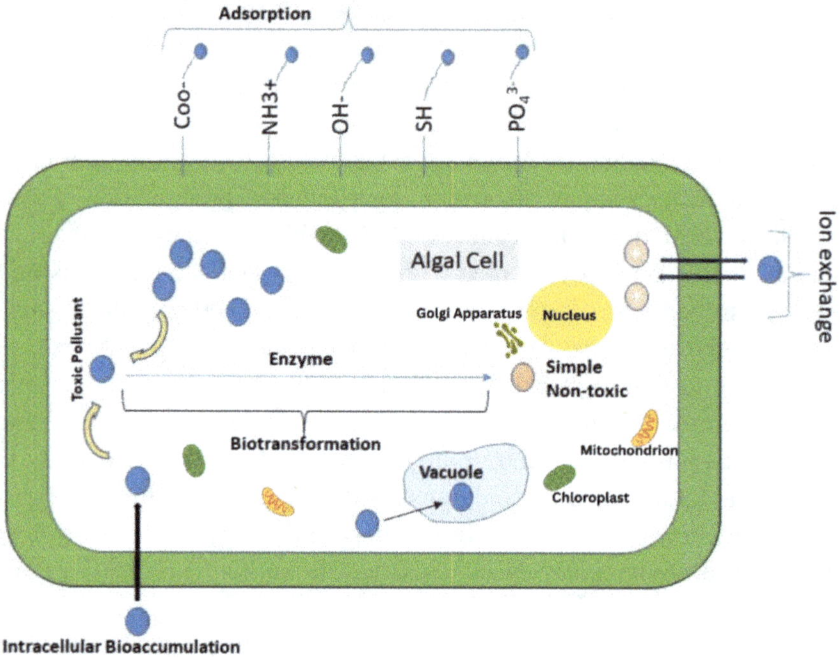

Passive biosorption- Passive biosorption is a process known to interact with polluted by means of dead algal cells through different functional group as mediator (Ayele et al., 2021). This process is metabolism independent and occur very instantly after 5 to 10 minutes interaction.

Active biosorption- Active biosorption is metabolically navigated, that requires living cells (Torres, 2020). It consists of two consecutive phases. Adsorption is the initial stage, which is comparable to passive biosorption. The metabolic process is quick, and the binding substances have no bearing on it. Transporting bound materials across the cell membrane is the next stage. It depends on the binding affinity and is a slow metabolic process.

Bioaccumulation- Bioaccumulation is completely a contrasting process than biosorption which follows a depending metabolic pathway. Bioaccumulation refers as the intracellular uptake of pollutant in living algae via cell membranes through passive and/or active transport mechanisms (Chojnacka, 2010). This process occurs in two sequential phases: initially, metal ions are rapidly and non-specifically adsorbed onto the cell wall through passive interactions, followed by their active and/or passive translocation across the cell wall and plasma membrane into the cytoplasm (M. Kumar et al., 2020). For instance, Cd((II) uptake in *Tetraselmis suecica* species is a biphasic process where, initial stage is adsorption to polysaccharides or protein later on by energy assisted accumulation to the cytosol (Pérez-Rama, 2002). Additionally, cations can be transported by the negatively charged groups of the cell surface to the intracellular compartment via active transport across the plasma membrane after binding to thiol molecules, mainly cysteine, when the concentration of metal in the extracellular environment is much higher than the intracellular concentration (Monteiro et al., 2012). Other amino acids like proline, glutamate, and histidine can also be very important for detoxification and metal chelation (Hayat et al., 2012). The molecular mechanisms of biosorption, biodegradation, and bioaccumulation outlined above underpin the design of targeted remediation strategies. Wastewater contaminated with heavy metals may rely on EPS-mediated biosorption, whereas xenobiotic-rich effluents require strains with robust oxidative enzyme systems.

5. DOWNSTREAMING PATHWAYS FOR ALGAL BIOMASS VALORIZATION

The interface of wastewater-based algal cultivation and subsequent downstreaming forms the critical juncture of the circular bioeconomy concept, where waste is converted into valuable bioresources. After the phycoremediation process, the algal masses are enriched with lipids, proteins, carbohydrates, and pigments. These need to be processed through several downstreaming steps to ensure maximum extraction of value, with minimum waste generation (Gupta et al., 2025; K. Mohanty, 2025).

5.1 Biochemical conversion pathways

Biochemical conversion routes rely mainly on biological catalysts like microorganisms or enzymes, to convert algal biomass into energy carriers and chemical intermediate products. This route is particularly suited for wet algal biomass, thus avoiding the need for energy-intensive drying steps, which are necessary for thermochemical conversion routes. Anaerobic digestion is one of the most prominent techniques in the conversion of algal biomass to biogas. The process involves four stages: hydrolysis, acidogenesis, acetogenesis, and methanogenesis. Co-digestion of microalgae with C-rich materials significantly increases the efficiency of this process. For example, in the co-digestion of *Chlorella vulgaris* with food waste enhanced methane yields and hence, resulted a balanced C/N ratio (Y. Zhang et al., 2019). Again, the biorefinery concept was employed in the production of microalgae in wastewater. Ruales et al., (2025) employing the concept of biorefinery found 86%

increase in methane yield upon co-digestion of carotenoid-extracted microalgae with primary sludge.

Another pathway is the fermentation of microalgae resulting in the production of bioethanol or biohydrogen. Microalgae have low levels of cellulose compared to lignocellulosic materials, while it contains significant starch levels in their composition. A microalgal consortium consisting mainly of *Chlorella* (40%), along with other algal species such as *Scenedesmus* and *Micractinium* were cultivated in hydroponic effluent using mixotrophic conditions. This boosted significant biomass productivity by approximately 55% (Chaudhuri & Balasubramanian, 2025). This was further subjected to offer a cost-effective approach for nutrient recovery from hydroponic effluent, along with producing lipid-rich microalgal biomass, offering both economic and environmental benefits.

5.2 Thermochemical conversion pathways

Thermochemical technologies include the applications of heat and pressure to decompose algal biomass to produce liquid, gas, and solid products. The advantages of these technologies include the treatment of whole biomass and the flexibility to accept variable compositions.

Hydrothermal technologies have gained significant attention due to their ability to process wet algal biomass containing 80-90% moisture, which does not require the energy-intensive step of dewatering the algae prior to conversion (Garcia Alba et al., 2012). Hydrothermal liquefaction (HTL) at 250-350°C and 5-20 MPa pressure, decomposes algal slurry to produce biocrude oil, aqueous phase nutrients, hydrochar, and gas. Recent advances in microwave-assisted pyrolysis (450-550°C) and catalytic gasification have demonstrated bio-oil yield improvements of 25-30% and syngas purity exceeding 80% compared to conventional methods (Reza et al., 2023). The aqueous phase nutrient-rich product can be recycled to cultivate algae, and the combined pretreatment and conversion process can produce 35-40% more lipid product compared to individual process applications.

While, pyrolysis technique involves thermal decomposition in the absence of oxygen to produce bio-oil, biochar, and syngas. While conventional pyrolysis requires dry feedstock, hydrothermal carbonization bridges this gap by processing wet biomass at lower temperatures (180-250°C) to produce hydrochar, a coal-like material suitable for soil amendment or solid fuel applications.

5.3 Extraction of high-value bioproducts

Before undergoing energy conversion, the extraction of high-value co-products is an important factor that improves the economics of biorefineries. Wastewater-based algal biomass can produce a variety of metabolites depending on culture conditions and applied stresses.

Pigments and lipids are primary targets for extraction during early stages. Carotenoids such as astaxanthin, lutein, and β-carotene are highly valuable due to their high market demand. Studies on microalgae cultivated on wastewater raceway pond microcosms revealed that high total carotenoids amount (21.81 mg g^{-1} DW) could be obtained as compared to the control (6.85 mg g^{-1} DW) (Dodangodage et al., 2026). Recent advancements in extraction processes, involving supercritical CO_2 and green solvents have shown impressive recoveries of 95% for lipid, 88% for protein, and 78% for polyphenols from mixed *Spirulina platensis* and fruit waste (Khalid et al., 2025). The green solvents used are free of toxic byproducts and can be used to produce food-grade products.

Another high-value compound is polyunsaturated fatty acids (PUFAs), which include eicosapentaenoic acid (EPA) and docosahexaenoic acid (DHA). The optimized mixotrophic culture of *Chlorella* sp. G049 in hydroponic effluent resulted in 23.78% lipid content with 27.78% PUFAs, which is suitable for both biodiesel and nutraceutical uses (Sriket et al., 2025). Protein extraction from algae also holds promise for the production of animal or human nutrition. After lipid extraction from the algae, the resulting protein-rich residue can be hydrolyzed to release amino acids with antioxidant activity. *Furcellaria lumbricalis*, a red macroalga, is a potential source of hydrocolloids such as carrageenans. It is reported to contain up to 50% carrageenans on a dry weight basis and thus holds promise for use in the food and pharmaceutical industries (Tuvikene et al., 2010).

5.4 Cascading biorefinery and zero-waste integration

The biorefinery concept involves the sequential extraction of multiple products to ensure maximum efficiency in the utilization of resources. The cascading method starts with the gentle extraction of high-value, low-volume products such as pigments and PUFAs, followed by the conversion of bulk materials to biofuels, and finally, the utilization of the left-over material for biofertilizers or bioenergy production.

One of the most compelling examples of zero-waste production through the application of the biorefinery concept was demonstrated in the cultivation of *Chlorella* sp. G049 in hydroponic effluent (Pekkoh et al., 2024). The production process involved the following steps: (1) mixotrophic cultivation with a maximum yield of 1.32 g/L, (2) lipid production for biodiesel with high heating value, (3) functional pigments exhibiting high antioxidant activities, and (4) biofertilizer production using the left-over material for improving soil quality and enhancing lettuce cultivation. The biorefinery process provided multiple revenue streams for a single biomass source. The techno-economic evaluation showed that the biorefinery process relies on the revenues generated from coproducts, as biofuels cannot generate profits on their own under the current market conditions.

5.5 Emerging technologies and process intensification

Innovations in the recent past have indicated promising possibilities in the efficiency of downstream processing. Carbon nano-ions derived from waste oil, when applied at 20 mg/L to *Chlorella pyrenoidosa* cultures, resulted in a 56.16% increase in chemical oxygen demand abatement compared to control cultures, while maintaining the lipid accumulation level (Asati et al., 2025). These nanomaterials may perform a dual role in cultivation enhancement and processing.

Hybrid enzymatic thermochemical pretreatment systems have indicated 40-50% reduction in tar formation during gasification, which is a major hindrance in syngas production (Issaka et al., 2025). Artificial intelligence and IoT monitoring technologies can optimize conversion conditions in real-time, which can bring down costs by 20-25% with improving energy efficiency (Geng et al., 2025). In-situ harvesting technologies, such as membrane photobioreactors combining cultivation with harvesting, can solve the dilute culture problem associated with algae. These technologies, in combination with metabolic engineering of feedstocks for specific conversion routes, form the technological frontier of algal biorefineries.

5.6 Challenges in the downstreaming process

Nevertheless, several challenges have been identified as barriers to the commercialization of algal downstreaming. For instance, the energy expenses of algal harvesting and dewatering is significant, contributing to 20-30% of total production costs. The quality of algal biomass is also variable due to wastewater sources. Moreover, scale-up challenges persist, with many of the promising technologies validated only on a laboratory scale. The research priorities in this regard include: (1) the design of low-energy algal harvesting technologies; (2) optimization of co-processing ratios of mixed feedstock sources; (3) understanding the structure-functional relationship of algal biopolymer materials to support algal-based products; (4) standardization of life cycle assessments and techno-economic analysis to facilitate cross-study comparability; and (5) design of supportive policy frameworks that recognize the co-benefits of algal wastewater biorefineries for both the environment and economy. The utilization of a multitude of conversion pathways within a cascading biorefinery framework, along with wastewater-based algal culture, presents a viable route to realizing a circular economy. The commercialization of this viable route is dependent on further advancements in extraction technologies. The multiplicity of conversion pathways from anaerobic digestion to HTL to pigment extraction, creates a strategic trade-off. Early-stage extraction of high-value co-products improves economic margins, whereas bulk conversion to biofuels maximizes volumetric throughput, with optimal configuration determined by the biochemical profile of the cultivated biomass.

6. SUSTAINABILITY AND ECONOMIC ANALYSIS OF INTEGRATED ALGAL SYSTEMS

The incorporation of microalgal cultivation and wastewater treatment has come up as a strategy to reinforced environmental impacts of the existing methods of wastewater treatments by using potential microalgae for nutrient recovery and removal, C capture and valorization of biomass within the circular bioeconomy boundary (Nagarajan et al., 2020). Life Cycle Assessment (LCA) and Techno Economic Analysis (TEA) are important framework tool to assess environmental impacts and economic feasibility of microalgae-based systems. LCA is a significant tool in order to increase sustainability and lower the environmental impacts. Expansion of LCA into bio-based materials and scaling of circular economy have led to profitable approach in solid waste management (Moutik et al., 2023). Recent study on TEA of microalgae-based biofertilizer using *Scenedesmus* sp. highlight the potential of microalgae-based systems as feasible perspective towards valorization of wastewater, which in turn indicates the need of scale up validation and merging existing treatment facilities that connects financial viability through photobioreactor installation and land requirements (Pérez Mesa et al., 2025).

6.1 Environmental sustainability assessment through life cycle assessment

LCA quantifies the environmental impacts of a product from conception to disposal throughout every stage of its lifecycle using ISO 14040 (2006) criteria. LCA of High-Rate Algal Pond (HRAP) systems for wastewater treatment revealed that microalgal biomass valorization for biogas production is found to be more environmentally suitable than biofertilizer production as compared with activated sludge systems in context with climate change, ozone layer depletion, photochemical oxidant formation and fossil resource depletion (Arashiro et al., 2018).

Reduction of environmental impacts depends on the variety of post-harvest microalgal biomass valorization tracks. Production of biostimulants denoted valuable improvements in many impact categories for efficient nutrient uptake (Crippa et al., 2025). Integration of solar technologies with closed-loop microalgal wastewater systems improves environmental performance by reducing greenhouse gas (GHG) emissions from over 50% to 99% and increases valorization of residual biomass as compared to open-loop systems, heading towards low-C circular bioeconomy (Kuppan et al., 2023). Further, microalgae-based wastewater systems are feasible for drug removal and energy recovery from urban wastewaters within the dimensions of circular economy (Avilia et al., 2025).

6.2 Techno-economic analysis

TEA is vital to assess economic feasibility of algal wastewater systems at scale by evaluating capital costs (CAPEX), operational costs (OPEX) and potential revenues from microalgal biomass valorization (Pérez Mesa et al., 2025). Prior studies on TEA have shown that electrocoagulation (EC) pretreatment of digestate effluents can enhance a significant level of algal biomass productivity at a lower production cost ($3.26/kg) as compared to traditional chemical pretreatment ($5.78/kg), specifying the role of pretreatment techniques that can hold the economic performance (Uludag-Demirer et al., 2022). Recently, TEA of wastewater treatment plant strategies via microalgae culture have pointed out that *Chlorella vulgaris* based tertiary treatment systems can immediately boost the ability of nutrient removal and create economic value. Such systems play vital role in resource recovery and goals of circular economy exceeding the regulatory rules for N and P (Nobre et al., 2024). Although wastewater utilization can minimize the input-related costs, but exhaustion of energy remains the main economic problem that affects the entire viability of the process (Dias et al., 2025).

6.3 Integrated environmental and economic sustainability

Now a days, comparative LCA and TEA outlooks are practiced in order to assess sustainability and economic achievement of environment friendly integrated algal wastewater systems. Combination of LCA and TEA methods highlighted that a source separated nutrient delivery approach, with integration of microalgae. This showed lower environmental impacts and increased the commercial feasibility of biofuel production using microalgae (Li et al., 2022). The identification of optimal places worldwide for biofuel production utilizing microalgae has been assessed through LCA and TEA approach within the aspects of environmental impacts, economic feasibility and biomass productivity. This analysis found Southeast Asia and Venezuela to be suitable regions that holds minimal algal biofuel prices ranges from $1.89–$2.15 per liter equivalent to gasoline. These results highlight the crucial role of geographical conditions and regional economic factors which regulate the practical existence of algal biofuel systems (Quiroz et al., 2023). Moreover, it has been found that replacement of Bold's Basal (BB) medium with greywater enhances the environmental and economic viability for the production of biofuel using microalgal systems. Such alternative growth media reduces energy demand, improves net energy recovery and minimizes the environmental impacts, operating costs and extra expenses, thereby increasing economic viability (Mohit et al., 2024). Therefore, the key to improve economic performance and environmental sustainability lies within incorporation of low-energy processing, strain optimization and recovery of high-value bioproduct (Geng et al., 2025).

6.4 Large-scale implementation challenges

6.4.1 Regulatory constraints and biomass safety

Microalgae cultivated on wastewater streams carry inherent safety risks that must be addressed before biomass can be valorized into food, feed, or agricultural products. Regulatory bodies require environmental risk assessments, biosafety approvals, and traceability protocols, particularly for biomass sourced from variable wastewater streams. The lack of harmonized international standards creates additional barriers for cross-border commercialization, with EU and US frameworks being the most stringent, while developing nations lack regulatory clarity and financial mechanisms to support implementation (Kumar & Mishra, 2025). Furthermore, the detection and mitigation of emerging contaminants (pharmaceuticals, microplastics, PFAS) in wastewater-cultivated biomass remains an unresolved analytical and regulatory challenge.

6.4.2 Economic competitiveness and cost structure

Economic assessments indicate that capital expenditure (CAPEX) dominates in photobioreactors, while operational expenditure (OPEX) prevails in high-rate algal ponds, driven by energy, nutrients, and harvesting costs (Calijuri et al. 2025). Similarly, a study by Hossain et al. (2026) found that strategies such as nutrient substitution with wastewater, renewable energy integration, and co-located biorefineries improve feasibility but remain limited by land requirements and market uncertainties. Furthermore, the wide variability in TEA/LCA results underscores uncertainty about the viability of microalgal systems due to low productivity, high resource demand, elevated costs, and environmental impacts.

Table 2. Major barriers affecting large-scale implementation of algal wastewater biorefineries.

Sl No.	Cost Component	Typical Share of Total Cost	Mitigation Strategies
1	Harvesting/dewatering	20-30%	Hybrid systems, membrane photobioreactors, flocculation
2	Nutrient input	Variable (50.3% reduction achievable with wastewater)	Wastewater substitution
3	Energy (cultivation + processing)	25–40%	Renewable energy integration, AI-optimized operations
4	CO_2 supplementation	10–20% of operational budget	Flue gas integration, direct air capture

6.4.3 Infrastructure retrofitting and practical application

Successful retrofitting of a microalgae-based tertiary treatment system can be integrated into existing waste water treatment plant (WWTP) infrastructure without major capital overhaul. A hybrid algal-bacterial pilot system can achieve less electricity consumption compared with conventional wastewater treatment systems, further supporting retrofit feasibility in decentralized contexts. For instances, a pilot-Scale retrofit project in UK demonstrated a 27 m^3 plant treating municipal tertiary wastewater system. This system achieved final effluent phosphate (0.27 mg/L) and ammonium (0.009 mg/L) within 12-hour HRT. A continuous flow-through system was subsequently retrofitted onto a small rural WWTP (1.5-2.2 m^3/h flow) operating for 12 months, achieving consistent reduction in energy consumption (López-Sánchez et al., 2022). Similarly, an outdoor pilot-scale hybrid algal-bacterial system in Egypt was evaluated for small communities with daily flows $\leq$5000 m^3/day. The system achieved 80% lower electricity consumption compared to conventional activated sludge processes, while meeting discharge standards for BOD/COD and nutrients (Mahapatra & Murthy, 2021).

7. FUTURE PERSPECTIVES AND CONCLUSION

The integration of algal-mediated processes into a wastewater circular bioeconomy framework indicates a paradigm shift from linear treatment to resource recovery. Algae possess a significant ability for efficient nutrient assimilation (N and P) and heavy metal via phycoremediation, which makes it feasible to transform waste resources into valuable bioresources such as proteins, lipids, and biopolymers. This process of valorization will help counter the environmental impacts of effluent discharge, while also providing feedstock for biofuels, biofertilizers, and animal feed, thus completing the material cycle. The integration of microalgae into wastewater treatment systems presents a series of unavoidable trade-offs that must be acknowledged and managed for successful implementation. These trade-offs include challenges such as productivity vs purity, CAPEX vs OPEX with respect to photobioreactors and single-product specialization vs multi-product biorefinery complexity.

The future researches should focus on overcoming systemic hurdles, moving beyond laboratory-scale demonstrations. The list of priorities includes the development of high-throughput omics and synthetic biology for the engineering of robust and high-yielding strains that can sustain themselves in wastewater matrices of varying composition. In addition, the development of low-cost and energy-efficient harvesting and downstream processing technologies is also a prerequisite for economic viability. Process integration with existing infrastructure and techno-economic analysis and life cycle analysis will play a pivotal role in ascertaining the environmental sustainability of this process and will help in its industrial-scale adoption. In conclusion, algal-mediated resource recovery provides an optimistic approach that aligns with bioeconomy goals, but its large-scale implementation requires interdisciplinarity to overcome existing technological and economic hurdles.

REFERENCES

Ahmed Bakhit, S. M., & Elhadi Sulieman, A. M. (2026). Algal enzymes and their applications. In *Microbial Enzymes* (pp. 87–107). Elsevier., DOI: 10.1016/B978-0-443-40352-1.00007-6

Ali Ch., B., Hussein, H. J., & Mushtak, T. J. (2022). *BOD: COD Ratio as Indicator for Wastewater and Industrial Water Pollution. 37.*

Ali, Z., Subeshan, B., Alam, M., Asmatulu, E., & Xu, J. (2023). Recent progress in extraction/transesterification techniques for the recovery of oil from algae biomass. *Biomass Conversion and Biorefinery*, *13*(4), 2553–2569. DOI: 10.1007/s13399-021-01326-y

Anh, H. T. H., Shahsavari, E., Bott, N. J., & Ball, A. S. (2021). Options for Improved Treatment of Saline Wastewater From Fish and Shellfish Processing. *Frontiers in Environmental Science*, *9*, 689580. DOI: 10.3389/fenvs.2021.689580

Apollo, S. (2023). A review of sludge production in South Africa municipal wastewater treatment plants, analysis of handling cost and potential minimization methods. *Physical Sciences Reviews*, *8*(11), 3847–3857. DOI: 10.1515/psr-2021-0234

Arashiro, L. T., Montero, N., Ferrer, I., Acién, F. G., Gómez, C., & Garfí, M. (2018). Life cycle assessment of high rate algal ponds for wastewater treatment and resource recovery. *The Science of the Total Environment*, *622–623*, 1118–1130. DOI: 10.1016/j.scitotenv.2017.12.051 PMID: 29890581

Arunakumara, K. K. I. U., & Zhang, X. (2008). Heavy metal bioaccumulation and toxicity with special reference to microalgae. *Journal of Ocean University of China*, *7*(1), 60–64. DOI: 10.1007/s11802-008-0060-y

Asati, H., Rana, M. S., Prajapati, S. K., & Tripathi, K. M. (2025). Water-Soluble Carbon Nano Onions-Augmented Microalgae for Coupled Wastewater Remediation and Biofuel Production: A Circular Economy Approach. *ACS Sustainable Resource Management*, *2*(10), 1957–1965. DOI: 10.1021/acssusresmgt.5c00276

Avilia, S., Pozzuoli, E., Iovinella, M., Ciniglia, C., & Papa, S. (2025). A Review on Innovative Strategies Towards Sustainable Drug Waste Management Through Algae-Based Systems. *Sci*, *7*(3), 92. DOI: 10.3390/sci7030092

Ayele, A., Suresh, A., & Benor, S. (2021). Phycoremediation of Heavy Metals, Factors Involved and Mechanisms Related to Functional Groups in the Algae Cell Surface – A Review. In Aravind, J., Kamaraj, M., Prashanthi Devi, M., & Rajakumar, S. (Eds.), *Strategies and Tools for Pollutant Mitigation* (pp. 269–289). Springer International Publishing., DOI: 10.1007/978-3-030-63575-6_13

Bhatt, P., Pandey, S. C., Joshi, S., Chaudhary, P., Pathak, V. M., Huang, Y., Wu, X., Zhou, Z., & Chen, S. (2022). Nanobioremediation: A sustainable approach for the removal of toxic pollutants from the environment. *Journal of Hazardous Materials, 427*, 128033. DOI: 10.1016/j.jhazmat.2021.128033 PMID: 34999406

Bora, J., Saha, I., Vaibhav, V., Singh, M., & Malik, S. (2023). Combined Applications of Physico-Chemical Treatments in Treatment of Industrial Wastewater. In Shah, M. P. (Ed.), *Advanced and Innovative Approaches of Environmental Biotechnology in Industrial Wastewater Treatment* (pp. 399–418). Springer Nature Singapore., DOI: 10.1007/978-981-99-2598-8_18

Boutin, C., & Eme, C. (2016). Domestic Wastewater Characterization by Emission Source. *13eme Congres Spécialisé IWA on Small Water and Wastewater Systems.*

Calijuri, M. L., Do Couto, E. D. A., Assemany, P. P., Ribeiro, V. J., Lorentz, J. F., Castro, J. D. S., Assis, L. R. D., Oliveira, A. P. D. S., Pereira, A. S. A. D. P., Marangon, B. B., Magalhães, I. B., Silva, T. A., Ferreira, J., Braga, M. Q., Gama, R. C. N. D., & Henriques, B. S. (2025). Microalgae-Based Wastewater Treatment and Biomass Valorization: Insights, Challenges, and Opportunities from 15 Years of Research. *ACS Omega, 10*(42), 49273–49299. DOI: 10.1021/acsomega.5c04364 PMID: 41179135

Chaudhuri, R., & Balasubramanian, P. (2025). Enhancing nutrient removal from hydroponic effluent with simultaneous production of lipid-rich biomass through mixotrophic cultivation of microalgae. *Journal of Applied Phycology, 37*(2), 957–971. DOI: 10.1007/s10811-024-03425-5

Choi, Y.-Y., Baek, S.-R., Kim, J.-I., Choi, J.-W., Hur, J., Lee, T.-U., Park, C.-J., & Lee, B. (2017). Characteristics and Biodegradability of Wastewater Organic Matter in Municipal Wastewater Treatment Plants Collecting Domestic Wastewater and Industrial Discharge. *Water (Basel), 9*(6), 409. DOI: 10.3390/w9060409

Chojnacka, K. (2010). Biosorption and bioaccumulation – the prospects for practical applications. *Environment International, 36*(3), 299–307. DOI: 10.1016/j.envint.2009.12.001 PMID: 20051290

Chugh, M., Kumar, L., Bhardwaj, D., & Bharadvaja, N. (2022). Bioaccumulation and detoxification of heavy metals. In *Development in Wastewater Treatment Research and Processes* (pp. 243–264). Elsevier., DOI: 10.1016/B978-0-323-85657-7.00013-4

Crippa, I., Dolci, G., Grosso, M., & Rigamonti, L. (2025). Life Cycle Assessment of Microalgal Biomass Valorization from a Wastewater Treatment Process. *Waste and Biomass Valorization, 16*(1), 525–541. DOI: 10.1007/s12649-024-02695-x

Da Fontoura, J. T., Rolim, G. S., Farenzena, M., & Gutterres, M. (2017). Influence of light intensity and tannery wastewater concentration on biomass production and nutrient removal by microalgae Scenedesmus sp. *Process Safety and Environmental Protection, 111*, 355–362. DOI: 10.1016/j.psep.2017.07.024

Das, P. P., Samanta, N. S., Dhara, S., & Purkait, M. K. (2023). Biofuel production from algal biomass. In *Green Approach to Alternative Fuel for a Sustainable Future* (pp. 167–179). Elsevier., DOI: 10.1016/B978-0-12-824318-3.00009-6

De La Mora-Orozco, C., González-Acuña, I. J., Saucedo-Terán, R. A., Flores-López, H. E., Rubio-Arias, H. O., & Ochoa-Rivero, J. M. (2018). Removing Organic Matter and Nutrients from Pig Farm Wastewater with a Constructed Wetland System. *International Journal of Environmental Research and Public Health, 15*(5), 1031. DOI: 10.3390/ijerph15051031 PMID: 29883370

De Philippis, R., Colica, G., & Micheletti, E. (2011). Exopolysaccharide-producing cyanobacteria in heavy metal removal from water: Molecular basis and practical applicability of the biosorption process. *Applied Microbiology and Biotechnology, 92*(4), 697–708. DOI: 10.1007/s00253-011-3601-z PMID: 21983706

Dias, R. R., Deprá, M. C., De Menezes, C. R., Zepka, L. Q., & Jacob-Lopes, E. (2025). Microalgae Cultivation in Wastewater: How Realistic Is This Approach for Value-Added Product Production? *Processes (Basel, Switzerland), 13*(7), 2052. DOI: 10.3390/pr13072052

Diep, P., Mahadevan, R., & Yakunin, A. F. (2018). Heavy Metal Removal by Bioaccumulation Using Genetically Engineered Microorganisms. *Frontiers in Bioengineering and Biotechnology, 6*, 157. DOI: 10.3389/fbioe.2018.00157 PMID: 30420950

Dodangodage, C. A., Gamage, G. N., Wijesekara, I. A., Kasturiarachchi, J. C., Perera, T. A., Rajapakshe, D., & Halwatura, R. U. (2026). Valorization of Canteen Wastewater Through Optimized Spirulina Platensis Cultivation for Enhanced Carotenoid Production and Nutrient Removal. *Phycology, 6*(1), 15. DOI: 10.3390/phycology6010015

EswaryDevi, T., Parthiban, R., Arun, J., & Gopinath, K. P.EswaryDevi. (2023). Processing of marine microalgae biomass via hydrothermal liquefaction for bio-oil production: Study on algae cultivation, harvesting, and process parameters. *Biomass Conversion and Biorefinery*, *13*(16), 15281–15294. DOI: 10.1007/s13399-022-03446-5

Gahlout, M., Prajapati, H., Tandel, N., & Patel, Y. (2021). Biosorption: An Eco-Friendly Technology for Pollutant Removal. In Panpatte, D. G., & Jhala, Y. K. (Eds.), *Microbial Rejuvenation of Polluted Environment* (Vol. 26, pp. 207–227). Springer Singapore., DOI: 10.1007/978-981-15-7455-9_9

Garcia Alba, L., Torri, C., Samorì, C., Van Der Spek, J., Fabbri, D., Kersten, S. R. A., & Brilman, D. W. F. (2012). Hydrothermal Treatment (HTT) of Microalgae: Evaluation of the Process As Conversion Method in an Algae Biorefinery Concept. *Energy & Fuels*, *26*(1), 642–657. DOI: 10.1021/ef201415s

Geng, Y., Shaukat, A., Azhar, W., Raza, Q.-U.-A., Tahir, A., Abideen, M. Z. U., Zia, M. A. B., Bashir, M. A., & Rehim, A. (2025). Microalgal biorefineries: A systematic review of technological trade-offs and innovation pathways. *Biotechnology for Biofuels and Bioproducts*, *18*(1), 93. DOI: 10.1186/s13068-025-02694-7 PMID: 40817213

Ghimire, U., Sarpong, G., & Gude, V. G. (2021). Transitioning Wastewater Treatment Plants toward Circular Economy and Energy Sustainability. *ACS Omega*, *6*(18), 11794–11803. DOI: 10.1021/acsomega.0c05827 PMID: 34056333

González-Camejo, J., Viruela, A., Ruano, M. V., Barat, R., Seco, A., & Ferrer, J. (2019). Effect of light intensity, light duration and photoperiods in the performance of an outdoor photobioreactor for urban wastewater treatment. *Algal Research*, *40*, 101511. DOI: 10.1016/j.algal.2019.101511

Gouda, S. A., & Taha, A. (2023). Biosorption of heavy metals as a new alternative method for wastewater treatment. *RE:view*.

Guedes, A. C., Amaro, H. M., Sousa-Pinto, I., & Malcata, F. X. (2019). Algal spent biomass—A pool of applications. In *Biofuels from Algae* (pp. 397–433). Elsevier., DOI: 10.1016/B978-0-444-64192-2.00016-0

Gupta, R., Mishra, N., Mukherjee, T., Mishra, S., & Bhattacharya, S. (2025). Microalgal biorefinery and circular bioeconomy model for wastewater treatment. In *Microbial Biotechnology: Integrated Microbial Engineering for B3 – Bioenergy, Bioremediation, and Bioproducts* (pp. 415–441). Elsevier. DOI: 10.1016/B978-0-443-23796-6.00016-X

Hasan, H. A., Muhamad, M. H., Ji, B., Nazairi, N. A., Jiat, K. W., Sim, S. I. S. W. A., & Poh, A. F. M. S. (2023). Revolutionizing wastewater treatment with microalgae: Unveiling resource recovery, mechanisms, challenges, and future possibilities. *Ecological Engineering*, *197*, 107117. DOI: 10.1016/j.ecoleng.2023.107117

Hayat, S., Hayat, Q., Alyemeni, M. N., Wani, A. S., Pichtel, J., & Ahmad, A. (2012). Role of proline under changing environments: A review. *Plant Signaling & Behavior*, *7*(11), 1456–1466. DOI: 10.4161/psb.21949 PMID: 22951402

Henze, M. (2008). *Biological wastewater treatment: Principles, modelling and design.*

Hossain, Md. S., Ahamed Anik, Md. S., Rafin, F.-A.-, Alam, S., Abir, A. A., Mourshed, M., & Ahmed, Md. R. (2026). Microalgae for sustainable energy: Techno-economic and life cycle insights into biodiesel and biohydrogen production. *Energy Conversion and Management: X*, *30*, 101840. DOI: 10.1016/j.ecmx.2026.101840

Huang, L., Jin, Y., Zhou, D., Liu, L., Huang, S., Zhao, Y., & Chen, Y. (2022). A Review of the Role of Extracellular Polymeric Substances (EPS) in Wastewater Treatment Systems. *International Journal of Environmental Research and Public Health*, *19*(19), 12191. DOI: 10.3390/ijerph191912191 PMID: 36231490

Hussain, A., Kumari, R., Sachan, S. G., & Sachan, A. (2021). Biological wastewater treatment technology: Advancement and drawbacks. In *Microbial Ecology of Wastewater Treatment Plants* (pp. 175–192). Elsevier., DOI: 10.1016/B978-0-12-822503-5.00002-3

Ijaz, S., & Ahmed, M. (n.d.). Harnessing Microalgae for Wastewater Treatment: Optimizing Cultivation, Pollutant Removal, and Photobioreactor Technologies for a Sustainable Future. In *Green Technologies for Wastewater Treatment and Bioenergy Production* (pp. 64–109).

Iman Shayan, S., Agblevor, F. A., Bertin, L., & Sims, R. C. (2016). Hydraulic retention time effects on wastewater nutrient removal and bioproduct production via rotating algal biofilm reactor. *Bioresource Technology*, *211*, 527–533. DOI: 10.1016/j.biortech.2016.03.104 PMID: 27038261

Issaka, E., Danso-Boateng, E., Fazal, A., Melville, L., & Adams, M. (2025). Algae biomass for energy generation: A comprehensive review of recent advances, mechanisms and bottlenecks. *Sustainable & Green Materials*, *1*(2–4), 32–94. DOI: 10.1080/29965292.2025.2515823

Kanamarlapudi, S. L. R. K., & Muddada, S. (2017). Characterization of Exopolysaccharide Produced by *Streptococcus thermophilus* CC30. *BioMed Research International*, *2017*, 1–11. DOI: 10.1155/2017/4201809 PMID: 28815181

Kato, S., & Kansha, Y. (2024). Comprehensive review of industrial wastewater treatment techniques. *Environmental Science and Pollution Research International, 31*(39), 51064–51097. DOI: 10.1007/s11356-024-34584-0 PMID: 39107648

Kehrein, P., Van Loosdrecht, M., Osseweijer, P., Garfí, M., Dewulf, J., & Posada, J. (2020). A critical review of resource recovery from municipal wastewater treatment plants – market supply potentials, technologies and bottlenecks. *Environmental Science. Water Research & Technology, 6*(4), 877–910. DOI: 10.1039/C9EW00905A

Khalid, S., Chaudhary, K., Aziz, H., Amin, S., Sipra, H. M., Ansar, S., Rasheed, H., Naeem, M., & Onyeaka, H. (2025). Trends in extracting protein from microalgae *Spirulina platensis*, using innovative extraction techniques: Mechanisms, potentials, and limitations. *Critical Reviews in Food Science and Nutrition, 65*(22), 4293–4309. DOI: 10.1080/10408398.2024.2386448 PMID: 39096052

Kumar, M., Singh, A. K., & Sikandar, M. (2020). Biosorption of Hg (II) from aqueous solution using algal biomass: Kinetics and isotherm studies. *Heliyon, 6*(1), e03321. DOI: 10.1016/j.heliyon.2020.e03321 PMID: 32042987

Kumar, V., & Mishra, S. K. (2025). Conclusion and Roadmap for Implementation. In Kumar Arya, S., Khatri, M., Singh, G., & Sahu, S. (Eds.), *Algal Bioengineering and Microbial Synergy to Green Remediation* (pp. 371–393). Springer Nature Singapore., DOI: 10.1007/978-981-96-8054-2_15

Kundu, P., Dutta, N., & Bhattacharya, S. (2024). Application of microalgae in wastewater treatment with special reference to emerging contaminants: A step towards sustainability. *Frontiers in Analytical Science, 4*, 1513153. DOI: 10.3389/frans.2024.1513153

Kuppan, P., Sudharsanam, A., Venkateswarlu, K., & Megharaj, M. (2023). Solar technology–closed loop synergy facilitates low-carbon circular bioeconomy in microalgal wastewater treatment. *NPJ Clean Water, 6*(1), 43. DOI: 10.1038/s41545-023-00256-8

Lee, C. S., Oh, H.-S., Oh, H.-M., Kim, H.-S., & Ahn, C.-Y. (2016). Two-phase photoperiodic cultivation of algal–bacterial consortia for high biomass production and efficient nutrient removal from municipal wastewater. *Bioresource Technology, 200*, 867–875. DOI: 10.1016/j.biortech.2015.11.007 PMID: 26595666

Li, P., Luo, Y., & Yuan, X. (2022). Life cycle and techno-economic assessment of source-separated wastewater-integrated microalgae biofuel production plant: A nutrient organization approach. *Bioresource Technology, 344*, 126230. DOI: 10.1016/j.biortech.2021.126230 PMID: 34732373

López-Sánchez, A., Silva-Gálvez, A. L., Aguilar-Juárez, Ó., Senés-Guerrero, C., Orozco-Nunnelly, D. A., Carrillo-Nieves, D., & Gradilla-Hernández, M. S. (2022). Microalgae-based livestock wastewater treatment (MbWT) as a circular bioeconomy approach: Enhancement of biomass productivity, pollutant removal and high-value compound production. *Journal of Environmental Management, 308*, 114612. DOI: 10.1016/j.jenvman.2022.114612 PMID: 35149401

Ma, Y., Rajkumar, M., Zhang, C., & Freitas, H. (2016). Inoculation of Brassica oxyrrhina with plant growth promoting bacteria for the improvement of heavy metal phytoremediation under drought conditions. *Journal of Hazardous Materials, 320*, 36–44. DOI: 10.1016/j.jhazmat.2016.08.009 PMID: 27508309

Mahapatra, D. M., & Murthy, G. S. (2021). Long term evaluation of a pilot scale multimodal algal bioprocess for treatment of municipal wastewater. *Journal of Cleaner Production, 311*, 127690. DOI: 10.1016/j.jclepro.2021.127690

Mahmood, T., Hussain, N., Shahbaz, A., Mulla, S. I., Iqbal, H. M. N., & Bilal, M. (2023). Sustainable production of biofuels from the algae-derived biomass. *Bioprocess and Biosystems Engineering, 46*(8), 1077–1097. DOI: 10.1007/s00449-022-02796-8 PMID: 36331626

Makuwa, S., Tlou, M., Fosso-Kankeu, E., & Green, E. (2020). Evaluation of Fecal Coliform Prevalence and Physicochemical Indicators in the Effluent from a Wastewater Treatment Plant in the North-West Province, South Africa. *International Journal of Environmental Research and Public Health, 17*(17), 6381. DOI: 10.3390/ijerph17176381 PMID: 32887228

Markai, S., Andrès, Y., Montavon, G., & Grambow, B. (2003). Study of the interaction between europium (III) and Bacillus subtilis: Fixation sites, biosorption modeling and reversibility. *Journal of Colloid and Interface Science, 262*(2), 351–361. DOI: 10.1016/S0021-9797(03)00096-1 PMID: 16256615

Martinez-Burgos, W. J., Bittencourt Sydney, E., Bianchi Pedroni Medeiros, A., Magalhães, A. I., De Carvalho, J. C., Karp, S. G., Porto De Souza Vandenberghe, L., Junior Letti, L. A., Thomaz Soccol, V., De Melo Pereira, G. V., Rodrigues, C., Lorenci Woiciechowski, A., & Soccol, C. R. (2021). Agro-industrial wastewater in a circular economy: Characteristics, impacts and applications for bioenergy and biochemicals. *Bioresource Technology, 341*, 125795. DOI: 10.1016/j.biortech.2021.125795 PMID: 34523570

Mehta, S. K., & Gaur, J. P. (2005). Use of Algae for Removing Heavy Metal Ions From Wastewater: Progress and Prospects. *Critical Reviews in Biotechnology, 25*(3), 113–152. DOI: 10.1080/07388550500248571 PMID: 16294830

Mohanty, K. (with Mohanty, S. S.). (2025). *A Biorefinery Approach to Algal Biomass Conversion for Biofuels and Bioproducts* (1st ed). Elsevier.

Mohit, A., Mishel, V. K., & Remya, N. (2024). Life cycle assessment and technoeconomic analysis of biofuels produced from polyculture microalgae cultivated in greywater. *Journal of Environmental Management*, *356*, 120711. DOI: 10.1016/j.jenvman.2024.120711 PMID: 38537468

Monteiro, C. M., Castro, P. M. L., & Malcata, F. X. (2012). Metal uptake by microalgae: Underlying mechanisms and practical applications. *Biotechnology Progress*, *28*(2), 299–311. DOI: 10.1002/btpr.1504 PMID: 22228490

Moutik, B., Summerscales, J., Graham-Jones, J., & Pemberton, R. (2023). Life Cycle Assessment Research Trends and Implications: A Bibliometric Analysis. *Sustainability (Basel)*, *15*(18), 13408. DOI: 10.3390/su151813408

Mukherjee, S., & Bairwa, D. H. (2025). Environmental Issues Arising from Urbanization: A Study on the Ecological Consequences of Rapid Urban Growth. *The Journal of Humanistic Education and Development*, *7*(3), 12–17. DOI: 10.22161/jhed.7.3.3

Munir, N., Sarwar, Z., Abideen, Z., Saleem, F., Hasanuzzaman, M., Siddiqui, Z. S., & El-Keblawy, A. (2025). Algae-based bioremediation of soil, water, and air: A solution to polluted environment. *Environmental Science and Pollution Research International*, *32*(36), 21338–21357. DOI: 10.1007/s11356-025-36880-9 PMID: 40952577

Nagarajan, D., Lee, D.-J., Chen, C.-Y., & Chang, J.-S. (2020). Resource recovery from wastewaters using microalgae-based approaches: A circular bioeconomy perspective. *Bioresource Technology*, *302*, 122817. DOI: 10.1016/j.biortech.2020.122817 PMID: 32007309

Nobre, M. L. F., Tavares, D., Fraga, C., Oliveira, B., Dias, M., Mesquita, S., Oliveira, C. M., & Pires, J. C. M. (2024). Techno-economic analysis of a circular microalgal approach for enhanced wastewater treatment and resource recovery in Northern Portugal. *Journal of Cleaner Production*, *434*, 140389. DOI: 10.1016/j.jclepro.2023.140389

Patil, A., Chakraborty, S., Yadav, Y., Sharma, B., Singh, S., & Arya, M. (2025). Bioremediation strategies and mechanisms of bacteria for resistance against heavy metals: A review. *Bioremediation Journal*, *29*(4), 448–480. DOI: 10.1080/10889868.2024.2375204

Pekkoh, J., Wichaphian, A., Kamngoen, A., Sriket, N., Zin, M. T., Lomakool, S., Maneechote, W., Chromkaew, Y., Pathom-aree, W., Cheirsilp, B., & Srinuanpan, S. (2024). Heterotrophic upcycling of hydroponic wastewater supplemented with glucose and indole-3-acetic acid into high-quality Chlorella biomass for zero-waste multiproduct microalgal biorefinery. *Environmental Technology & Innovation, 36*, 103813. DOI: 10.1016/j.eti.2024.103813

Pérez Mesa, A., Céspedes Grattz, P. A., Vidal Vargas, J. J., Ríos, L. A., & Ocampo Echeverri, D. (2025). Techno-Economic Assessment of Microalgae-Based Biofertilizer Production from Municipal Wastewater Using Scenedesmus sp. *Water (Basel), 17*(20), 2941. DOI: 10.3390/w17202941

Pérez-Rama, M. (2002). Cadmium removal by living cells of the marine microalga Tetraselmis suecica. *Bioresource Technology, 84*(3), 265–270. DOI: 10.1016/S0960-8524(02)00045-7 PMID: 12118704

Priyadarshini, S., Jagatee, S., Tudu, J., Kumbhar, M. M., & Das, A. P. (2026). Microbe-mediated wastewater treatment: Harnessing nature's engineers for sustainable remedies. In *Decarbonization of Wastewater Pollutants as a Sustainable Solution* (pp. 103–136). Elsevier., DOI: 10.1016/B978-0-443-33062-9.00008-4

Qian, W., Yang, Y., Chou, S., Ge, S., Li, P., Wang, X., Zhuang, L.-L., & Zhang, J. (2024). Effect of N/P ratio on attached microalgae growth and the differentiated metabolism along the depth of biofilm. *Environmental Research, 240*, 117428. DOI: 10.1016/j.envres.2023.117428 PMID: 37875171

Quiroz, D., Greene, J. M., Limb, B. J., & Quinn, J. C. (2023). Global Life Cycle and Techno-Economic Assessment of Algal-Based Biofuels. *Environmental Science & Technology, 57*(31), 11541–11551. DOI: 10.1021/acs.est.3c02892 PMID: 37499260

Ramírez Mérida, L. G., & Rodríguez Padrón, R. A. (2023). Application of microalgae in wastewater: Opportunity for sustainable development. *Frontiers in Environmental Science, 11*, 1238640. DOI: 10.3389/fenvs.2023.1238640

Ranjit, P., Jhansi, V., & Reddy, K. V. (2021). Conventional Wastewater Treatment Processes. In Maddela, N. R., García Cruzatty, L. C., & Chakraborty, S. (Eds.), *Advances in the Domain of Environmental Biotechnology* (pp. 455–479). Springer Singapore., DOI: 10.1007/978-981-15-8999-7_17

Reza, M., Iskakova, Z., Afroze, S., Kuterbekov, K., Kabyshev, A., Bekmyrza, K., Kubenova, M., Bakar, M., Azad, A., Roy, H., & Islam, M. (2023). Influence of Catalyst on the Yield and Quality of Bio-Oil for the Catalytic Pyrolysis of Biomass: A Comprehensive Review. *Energies, 16*(14), 5547. DOI: 10.3390/en16145547

Roy, A., Ganguly, N., Banerjee, S., & Pandit, S. (2024). From Linear Economy to Circular Bio-economy: A Paradigm Shift in Wastewater Management. In Pandit, S., Kumar, L., Mathuriya, A. S., Jadhav, D. A., & Ray, S. (Eds.), *Advancements in Bio-systems and Technologies for Wastewater Treatment* (Vol. 118, pp. 135–154). Springer International Publishing., DOI: 10.1007/978-3-031-58331-5_8

Ruales, E., Bellver, M., Álvarez-González, A., Carvalho Fontes Sampaio, I., Garfi, M., & Ferrer, I. (2025). Carotenoids and biogas recovery from microalgae treating wastewater. *Bioresource Technology*, *428*, 132427. DOI: 10.1016/j.biortech.2025.132427 PMID: 40120990

Schiewer, S., & Wong, M. H. (2000). Ionic strength effects in biosorption of metals by marine algae. *Chemosphere*, *41*(1–2), 271–282. DOI: 10.1016/S0045-6535(99)00421-X PMID: 10819211

Selvaraj, D., Dhayabaran, N. K., & Mahizhnan, A. (2022). An insight on pollutant removal mechanisms in phycoremediation of textile wastewater. *Environmental Science and Pollution Research International*, *30*(60), 124714–124734. DOI: 10.1007/s11356-022-21307-6 PMID: 35708812

Singh, R., Singh, A., Balomajumder, C., & Vidyarthi, A. K. (2025). Assessment of industrial effluent discharges contributing to Ganga water pollution through a multivariate statistical framework: Investigating the context of Indian industries. *Environmental Science and Pollution Research International*, *32*(59), 31625–31643. DOI: 10.1007/s11356-024-35823-0 PMID: 39760835

Singh, U. B., & Ahluwalia, A. S. (2013). Microalgae: A promising tool for carbon sequestration. *Mitigation and Adaptation Strategies for Global Change*, *18*(1), 73–95. DOI: 10.1007/s11027-012-9393-3

Sivakumar, S., Nurgat, M. Z., Srinivasan, L. V., Bora, A., Wagh, M. S., Kungwani, N. A., & Narayan, M. (2025). Bioengineered Nano-sorbents: Microbial Allies in Pollutant Sequestration. In Mohanta, Y. K., Mishra, B., & Pudake, R. N. (Eds.), *Nano-microbiology for Sustainable Development* (pp. 331–372). Springer Nature Switzerland., DOI: 10.1007/978-3-031-78845-1_13

Sonune, M., Mungal, M., & Kamble, S. (2015). *Study of physico-chemical characteristics of domestic wastewater in Vishnupuri*. Nanded.

Soo, A., & Shon, H. K. (2024). A nutrient circular economy framework for wastewater treatment plants. *Desalination*, *592*, 118090. DOI: 10.1016/j.desal.2024.118090

Sriket, N., Wichaphian, A., Kamngoen, A., Pekkoh, J., Pathom-aree, W., Chromkaew, Y., Cheirsilp, B., Maneechote, W., Leong, Y. K., & Srinuanpan, S. (2025). Converting hydroponic effluent into high PUFA-rich biomass by mixotrophically cultivated microalga Chlorella sp. Through light regulation for sustainable production of bioingredient and bioenergy. *Journal of Environmental Chemical Engineering*, *13*(3), 117187. DOI: 10.1016/j.jece.2025.117187

Swaminaathan, P., Shaji, A., Saravanan, A., & Yaashikaa, P. R. (2024). Innovative Approaches in Extremophile-Mediated Remediation of Toxic Pollutants: A Comprehensive Review. *Water Conservation Science and Engineering*, *9*(2), 39. DOI: 10.1007/s41101-024-00274-8

Torres, E. (2020). Biosorption: A Review of the Latest Advances. *Processes (Basel, Switzerland)*, *8*(12), 1584. DOI: 10.3390/pr8121584

Touliabah, H. E.-S., El-Sheekh, M. M., Ismail, M. M., & El-Kassas, H. (2022). A Review of Microalgae- and Cyanobacteria-Based Biodegradation of Organic Pollutants. *Molecules (Basel, Switzerland)*, *27*(3), 1141. DOI: 10.3390/molecules27031141 PMID: 35164405

Tuvikene, R., Truus, K., Robal, M., Volobujeva, O., Mellikov, E., Pehk, T., Kollist, A., Kailas, T., & Vaher, M. (2010). The extraction, structure, and gelling properties of hybrid galactan from the red alga Furcellaria lumbricalis (Baltic Sea, Estonia). *Journal of Applied Phycology*, *22*(1), 51–63. DOI: 10.1007/s10811-009-9425-x

Uludag-Demirer, S., Bustamante, M., Liu, Y., & Liao, W. (2022). Techno-economic analysis and life cycle assessment of algal cultivation on liquid anaerobic digestion effluent for algal biomass production and wastewater treatment. In *Integrated Wastewater Management and Valorization Using Algal Cultures* (pp. 149–164). Elsevier., DOI: 10.1016/B978-0-323-85859-5.00009-9

Upadhyay, A. K., Singh, R., & Singh, D. P. (2019). Phycotechnological Approaches Toward Wastewater Management. In Bharagava, R. N., & Chowdhary, P. (Eds.), *Emerging and Eco-Friendly Approaches for Waste Management* (pp. 423–435). Springer Singapore., DOI: 10.1007/978-981-10-8669-4_18

Van Ho, A., Ward, D. M., & Kaplan, J. (2002). Transition Metal Transport in Yeast. *Annual Review of Microbiology*, *56*(1), 237–261. DOI: 10.1146/annurev.micro.56.012302.160847 PMID: 12142483

Wang, T., Liu, W., Xiong, L., Xu, N., & Ni, J. (2013). Influence of pH, ionic strength and humic acid on competitive adsorption of Pb(II), Cd(II) and Cr(III) onto titanate nanotubes. *Chemical Engineering Journal*, *215–216*, 366–374. DOI: 10.1016/j.cej.2012.11.029

Winkler, M.-K. H., & Van Loosdrecht, M. C. M. (2022). Intensifying existing urban wastewater. *Science, 375*(6579), 377–378. DOI: 10.1126/science.abm3900 PMID: 35084964

Zabochnicka, M., Krzywonos, M., Romanowska-Duda, Z., Szufa, S., Darkalt, A., & Mubashar, M. (2022). Algal Biomass Utilization toward Circular Economy. *Life (Basel, Switzerland), 12*(10), 1480. DOI: 10.3390/life12101480 PMID: 36294915

Zhang, S., Sun, Z., Zheng, T., He, C., & Lin, D. (2024). Nanoplastics increase algal absorption and toxicity of Cd through alterations in cell wall structure and composition. *Water Research, 254*, 121394. DOI: 10.1016/j.watres.2024.121394 PMID: 38442610

Zhang, Y., Caldwell, G. S., Zealand, A. M., & Sallis, P. J. (2019). Anaerobic co-digestion of microalgae Chlorella vulgaris and potato processing waste: Effect of mixing ratio, waste type and substrate to inoculum ratio. *Biochemical Engineering Journal, 143*, 91–100. DOI: 10.1016/j.bej.2018.12.021

Chapter 2
Microalgal Consortia for Nutrient Removal and Pollutant Biotransformation

S. Harini
http://orcid.org/0009-0008-0300-5081
Sacred Heart College, India

G. Anusuya
http://orcid.org/0009-0007-2211-4980
Sacred Heart College, India

A. Mubarak Ali
http://orcid.org/0009-0009-6367-7864
Sacred Heart College, India

T. Elavarasi
http://orcid.org/0009-0003-0792-4484
Sacred Heart College, India

Yuvalakshmi Logesan
http://orcid.org/0009-0005-4579-1170
Sacred Heart College, India

ABSTRACT

Eutrophication, which is usually caused by the disposal of wastewater containing nutrients such as those present in hydroponic effluent, is known to have a negative effect on the quality and balance of water. Therefore, the optimization of the treatment

DOI: 10.4018/979-8-2600-3080-6.ch002

of wastewater in an economical manner is necessary. Phosphorus and nitrogen are nutrients which can be extracted and recovered from wastewater with the help of bacterial and microalgal consortia. This is in accordance with the concept of the circular bio-economy, which is the production of goods from waste. The concept of the use of microalgae-bacteria consortia in the bioremediation of wastewater will be presented briefly in this chapter. The principle of symbiotic connections states that the use of microalgae-bacteria consortia is better than the use of single species in the bioremediation of wastewater. These consortia collect lipids, proteins, and carbohydrates which can be converted into biofuel and other beneficial products, aside from the removal of contaminants from wastewater.

1. INTRODUCTION

The environment is in a critical condition of being threatened by the rising need for water at the global level, the decreasing supply of water, and the pollution of water. Eutrophication is considered the major issue that threatens the quality of water. Eutrophication occurs when wastewater is discharged into the water, containing nutrients. This issue is further complicated when wastewater from domestic water sources, as well as water from future water practices of hydroponics, is discharged into the water. Even though the traditional practices of water treatment have proven successful, they have also proven unsustainable because of the energy required, the waste produced, and the greenhouse gases they emit (Jia & Yuan, 2018). The process of bioremediation, which utilizes bio-agents to clean the environment, is a more sustainable way to address the problem. The use of the microalgae method has gained more interest from the research community compared to other bioremediation processes due to the inherent capacity of the method to allow the treatment of wastewater (Chia et al., 2020). Various species of microalgae can thrive in different wastewater systems or environments. Microalgae have shown their potential to remove nitrogen, phosphorus, and organic compounds as they remove different contaminants present in wastewater, such as Chemical Oxygen Demand (COD) and bio-pollutants. They are a sustainable source of chemicals for industrial use (Li et al., 2021a). The potential is multiplied multiple times when bacteria and microalgae are used in symbiotic consortia (Mondal & Palit, 2022). Compared to using them separately, the symbiotic interaction between microalgae and bacteria can improve the removal of both contaminants and nutrients. The term microalgae-bacteria symbiosis describes naturally occurring interactions that help wastewater treatment ponds reduce their biological oxygen consumption. Active contact between the species is necessary for microalgae-bacteria consortia to be effective (Cao et al., 2020). The incorporation of this biotechnology into a circular bioeconomy is particularly noteworthy. Microalgae-bacteria consortia not only remove nutrients and detoxify pollutants but also mitigate CO_2 while generating biomass and valuable metabolites (Dongyu et al., 2022). In this approach, circular resource recovery systems replace linear waste treatment concepts. To further improve the use of these consortia, however, a thorough comprehension of intricate inter-kingdom interactions and system optimisation would be required, particularly for particular waste streams such nutrient-rich hydroponic effluent (Zhibei et al., 2022). At the same time, it is expected that this chapter will provide a comprehensive overview regarding the prospects and challenges associated with integrating wastewater resources, such as hydroponic waters, with sustainability.

2. TYPES AND APPLICATIONS OF MICROALGAL CONSORTIA IN WASTEWATER TREATMENT

2.1. Naturally Occurring Consortia

Microbial consortia exist in almost all natural environments. Such consortia colonize diverse environments, which range from soil and water to extreme environments such as hot springs, ocean vents, and polar regions. The early 20th century witnessed the first study on the characterisation of the roles of microorganisms in their natural habitat by the Russian scientist Winogradsky, which gave rise to the science of microbial ecology (Ho et al., 2017). Since the cow and its relatives have the ability to consume food that cannot be digested by other animals, including humans, the rumen of the cow is one of the most studied microbial consortia (Daverey et al., 2019). The microbial consortia have been the subject of much study, which has come up with innovative methods of producing biofuels and other bio-products. Over the past ten years, there has been an increased focus on the potential applications of microbial consortia in the field of biotechnology remediation of the environment, and human health maintenance. For example, consortia have been used to bio-clean contaminated environments, to produce probiotics, and to clean wastewater. The natural occurrence of microbial consortia has remained a dynamic field with the potential to change various fields of science and technology (Tournier et al., 2020).

2.2. Synthetic/Engineered Consortia

Using the knowledge obtained from natural systems, synthetic consortia have been created to perform specific functions. Synthetic consortia offer advantages such as division of labor and metabolic complementarity (Abbas & Halog, 2021). To provide the highest performance in particular tasks, such as the production of metabolites and the breakdown of contaminants, synthesised consortia have been developed. By using the best strains, synthetic consortia can offer the best performance and stability compared to natural consortia. For example, synthetic consortia composed of microalgae and hydrocarbon-degrading bacteria have been created to enhance the efficiency of petroleum compound degradation in water. In agriculture, synthetic consortia have been created to offer the best performance in the field of biocontrol. For example, Schlembach et al. (2021) characterized a microbial consortium developed by Koppert Biological Systems for use in various biocontrol strategies. The authors further screened microbial libraries to identify effective consortia against plant pathogens.

3. TARGET POLLUTANTS: REMOVAL MECHANISMS AND EFFICIENCIES

3.1. Nutrients

3.1.1. Nitrogen

Nitrogen compounds present in wastewater may be either organic or inorganic, which include ammonium (NH_4^+), nitrite (NO_2^-), and nitrates (NO_3^-). Higher levels of these compounds affect the quality of water and lead to eutrophication (Kumar et al., 2018). Microalgae need nitrogen to produce protein, enzymes, and nucleic acids. Nitrogen removal in algal-bacterial systems occurs through four primary pathways follows assimilation (anabolism), ammonia stripping, nitrification, and denitrification. Microalgae directly assimilate ammonium, nitrate, and nitrite from wastewater (Kube et al., 2018).

3.1.2. Phosphorus

The release of wastewater poses a risk of eutrophication due to elevated phosphorus levels. During microalgae metabolism, the majority of phosphorus assimilation can be found in HPO_4^- (phosphoric acid) and HPO_4^{2-} (hydrogen phosphate). Phosphorus (P) is a macronutrient essential for the synthesis of nucleic acids, proteins, and lipids, and it plays a key role in energy metabolism (Wu et al., 2025). In general, both biotic and abiotic assimilation are the methods that are employed in the elimination process. Similar to this, soluble P and polyvalent cations found in wastewater are combined in the abiotic orthophosphate precipitation process, also referred to as auto flocculation. This process states that P is transformed into solid P at higher pH values, which is then readily precipitated out. However, as P is also thought to be a limiting nutrient for the cellular mass of algae, absorption is the primary process used to remove P from wastewater. Therefore, phosphorus can be efficiently removed from textile wastewater under appropriate conditions. Algae demonstrated a superior mechanism for extracting both N and P from effluent (Jayabal., 2025).

3.2. Heavy Metals

Heavy metals (HMs) biosorption has been studied by many researchers using algae-bacteria systems. Research indicates that because permease proteins transfer heavy metals from the external environment into cells, they are critical for biosorption. The efficiency of this process is determined by the shape of the cell membrane and the formation of the protein complexes that sequester the metals (Shekharaiah et al., 2020). Extracellular polymeric substances (EPS) and siderophores made from bacteria or algae are used to complex heavy metals into a less toxic and soluble form using an enzyme that starts oxidation-reduction reactions and biotransformation in order to biologically remove HMs from their medium. Consequently, it was suggested that an algae-bacteria consortium perform a biological detoxifying system that included Cadmium (II), nitrate (II), copper (II), and zinc (II) ions in wastewater. As a result, HMs might be bio observed onto the cell surfaces in a gradual phase and then slowly absorbed by the bacteria. Compared to algae-only systems, growth of algae in algae-bacterial systems is frequently lower. This suggests that although higher removal rates of heavy metals (HM) are encouraged in algae-bacteria systems, this may not always be because the algae absorb more HM rather, it could be a result of the bacteria and algae working together in a synergistic manner. It has been demonstrated that algae-bacteria biofilms display facultative symbiotic associations in some harsh aquatic environments, such as acid mine drainage (AMD), where algae-bacteria consortia may be beneficial. This is because they provide heterotrophic bacteria with the necessary metabolites required for growth as well as a mechanism of removing heavy metals in AMD (Abinandan et al., 2018).

3.3. Organic Pollutants

3.3.1. Pharmaceuticals and Personal Care Products (PPCPs)

The removal of PPCPs and their constituents is a matter of concern, as they exhibit a high level of biological persistency. The five mechanisms, namely volatilisation, biosorption, bioaccumulation, biodegradation, and photodegradation, through which microorganisms remove pharmaceuticals and personal care products and other similar compounds from the environment. Algae remove pharmaceuticals and personal care products through intracellular and extracellular biodegradation, bioadsorption, and bioaccumulation. The bioremediation of pharmaceuticals and personal care products can take place through other enzymatic mechanisms, too, which are found to be active in algae, namely, hydroxylation, carboxylation, an oxidisation, hydrogenation, a glycosylation, a demethylation, as well as a breaking, a decarboxylation, a dihydroxylation, and a bromination. The group of antibiotics is considered to be a major group of developing water pollutants under pharmaceuticals and personal care products. The identification of certain antibiotic residues found in any medium is considered a critical issue, as the level of toxicity associated with these residues can be very high. *Chlamydomonas* sp. is found to be effective in the removal of antibiotics from water. The microalga *Chlamydomonas* sp. Tai-03 showed 100% removal of ciprofloxacin (CIP) and 54.53% removal of sulfadiazine (SDZ) (Xie et al., 2020). It should be mentioned that removal of CIP was mainly through biodegradation, while removal of SDZ was through photolysis. Smaller molecular weight (MW) polysaccharides are produced by algae in algae-bacteria granules in response to tetracycline, according to recent observations. Tetracycline is therefore removed more quickly as a result of the increased concentration of these polysaccharides in EPS. Algae-bacteria systems have also been observed to improve removing of the sulfamethoxazole (SMX) compounds from WWTP effluent. Even though high levels of bisphenol A (BPA) are detrimental to algae growth, bacterial proliferation in an algae-bacteria system promotes BPA breakdown and decreases the BPA inhibitory impact, making algae-bacteria systems useful for BPA removal (Wang et al., 2020a).

3.3.2. Phenolics

One of the main causes of environmental contamination nowadays is phenolic compounds and their derivatives. Although there were few research examining the use of algal-bacterial symbiosis in phenolic compound elimination, the results demonstrated that systems made up of bacteria and algae had a high potential for effective removal of phenolic compounds. For example, the algae *Chlorella vulgaris* is a highly effective bio-cleaner that may be used to recover phenolic chemicals from waste materials (Xiao et al., 2019). This suggests that algae have a significant ability to extract phenol at concentrations lower than 300 mg/L. In a stable algal-bacterial system, the microbial population can efficiently eliminate the phenolic chemicals. Numerous elements define the microbial community's responsibilities, which have a substantial impact on the elimination process. This lends credence to the theory that nitrogen heterocyclic chemicals such as quinoline and indole are better removed by the algae-bacteria combination. Additionally, it suggests that the enrichment of *Chlorobaculum, Aquabacterium, Levilinea*, and *Longilinea* has a major impact on the breakdown of the phenolic compounds.

3.3.3. Dyes

The colour of the material is one of its most striking features. The textile sector uses more than 70% of all dyes. In recent years, the use of synthetic dyes in dying textiles has increased dramatically. These companies' exhaust emissions and wastewater were the primary sources of environmental pollution. Numerous chemical groups can be found in dye compounds (Chu & Phang, 2019). The auxochrome, which improves the dye's binding affinity to the fabric, and the chromophore-one, which gives the fabric its colour, are two distinct components. The charge types and structures of the dyes were used to group them. Algae remove dyes by a variety of natural processes, including biosorption and bioconversion/biodegradation. The functional groups in the algal cell wall allow dye molecules to adsorb and collect on the surface of the algal polymer. It has been demonstrated that the biomass of many microalgae species, including *Spirogyra, Caulerpa vulgaris*, *Caulerpa lentillifera,* and *Caulerpa scalpelliformis*, can effectively eliminate reactive colours. The algae use a variety of enzymes to biodegrade the dyes, and depending on the species, they biodegrade the chemical molecule into a different chemical form. Azoreductase is the enzyme that catalyses the conversion of azo groups into aromatic amine compounds in this process (Lakshmana Prabu et al., 2021). Due to their extreme toxicity, aromatic amine compounds undergo further conversion into simple organic forms, which produces CO_2 + H_2O. The oxidation of various aromatic dye compounds such as phenolic compounds and reduction of O_2 to H_2O by the laccase enzyme are significant processes. Another enzyme of the oxidoreductase family of enzymes that have the oxidation process of phenolic compounds, aromatic amine compounds, and non-phenolic compounds incorporated in their structure is polyphenol oxidase. In addition, the biosorption process of reactive color chemicals from textile wastewaters by *Spirogyra* sp. (nonviable biomass) and *Chlorella* sp. (active biomass) was documented. The biosorption process was employed for decolorizing textile wastewaters by using *Chlorella vulgaris* as a non-viable immobilized microalga. In addition, the biosorption of monoazo dye in the bioconversion process of textile yellow 2G was carried out by *C. vulgaris* (Behera et al., 2021).

3.3.4. Naphthenic Acids

Certain algae have been shown to facilitate the biotransformation or mineralization of recalcitrant organic molecules, including the degradation of naphthalene by blue-green, red, diatomaceous, and green algae. Naphthenic acids (NAs) are recalcitrant carboxylic acids that have been identified as particularly harmful pollutants in aquatic ecosystems (Fan et al., 2022). Demonstrated that electro-oxidation followed by aerobic biodegradation effectively degrades naphthenic acids and completely removes toxicity towards *Vibrio fischeri*, highlighting the synergistic potential of combined approaches for NA detoxification. While that study focused on electrochemical pre-treatment coupled with bacterial degradation, it underscores the importance of integrating biological processes whether bacterial or algal-bacterial—for the remediation of persistent organic pollutants (Abdalrhman et al., 2019).

3.4. Emerging Pollutants: Microplastics and Antibiotic Resistance Genes

As their role in combating emerging contaminants such as microplastics and antibiotic resistance genes is an important and rising field of study, the examined microalgae-bacteria consortia are effective in detoxifying conventional pollutants. Because of their persistence, mobility, and tendency to bioaccumulate, emerging pollutants represent serious risks to aquatic ecosystems and human health. Algal-bacterial biofilms and flocs have the ability to sequester and bioadsorb microplastics (particles smaller than 5 mm). Microplastics can be trapped and consolidated by the sticky EPS produced by consortia, which will aid in their sedimentation and eventual removal from the water column. For instance, the bacterial strains present in the consortia could have the enzyme potential to degrade certain polymers however, this is still under study (Xie et al., 2020). Consortia may also have an effect on the fate of antibiotic resistance genes (ARGs), which have been increasingly recognized to be a significant contaminant of wastewater. The presence of algae may reduce the risk of horizontal gene transfer through the modification of the microbial population structure, the production of antibacterial compounds, or the removal of selective pressure through the effective removal of antibiotics, although bacteria may have antibiotic resistance genes (ARGs) (Zhang et al., 2021).

Figure 1. The Microalgae-Bacteria Symbiotic Cycle in Wastewater Treatment

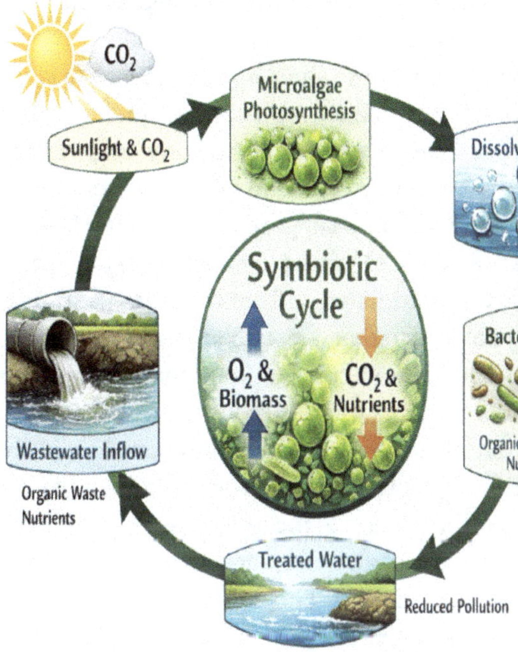

Nevertheless, the risk of ARG transfer among the consortia should be evaluated. The application of the metagenomics and transcriptomics of the evolution of ARGs in Consortia provides a novel perspective on the ecology and genetics of Consortia, which is important for these dynamic pollutants. Some of the promising applications of Consortia include biopolymer degradation, plastics adhesion, and competitive elimination of antibiotic-resistant bacteria. As depicted in Figure 1, the symbiotic relationship between the bacteria and microalgae is the major driving force behind the effectiveness of the consortia. In the consortia, microalgae have the ability to produce oxygen through photosynthesis. This diagram illustrates the mutually beneficial relationship between bacteria and microalgae in wastewater treatment. Through the process of photosynthesis, microalgae generate oxygen, which bacteria use to break down organic waste. In exchange, germs release carbon dioxide and nutrients like phosphate and ammonium, which microalgae utilise to thrive.

4. CO_2 CAPTURE AND SEQUESTRATION BY THE CONSORTIA MITIGATION OF GREENHOUSE GAS

It is constantly necessary to grow these algae in order to meet a variety of purposes, such as fixing CO_2, producing biomass for biogas, producing fine chemicals and extracellular vesicles, and purifying wastewater. An early study from the Carnegie Institution in Washington explored large-scale algal cultivation for CO_2 abatement (Zhibei et al., 2022). One of the greenhouse gases is CO_2. One immediate result of global warming is the need to cut CO_2 emissions. Microalgae and cyanobacteria provide a viable approach to CO_2 mitigation and contribute to the fight against global warming since they fix CO2 at a pace that is between 10–50 times quicker than that of terrestrial plants. It takes about 1.83 kg of CO_2 to produce 1 kilogram of dry algal biomass. Three sources of CO_2 have historically been used in algal cultivation. CO_2-enriched air in closed photobioreactors, up to 1.0%, ambient CO_2 at about 380 ppm, and flue gas from coal-fired power plants, which normally includes 10 to 20% CO_2. Because the Kyoto Protocol offers financial incentives in the form of carbon credits, the potential of algal-bacterial consortia to trap and sequester CO_2 will become increasingly significant (Zhibei et al., 2022). This interaction of the two organisms, in which the algae would be providing the oxygen and the bacteria would be providing the CO_2, would also increase the efficiency of the mechanism of fixation of the CO_2. The generation of renewable material, which would be utilised as fertiliser and act as a biosorbent in order to remove the metals, would help mitigate the CO_2 from biomass-based material, which is one of this method's environmental benefits. The energy savings from reduced aeration and the valorisation of algal biomass provide additional economic and environmental incentives. Future research priorities include selecting bacterial and algal strains based on their CO_2 fixation rates, tolerance to elevated CO_2 and temperature, and compatibility designing optimized bioreactors and developing strains that can thrive under high CO_2 concentrations.

5. OMICS APPROACHES FOR ENHANCEMENT OF BIOREFINERY CAPABILITIES OF MICROALGAE

Considering algae can create a wide range of products and are used in many scientific domains, recent advances in transcriptomics, metabolomics, proteomics, genomics, and other fields have generated new interest in algae. The application of omics techniques has helped us comprehend a range of biological alterations. These include the detection of complex formation, traits of biomolecules, mutational alterations, stress caused by nitrogen deprivation, etc (Lee et al., 2020). While omics techniques are well established in plant and human research, their application to microalgae is still emerging. Understanding the transcriptomic and proteomic responses of microalgae under different conditions is essential for strain improvement and product development. More well-known is the ability of the oleaginous microalgae strain *C. vulgaris* to increase lipid production, which in turn increases the production of biofuel. Important targets for boosting lipid synthesis are revealed by the proteome analysis of the oleaginous strain *C. vulgaris*, both with and without nitrogen. This lack of nitrogen led to the synthesis of fatty acids and the development of lipids, as well as a number of regulators related to transcription factors, the cell cycle, and cell signalling. Thus, the abundance of the product in a nitrogen-deficient state can contribute to the overexpression of the product. If the proteins expressed in the presence and absence of nitrogen in *Tisochrysis lutea* were examined, it was found that 37 proteins were identified as differentially expressed, 17 of which were related to lipid accumulation and N-starvation. In order for the production of biofuel to be enhanced, the green algae species such as *C. pyrenoidosa* and *A. pyrenoidosa* were analyzed under various concentrations of nitrate and copper. The species of algae known as *D. salina* is essential for the synthesis of biosynthetic materials such as biofuel and glycerol since it is capable of adapting to nitrogen starvation, osmotic stress, and oxidative stress. The influence of nitrogen on the transcriptome and proteome analysis of *A. protothecoides* resulted in enhanced biofuel production, especially when combined with the presence of oil and phosphorus starvation. Another type of algae known as *Chromochloris zofingiensis* is also known for the production of biofuel like fatty acids and carotenoids. Research has indicated that the application of the transcriptome and proteome of a type of algae known as *N. gaditana* at different concentrations of light and nitrogen is capable of increasing the production of biofuel (Ahmad et al., 2017, Cao et al., 2020).

6. BIOREACTOR SYSTEMS AND OPERATIONAL STRATEGIES

The type of bioreactor also significantly impacts the microalgae-bacteria consortia's performance. The open systems, such as the raceway pond, provide the following advantages: they are cost-effective and scalable. The main drawback of using the systems could be the difference in environmental changes. On the other hand, by making regular adjustments, the tube, panel, and column types of photobioreactor systems might ensure a greater response in biomass productivity. Hybrid and biofilm reactors may provide more operating stability and make harvesting easier by efficiently improving biomass retention. Controlling important factors including temperature, pH, light intensity, hydraulic holding time, and oxygen rate is necessary for optimal efficiency. Key considerations for practical implementation include scalability, harvesting, dewatering, and energy balance (Daverey et al., 2019) (Table 1).

6.1. AI and IoT Integration for Smart Monitoring and Control

In the case of the microalgae-bacteria systems, artificial intelligence (AI), machine learning (ML), and the Internet of Things (IoT) can be highly beneficial in terms of their contribution to the scalability of the systems. These systems present good opportunities for real-time bioreactor management with the provision of optimal conditions of growth, pollutant removal, and biomass productivities with minimal electrical requirements and human supervision. IoT networks with integrated sensors continuously monitor critical parameters such as temperature, turbidity, algal biomass density, pH, dissolved oxygen, and nutrient concentrations (nitrogen (N) and phosphorus (P)). Machine learning algorithms would be used to improve the quality of the stream of data from this source, which would be used to automate adjustments, identify areas of inefficiency, as well as predict the behavior of the system. The real-time feedback adaptive control systems would adjust the light intensity, aeration rate, or the rate of nutrition to keep the consortia activity optimal. The AI would be used to predict the time of harvesting to get the maximum value of the biomass, as depicted by the growth curve and lipid accumulation patterns. The variables can be simulated and adjusted before actual operation through the use of the use of digital twins which makes it possible to create virtual replicas of actual bioreactors. This removes the need to use the trial-and-error method, which is a part of the scale-up operation (Cheng et al., 2023). Cloud platforms can be used to manage the presence of decentralized water treatment systems, which allows intelligent water management systems to be used in situations of low resources or in a rural setting. Another paradigm shift is the fusion of biotechnology with digital technologies for precise bioremediation and the incorporation of consortia into intelligent monitoring/control systems. With the advancement of edge computing and sensor technologies, intelligent systems will be more accessible in the near future, thus widening their potential application for microalgae-bacteria consortia-based bioremediation in various wastewater systems (Kong et al., 2024).

7. MECHANISMS OF SYNERGISTIC ACTION IN ALGAL-BACTERIAL CONSORTIA

The enhanced performance of microalgae-bacterial consortia in wastewater treatment can be attributed to the various synergistic interactions that make the process stable and efficient. Therefore, this shows that the two interact with each other through their metabolic processes and cannot be described easily (Madhav, 2018).

7.1. Metabolic Exchange and Nutrient Cycling

This basic mechanism of exchange of gases and nutrients is two-way. By giving back the oxygen to the aquatic medium through the process of photosynthesis, the microalgae assist the aerobic bacteria in degrading the organic substances present in the water and reducing the biological oxygen demand (BOD) levels. The algae assimilate the carbon content, which is reduced through the respiration of the bacteria and the carbon dioxide produced through their metabolism. The complex forms of nitrogen and phosphorus present in the water are converted into simpler forms like phosphate and ammonium through the metabolism of bacteria, which can be assimilated directly by the algae (Kasavan et al., 2021).

7.2. Biofilm Formation and Co-aggregation

The biofilm or granule form is used to obtain and sustain the tight aggregation of the bacteria and algae, which often occurs in the case of a consortium. The process by which the bacteria and algae produce and release the Extracellular Polymeric Substances (EPS) leads to the tight aggregation. The advantages of the biofilm include the higher biomass concentration, the gradient effect, and the protective effect of biofilms against environmental stressors. The synergistic effect on the degradation process is a result of the association between the biofilms (Richa et al., 2020).

7.3. Role of Extracellular Polymeric Substances (EPS)

EPS is a vital component of the consortium matrix, which comprises mainly lipids, proteins, polysaccharides, and nucleic acids. The functions of EPS can be listed as:

i. **Entrapment of Pollutants and Biosorption:** The carboxyl, hydroxyl, and other amino groups, which build up the composition of EPS, exhibit a trend to retain heavy metals, dyes, and other organic pollutants through ionic, precipitation, and complexation reactions, respectively.
ii. **Protective Barrier:** The microbial cell is shielded from harmful substances, e.g., antibiotics and phenolic compounds, which have limited bioavailability because of the presence of EPS.
iii. **Enzyme Stabilization:** The functionality of enzymes necessary to degrade complex biofilm structures, as well as the retention of extracellular enzymes

secreted by microorganisms, such as laccase, phosphatase, and azoreductase enzymes, is preserved by EPS.

7.4. Signal-Mediated Interactions and Community Stability

Recent research has shown that signaling compounds from other kingdoms of life, e.g., algal exudates or quorum sensing peptides, can control population dynamics or functional responses. The consortium can respond to a range of wastewater loads because these compounds have a significant effect on biofilm balance, enzyme production, and even stress responses. To achieve a stable and long-term operation with a range of wastewater loads, e.g., hydroponic or industrial waste, it is important to note that a stable and robust consortium is not a random mixture of microbes, but rather a microbial network of interaction (Zhou et al., 2020).

8. MECHANISMS OF POLLUTANT REMOVAL

There are three main processes involved in the removal of pollutants by algae, which are biosorption, bioaccumulation, and biodegradation. These mechanisms are often simultaneous, thus improving the overall efficiency of treatment. This is a schematic diagram illustrating the three major mechanisms whereby microalgae-bacteria consortia remove contaminants from wastewater. Biosorption is the binding of contaminants to the surface of the cells. Bioaccumulation is the uptake of contaminants into the cells, which are stored. Biodegradation is the enzymatic degradation of contaminants to simpler, less toxic substances. These mechanisms are often simultaneous, thus improving the overall treatment efficiency.

8.1. Biodegradation

Biodegradation is the breakdown of organic pollutants by naturally occurring microorganisms. Like bacteria, algae produce enzymes that cleave bonds in complex organic molecules, converting them into simpler, less harmful compounds (Huy et al., 2018). Some of the enzymes in this category include reductase, oxidoreductase, and oxidative. Indeed, further research is necessary to acquire more information in terms of biochemistry related to dye compounds, as different compounds result in different products.

8.2. Biosorption

Biosorption is a physicochemical process in which pollutants (sorbates) are removed from solution by biological materials (sorbents). These sorbents can be living or dead biomass, and may be organic or inorganic in nature. The process does not require metabolic activity and is often rapid (Li et al., 2021).

8.2.1. Passive Biosorption

The chemical, nutrient, or contaminant present in the wastewater interacts or reacts with the dead biomass cell wall to start the passive biosorption process. This is attributable to the functional groups mentioned above. This is a fast process, and its time span is five to ten minutes without the need for a metabolic response.

8.2.2. Active Biosorption

The process involved in the active biosorption, and where, in some cases, biosorption is equivalent to bioaccumulation, is the one that occurs in the wet biomass or cells. Adsorption is the initial phase, and it also implies the process of passive biosorption, which is part of the two-part process involved in active biosorption. The type of metabolism involved in the process is called fast-rate metabolism, and the bonding agent is not needed during the time these reactions are occurring. The transporting of the agents, or bonding, involved in the slow-rate metabolism, forms the second part of the process involved in the above-discussed phenomenon (Kasavan et al., 2021).

9. INTEGRATION WITH OTHER WASTE-TO-ENERGY TECHNOLOGIES

It was found that adding waste-to-energy techniques, particularly anaerobic digestion (AD) and microbial fuel cells (MFCs), to wastewater treatment processes based on microalgae and bacteria might significantly improve the process's sustainability and energy balance (Jayabal., 2025). This synergy forms the basis for integrating algae-bacteria systems with anaerobic digestion and microbial fuel cells. Algal biomass from WWTPs can be utilised as feedstock for AD in an integrated system that combines algae and AD. Digestate and biogas, which includes methane, are produced by AD. The biogas that is generated can be used as fuel, but the CO2 that is present in the flue gas can be recycled at the location where the algae growth occurs, increasing photosynthesis. The generated AD can also be used to supply nutrients to algae, which are simultaneously recovered from the WWTPs as phosphate and ammonia content (Schlembach et al., 2021). Alternatively, direct energy and clean wastewater can be produced at the same time using microbial fuel cells (MFCs) and microalgae-bacteria consortia. In this case, the bacteria oxidise the organic compounds in the anode to produce electrons that travel to the cathode via an external circuit. There, algae (or algae-powered oxygen production) aid in the reduction reaction. Algal photosynthesis at the cathode boosts the oxygen supply and improves MFC performance. These algal-MFC systems have the potential to clean wastewater, create renewable electricity, and produce algae biomass for later use (Wu et al., 2025). These integrations are excellent illustrations of the circular bioeconomy in action because they transform linear waste flows into interrelated cycles of energy, nutrients, and clean water. Further research should focus on managing scale-up difficulties, optimising system design, and conducting comprehensive life-cycle assessments to validate the overall economic and environmental benefits of such hybrid setups (Li et al., 2020).

10. ECONOMIC AND LIFE-CYCLE ANALYSIS

The economic feasibility and environmental impact of microalgae-bacteria-based systems should be carefully evaluated for larger-scale development. According to techno-economic analysis, these systems do have certain expenses, but they can be offset by the revenue generated from the final product, the value of biomass, biofuels, animal husbandry, fertilizers, and high-value biochemicals that are produced (Pandey & Sinha, 2022). Microalgae-bacteria systems have a lower environmental impact than conventional processes due to carbon sequestration, reduced chemical use, and energy recovery that generates valuable end products A comparative analysis of conventional WWTPs and microalgae-bacteria systems indicates that there is a wide gap between them in terms of economic and environmental viability. Table 1 presents a comparison of some of the important parameters associated with conventional WWTPs and microalgae-bacteria systems, along with their economic and environmental feasibility. This is another factor that reinforces global sustainability strategies and provides a new pathway towards a sustainable bio-based economy, as proposed by Jiang et al. (2021).

Table 1. Economic and Environmental Impact: Conventional vs. Algal Systems

Parameter	Conventional WWTP	Microalgae-Bacteria System
Energy Use	High (aeration, sludge handling)	Low to moderate (solar-driven)
Sludge Production	High (waste sludge)	Low (valuable biomass)
Carbon Footprint	High (CO_2 emissions)	Negative (CO_2 sequestration)
Operational Cost	High	Moderate (offset by products)
Revenue Streams	None / tipping fees	Biofuel, feed, fertilizer, credits
Chemical Use	High (coagulants, disinfectants)	Minimal (biological process)

11. CASE STUDIES AND REAL-WORLD APPLICATIONS

The mentioned increase in the number of expanding microalgae-bacteria consortium projects worldwide is evidence of the practical application of the microalgae-bacteria consortium technology. It has reportedly been successful in treating municipal wastewater using the microalgae-bacteria consortium method. Successful applications have been reported in Spain, New Zealand, and the United States (Abbas & Halog, 2021). The practical application of microalgae-bacteria consortium technology is required in order to treat wastewater containing PPCPs from textile colors, medicines, as well as wastewater rich in phosphate and nitrate (Abbas & Halog, 2021). Hydroponics is one notable example of the application of microalgae-bacteria consortium technology. The feasibility of the microalgae-bacteria consortium technology is further supported by the fact that there is already a large number of pilot plants and full-scale plants for different types of wastewater. Table 2 shows an overview of a number of case studies on different types of wastewater, such as municipal, industrial, and agricultural. Decentralised wastewater treatment is another potential application for microalgae-bacteria consortium technology (Zhibei et al., 2022). Apart from the proposed application, there are a number of different applications for microalgae-bacteria consortium technology, especially for decentralised systems.

Table 2. Summary of Selected Pilot and Full-Scale Case Studies

Application Context	Key Performance	Valorization Output
Municipal WWTP (New Zealand)	>90% N & P removal	Biomethane, biofertilizer
Textile Effluent (India)	95% dye removal, detoxification	Biogas, reusable water
Hydroponic Integration (Netherlands)	75% water saving, nutrient recycling	Liquid fertilizer, irrigation reuse

Application Context	Key Performance	Valorization Output
Acid Mine Drainage (USA)	80% heavy metal removal	Metal recovery, biomass pellets
Agricultural Runoff (Spain)	85% nitrate reduction	Biodiesel, animal feed

12. CHALLENGES AND FUTURE PROSPECTS

12.1. Technical and Operational Challenges

Despite the promising potential offered by microalgae-bacteria consortiums, there have been a number of challenges which have slowed down the full exploitation and usage of the technology. First off, the costly and energy-intensive nature of the harvesting and dewatering process of the microalgae poses a challenge that needs to be worked on. In addition, the seasonal changes in temperature and light availability may have some impact on the efficiency of the PBRs designed for outdoor applications. Thirdly, the growth of the consortia in a stable environment over a long period of time is a challenge. This is because of the dynamic nature of the consortia and the possible invasion by impure microalgae species. Therefore, to overcome the challenges, new methods need to be developed for the harvesting of the microalgae, new strains need to be developed which will withstand the seasonal changes, and finally, controlled environment bioreactors need to be designed to optimize the growth of the microalgae.

12.2. Future Prospects and Interdisciplinary Convergence

This would be achieved through the development of interdisciplinary synergies, including the development of new tools in the areas of synthetic biology and metabolic engineering, which would facilitate the development of specific-to-the-task consortia of microbes, which would be optimised to degrade a specific pollutant or produce a specific product of interest. Another area that would be explored in the future is materials science, which would allow the development of new, less expensive biofilm supports or harvesting materials. Finally, as was mentioned in Section 6.1, the development of new tools in the field of digital technologies would allow the optimization of biological treatment systems to an unprecedented extent (Yang et al., 2023). In order to use this technology to its fullest potential, however, synergistic interactions of the communities, the economy, the government, and the sciences would be required. In order to advance from the pilot-scale successes to the global level, the proper alignment of environmental goals with the economy and society would be required (Cheng et al., 2023).

13. CONCLUSION

Consortia of bacteria and microalgae have the potential to revolutionise wastewater treatment by shifting the focus from pollution reduction to the creation of opportunities for sustainable resource recovery. This is because the microalgae-bacteria consortia have the ability to sequester carbon, efficiently eliminate pollutants and nutrients from wastewater through natural mutually beneficial processes, and use waste water as a resource to generate valuable biomass that can be used to produce feed, energy, and other bioproducts. The path forward in the development of these biotechnologies is still marked by a number of obstacles that must be addressed in the future, even with the significant advancements in the design and development of bioreactors and the process optimisation achieved through the use of omics. The future of water management is to be seen as part of an integrated vision which includes the creation of closed cycles of water, waste, and energy through biological innovation. And to create a sustainable future, this is a must, not an option.

REFERENCE

Abbas, S., & Halog, A. (2021). Analysis of Pakistani textile industry: Recommendations towards circular and sustainable production. In *Circular Economy and Sustainability* (pp. 77–111). Elsevier. DOI: 10.1007/978-981-16-3698-1_3

Abdalrhman, A. S., Zhang, Y., & Gamal El-Din, M. (2019). Electro oxidation by graphite anode for naphthenic acids degradation, biodegradability enhancement and toxicity reduction. *The Science of the Total Environment*, *671*, 270–279. DOI: 10.1016/j.scitotenv.2019.03.262 PMID: 30928756

Abinandan, S., Subashchandrabose, S. R., Venkateswarlu, K., & Megharaj, M. (2018). Microalgae–bacteria biofilms: A sustainable synergistic approach in remediation of acid mine drainage. *Applied Microbiology and Biotechnology*, *102*(3), 1131–1144. DOI: 10.1007/s00253-017-8693-7 PMID: 29260261

Ahmad, A., Srivastava, S., & Ali, B. M. J. (2017). A FRET based optical sensor for selective detection of heavy metal ions. *Algal Research*, *26*, 354–361.

Behera, M., Nayak, J., Banerjee, S., Chakrabortty, S., & Tripathy, S. K. (2021). A review on the treatment of textile industry waste effluents towards the development of efficient mitigation strategy: An integrated system design approach. *Journal of Environmental Chemical Engineering*, *9*(4), 105277. DOI: 10.1016/j.jece.2021.105277

Cao, X., Li, H., Zhou, Y., & Song, C. (2020). The shift of mutualistic relationships among algae, free living and attached bacteria through different nutrient addition mode: A mesocosm study. *Journal of Freshwater Ecology*, *35*(1), 535–548. DOI: 10.1080/02705060.2020.1858984

Cheng, J., Li, K., & Yang, Z. (2023). Digital twins for photobioreactor optimization: From lab to pilot scale. *Algal Research*, *69*, 102943. DOI: 10.1016/j.algal.2022.102943

Chia, W. Y., Ying Tang, D. Y., Khoo, K. S., Kay Lup, A. N., & Chew, K. W. (2020). Nature's fight against plastic pollution: Algae for plastic biodegradation and bioplastics production. *Environmental Science and Ecotechnology*, *4*, 100065. DOI: 10.1016/j.ese.2020.100065 PMID: 36157709

Chu, W. L., & Phang, S. M. (2019). Biosorption of heavy metals and dyes from industrial effluents by microalgae. In Alam, M., & Wang, Z. (Eds.), *Microalgae Biotechnology for Development of Biofuel and Wastewater Treatment* (pp. 599–634). Springer., DOI: 10.1007/978-981-13-2264-8_23

Daverey, A., Pandey, D., Verma, P., Verma, S., Shah, V., Dutta, K., & Arunachalam, K. (2019). Recent advances in energy efficient biological treatment of municipal wastewater. *Bioresource Technology Reports*, *7*, 100252. DOI: 10.1016/j.biteb.2019.100252

Dongyu, C., Lingchao, K., Yi, W., Yuanqing, Z., & Chuanlun, Z. (2022). In situ identification of environmental microorganisms with Raman. *Environmental Science and Ecotechnology*, *11*, 100187. DOI: 10.1016/j.ese.2022.100187 PMID: 36158754

. European Commission. (2020). *A new Circular Economy Action Plan*. COM(2020) 98 final.

Fan, H., Wang, Y., Wang, Y., & Coyte, P. C. (2022). The impact of environmental pollution on the physical health of middle aged and older adults in China. *Environmental Science and Pollution Research International*, *29*(3), 4219–4231. DOI: 10.1007/s11356-021-15832-z PMID: 34403062

Ho, A., Di Lonardo, D. P., & Bodelier, P. L. E. (2017). Revisiting life strategy concepts in environmental microbial ecology. *FEMS Microbiology Ecology*, *93*(3), fix006. Advance online publication. DOI: 10.1093/femsec/fix006 PMID: 28115400

Huy, M., Kumar, G., Kim, H., & Kim, S. (2018). Photoautotrophic cultivation of mixed microalgae consortia using various organic waste streams towards remediation and resource recovery. *Bioresource Technology*, *247*, 576–581. DOI: 10.1016/j.biortech.2017.09.108 PMID: 28982087

Jayabal, R. (2025). Microalgae-bacteria systems in wastewater treatment and biofuel production: Challenges, advances, and future perspectives. *Algal Research*, *91*, 104228. DOI: 10.1016/j.algal.2025.104228

Jia, H., & Yuan, Q. (2018). Ammonium removal using algae bacteria consortia: The effect of ammonium concentration, algae biomass, and light. *Biodegradation*, *29*(2), 105–115. DOI: 10.1007/s10532-017-9816-7 PMID: 29236193

Jiang, R., Qin, L., Feng, S., Huang, D., Wang, Z., & Zhu, S. (2021). The joint effect of ammonium and pH on the growth of *Chlorella vulgaris* and ammonium removal in artificial liquid digestate. *Bioresource Technology*, *325*, 124690. DOI: 10.1016/j.biortech.2021.124690 PMID: 33465643

Kasavan, S., Yusoff, S., Guan, N. C., Zaman, N. Q. K., & Fakri, M. F. R. (2021). Global trends of textile waste research from 2005 to 2020 using bibliometric analysis. *Environmental Science and Pollution Research International*, *28*(33), 44780–44794. DOI: 10.1007/s11356-021-15303-5 PMID: 34235692

Kong, W., Kong, J., Feng, S., Yang, T., Xu, L., Shen, B., Bi, Y., & Lyu, H. (2024). Cultivation of microalgae-bacteria consortium by waste gas-waste water to achieve CO_2 fixation, wastewater purification and bioproducts production. *Biotechnology for Biofuels and Bioproducts*, *17*(1), 26. Advance online publication. DOI: 10.1186/s13068-023-02409-w PMID: 38360745

Kube, M., Jefferson, B., Fan, L., & Roddick, F. (2018). The impact of wastewater characteristics, algal species selection and immobilisation on simultaneous nitrogen and phosphorus removal. *Algal Research*, *31*, 478–488. DOI: 10.1016/j.algal.2018.01.009

Kumar, G., Huy, M., & Kim, S. (2018). Evaluation of gradual adaptation of mixed microalgae consortia cultivation using textile wastewater via fed batch operation. *Biotechnology Reports (Amsterdam, Netherlands)*, *20*, e00289. DOI: 10.1016/j.btre.2018.e00289 PMID: 30416980

Lakshmana Prabu, S., Umamaheswari, A., & Puratchikody, A. (2021). Bioremediation approaches for degradation of textile dye effluents. In Marimuthu, P. D., Sundaram, R., Jeyaseelan, A., & Kaliannan, T. (Eds.), *Bioremediation and Green Technologies* (pp. 113–133). Springer., DOI: 10.1007/978-3-030-64122-1_9

Lee, J., Hyeon, D. Y., & Hwang, D. (2020). Single cell multiomics: Technologies and data analysis methods. *Experimental & Molecular Medicine*, *52*(9), 1428–1442. DOI: 10.1038/s12276-020-0420-2 PMID: 32929225

Li, S., Li, X., & Ho, S.-H. (2021). Microalgae as a solution of third world energy crisis for biofuels production from wastewater toward carbon neutrality: An updated review. *Chemosphere*, *291*, 132863. DOI: 10.1016/j.chemosphere.2021.132863 PMID: 34774903

Li, X., Wang, X., Zhang, Y., Ding, N., & Zhou, Q. (2020). Algal bacterial consortia for enhanced bioelectricity generation and wastewater treatment in microbial fuel cells. *The Science of the Total Environment*, *727*, 138691. DOI: 10.1016/j.scitotenv.2020.138691

Li, Y., Xu, Y., Fu, Z., Li, W., Zheng, L., & Li, M. (2021). Assessment of energy use and environmental impacts of wastewater treatment plants in the entire life cycle: A system meta analysis. *Environmental Research*, *198*, 110458. DOI: 10.1016/j.envres.2020.110458 PMID: 33188763

Madhav, S., Ahamad, A., Singh, P., & Mishra, P. K. (2018). A review of textile industry: Wet processing, environmental impacts, and effluent treatment methods. *Environmental Quality Management*, *27*(3), 31–41. DOI: 10.1002/tqem.21538

Mondal, S., & Palit, D. (2022). Challenges in natural resource management for ecological sustainability. In *Natural Resources Conservation and Advances for Sustainability* (pp. 29–59). Elsevier., DOI: 10.1016/B978-0-12-822976-7.00004-1

Pandey, A., & Sinha, A. (2022). IoT enabled smart algal bioreactors for wastewater treatment and bioenergy production. *Journal of Cleaner Production, 365*, 132675. DOI: 10.1016/j.jclepro.2022.132675

Richa, A., Touil, S., Fizir, M., & Martinez, V. (2020). Recent advances and perspectives in the treatment of hydroponic wastewater: A review. *Reviews in Environmental Science and Biotechnology, 19*(4), 945–966. DOI: 10.1007/s11157-020-09555-9

Schlembach, I., Grünberger, A., Rosenbaum, M. A., & Regestein, L. (2021). Measurement techniques to resolve and control population dynamics of mixed culture processes. *Trends in Biotechnology, 39*(10), 1093–1109. DOI: 10.1016/j.tibtech.2021.01.006 PMID: 33573846

Shekharaiah, P. C., Sanyal, D., Dasgupta, S., Sapre, A., & Banik, A. (2020). Heavy metal mitigation with special reference to bioremediation by mixotrophic algae bacterial protocooperation. In *Cellular and Molecular Phytotoxicity of Heavy Metals* (pp. 305–334). Springer., DOI: 10.1007/978-3-030-45975-8_15

Tournier, V., Topham, C. M., Gilles, A., David, B., Folgoas, C., Moya-Leclair, E., & Pelloux-Prayer, H. (2020). An engineered PET depolymerizing enzyme breaks down PET plastics. *Nature Chemical Biology, 16*(7), 776–782. DOI: 10.1038/s41589-020-0550-4 PMID: 32367018

Wang, S., Ji, B., Zhang, M., Ma, Y., Gu, J., & Liu, Y. (2020). Defensive responses of microalgal bacterial granules to tetracycline in municipal wastewater treatment. *Bioresource Technology, 312*, 123605. DOI: 10.1016/j.biortech.2020.123605 PMID: 32504951

Wu, Z., Shi, W., Yuan, W., Chen, Z., Xie, Y., Lv, Z., Xu, J., Amadu, A. A., Qiu, S., & Ge, S. (2025). Development and operation of indigenous microalgal-bacterial consortium system treating eutrophic lake water: Consortium identification and system demonstration. *Bioresource Technology, 132462*, 132496. Advance online publication. DOI: 10.1016/j.biortech.2025.132496 PMID: 40204026

Xiao, M., Ma, H., Sun, M., Yin, X., Feng, Q., Song, H., & Gai, H. (2019). Characterization of cometabolic degradation of p-cresol with phenol as growth substrate by *Chlorella vulgaris*. *Bioresource Technology, 281*, 296–302. DOI: 10.1016/j.biortech.2019.02.079 PMID: 30826515

Xie, P., Chen, C., Zhang, C., Su, G., Ren, N., & Ho, S.-H. (2020). Revealing the role of adsorption in ciprofloxacin and sulfadiazine elimination routes in microalgae. *Water Research, 172*, 115475. DOI: 10.1016/j.watres.2020.115475 PMID: 31972413

Yang, Y., Ge, S., Pan, Y., Qian, W., Wang, S., Zhang, J., & Zhuang, L.-L. (2023). Screening of microalgae species and evaluation of algal lipid stimulation strategies for biodiesel production. *The Science of the Total Environment, 857*, 159281. DOI: 10.1016/j.scitotenv.2022.159281 PMID: 36216060

Zhang, Y., Li, A., Dai, T., & Wang, Y. (2021). Fate of antibiotic resistance genes in an algal bacterial system treating wastewater: Insights from metagenomic analysis. *Environmental Science & Technology, 55*(14), 9910–9919. DOI: 10.1021/acs.est.1c01471

Zhibei, C., Wenlong, Y., Mingzhu, D., & Yingjin, Y. (2022). Construction of microbial consortia for microbial degradation of complex compounds. *Frontiers in Bioengineering and Biotechnology, 10*, 1051233. DOI: 10.3389/fbioe.2022.1051233 PMID: 36561050

Zhou, J., Lyu, Y., Richlen, M. L., Anderson, D. M., & Cai, Z. (2020). Quorum sensing mediated microbial interactions in algal bacterial consortia: A review. *Bioresource Technology, 312*, 123589. DOI: 10.1016/j.biortech.2020.123589

Chapter 3
Algal-Bacterial Interactions for Advanced Wastewater Treatment and Resource Recovery

Ifrah Javed
http://orcid.org/0009-0006-3735-7830
University of Agriculture, Faisalabad, Pakistan

Muhammad Salman Javed
Independent Researcher, Malaysia

ABSTRACT

The increasing volume of emerging contaminants in wastewater poses a major threat to the environment. This study explores the ground-breaking potential of the algal-bacterial interacting system as an eco-friendly and sustainable alternative for water purification and resource recovery. This study explains the basic mechanism of algal-bacterial interactions, including symbiosis, gene transfer, and quorum sensing, by enhancing the system efficiency and resilience. These systems demonstrate high proficiency in eliminating pollutants through biodegradation. The harvested algal-bacterial biomass is utilized as a feedstock for the circular economy by enabling the synthesis of biogas, biodiesel, bioplastic, and biofertilizers. Algal-bacterial interactions face some challenges during wastewater treatment and resource recovery. This study connects the microbiology, economics, and environmental studies for ABS as an integrated solution of transforming wastewater treatment to a resource-generating platform by aligning with global sustainability goals.

DOI: 10.4018/979-8-2600-3080-6.ch003

1. INTRODUCTION

Water is a vital component for all living beings on Earth, as it covers almost 71% of Earth's surface. Recently, water shortage has become a considerable global issue due to the overexploitation of natural water resources (Oruganti *et al.*, 2022). Wastewater is polluted water produced from rainwater runoff and human activities, including domestic, industrial, and commercial uses. Wastewater is composed of approximately 99.9% water, with the remaining 0.1% being what has been removed. This 0.1% includes organic matter, nutrients, microorganisms, and various inorganic compounds (Aleruchi, Fubara, & Otto, 2023).

With the increase in urbanization, agricultural activities and industrialization, the volume of wastewater generated rises, containing harmful pollutants that contaminate the water (Saravanan *et al.,* 2021). Among the various contaminants in wastewater are inorganic and organic compounds, toxic heavy metals, pharmaceuticals, and fluorinated compounds. They pose serious threats to human health and other living organisms due to their toxicity and acquisitive nature. Untreated wastewater causes various human health problems, including dysentery, cholera outbreaks and typhoid. By discharging into lakes, rivers, and oceans, decaying organic matter and debris in wastewater deplete dissolved oxygen, which is crucial for aquatic organisms to survive (Silva, 2023). Its presence in aquatic habitats deteriorates the downstream water quality.

Water contamination poses significant challenges, e.g., water shortage for drinking and all other essential life activities in households, agriculture, and industries. Wastewater treatment plays a crucial role in safeguarding the environment, protecting public health, and ensuring the sustainable management of all water resources (Karimi-Maleh *et al.,* 2020).

At present, wastewater treatment faces significant pressure globally to reduce carbon emissions, conserve energy, and recover resources. Traditional physical, biological, and chemical techniques for wastewater purification have been widely used, but they have limitations in operational and technical aspects. They are very costly, energy-intensive, labor-intensive, and produce secondary waste that requires additional attention. Sometimes these techniques are not suitable for all types of wastewater (Garrido-Baserba *et al.*, 2024). Alternative processes are urgently needed to swiftly eliminate wastewater contaminants without harming the environment.

Biological processes have received considerable attention due to their cost-effectiveness, improved efficiency, lack of toxic by-product generation, and environmental sustainability, achieved through the use of microorganisms (Oruganti *et al.*, 2022). Algal-bacterial systems have been a highly promising approach in recent years due to their unique capabilities and environmental benefits, including high pollutant removal efficiency, carbon capture, bioresource production, and greenhouse gas sequestration. They can remove nutrients and emerging contaminants from wastewater. Besides, algae also show good efficiency in removing all pathogens. The microalgal-bacterial wastewater treatment system (MBWTS) is a sustainable alternative for achieving high CO_2 fixation and adding value by removing P and N from wastewater without providing additional carbon sources. This mutualistic symbiosis creates a highly efficient, low-energy treatment system that aligns perfectly with principles of the circular economy (Saravanan *et al.*, 2021).

Bacteria use the oxygen produced by algae during photosynthesis, whereas algae use the carbon dioxide produced by bacteria, thereby establishing a symbiotic relationship. A large number of bacteria in the water ecosystem interact with algae through a specific phenomenon, thereby acting as natural water purifiers. Single-algae systems exhibit much lower removal efficiency than algal-bacterial systems. *Euglena, Oscillatoria, Chlamydomonas, Scenedesmus, Chlorella, Nitzschia, Navicula,* and *Stigeoclonium* are highly tolerant algal genera found in polluted water. The symbiotic interaction between algae and bacteria can be described as commensalism, mutualism, and parasitism. Algae use nutrients from wastewater as a renewable energy source. Microbes linked with wastewater interrelate directly or indirectly with the algae through a symbiotic relationship, which can hinder the growth of the species involved (Abdel-Raouf, Al-Homaidan, & Ibraheem, 2012).

Algal-bacterial consortia offer a promising technology for wastewater purification and resource recovery. They capture the sources, including nutrients and carbon, during treatment. Bacteria-algal interacting systems reduce the emission of carbon and promote the recycling of resources in wastewater purification. It provides valuable biofuels and makes it a highly sustainable option for resource recovery from the environment (S. Liu *et al.*, 2024).

This chapter investigates the fundamental biochemical mechanisms underlying algal-bacterial interactions in wastewater and the system's proficiency in removing nutrients, degrading organic carbon, and breaking down emerging contaminants. This study explores the biological system within the circular economy model, focusing on pathways for decontaminating and converting harvested algal-bacterial biomass into biofuels (e.g., biodiesel), bioplastics (e.g., PHA), and biofertilizers. This chapter aims to connect microbiology, environment, engineering, and economics to guide future research and applications.

2. FUNDAMENTALS OF ALGAL-BACTERIAL INTERACTIONS

Algal-bacterial interactions in wastewater treatment are based on a mutualistic relationship. They interact through symbiosis, signalling transduction, and gene transfer mechanisms to increase their efficiency. The algal-bacterial system exhibits a highly efficient design and operation, enabling faster removal rates and lower operational costs. For example, *Chlamydomonas reinhardtii and Saccharomyces cerevisiae* can metabolize and interact with one another by exchanging available nitrogen and carbon sources. *Chlorella vulgaris* & *Microcystis aeruginosa* also form efficient algal-bacterial co-cultures (Xiong *et al.*, 2023).

2.1. Symbiosis

Algae and bacteria exhibit symbiosis, in which both partners get benefit. This relationship has a positive impact on algal-bacterial structure, function, and metabolism, which is crucial for altering the ecological functions and diversity of the microbial community. To attain an ideal wastewater treatment system, algae-bacteria symbiosis (ABS) is an emerging technique that has received widespread attention in recent years. This system primarily focuses on biomass growth and the removal of toxic heavy metals and nutrients, requiring very low energy and offering great potential for resource recycling (Ren *et al.,* 2024).

Nutrient Exchange

Algae release Dissolved Organic Carbon (DOC), which bacteria take up to feed on. Algae use CO_2 for photosynthesis, take up nutrients (N, P), and release O_2.

$$6CO_2 + 6H_2O + \text{Light}\ C_6H_{12}O_6 + 6O_2$$

Bacteria, in return, utilize the O_2 from algae for metabolism to break down ammonia and organic matter. During their metabolism, bacteria release inorganic carbon, nitrogen, and phosphorus as byproducts, which are utilized by algae. They also provide growth factors (like B vitamins) for algae (Abinandan & Shanthakumar, 2015).

Figure 1. Schematic representation of algal-bacterial symbiotic interaction (Oruganti et al., 2022)

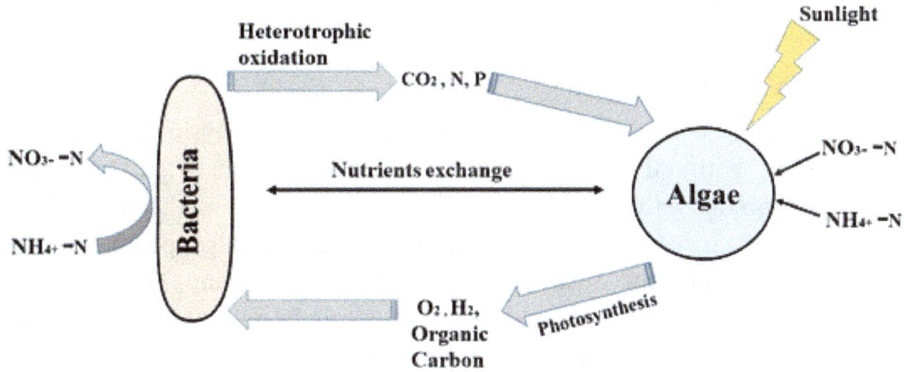

A consortium of *Pseudomonas putida* (a Gram-negative bacterium) and *Chlorella vulgaris* (a Green microalgae), *Agrobacterium sp.*, and *Ensifer adherens* symbionts with *Chlorella vulgaris* can effectively degrade heavy metals from wastewater. Algae–bacteria symbiotic systems are used to remove low-carbon, efficient nitrogen from wastewater and to degrade organic matter.

Bio Flocculation

Microalgal surfaces act as a habitat for bacteria. Algal growth protects the bacteria from optimal contamination. The stability of the algae and bacteria depends on microbial adhesion and fixation to form biomass aggregates. Algae and bacteria stick to the extracellular polymeric substances (EPS), a sticky matrix that binds them into cohesive clusters. Microalgae and bacteria aggregate to form bioflocs using extracellular polymeric substances (EPS) produce by themselves, which is called bioflocculation. It can minimize mechanical disturbance, including excessive aeration, which is crucial for allowing aggregates to form and remain stable. They are formed with a smaller volume and lower stability than the activated sludge granules. Bioflocculation is usually the last stage of cultivation (H. Wang, Deng, Qi, & Wang, 2022).

2.1.1 Symbiotic Forms of Algae and Bacteria

The symbiotic interaction between algae and bacteria results in various forms, including Suspended Algal-Bacterial Consortia, attached growth systems, and algal-bacterial biofilms.

a) Suspended Algal-Bacterial Consortia

A mixed culture of algal-bacterial consortium grows freely in the wastewater stream and floats within the treatment medium. Bacteria attach to the algal surface and form clumps (flocs) that settle readily. Suspended growth method of ABS system is mostly performed in the open ponds, such as high-rate algal ponds (HRAPs), photo bioreactors, and active algae treatment systems (Jin, Zhang, Zhou, Zhang, & Wu, 2024). Open bioreactors are pond systems that can be natural or artificial. Suspended growth processes can successfully complete their mixing stage by promoting cross-linking between the microorganisms and wastewater treatment. Biomass is kept in suspension to prevent precipitation.

b) High-Rate Algal Ponds (HRAPs)

Shallow, open bioreactors are typically formed by mixing microalgae and bacteria in traditional biological stabilization ponds (TBSPs). They possess short treatment times (4-10 days), low cost, low energy consumption, and are ideal for sunny regions as compared to other sludge treatment techniques. They can produce valuable biomass for resource recovery. HRAP design features a raceway-style pond with a shallow water depth (0.2-1.0m). A paddle wheel is used to gently mix the system with horizontal flow velocity (0.15-0.3ms^{-1}). The proliferation stage of microalgae can be observed here, leading to nutrient removal through assimilation into biomass (Craggs, Heubeck, Lundquist, & Benemann, 2011). Microalgae production and wastewater treatment in HRAPs are enhanced by optimizing light penetration by adjusting pond depth or hydraulic retention time (Park, Craggs, & Shilton, 2011). The drawbacks of this system are the difficulty in separating the biomass after treatment and the requirement for a large operating area.

c) Attached Growth System

Biomass is attached to a support medium by increasing algae density, making it easier to collect. They are represented by biofilm reactors. Wastewater flows through the bioreactor, resulting by taking long residence times than HRTs for algae and bacteria. This growth process exhibits a higher degradation rate of all nutrients by using shorter HRTs because communities are remained in the reactor at a slower growth rate. It can operate to reduce the reactor footprint at great depths. Cell entrapment and algal turf scrubbers are main applications of this system.

d) Algal-Bacterial Granular Sludge (ABGS)

Algae and bacteria can form dense, fast-settling granules (small balls) by living in close proximity to each other. They can be cultivated from regular sludge under light. Microalgae have the ability to attach to the surface of activated sludge, forming small granules in the initial cultivation stage. In this structure, activated sludge has bacteria present inside, while microalgae are outside. Microalgae-activated sludge granules have a size range from μm to mm (Wang *et al.,* 2022). This system has garnered considerable attraction due to the exceptional settling ability. This ability is mainly developed due to the mutualism between algae and bacteria. Formation of this symbiotic form can be accelerated by adding fungi or encapsulating algae/bacteria in alginate. They solve the separation problem of suspended systems. This system can remove more than 90% of nutrients (Nitrogen & Phosphorus) without external aeration. It can produce net energy gain and reduce greenhouse gas emissions. During this system, long-term stability is a main challenge in the real world (L. Tang, Gao, Liang, Wang, & Wang, 2024).

2.1.2 Photobioreactors

Photobioreactors (PBRs) can provide controlled, optimal environmental conditions, including nutrients, light, temperature, and mixing, for the growth of algae. In open cultivation systems, the monoculture strains are difficult to maintain due to the risk of attack by other species.

Closed photobioreactors offer a small land footprint, higher harvesting efficiency, lower water evaporation, and greater photosynthesis than high-rate algal ponds. They can improve mass transfer, reduce contamination risk, and better control to study specific algae species and their removal abilities (Oruganti *et al.*, 2022).

Flat-plate bioreactors use low energy, are easy to maintain, and achieve high biomass yields. They consist of a flat, thin surface medium to attach the microalgae. They have a high surface-to-volume ratio and require minimal mechanical parameters, but they struggle to control temperature. *Scenedesmus Obliquus* and *Chlorella vulgaris cultivated in vertical flat-plate photobioreactors remove more than 99% of* nutrients at a retention time of 8 days to treat municipal wastewater (I.-S. Yang et al., 2016).

Column PBR has excellent gas-liquid mass transfer and promotes bacteria-algae granulation. A major challenge for suspended systems is separating tiny, water-like-density algal biomass from treated water. Magnetic nanoparticles can be used to attach bacteria and algae, which allows easy separation in a magnetic field for large-scale applications (Ren *et al.,* 2024)

a) Hybrid Biofilm Photobioreactors

This photobioreactor uses the symbiotic relationship between algae and bacteria to purify wastewater. Biofilm structure offers a stable environment for both organisms, enhancing nutrient uptake and pollutant removal. Hydraulic retention time (HRT) of less than 24 hours can be achieved using 3-dimensional media as a biofilm carrier. Polyurethane sponge cubes are used in the trickling filter for algae and bacteria to treat domestic wastewater, achieving 90%, 24%, and 37% removal of C, N, and P within 8 hours of HRT (Katam, Shimizu, Soda, & Bhattacharyya, 2020).

b) Algal Turf Scrubber (ATS)

ATS is used to promote natural methods for degrading pollutants in wastewater. This technique has long used inclined beds to support biofilm synthesis. This biofilm has a structure of a mixed community, including filamentous microalgae (periphyton and epiphytic diatoms), aerobic bacteria, and fungi. In this system, wastewater flows from the biofilm, and nutrients are removed by assimilation into biomass. Cyanobacteria (particularly *Oscillatoria sp.,*) and diatoms (*Nitzschia sp., Navicula sp., and Cyclotella sp.*) are predominant species in this scenario. ATS technique can solve many issues related to the microalgal growth and harvesting process. In this context, the Benthic algae type that grows on turf scrubbers can be scraped once to make harvesting easier. It also reduces the total production cost of biomass. Harvesting rates play an important role in the ATS technique. The hybrid ATS system has increased total nitrogen elimination through constructed wetlands (Oruganti *et al.*, 2022).

c) Membrane Aerated Biofilm Reactor (MABR)

MABR is a highly innovative treatment system for removing wastewater pollutants using a hydrophobic membrane. This membrane serves as a biofilm carrier and can function as a bubble-free air diffuser. Nitrogen removal efficiency is usually increased by using the membrane-aerated algal-bacterial biofilm reactor (MABAR). This advantage of increasing the removal rate of pollutants stems from algal assimilation.

2.1.3. Algal-Bacterial Symbiotic Biofilms

An algal-bacterial biofilm is a symbiotic growth of these organisms after they attach to a solid carrier. They are complex, ecologically significant, three-dimensional structures of microbial communities embedded in the self-produced matrix called extracellular polymeric substance (EPS). This synergistic relationship ensures biofilm stability and enhances its efficiency, making algae-bacteria systems more effective for wastewater treatment and other environmental applications (Wen, Sun, & Liu, 2025). It also facilitates downstream processing recovery. These biofilms exhibit superior removal of organic matter and nutrients compared to standard biofilms. An algal-bacterial biofilm offers high resistance to shock loads and higher nutrient removal, at lower operating costs. EPS secreted by bacteria can promote attachment by protecting organisms from high salinity, pH, and high temperature (Chen, Xie, Qiu, Li, & Ge, 2024). There are four stages present in the formation process of an algal-bacterial biofilm.

Formation Stages

1. Suspended bacteria and algae transfer to the surface of the membrane carrier by adsorption, gravitational hydrodynamics, or phototropism of algae. Bacteria first stick to the surface and produce a glue-like substance, extracellular polymeric substances (EPS).
2. During the irreversible attachment, Algae cells attach to this sticky layer under the influence of bacterial EPS. This is an early growth phase. This attachment is secured by the EPS.
3. At the growth stage of biofilm, the growth and reproduction rate of bacteria usually become higher than that of microalgae. Here, a large amount of bacterial EPS is present. Different varieties of microalgae, cyanobacteria, and algae begin to grow in large quantities on EPS substrates by attaching to the carrier. During the development phase, algal cells proliferate by exchanging nutrients with bacteria within the EPS matrix. Here, a symbiotic relationship starts to develop between algae and bacteria.
4. During the mature stage, the algae-to-bacteria ratio stabilizes on the biofilm, and a structurally stable multilayer, good symbiotic relationship is achieved. A mature algal-bacterial system forms a biofilm with a three-dimensional reticulation structure cross-linked by EPS under nutrient-rich conditions (Chen *et al.*, 2024).

2.2. Gene Transfer

Algae and bacteria can exchange genes directly because they live in close proximity. This process is called horizontal gene transfer. The mechanism of gene transfer between microorganisms also relates to coevolution. This method, involving algae and bacteria, occurs mostly during long-term culture. Genes of bacteria favour the microalgal growth after entering the inner side of algal cells in a special manner.

Gene transfer from bacteria increased the nitrogen uptake rate of microalgae by stimulating their growth rate under unfavourable conditions.

For example, the algae Seminavis robusta has some chloroplast genes that are derived from *Clostridia* bacteria. Bacteria can deliver genes encoding *ornithine cyclodeaminase* and *agmatinase* to nitrogen-restricted microalgae, helping them better utilize nitrogen and grow. In wastewater treatment, ABS genes encoding cyanophycin synthase and other nitrogen-cycle genes have been actively transferred to enhance the system's resilience and biomass production (H. Liu *et al.*, 2024). Future research must focus on identifying the exact mechanisms and environmental factors that drive gene transfer in wastewater treatment.

2.3. Signal Transduction (Chemical Communication)

Bacteria release chemicals acting as signals to communicate with algae. These signals can transfer algal genes, regulate biological activities, or change their physical development. For example, Bacteria can alter the structure (morphogenesis) of algae through these signals (Saravanan *et al.,* 2021).

Quorum Sensing (QS) is a type of signal transmission between bacteria and algae. It is a cell-to-cell communication mechanism involving the production, release, and detection of signalling molecules. QS can operate at the intraspecific (within the same species) and interspecific (between different species) levels of communication. N-acyl-homoserine lactones (AHLs) and Autoinducing peptides (AIPs) are the most commonly used QS molecules (Yu *et al.,* 2024). Gram-negative bacteria mostly secrete small sensing molecules, such as AHLs, whereas gram-positive bacteria secrete AIPs. The interaction of signalling molecules usually occurs in the phycosphere, the outer surface of the algal cell that is permeable to diffusion. Microalgae provide nutrients to bacteria and stimulate their proliferation by secreting substances that increase the biofilm formation (Y. Zhang *et al.,* 2021). The bacterial QS molecules include siderophores, autoinducing oligopeptides, and N-3-oxohexanoyl-L-homoserine lactone.

3. FACTORS AFFECTING ALGAL-BACTERIAL INTERACTIONS

Various factors influence the efficiency and stability of algal-bacterial interactions, including temperature, pH, light, hydraulic retention time, aeration, nutrient load, and mixing conditions. Algal-bacterial consortia play a very important role in wastewater treatment by enhancing nutrient removal, organic matter degradation, and oxygen balance. There is a great need to precisely control these factors to improve the system efficiency (Zhao *et al.,* 2025).

3.1. Temperature

Temperature directly influences the growth and metabolic activity rates of microalgae and bacteria. Each species has a specific optimal temperature range for growth. Outside this range, biomass growth rate slows and then abruptly stops. An optimal temperature range (e.g., 30–35°C) can maximize algal oxygen release while meeting the elevated oxygen demand of heterotrophic bacteria during COD and ammonia nitrogen degradation. This thermal optimization enables highly efficient wastewater treatment with minimal external energy input (Siew Moi, 2004). The temperature range between 25-35°C is maintained mostly in laboratory and pilot research to obtain satisfactory performance. The optimal growth temperatures are 27.2°C for cyanobacteria and 26.3°C for green algae (Peng, Liu, & Huang, 2020).

3.2. Effect of pH

pH significantly influences the nutrient speciation, microbial metabolism, and overall system stability. Algae indirectly consume CO_2 to increase pH. It can enhance the ammonia stripping and phosphorus precipitation, but may also inhibit bacterial nitrification at excessively high pH levels. Most algal-bacterial systems perform optimally at pH 7-9. Extreme pH conditions can damage cell integrity, reduce microbial diversity, and destabilize mutualistic interactions. Therefore, it is necessary to adjust operating pH conditions to maintain the system stability (Jin *et al.*, 2024).

3.3. Light Intensity

Microalgae synthesize the organic matter in the presence of sunlight by integrating nutrients and dissolved inorganic carbon. Light availability is essential for algal photosynthesis. It is important to provide oxygen to heterotrophic bacteria for aerobic degradation of the organic pollutants. Higher light intensities can cause some photo-inhibition in the microalgae and nitrifying bacteria. Insufficient light can limit the photosynthetic activity, reduce oxygen supply, and affect system efficiency. Light intensity and light/dark cycle should be optimized for real-scale application of algae-bacterial interacting systems to treat wastewater (Oruganti *et al.*, 2022).

3.4. Nutrient Availability

Nutrient availability, especially nitrogen, phosphorus, and carbon, directly influences the growth and interaction of algae and bacteria. Algae take in inorganic nutrients for biomass production and produce oxygen through photosynthesis for bacteria. Bacteria supply the required metabolites and inorganic carbon source for algal growth. They benefit from algal-derived oxygen and organic exudates. An imbalanced nutrient ratio can shift community composition, leading to dominance of either algae or bacteria and reduced treatment efficiency. Adequate and balanced nutrient availability promotes synergistic interactions, improving nutrient removal and biomass productivity (Oruganti *et al.*, 2022).

3.5. Reactor Selection

ABSS are mainly grown in photobioreactors, including tubular, flat-plate, and column reactors. Flat plate photobioreactors use high-energy light and high temperatures. Light and dark conditions using this reactor are difficult to control. Tubular photobioreactors require a high light-to-volume ratio. This condition favours the growth of algae and large-scale construction under sunlight. But the presence of a long tube hinders the mixing of liquid and gas. The aerobic-anaerobic cascade linkage in the reactor is regulated by controlling the light cycle time. Under sufficient light conditions, algae cannot grow fast. It is difficult for algae to provide sufficient electron acceptors under these conditions, leading to a decrease in anode removal rate. The photobioreactors equipped with external solar photovoltaic capacitors effectively increase the problems. At night, the capacitor stores solar energy, and the removal rate increases. Reactor designs, including flat plate, column, tubular, and hybrid configurations, affect biomass yield and pollutant removal (Vo *et al.*, 2019).

3.6. Hydraulic Retention Time

HRT determines the contact time between wastewater constituents and microbial communities. Wastewater treatment plants with large treatment capacity have longer HRT, but treatment efficiency declines, ultimately reducing economic efficiency. Short HRT and non-aeration can enhance assimilatory metabolism, particularly in algal-bacterial biofilms. Excessive HRT may reduce system productivity and increase operational costs. An optimal HRT supports sufficient algal growth, bacterial metabolism, and stable biomass retention by enhancing nutrient removal efficiency and system resilience (Jin *et al.*, 2024).

4. ALGAL-BACTERIAL SYSTEM IN WASTEWATER TREATMENT

4.1. Mainstream Wastewater

Municipal sewage is a large wastewater system that allows observation of algal-bacterial interactions during purification. Mainstream wastewater contains several nutrients from domestic sewage and toxic contaminants from industrial wastewater. Algal-bacterial biofilms are more efficient at removing nitrogen and phosphorus from this water. A continuous-flow algal-bacterial biofilm is primarily used to remove C/N from municipal wastewater. This system removes organic matter by producing the biomass that can settle quickly (Hammond *et al.,* 2025). The ABGS system has been shown to be more efficient for removing antibiotics from main-stream wastewater (Wang *et al.*, 2020).

4.2. Agricultural and Aquaculture Wastewater

Agricultural wastewater contains many chemicals, making it more complex to treat. Very high ammonia levels, antibiotics, and toxic fertilizers are present in this water, which need to be removed more carefully. Heavy metals are also present in agricultural wastewater. Algal-bacterial symbiotic biofilm system has been proven more useful with the best C: N: P ratio (106:6:1) to treat swine wastewater with the removal percentage of (COD-99%, NH_4^+-92%, TP-99%). Interestingly, ABSS interactions are more reliable and effective for wastewater at low antibiotic concentrations. Antibiotics stimulate the microbial electron transfer and gene expression during degradation (Zhang *et al.,* 2024). Moreover, investigating the more suitable species to strengthen the coupling mechanism between algae and bacteria is more beneficial. It helps them achieve high resource recovery from agricultural wastewater.

4.3 Industrial Wastewater

The primary component of dye wastewater is azo dyes, which can be absorbed by the human body and cause carcinogenic and mutagenic effects. The azo group introduces high chromaticity in wastewater, and its aromatic group, with its conjugated bonds, is highly stable, making it difficult to degrade during wastewater purification. Anthraquinone dyes are first oxidized by *Chlorella vulgaris*. Bacteria break down microalgal metabolites and accomplish mineralization. The symbiotic interaction between algae and bacteria is a highly cost-effective, eco-friendly, and efficient system for removing nitrogen and phosphorus from printing and dye wastewater compared to either alone. These organisms form aggregates to enhance their tolerance to salinity and dyes. Symbiotic bacteria stimulate the growth of algae and the metabolism of nitrogen, phosphorus, and carbon. While algae increases the transformation of dyes, it also supports bacterial metabolism (Tang et al., 2019).

4.4 Pharmaceutical Wastewater

This type of wastewater from industrial sector include organic solvents, intermediates, and catalysts, as well as the pharmaceutical ingredients. Algal-bacterial system for wastewater treatment improve the removal efficiency of pollutants and increase the production of algal biomass, which can be used to generate biofuels (Narayanan *et al.*, 2025). ABSS can remove 97% of antibiotics from wastewater compared to other systems. *Amoxicillin (AMX)*-degrading microalgae were selectively cultured from the activated sludge. They reported that the AMX degradation percentage exceeded 90%. It has been achieved by using the ABS system, which couples cultured microalgae with two other bacteria. Microbiological studies recommended that the ABS system enhance microalgae's tolerance and promote biopolymer production by upregulating precursor molecules (Jin *et al.*, 2024).

5. REMOVAL OF POLLUTANTS USING ALGAL-BACTERIAL SYSTEMS

5.1. Biodegradation

Biodegradation is widely used to remove various pollutants from wastewater due to its cost-effectiveness, high contaminant-removal efficiency, and eco-friendliness. Biological methods break down organic matter in wastewater using microorganisms, including algae, bacteria, fungi, and yeast. They degrade pollutants aerobically or anaerobically into water, CO_2, and CH_4, releasing energy in the process. Some waste can also be transformed into fuels and other valuable chemicals (Wu *et al.,* 2021).

Role of algae and bacteria in wastewater treatment

Wastewater contains various organic matter, including total nitrogen (TN), total phosphorus (TP), and other chemicals. Algae and bacteria interact in different ways to help treat wastewater (municipal, industrial, agricultural, and domestic) by removing emerging contaminants. Wastewater can support this co-culture system without supplemental oxygen. Simultaneously, nutrients can be converted into biomass, and CO_2 emissions into the atmosphere will be reduced. The oxygen produced by microalgae during photosynthesis supports the growth and respiration of the bacteria that degrade pollutants in the wastewater. It substantially reduces the costs and greenhouse gas emissions associated with mechanical aeration in activated sludge systems (Alcántara et al., 2015).

5.2 Nitrogen Removal

Nitrogen is an essential nutrient for the growth of many organisms in water. Excess nitrogen compounds can degrade water quality and cause eutrophication by discharging into natural waterways. Biological methods are mostly used to remove nitrogen from municipal wastewater. Algae and bacteria interaction removes nitrogen through nitrification, denitrification, and assimilation. 99.51% nitrogen removal can be achieved without additional aeration through this system (Sial *et al.*, 2021).

Nitrification (Aerobic Process)

This process is performed by ammonia-oxidizing bacteria (AOB) and Nitrite-oxidizing bacteria (NOB). Algal-bacterial interactions provide energy-intensive aeration for nitrification by converting ammonia to nitrate. The algae, as photosynthetic organisms, provide free oxygen for bacterial nitrification. Nitrogen, as ammonia, is biologically oxidized, initiating nitrogen removal.

$$\text{Ammonia } (NH_4^+\text{-N}) \rightarrow \text{Nitrite } (NO_2^-\text{-N}) \rightarrow \text{Nitrate } (NO_3^-\text{-N})$$

Denitrification (Anoxic Process)

This is the biological reduction of nitrate to nitrogen gas (N_2), which is released into the atmosphere. This reaction is performed by denitrifying bacteria under anoxic or low-oxygen conditions. They can add external carbon sources for denitrification by converting nitrate to nitrogen gas. Algae supply organic carbon (through exudates) for bacterial denitrification (Nguyen *et al.*, 2020).

$$\text{Nitrate} \rightarrow NO \ \ \text{(nitrite)} \rightarrow NO \rightarrow N_2O \rightarrow N_2$$

In the integrated algal-bacterial system, algae help to create a microenvironment with alternating aerobic and anoxic conditions facilitating nitrification and denitrification. They enhance nitrogen removal and improve water quality. The removal efficiency of Nitrogen and other pollutants can be calculated by following equation.

$$\textit{Removal efficiency} = \frac{C_{in} + C_e}{C_{in}} \times 100\%$$

where C_{in} indicates pollutant concentration in influent while C_e is concentration in effluent (Sharma and Bhatti, 2022). *C. vulgaris* and wastewater bacteria removed up to 97% N from municipal wastewater. Another study determined the simultaneous removal of organics, nitrate and phosphate through photosynthetic bacteria (*Rhodobacter*) and algae (*C. sorokiniana*) (Kim *et al.*, 2014).

5.3 Phosphorus Removal From Wastewater

Phosphorus released into the environment by various industries is a major issue due to its significant contribution to the eutrophication of water bodies. Algae have a high demand for phosphorus, as it is an important nutrient for algal growth. Phosphorus can be efficiently removed from wastewater through biological processes using an algal-bacteria interacting system. In biological mechanisms for removing phosphorus, it is assimilated into biomass through phosphorylation (Bezerra et al., 2025). Algae can grow healthy on bacteria present in wastewater using a photobioreactor (PBR). The batch experiment can be mostly used for the co-cultivation of the algae and wastewater-borne bacteria (*Acinetobacter* sp.).

The co-culture of algae and this bacterium has been shown to be highly efficient, removing 98.78% of total phosphorus from wastewater. Synergistic cooperation has been observed in growth of algae and *Acinetobacter sp*. The exchange of CO_2 and O_2, which exhibited a promising way to treat wastewater to improve nutrient removal and the production of biomass (Liu *et al.*, 2017).

5.4. Removal of Emerging Contaminants

Microalgae and bacterial consortiums proved more reliable system in removing organic and inorganic pollutants from wastewater including heavy metals, total phosphorus, nitrogen and dissolved organic carbon. The consortium system provides very intricate interaction and also substrate exchange. It is able to provide robust growth, very stable environment by preventing the pathogenic microbial invasion. The symbiosis interaction between *Chlorella* and bacteria exhibited predominant efficiency to treat wastewater both by increasing biomass productivity and degrading the organic and inorganic contaminants (Adhithya et al., 2025).

5.4.1 Heavy Metals Removal

Heavy-metal-resistant bacteria use various detoxification strategies, including metal exclusion, biochelation, sequestration, and enzymatic transformation. In the algal-bacterial symbiotic system, these bacteria mitigate HM toxicity towards algae and benefit from the microenvironment created by algal growth (Zhao *et al.,* 2025). The *Pseudomonas putida* and *Chlorella vulgaris* consortium neutralizes Cd(II) and Pb(II) toxicity via bacterial activity, enabling algal survival under metal stress. In this system, AHL-producing bacteria can improve HM removal efficiency mainly by stimulating biomass production through increased protein synthesis. (Yu *et al.*, 2024). The ABSS system has been proven efficient to remove Cr(VI) and Cd(II). They depend on biosorption and intracellular accumulation (Huang *et al.*, 2025).

Industrial wastewater has been treated to remove manganese, cobalt, cadmium and chromium metals using bioremediation with algal-bacterial (Bashir *et al.,* 2023). Both algae and bacteria have cell walls rich with functional groups, including carboxyl, hydroxyl, phosphate, amino, and sulfate groups. The negatively charged groups can bind the positively charged Cd^{2+} ions using ion exchange and electrostatic attraction. It does not require metabolic activity and makes it effective with dead biomass (Parmar and Patel, 2025).

Synergistic Interactions

Algae supply oxygen molecules and other organic matter through photosynthesis to support bacterial metabolism. Bacteria, in return, release carbon dioxide, vitamins, and all other nutrients to enhance algal growth. This mutualistic behaviour of organisms is necessary to improve system stability and metal removal efficiency from wastewater. Algal–bacterial interactions provide a sustainable and effective approach to remove metals from wastewater by combining natural processes with minimal chemical and energy inputs.

5.4.2 Surfactants

Surfactants are significant portion of domestic wastewater pollutants. Linear Alkylbenzene Sulfonate (LAS) anionic surfactant is an important component in detergents and cleansing agents. Biological treatment is a feasible option for removing it from wastewater using a photobioreactor system with an algal-bacterial consortium. LAS and caffeine removal from wastewater using this biological system with photobiodegradation, light, and dark reactors has already been studied. Synthetic wastewater containing the target pollutants was fed into the reactors, achieving removal efficiencies of 99% for LAS and 96% for caffeine (Katam et al., 2020).

5.4.3 Antibiotics

Antibiotics present in wastewater from various sources have a substantial influence on the environment. They disrupt the wastewater treatment procedure and negatively impact the ecosystem. (Eheneden et al., 2023). Biological removal mechanisms of typical antibiotics, including bioadsorption, bioaccumulation, and biodegradation, are also used in this approach. Bacteria, supported by algal oxygen, can biodegrade antibiotics via enzymatic pathways better than a single algal procedure. The use of microalgae and bacteria has emerged as a sustainable and very promising technique due to their non-target status, rapid growth, and carbon recovery abilities. This technique involves the extracellular polymeric substances, pigments, antioxidant enzyme systems, signalling molecules, and metabolic pathways, and analyses the role of treatment facilities in facilitating algae–bacteria symbiosis, including membrane bioreactors and bioelectrochemical systems. *Tetadesmus obliqus (TO)*, when symbiotically interacting with Alcaligenes faecalis bacteria, has been used to remove tetracycline, achieving a 93.87 % removal rate within 1.25 days by improving degradation efficiency over a short time. Photosynthetic oxygenation by microalgae sustains bacterial respiration in this system. Bacterial CO_2 fuels algal carbon fixation, creating an oxygen-carbon loop (Wang *et al.,* 2025).

Algal-bacteria Interactions in Antibiotic Environment

To cope with antibiotic stress, the interaction between algae and bacteria alters the cellular response system to adapt to external environmental pressures. These changes involve EPS secretions within cells, synthesis of pigments, alterations in signaling and metabolic pathways, regulation of gene activity, and synthesis and degradation of proteins.

- Algae–bacteria symbiosis systems are used to increase the EPS secretion, improve uptake of antibiotics, and enhance their activity. To deal with antibiotics, algae and bacteria increase their levels of chlorophyll a and b. In this system, the activity of intracellular pigments helps to reduce antibiotic damage.
- Algal-bacterial symbiosis also affects the metabolic processes and can disturb the whole system by elevating oxidative stress, altering gene expression, and afterwards influence the production and function of antioxidant enzymes.
- Using gene transfer and signal transduction mechanisms during antibiotic stress, algae and bacteria activate or suppress the special signal pathways. They lead to changes in the expression of signalling elements, including gene transcription factors. They help to control the cellular responses to antibiotics (Xiao *et al.,* 2024).
- Bacteria and algae have the ability to modify their metabolic functions in the presence of antibiotics by changing associated enzymes, gene transcriptions, and translation methods.

It is necessary to employ advanced, efficient tools to deepen our understanding of the effects of algal-bacterial interactions on cellular and biological responses to antibiotic stress in the environment.

Table 1. Algal-bacterial System with pollutants removal efficiency

Process type	Co-culture (Algae-Bacteria)	pH	Pollutants removal efficiency	References
PBR	*Leptolyngbya sp. XZ1 and Bacillus sp. S1*	5	Cd (II), (78.1%)	(Ri *et al.*, 2023)
HB photobioreactor	*Ulothrix zonata* and the natural biofilm	3.5	Cu(II) >50% Mn(II)>45% Zn(II) >30% Ni(II) >45%	(Orandi et al., 2012)
HRAR	*Chlorella vulgaris* and *Azotobacter S8*	7	Cr (VI), 18.68%	(Tangahu et al., 2020)
SBR	*Comamonadaceae* and *Nitrosomonadaceae* and *Navicula* and *Stigeoclonium*	7 - 8.4	61–80% (TN)	(Meng *et al.*, 2019)
MABAR	*Proteobacteria, Chloroflexi* and *Cyanobacteria, Bacteroidetes, Acidobacteria* and *Actinobacteria*	8.3	90%/75%(NH -N)	(Li et al., 2022)
Signal transduction (AHLs and IAA)	*Chlorella pyrenoidosa FACHB-5* and activated sludge	7	TN (95.8%), TP (68.1%), COD (96.1%)	(K.-Y. Li *et al.*, 2025)

6. RESOURCE RECOVERY FROM ALGAL-BACTERIAL SYSTEMS

Resource recovery refers to the process of harnessing the symbiotic relationship between algae and bacteria to capture and convert pollutants into valuable products. In wastewater treatment, some valuable resources need to be recovered economically. Many valuable products, e.g., biogas and biofuels, can be produced from the biomass generated during wastewater treatment. This conversion depends on the technology used and the end use to add economic benefits. Methane, hydrogen, and clean biofuels are generated using biomass by the process of anaerobic digestion. This process mainly relies on algae and bacteria to recover valuable resources, such as lipids for biodiesel and biomass for energy (Liu *et al.,* 2024).

6.1 Biodiesel Production

The synergistic interaction between algae and bacteria instantaneously provides lipid-rich biomass for biodiesel production during wastewater treatment (Ravikumar, 2025). Biodiesel is produced from algal lipid-containing biomass using the transesterification process. This section highlights the essential components of lipid-accumulation-based production and process integration. Microalgal biomass is grown on wastewater with a lipid content of 10-30% on a dry-weight basis. Wastewater media provide favourable conditions for cyanobacteria and microalgae, enabling them to grow and reproduce rapidly in this scenario. The mechanism involves basic steps, including obtaining the biomass (cultured or natural), followed by pre-treatment (dewatering, thickening, and drying), lipid extraction, lipid transesterification, and the biodiesel separation.

Figure 2. Biodiesel production using algae and bacteria

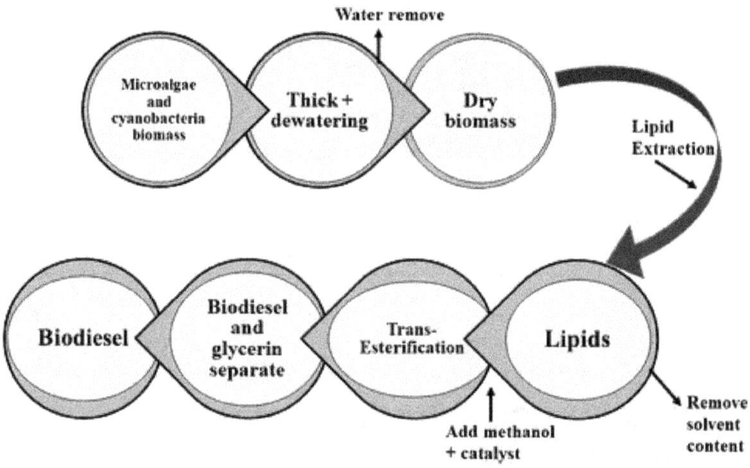

6.1.1. Pre-Treatment Method

The pre-treatment process for microalgae/cyanobacteria biomass includes harvesting, concentrating, and drying to prepare it for efficient lipid extraction. Biomass is dried under solar light or in an oven, then ground into a fine powder. Here, cell barriers break, and particle size is reduced by minimizing water content.

6.1.2. Lipid Extraction

For lipid extraction from microalgae and cyanobacteria, chemical or mechanical methods are most commonly used. Chemical processes include solvent extraction, Soxhlet extraction, and supercritical CO_2 extraction. In laboratory-scale synthesis, chemical-based extraction is mostly used. This process is not only lipid-specific;

it also requires reducing the co-extraction of non-lipid contaminants, e.g., proteins and carbohydrates. They choose based on the desired neutral lipid fractions.

6.1.3. Trans-Esterification

Methanol reacts with triglycerides to form glycerol and a fatty acid methyl ester in the presence of catalysts or enzymes. Alkali-catalyzed saponification of FFA during transesterification alters fuel properties. This reaction produces methyl esters (biodiesel) and glycerol (byproduct) (V. Sharma & Gupta, 2022).

6.2 Biogas Generation

After the lipid extraction, residual biomass is subjected to alcoholic fermentation to produce ethanol or to anaerobic digestion to generate methane. Anaerobic digestion for the biomass and then converting it into the combustible CH_4 gas is a very promising technique to recover the resources from wastewater. The different operating conditions of algal-bacterial systems used to treat wastewater also affect the biochemical methanogenic potential (BMP) of the biomass (Passos *et al.,* 2017).

6.2.1. Biohydrogen Production

Production of this resource from algal-bacterial biomass is a potentially very sustainable approach, as it recovers energy from the phenomenon. Microalgae are able to produce biohydrogen through fermentation or photolysis. The high contents of carbohydrates, lipids, proteins, and other nutrients in algal biomass make it a very promising feedstock for biohydrogen production. Biohydrogen production during fermentation relies on the availability of fermentative organisms and on biomass pretreatment. Feasibility of coupling the urban wastewater treatment during the biohydrogen production in an integrated approach. Biohydrogen production yielded the 56.8 mL/g volatile solids from using the dark fermentation process of *Scenedesmus obliquus* grown on the wastewater (Batista *et al.*, 2015).

6.3 Production of Polyhydroxyalkanoates

Excessive use of plastics causes major pollution issues in the environment and harms aquatic flora and fauna. Nowadays, plastics are usually replaced by the sustainable, eco-friendly bioplastics produced from various the biopolymers obtained from the living organisms. These biopolymers are produced from natural substances under nitrogen-limited conditions. They have thermal and mechanical properties with the extra advantage of being biodegradable, like petroleum-based polymers.

Algal-bacterial symbiotic systems in wastewater purification offer a sustainable platform for pollutant removal and the production of biodegradable plastics, primarily polyhydroxyalkanoates (PHA).

Microalgae produce oxygen and carbon during wastewater treatment, while bacteria use that carbon for growth. Under conditions of nutrient imbalance (limited nitrogen and phosphorus but excessive carbon), bacteria shift their metabolism to store carbon intracellularly as PHA granules.

PHA are linear polyesters naturally synthesized by many bacteria as energy and carbon storage materials. They are biodegradable materials in the marine and soil environments. They have similar material properties to conventional plastics like polypropylene. Producing PHA from algal-bacterial systems is a sustainable approach to addressing environmental pollution, paving the way for a circular bio-economy. Polyhydroxyalkanoates obtained from cyanobacteria and microalgae serve as a very cost-effective and eco-friendly alternative to boost the competitiveness of biological-based biopolymers (Anand *et al.,* 2023)

6.4 Biofertilizer Production

The algal biomass used in wastewater treatment can be applied as a biofertilizer or soil additive to increase soil phosphate and nitrogen levels. These biomasses can improve soil levels of calcium, potassium, iron, and magnesium. However, there is a risk of heavy metals, which must be avoided. Additionally, slow-release biofertilizers derived from microalgae can be utilized to increase the amount of organic matter in the soils. Additionally, they may add various plant-stimulating substances to the soil, along with potentially harmful and other micropollutants, raising questions about their use as biofertilizers (Anand *et al.,* 2023). Biofertilizer is produced from the cultivation and harvesting process of microalgae. That converts algae into biomass, and with the help of the digestion process, biofertilizers can be produced.

Figure 3. Production of biofertilizer from Microalgal cultivation (Yadavalli & Heggers, 2013)

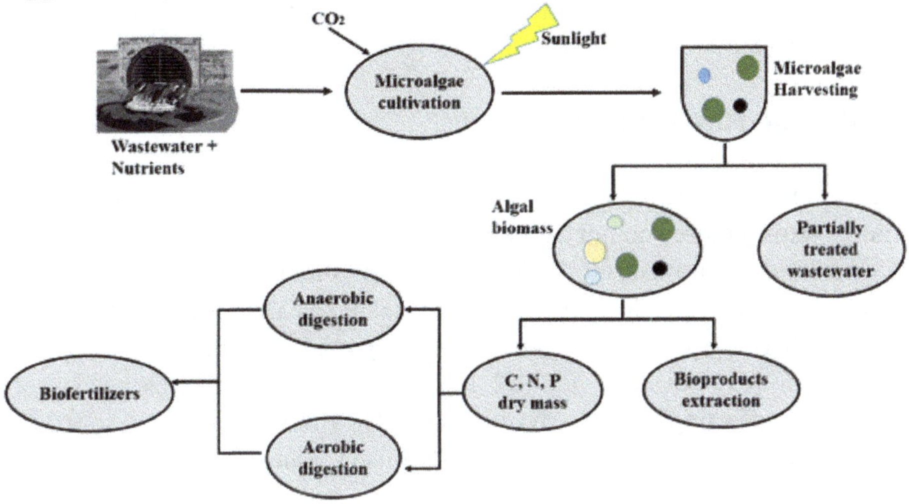

Figure 3: Production of biofertilizer from Microalgal cultivation

Cyanobacteria are able to assimilate more nitrogen than inorganic fertilizers, making it more suitable for rice plant cultivation. Biofertilizers obtained from the immobilized *Chlorella pyrenoidosa* grown in dairy wastewater also increased the growth in paddy (Yadavalli & Heggers, 2013).

7. CHALLENGES IN ALGAL-BACTERIAL SYSTEM

Algal-bacterial interactions offer significant benefits in wastewater treatment, including nutrient recovery, lower operational costs, and reduced aeration requirements. However, this system faces operational and technical challenges that must be addressed to compete with conventional wastewater treatment systems, such as activated sludge. Light intensity, nutrient supply (C: N: P), temperature, pH, Hydraulic retention time, and solid retention time need to be adjusted. The selection of appropriate algal and bacterial strains is also very important.

7.1. Light Intensity

Provision of sufficient light intensity during photosynthesis is the major limitation in this system. The interaction between algae and bacteria for removing wastewater pollutants requires a precise wavelength. Light intensity should be sufficient for organisms to carry out the nutrient exchange. High biomass concentrations (>1 g/L) cause shading and inhibit photosynthesis. Existing outdoor activated sludge plants lack the submerged illumination required to support significant algal growth (Bilad *et al.,* 2014).

7.2. Hydraulic Retention Time

HRT is the average time in which wastewater stays inside the reactor. Algal-bacterial systems require enough contact time for algae to consume most nutrients and oxygen production for bacteria, for this reactor should not ne inefficient and expensive. Hydraulic retention time should be treated as a key variable to operate the system rather than a fixed constant. Typical hydraulic retention times (HRTs) for most systems have been reported as 2-10 days to purify water. It has been a very long time since affecting the system efficiency. They must be reduced to 0.2-0.4 days to match other conventional system footprints (J. Yang *et al.,* 2018).

7.3. Consortium Stability

Consortium stability means maintaining a balanced, functioning partnership between algal and bacterial populations rather than allowing one side to dominate. The symbiotic relationship is complex and vulnerable. Stability relies on suitable metabolite exchange and division of labour. It can be disrupted by competition for nutrients. This stability depends on the microalgal-bacterial ratio, light availability, carbon source, C/N/P ratio, aeration, dissolved oxygen, and temperature. By changing these conditions, competition can replace collaboration, and treatment performance drops (Qi *et al.,* 2024).

7.4. Strain Selection

The use of robust algal strains in the reaction system that will be able to thrive in low-light and other harsh conditions (e.g., certain diatom species) is also very important. Strain selection poses another important challenge in the algal-bacterial system that warrants discussion.

7.5. Scale-Up the System

Most research experiments are lab-scale and short-term. There is a need to operate the process at a large scale with long-term studies. Large-scale, long-term studies of this interacting system to match real-world stability and performance are also required.

7.6. Low-Cost Bioproducts

Extensive research is required to develop efficient biomass harvesting and conversion processes to produce low-cost biofuels and bioproducts. Low-coat bioproducts are beneficial outputs that can make the system more economical after wastewater treatment. Biomass for biofertilizers, biogas, biofuels, and other useful biochemicals is considered in this scenario.

8. CHALLENGES IN ENABLING THE CIRCULAR ECONOMY

The struggle to create a circular economy is facing several challenges in various aspects of the wastewater treatment sector. The large quantity, difference in composition, and appearance of new pollutants make it difficult to choose appropriate treatment methods. Additionally, selecting different approaches to wastewater treatment and water resource management creates a complex decision-making process. Furthermore, by selecting a suitable treatment technology from a technical perspective, there are other challenges that need to be overcome now, including regulatory obstacles, customer agreement, market penetration, and public perception goals. However, marketability, social, economic, and political factors play an important role in enabling a circular route for recovering resources from wastewater. There is an urgent need to solve these challenges regarding technology, infrastructure, cost consideration policy, and regulation.

This process addresses the algae-based wastewater treatment and also discusses the problems of nutrient pollution, water scarcity, and environmental degradation. These problems need to be solved by optimizing operational parameters and conducting economic analysis. Using a light-emitting diode (LED) with a specific wavelength tailored to algal growth requirements enables precise control over light specifications and improves cultivation performance. There is a need to investigate the effects of nutrient availability and stoichiometric ratios on algal growth to optimize pollutant removal efficiency and biomass composition. By advancing the scientific knowledge, innovating in current techniques, and engaging stakeholders, researchers, policy makers, and practitioners can accelerate the transition towards wastewater management practices and resource preservation efforts (Ugwuanyi, Nwokediegwu, Dada, Majemite, & Obaigbena, 2024).

9. FUTURE DIRECTIONS

Future research on algal-bacterial systems (ABS) should transition from laboratory-scale proofs-of-concept to robust industrial applications by prioritizing the following three areas:

1. **Reactor Intensification and Scalability:** Hydraulic retention time (HRT)should be reduced using various innovative reactor configurations. Granular sludge systems and hybrid biofilm photobioreactors are the most common and essential for increasing biomass retention, enhancing operational speed, and stabilizing consortia against environmental fluctuations in temperature and contaminant loads.
2. **Genetic and Metabolic Engineering:** Use of flourishing strategies, including CRISP or synthetic biology, to modify algal-bacterial strains for enhancing the degradation of complex emerging pollutants. Engineering characters for improved transmission of light in turbid media and metabolic pathways for high-value bioproducts will considerably enhance the techno-economic viability of the biorefinery concept.
3. **Systems Integration and Circularity:** Detailed Life Cycle Assessments (LCAs) and Techno-Economic Analyses (TEAs) for the algal-bacterial system are required to optimize the "wastewater-to-wealth" pipeline. Research should explore the direct integration of ABS with industrial flue gas for CO_2 fixation, and the standardization of protocols for high-value resource recovery (e.g., bioplastics and antioxidants) to establish wastewater treatment plants as sustainable urban biorefineries.

CONCLUSION

The algal-bacterial system presents a standard modification in wastewater treatment and management, transitioning from the removal of various contaminants to a circular framework of nutrient resource recovery from the environment. The mutualistic interactions comprising nutrient exchange, gene transfer, and chemical signalling that underlie these continually-maintained biochemical loops are complex. Minimization of external aeration, the primary energy sink in traditional treatment processes, is achieved through enhanced exchange of CO_2, O_2, and essential nutrients. Due to its versatility, the system can treat a variety of high-strength municipal, agricultural, and industrial streams by integrating biosorption, assimilation, and biodegradation. Its integration with the circular bioeconomy is one of its major strengths; the biomass produced becomes an asset, rather than a waste.

Such biomass can be converted into a wide array of bioproducts, such as biodiesel, biogas, and also biodegradable plastics. Resource recovery coupled with water purification is an attractive economic driver, leading to both lower operational costs and improved infrastructure sustainability. Future research must focus on reducing Hydraulic Retention Time (HRT) and on developing innovative designs, such as hybrid biofilm photobioreactors or granular sludge systems, to improve biomass retention. Wastewater treatment plants can be redesigned as urban biorefineries to integrate microbial ecology with process engineering. The success of this transition depends on interdisciplinary collaboration and a dedicated shift toward valuing waste products as integral components of a resource-efficient future. Ultimately, the algal-bacterial system offers a sustainable path to reducing greenhouse gas emissions and achieving global circular-economy goals.

REFERENCES

Abdel-Raouf, N., Al-Homaidan, A. A., & Ibraheem, I. B. (2012). Microalgae and wastewater treatment. *Saudi Journal of Biological Sciences, 19*(3), 257–275. DOI: 10.1016/j.sjbs.2012.04.005 PMID: 24936135

Abinandan, S., & Shanthakumar, S. (2015). Challenges and opportunities in application of microalgae (Chlorophyta) for wastewater treatment: A review. *Renewable & Sustainable Energy Reviews, 52*, 123–132. DOI: 10.1016/j.rser.2015.07.086

Adhithya, S., Nithya, K., Sathish, A., & Kumar, V. (2025). Uncovering the feasibility of using live Chlorella microbiomes in domestic and industrial wastewater treatment: Insights into monoculture and synergistic mixed co-cultured system. *Journal of Industrial and Engineering Chemistry, 149*, 140–156. DOI: 10.1016/j.jiec.2025.02.006

Alcántara, C., Posadas, E., Guieysse, B., & Muñoz, R. (2015). Microalgae-based Wastewater Treatment. In Kim, S.-K. (Ed.), *Handbook of Marine Microalgae* (pp. 439–455). Academic Press. DOI: 10.1016/B978-0-12-800776-1.00029-7

Aleruchi, O., Fubara, E., & Otto, B. (2023). Wastewater Quality-It's Impact on the Environment and Human Physiology: A Review. *International Journal of Advance Research and Innovation, 9*(4), 43–58. DOI: 10.51976/ijari.942107

Anand, U., Dey, S., Parial, D., Federici, S., Ducoli, S., Bolan, N. S., Dey, A., & Bontempi, E. (2023). Algae and bacteria consortia for wastewater decontamination and transformation into biodiesel, bioethanol, biohydrogen, biofertilizers and animal feed: A review. *Environmental Chemistry Letters, 21*(3), 1585–1609. DOI: 10.1007/s10311-023-01562-w PMID: 36785620

Bashir, K., Khan, S., Ali, R., Yasmin, H., Gaafar, A.-R. Z., Khilgee, F. E. A., Butt, S., & Ullah, A. (2023). Bioremediation of Metal-Polluted Industrial Wastewater with Algal-Bacterial Consortia: A Sustainable Strategy. *Sustainability (Basel), 15*(19), 14056. DOI: 10.3390/su151914056

Batista, A. P., Ambrosano, L., Graça, S., Sousa, C., Marques, P. A. S. S., Ribeiro, B., Botrel, E. P., Castro Neto, P., & Gouveia, L. (2015). Combining urban wastewater treatment with biohydrogen production – An integrated microalgae-based approach. *Bioresource Technology, 184*, 230–235. DOI: 10.1016/j.biortech.2014.10.064 PMID: 25453433

Bezerra, S. S., Fontana, L., Arantes, C. C., & de Jesus, T. A. (2025). Phosphorus removal from wastewater by microalgal cultivation in photobioreactors: A systematic literature review and multivariate analysis. *Environmental Monitoring and Assessment*, *197*(11), 1182. DOI: 10.1007/s10661-025-14524-2 PMID: 41055785

Bilad, M. R., Discart, V., Vandamme, D., Foubert, I., Muylaert, K., & Vankelecom, I. F. (2014). Coupled cultivation and pre-harvesting of microalgae in a membrane photobioreactor (MPBR). *Bioresource Technology*, *155*, 410–417. DOI: 10.1016/j.biortech.2013.05.026 PMID: 24559585

Chen, Z., Xie, Y., Qiu, S., Li, M., & Ge, S. (2024). Enriched functional exoproteins and increased exopolysaccharides with altered molecular conformation mutually promoted indigenous microalgal-bacterial consortium biofilm growth under high light intensity. *Chemical Engineering Journal*, *480*, 148056. DOI: 10.1016/j.cej.2023.148056

Craggs, R. J., Heubeck, S., Lundquist, T. J., & Benemann, J. R. (2011). Algal biofuels from wastewater treatment high rate algal ponds. *Water Science and Technology*, *63*(4), 660–665. DOI: 10.2166/wst.2011.100 PMID: 21330711

Eheneden, I., Wang, R., & Zhao, J. (2023). Antibiotic removal by microalgae-bacteria consortium: Metabolic pathways and microbial responses. *The Science of the Total Environment*, *891*, 164489. DOI: 10.1016/j.scitotenv.2023.164489 PMID: 37279806

Garrido-Baserba, M., Sedlak, D. L., Molinos-Senante, M., Barnosell, I., Schraa, O., Rosso, D., Verdaguer, M., & Poch, M. (2024). Using water and wastewater decentralization to enhance the resilience and sustainability of cities. *Nature Water*, *2*(10), 953–974. DOI: 10.1038/s44221-024-00303-9

Hammond, C. R., Hernández, M. S. G., & Loge, F. J. (2025). Microalgal-bacterial aggregates for wastewater treatment: Origins, challenges, and future directions. *Water Environment Research : A Research Publication of the Water Environment Federation*, *97*(2), e70018. DOI: 10.1002/wer.70018 PMID: 39900561

Huang, Y., Yu, J., Huang, X., Sun, J., Tang, W., Wei, M., Yang, G., & Fu, Y. (2025). Algal-bacterial symbiotic system for treatment of heavy metal containing wastewater: Performance, mechanisms and applications. npj. *NPJ Clean Water*, *8*(1), 86. DOI: 10.1038/s41545-025-00475-1

Jin, D., Zhang, X., Zhou, L., Zhang, X., & Wu, P. (2024). Emerging applications and mechanisms of algal-bacterial symbiosis on sustainable wastewater treatment: A comprehensive review. *Journal of Water Process Engineering*, *65*, 105912. DOI: 10.1016/j.jwpe.2024.105912

Karimi-Maleh, H., Shafieizadeh, M., Taher, M. A., Opoku, F., Kiarii, E. M., Govender, P. P., Ranjbari, S., Rezapour, M., & Orooji, Y. (2020). The role of magnetite/graphene oxide nano-composite as a high-efficiency adsorbent for removal of phenazopyridine residues from water samples, an experimental/theoretical investigation. *Journal of Molecular Liquids, 298*, 112040. DOI: 10.1016/j.molliq.2019.112040

Katam, K., Shimizu, T., Soda, S., & Bhattacharyya, D. (2020). Performance evaluation of two trickling filters removing LAS and caffeine from wastewater: Light reactor (algal-bacterial consortium) vs dark reactor (bacterial consortium). *The Science of the Total Environment, 707*, 135987. DOI: 10.1016/j.scitotenv.2019.135987 PMID: 31869619

Kim, K.. (2014). Simultaneous removal of organics and nutrients from municipal wastewater using a photosynthetic bacterial and microalgal consortium. *Journal of Microbiology and Biotechnology, 24*(11), 1548–1554.

Li, J., Ou, R., Liao, H., Ma, J., Sun, L., Jin, Q., He, D., & Wang, Q. (2022). Natural lighting enhancing the algae proliferation and nitrogen removal in membrane-aerated bacterial-algal biofilm reactor. *The Science of the Total Environment, 851*(Pt 1), 158063. DOI: 10.1016/j.scitotenv.2022.158063 PMID: 35981577

Li, K.-Y., Zhou, J.-L., Vadiveloo, A., Li, J.-N., Zhou, D., Wang, J.-M., Zhang, C., Xie, Y.-L., Chen, D.-Z., & Gao, F. (2025). Signal molecule-mediated algae-bacteria interactions and membrane-enabled separation synergize wastewater bioremediation. *Chemical Engineering Journal, 523*, 168368. DOI: 10.1016/j.cej.2025.168368

Liu, H., Al-Dhabi, N. A., Jiang, H., Liu, B., Qing, T., Feng, B., Ma, T., Tang, W., & Zhang, P. (2024). Toward nitrogen recovery: Co-cultivation of microalgae and bacteria enhances the production of high-value nitrogen-rich cyanophycin. *Water Research, 256*, 121624. DOI: 10.1016/j.watres.2024.121624 PMID: 38669903

Liu, H., Lu, Q., Wang, Q., Liu, W., Wei, Q., Ren, H., Ming, C., Min, M., Chen, P., & Ruan, R. (2017). Isolation of a bacterial strain, Acinetobacter sp. from centrate wastewater and study of its cooperation with algae in nutrients removal. *Bioresource Technology, 235*, 59–69. DOI: 10.1016/j.biortech.2017.03.111 PMID: 28364634

Liu, S., Kong, Z., Guo, H., Zhang, Y., Han, X., Gao, Y., Daigger, G. T., Zhang, G., Li, R., Liu, Y., Zhang, P., & Song, G. (2024). Performance, mechanism regulation and resource recycling of bacteria-algae symbiosis system for wastewater treatment: A review. *Environmental Pollution, 362*, 125019. DOI: 10.1016/j.envpol.2024.125019 PMID: 39326826

Meng, F., Xi, L., Liu, D., Huang, W., Lei, Z., Zhang, Z., & Huang, W. (2019). Effects of light intensity on oxygen distribution, lipid production and biological community of algal-bacterial granules in photo-sequencing batch reactors. *Bioresource Technology*, *272*, 473–481. DOI: 10.1016/j.biortech.2018.10.059 PMID: 30390540

Narayanan, I., Rajamanickam, R., Singh, R. K., Singh, P., Prabhu, A. A., Pandey, A., Pradhan, N., Nancharaiah, Y. V., Kumar, V., Selvasembian, R., & Mal, J. (2025). Microalgae-based remediation of pharmaceutical contaminants: Emerging strategies and technological synergies. *Discover Applied Sciences*, *7*(11), 1288. DOI: 10.1007/s42452-025-07226-4

Nguyen, T.-T.-D., Nguyen, T.-T., An Binh, Q., Bui, X.-T., Ngo, H. H., Vo, H. N. P., Andrew Lin, K.-Y., Vo, T.-D.-H., Guo, W., Lin, C., & Breider, F. (2020). Co-culture of microalgae-activated sludge for wastewater treatment and biomass production: Exploring their role under different inoculation ratios. *Bioresource Technology*, *314*, 123754. DOI: 10.1016/j.biortech.2020.123754 PMID: 32650264

Orandi, S., Lewis, D. M., & Moheimani, N. R. (2012). Biofilm establishment and heavy metal removal capacity of an indigenous mining algal-microbial consortium in a photo-rotating biological contactor. *Journal of Industrial Microbiology & Biotechnology*, *39*(9), 1321–1331. DOI: 10.1007/s10295-012-1142-9 PMID: 22644382

Oruganti, R. K., Katam, K., Show, P. L., Gadhamshetty, V., Upadhyayula, V. K. K., & Bhattacharyya, D. (2022). A comprehensive review on the use of algal-bacterial systems for wastewater treatment with emphasis on nutrient and micropollutant removal. *Bioengineered*, *13*(4), 10412–10453. DOI: 10.1080/21655979.2022.2056823 PMID: 35441582

Park, J. B. K., Craggs, R. J., & Shilton, A. N. (2011). Wastewater treatment high rate algal ponds for biofuel production. *Bioresource Technology*, *102*(1), 35–42. DOI: 10.1016/j.biortech.2010.06.158 PMID: 20674341

Parmar, K. S., & Patel, K. M. (2025). Biosorption and bioremediation of heavy metal ions from wastewater using algae: A comprehensive review. *World Journal of Microbiology & Biotechnology*, *41*(7), 262. DOI: 10.1007/s11274-025-04424-5 PMID: 40640643

Passos, F., Gutiérrez, R., Uggetti, E., Garfí, M., García, J., & Ferrer, I. (2017). Towards energy neutral microalgae-based wastewater treatment plants. *Algal Research*, *28*, 235–243. DOI: 10.1016/j.algal.2017.11.006

Peng, Z., Liu, G., & Huang, K. (2020). Cold Adaptation Mechanisms of a Snow Alga Chlamydomonas nivalis During Temperature Fluctuations. *Frontiers in Microbiology*, *11*, 611080. DOI: 10.3389/fmicb.2020.611080 PMID: 33584575

Qi, L., Liu, X., Gao, Y., Wang, Z., Zhang, N., & Su, X. (2024). Interactions between bacteria and microalgae in microalgal-bacterial symbiotic wastewater treatment systems: Mechanisms and influencing factors. *Aquatic Microbial Ecology, 90,* 41–60. DOI: 10.3354/ame02007

Ren, Z., Li, H., Sun, P., Fu, R., Bai, Z., Zhang, G., Sun, L., & Wei, Y. (2024). Development and challenges of emerging biological technologies for algal-bacterial symbiosis systems: A review. *Bioresource Technology, 413,* 131459. DOI: 10.1016/j.biortech.2024.131459 PMID: 39255948

Ri, C., Tao, Y., Tu, J., Li, X., She, S., Hou, L., Fu, Y., & Chen, L. (2023). Improvement of the Cd removal efficiency of a filamentous cyanobacterium Leptolyngbya sp. XZ1 through co-culture with Bacillus sp. S1. *Journal of Applied Phycology, 35*(6), 2935–2944. DOI: 10.1007/s10811-023-03078-w

Saravanan, A., Kumar, P. S., Varjani, S., Jeevanantham, S., Yaashikaa, P. R., Thamarai, P., Abirami, B., & George, C. S. (2021). A review on algal-bacterial symbiotic system for effective treatment of wastewater. *Chemosphere, 271,* 129540. DOI: 10.1016/j.chemosphere.2021.129540 PMID: 33434824

Sharma, A., & Bhatti, M. S. (2022). Simultaneous Removal of Organic Matter and Nutrients from High Strength Organic Wastewater Using Sequencing Batch Reactor (SBR). *Processes (Basel, Switzerland), 10*(10), 1903. https://www.mdpi.com/2227-9717/10/10/1903. DOI: 10.3390/pr10101903

Sharma, V., & Gupta, R. (2022). Biomass Digestion for Production of Electrical Energy. In (pp. 151-178). DOI: 10.1007/978-981-16-9135-5_7

Sial, A., Zhang, B., Zhang, A., Liu, K., Imtiaz, S. A., & Yashir, N. (2021). Microalgal–Bacterial Synergistic Interactions and Their Potential Influence in Wastewater Treatment: A Review. *BioEnergy Research, 14*(3), 723–738. DOI: 10.1007/s12155-020-10213-9

Siew Moi, P. (2004). Handbook of Microalgal Culture. Biotechnology and Applied Phycology. *Journal of Applied Phycology, 16*(2), 159–160. DOI: 10.1023/B:JAPH.0000044943.45511.5e

Silva, J. A. (2023). Wastewater Treatment and Reuse for Sustainable Water Resources Management: A Systematic Literature Review. *Sustainability (Basel), 15*(14), 10940. https://www.mdpi.com/2071-1050/15/14/10940. DOI: 10.3390/su151410940

Tang, L., Gao, M., Liang, S., Wang, S., & Wang, X. (2024). Enhanced biological phosphorus removal sustained by aeration-free filamentous microalgal-bacterial granular sludge. *Water Research*, *253*, 121315. DOI: 10.1016/j.watres.2024.121315 PMID: 38382289

Tang, W., Xu, X., Ye, B.-C., Cao, P., & Ali, A. (2019). Decolorization and degradation analysis of Disperse Red 3B by a consortium of the fungus Aspergillus sp. XJ-2 and the microalgae Chlorella sorokiniana XJK. *RSC Advances*, *9*(25), 14558–14566. DOI: 10.1039/C9RA01169B PMID: 35519313

Tangahu, B., Berlianto, M., & Kartika, A. (2020). Deconcentration of Chromium Contained in Wastewater Using a Bacteria and Microalgae Consortia with a High Rate Algal Reactor System. *Journal of Ecological Engineering*, *21*(8), 272–284. DOI: 10.12911/22998993/126878

Ugwuanyi, E., Nwokediegwu, Z., Dada, M., Majemite, M., & Obaigbena, A. (2024). The role of algae-based wastewater treatment systems: A comprehensive review. *World Journal of Advanced Research and Reviews*, *21*(2), 937–949. DOI: 10.30574/wjarr.2024.21.2.0521

Vo, H. N., Ngo, H. H., Guo, W., Nguyen, T., Liu, Y., Liu, Y., Nguyen, D. D., & Chang, S. W. (2019). A critical review on designs and applications of microalgae-based photobioreactors for pollutants treatment. *The Science of the Total Environment*, *651*(Pt 1), 1549–1568. DOI: 10.1016/j.scitotenv.2018.09.282 PMID: 30360283

Wang, H., Deng, L., Qi, Z., & Wang, W. (2022). Constructed microalgal-bacterial symbiotic (MBS) system: Classification, performance, partnerships and perspectives. *The Science of the Total Environment*, *803*, 150082. DOI: 10.1016/j.scitotenv.2021.150082 PMID: 34525774

Wang, J., Chen, X., Sun, H., Li, X., Lu, Q., Qin, P., Yang, Y., Lai, D., Luo, L., Peng, X., Yang, Y., & Wu, Z. (2025). Accelerated removal and mechanism of tetracycline from water using immobilized bacteria combined with microalgae. *Journal of Environmental Chemical Engineering*, *13*(2), 115424. DOI: 10.1016/j.jece.2025.115424

Wang, S., Ji, B., Zhang, M., Ma, Y., Gu, J., & Liu, Y. (2020). Defensive responses of microalgal-bacterial granules to tetracycline in municipal wastewater treatment. *Bioresource Technology*, *312*, 123605. DOI: 10.1016/j.biortech.2020.123605 PMID: 32504951

Wen, X., Sun, X., & Liu, X. (2025). Synergizing mechanistic and AI models for deeper insights into algal-bacterial systems in sustainable wastewater treatment. *Journal of Water Process Engineering*, *76*, 108169. DOI: 10.1016/j.jwpe.2025.108169

Wu, B., Deng, J., Niu, H., Liang, J., Arslan, M., Gamal El-Din, M., Wang, Q., Guo, S., & Chen, C. (2021). Establishing and Optimizing a Bacterial Consortia for Effective Biodegradation of Petroleum Contaminants: Advancing Classical Microbiology via Experimental and Mathematical Approach. *Water (Basel)*, *13*(22), 3311. https://www.mdpi.com/2073-4441/13/22/3311. DOI: 10.3390/w13223311

Xiao, Z., Meng, H., Li, S., Ning, W., Song, Y., Han, J., Chang, J.-S., Wang, Y., & Ho, S.-H. (2024). Insights into the removal of antibiotics from livestock and aquaculture wastewater by algae-bacteria symbiosis systems. *Environmental Research*, *257*, 119326. DOI: 10.1016/j.envres.2024.119326 PMID: 38849002

Xiong, W., Wang, S., Jin, Y., Wu, Z., Liu, D., & Su, H. (2023). Insights into nitrogen and phosphorus metabolic mechanisms of algal-bacterial aerobic granular sludge via metagenomics: Performance, microbial community and functional genes. *Bioresource Technology*, *369*, 128442. DOI: 10.1016/j.biortech.2022.128442 PMID: 36470490

Yadavalli, R., & Heggers, G. R. V. N. (2013). Two stage treatment of dairy effluent using immobilized Chlorella pyrenoidosa. *Journal of Environmental Health Science & Engineering*, *11*(1), 36. DOI: 10.1186/2052-336X-11-36 PMID: 24355316

Yang, I.-S., Salama, E.-S., Kim, J.-O., Govindwar, S. P., Kurade, M. B., Lee, M., Roh, H.-S., & Jeon, B.-H. (2016). Cultivation and harvesting of microalgae in photobioreactor for biodiesel production and simultaneous nutrient removal. *Energy Conversion and Management*, *117*, 54–62. DOI: 10.1016/j.enconman.2016.03.017

Yang, J., Gou, Y., Fang, F., Guo, J., Ma, H., Wei, X., & Shahmoradi, B. (2018). Impacts of sludge retention time on the performance of an algal-bacterial bioreactor. *Chemical Engineering Journal*, *343*, 37–43. DOI: 10.1016/j.cej.2018.02.118

Yu, Q., Chen, J., Ye, M., Wei, Y., Zhang, C., & Ge, Y. (2024). N-acyl homoserine lactones (AHLs) enhanced removal of cadmium and other pollutants by algae-bacteria consortia. *Journal of Environmental Management*, *366*, 121792. DOI: 10.1016/j.jenvman.2024.121792 PMID: 39002459

Zhang, H., Wu, T., Sun, L., Han, Z., Wang, Q., & Li, D. (2024). The construction of a microalgal-bacterial biofilm reactor for enhanced swine wastewater treatment. *Algal Research*, *79*, 103494. DOI: 10.1016/j.algal.2024.103494

Zhang, Y., Zheng, L., Wang, S., Zhao, Y., Xu, X., Han, B., & Hu, T. (2021). Quorum Sensing Bacteria in the Phycosphere of HAB Microalgae and Their Ecological Functions Related to Cross-Kingdom Interactions. *International Journal of Environmental Research and Public Health*, *19*(1), 163. Advance online publication. DOI: 10.3390/ijerph19010163 PMID: 35010421

Zhao, W., Tian, K., Zhang, L., Tang, Y., Chen, R., Zheng, X., & Zhao, M. (2025). Harnessing an Algae–Bacteria Symbiosis System: Innovative Strategies for Enhancing Complex Wastewater Matrices Treatment. *Sustainability (Basel)*, *17*(15), 7104. https://www.mdpi.com/2071-1050/17/15/7104. DOI: 10.3390/su17157104

Chapter 4
Synergistic Algal–Bacterial Interactions in Wastewater Bioremediation

Sakibul Islam Ratul
http://orcid.org/0009-0000-7217-5190
Daffodil International University, Dhaka, Bangladesh

Md. Shoeab Akhter
http://orcid.org/0009-0004-4582-2937
International University of Business Agriculture and Technology, Dhaka, Bangladesh

Sakik Rahman
Daffodil International University, Dhaka, Bangladesh

Md. Ratul Bhuyan
Daffodil International University, Dhaka, Bangladesh

Mohammed Ataur Rahman
International University of Business Agriculture and Technology, Dhaka, Bangladesh

Meshabul Ferdous Mithun
Daffodil International University, Dhaka, Bangladesh

Tasnim Taranni Shuchi
Daffodil International University, Dhaka, Bangladesh

ABSTRACT

Microalgae-bacteria consortia are a new and energy-efficient system of wastewater treatment and recovery of resources. This chapter emphasizes the metabolic interactions that are synergistic between algal photosynthesis and bacterial respiration to promote carbon, nitrogen, and phosphorus cycles. Co-cultures involving Chlorella vulgaris and Pseudomonas putida were studied on a laboratory scale with controlled conditions of light and aeration in order to determine the efficiency in

DOI: 10.4018/979-8-2600-3080-6.ch004

the removal of pollutants. The consortia clearly showed up to a 90 percent reduction in chemical oxygen demand and an improvement on ammonia and phosphate removal over monocultures. Metagenomic and metabolomic studies established a nutritional complementation and symbiotic control of growth. The results show that algal-bacterial systems are a sustainable and innovative approach to valorizing wastewater, making it cheaper in terms of energy and yielding biofertilizers and biomass that can be used to create biofuels.

1. INTRODUCTION

Blistering urbanization, population increase and industrial development have augmented the world issues on wastewater creation and nutrient contamination. The industrial and municipal wastewaters have a high level of organic matter, nitrogen, and phosphorus, which, when discharged without proper treatment, cause eutrophication, depletion of oxygen, and degradation of aquatic ecosystems (Qin et al., 2015). The other traditional methods of wastewater treatment, such as activated sludge and physicochemical systems, such as chemical precipitation and aeration, are good, but they consume a lot of energy, are expensive, and are not so effective in recovering useful nutrients (Peters et al., 1984). In addition, such systems produce huge amounts of sludge and produce greenhouse gases, which are inconsistent with the aim of global sustainability and mitigation of climate.

Biological treatment technologies have become better in terms of environmental performance, but monoculture-based microbial systems are often unstable, with low nutrient recovery and high operation requirements (Luo et al., 2020). These constraints have given rise to pursuing the nature-based and low-energy options that could combine wastewater treatment with resource recovery. Microalgae-bacteria consortia in this regard have been viewed as a viable solution. These systems replicate natural aquatic ecosystems in which photosynthetic and heterotrophic microorganisms coexist with strictly coupled metabolic interactions between each other.

Chlorella vulgaris, *Scenedesmus obliquus*, and *Chlamydomonas reinhardtii* are examples of microalgae that utilize light energy to fix carbon dioxide and incorporate inorganic nitrogen and phosphorus, giving out oxygen as a by-product (Shen, 2014). This oxygen helps in supporting aerobic bacterial respiration thus allowing the bacteria to effectively break down organic pollutants like *Pseudomonas putida*, *Bacillus subtilis* and *Nitrosomonas spp*. Bacterial metabolism, in its turn, recycles carbon dioxide, ammonium, and phosphate, which serve to support the growth of algae. Such a symbiotic relationship increases the efficiency of pollutant removal and decreases the outside aeration and chemical application.

In addition to the elimination of pollutants, algal-bacterial consortia is directly applicable in the field of carbon sequestration, nutrient recovery, and circular bioeconomy schemes (Viswanaathan et al., 2022). The algal biomass produced in the process of wastewater treatment can be utilized as biofertilizers, bioenergy feedstock, or bioproduct precursors and convert wastewater as a source of environmental liability to a resource. These systems are in line with the world's efforts to achieve carbon-neutral wastewater treatment and sustainable management of resources.

This chapter has the following objectives:

- To explain the biological and metabolic interactions that were inherent in algal-bacterial symbiosis in wastewater systems.
- To analyze the carbon, nitrogen and phosphorus cycling in algal-bacterial consortia and their contribution to the removal of pollutants.
- To assess the opportunity of an algal-bacteria system to valorize wastewater as part of a circular bioeconomy.

2. PRINCIPLES OF ALGAL-BACTERIAL SYMBIOSIS

Metabolic complementarity and ecological cooperation regulate algal-bacterial symbiosis in wastewater systems. Microalgae are primary producers capable of fixing carbon and accumulating nutrients and bacteria are decomposers capable of mineralizing organic matter and reproducing organic nutrients in inorganic forms. Oxygen-carbon exchange is the way to stabilize the interaction, as well as the nutrient recycling and microbial signaling (Lock & Hoyer, 1974). The outputs of these processes include increased system resilience, increased nutrient-use efficiency, and decreased energy requirements of the operation, which makes algal-bacterial consortia a solid biological platform to achieve sustainable wastewater treatment. The Figure 1 provides a conceptual overview of algal–bacterial wastewater bioremediation, illustrating wastewater inflow, metabolic interactions between microalgae and bacteria, reduced aeration demand, treated effluent generation, and biomass harvesting.

Figure 1. Conceptual overview of algal–bacterial wastewater bioremediation

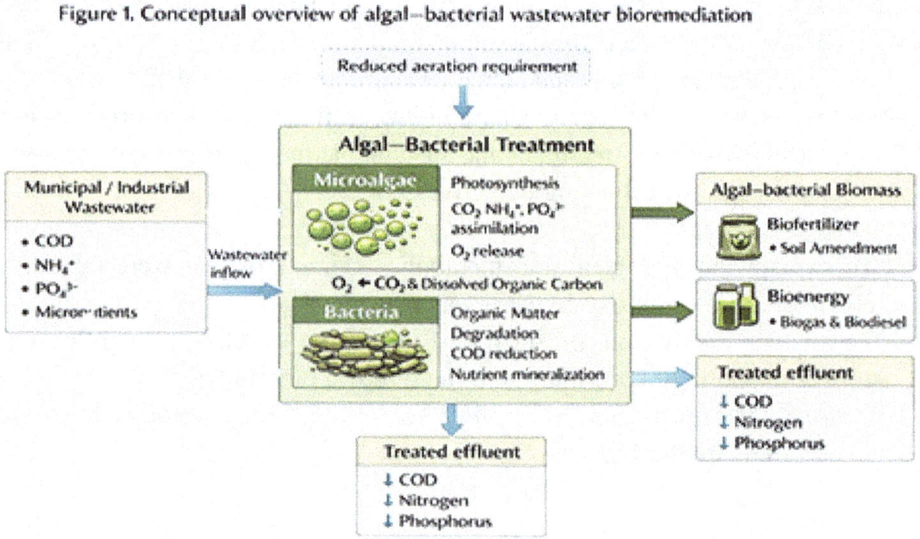

2.1 Microalgal Wastewater Treatment Systems

The microalgae are of key importance in the process of the treatment of waste water due to their metabolic characteristic of photosynthesis and their ability to absorb the nutrients. Other species like *Chlorella vulgaris* and *Scenedesmus obliquus* use the sunlight energy to transform the carbon dioxide into the organic biomass through oxygenic photosynthesis and directly contribute towards the process of carbon sequestration (Ramjee Chaudhary et al., 2018). Wastewater is an excellent source of dissolved inorganic carbon, ammonium, nitrate, and phosphate that microalgae easily absorb to grow the cells. The uptake of nitrogen takes place as ammonium (NH_4^+), but in some cases nitrate (NO_3^-) and nitrite (NO_2^-) can be used as well based on the species and environmental factors. Phosphorus will be absorbed as orthophosphate, and it will be integrated into nucleic acids, phospholipids, and energy containing molecules like ATP. The capacity to remove nutrients such as nitrogen and phosphorus simultaneously makes microalgae especially useful in the prevention of the release of nutrients as well as eutrophication. An important ecological role of microalgae is releasing oxygen as a result of photosynthesis (Masojídek et al., 2013). This oxygen augments aerobic metabolism of bacteria and minimizes or does away with the necessity of mechanical aeration. Moreover, there are algal exudates, such as carbohydrates and organic acids, which supply the carbon substrates that promote the activity of bacteria. Microalgae, due to these functions, are biological agents of wastewater purification and produce biomass suitable to the downstream valorization process.

2.2 Bacterial Functional Roles

Bacteria are essential companions in the algal-bacterial treatment process of wastewater because of their contribution to the degradation of organic matter and conversion of nutrients. The heterotrophic bacteria like *Pseudomonas putida* and *Bacillus subtilis* decompose dissolved and particulate organic matter and thus decrease that of chemical oxygen demand (COD) (Djaoudi et al., 2020). It is an activity of respiration that can convert organic carbon to carbon dioxide that is easily utilized by algal photosynthesis. There is also transformation of nitrogen through bacteria. The conversion of organic nitrogen into ammonium is done through the process of ammonification, and further oxidation of ammonium into nitrate is done by nitrifying bacteria like *Nitrosomonas* and *Nitrobacter*. Nitrate is also reduced to gaseous forms of nitrogen by the denitrifying bacteria in microzones where there is no oxygen, thus assimilating nitrogen (Rajta et al., 2020). These successive processes increase the efficiency of nitrogen cycling in consortia. Bacteria that solubilize phosphate do so by means of the release of bioavailable phosphate by insoluble mineral or organic phosphate by means of organic acids and phosphatases. On the whole, the collective response of bacteria leads to nutrient regeneration, microbial community stabilization, and the overall enhancement of the efficiency of wastewater treatment systems.

2.3 The foundation of the Synergistic Interactions

The oxygen-carbon cycle mainly causes the synergy between microalgae and bacteria. The carbon fixation of algal photosynthesis is sustained by oxygen produced by photosynthesis, and the fixation of CO_2 voluntarily by aerobic bacteria is transported most of the time by oxygen (Shen, 2014). The mutualism relationship also reduces the need to aerate the water and improves the efficiency of the systems. The collaboration is also enhanced by mutualistic nutrient cycling. Ammonium and phosphate are recreated in processes of bacterial mineralization and are quickly utilised by microalgae (Zhang et al., 2024). The algal exudates, in their turn, constitute easily accessible organic material, triggering the growth of bacteria. These feedback systems encourage an equal population growth and inhibit the prevalence of one group. There are also synergistic interactions that facilitate growth stimulation and system stability. The association of algae and bacteria is associated with enhanced ability to endure environmental changes, like the nutrient loading variations or light. The biofilm formation and microbial signaling are associated with functional redundancy and resilience which make these systems reliable as compared with monoculture-based treatments.

3. METABOLIC COUPLING AND BIOGEOCHEMICAL CYCLES

Carbon, nitrogen, and phosphorus cycling in algal-bacterial consortia Metabolism Algal bacterial consortia combine metabolic coupling to form a single biogeochemical web (Yu et al., 2024). Photosynthetic fixation of carbon, bacterial respiration, mineralization of nutrients, and assimilation is an ongoing process, and it becomes possible to remove the pollutants efficiently and recover the resources. These closely interconnected cycles are the basis of the high performance of consortia as opposed to traditional biological treatment systems.

3.1 The Cycling of Carbon Between Algal and Bacterial

The characteristics of algal-bacterial systems of wastewater include carbon cycling. Microalgae absorb inorganic carbon by making use of photosynthesis that transforms CO_2 into organic biomass. Part of this carbon is being emitted as dissolved organic substances such as sugars and organic acids which act as food to the heterotrophic bacteria. The plants contribute to pollution with wastewater in the form of exudates and organic substances which can be used by bacteria, which then oxidize them via respiration. This reaction inhibits the chemical oxygen requirement and returns the CO_2 in the system, which perpetuates the growth of algae (Singh & Singh, 2014). The interchange of the carbon between the algal and bacterial partners is always a closed-loop system that improves the efficiency of the treatment. Combining carbon fixation with organic matter degradation, algal-bacterial consortia attain a high COD reduction, and at the same time, fix carbon in biomass. This coupling capability is one of the differences that separates them in comparison to traditional treatment systems and contributes towards their involvement in climate-sensitive wastewater treatment.

3.2 Nitrogen Conversion Reaction Systems

In algal-bacterial systems, the elimination of nitrogen takes place using successive complementary assimilation and conversion pathways (Saravanan et al., 2021). Ammonium is more likely to be assimilated by microalgae because of its lower cost of energy, which includes nitrogen in proteins and nucleic acids. Such direct uptake lowers the amounts of nitrogen in treated effluents. Nitrogen removal occurs further through the action of bacteria. Organic nitrogen is converted into ammonium through ammonification, and nitrate is produced by oxidizing ammonium to nitrates through the process of nitrification in the presence of oxygen. The process of nitrogen removal is complete because in micro-anoxic areas, denitrifying bacteria break down nitrate to gaseous nitrogen (James & Vijayanandan, 2023). The algal-bacterial consortia are more effective in removing nitrogen compared to monocultures because there are assimilation and transformation processes that are simultaneously taking place in the same system.

3.3 Phosphorus Intake and Recycling

The phosphorus extraction is done by a combination of assimilation of algae and mobilization of bacteria (Shen et al., 2017). Microalgae absorb the phosphate to be used as cell energy, and it is built into the structural and energy-related product. Bacterial phosphate-solubilizing processes increase the supply of phosphate by transforming insoluble mineral and organic phosphorus to bioavailable phosphates. Enzymes secreted by bacteria, such as organic acids and phosphatase, are important in the process of mineralizing phosphorus. This recycling ensures that they do not accumulate phosphorus in the effluents and avoid incidence of eutrophication. The combination of algal uptake and bacteria solubilization is effective in the recovery of phosphorus as well as biomass generation and long-term control of nutrients. Table 1 summarizes the complementary functionalities of the microalgae and bacteria in the treatment and cycling of nutrients in wastewater.

Table 1. Functional roles of microalgae and bacteria in algal–bacterial wastewater treatment systems

Component	**Representative organisms**	**Primary function**	**System-level benefit**
Microalgae (R. Chaudhary et al., 2018)	*Chlorella vulgaris, Scenedesmus obliquus*	CO_2 fixation, N and P uptake, O_2 release	Carbon sequestration, nutrient removal
Heterotrophic bacteria (Safitri et al., 2015)	*Pseudomonas putida, Bacillus subtilis*	Organic matter degradation	COD reduction
Nitrifying/denitrifying bacteria (Han et al., 2022)	*Nitrosomonas, Nitrobacter* spp.	Nitrogen transformation	Enhanced nitrogen removal
Phosphate-solubilizing bacteria (do Carmo et al., 2019)	*Bacillus, Pseudomonas* spp.	Phosphorus mobilization	Phosphorus recovery

3.4 Emerging Contaminants Removal Potential

In addition to conventional pollutants, wastewater increasingly contains emerging contaminants such as pharmaceuticals, personal care products, microplastics, and endocrine-disrupting compounds. Algal–bacterial consortia show potential in removing these contaminants through mechanisms including biodegradation, biosorption, and photodegradation. Microalgae can uptake certain pharmaceuticals, while bacteria contribute to enzymatic breakdown of complex organic compounds. Although promising, further research is needed to optimize removal efficiency and understand long-term environmental impacts.

4. MODEL ALGAL–BACTERIAL CONSORTIA

Synergistic mechanisms and how to optimize the performance of the wastewater treatment are studied using model algal-bacterial consortia under controlled conditions. Co-culture of microalgae and heterotrophic bacteria has become a popular choice among other combinations because this type of culture is robust, adaptable, and complementary with respect to metabolism (Han et al., 2016). The choice of species is essential in the realization of stable population dynamics, effective nutrient removal, and high biomass productivity. *Chlorella vulgaris* and *Pseudomonas putida* are also an effective and well-known consortium which is widely used in the laboratory-scale testing of an algal-bacterial wastewater treatment and resource recovery system.

4.1 *Chlorella vulgaris* as one of the Core Microalgal Species

Chlorella vulgaris is among the most commonly used microalgal species in the treatment of wastewater because it has physiological strength and high metabolic efficiency. It has great resistance to changes in nutrient levels, pH, and light levels, and can be used in a variety of streams of wastewater (Khalili et al., 2015). The species boasts of high growth rates and photosynthetic capacity that allows it to fix carbon dioxide and assimilate nitrogen and phosphorus that are inorganic in nature.

Within wastewater systems, *C. vulgaris* selectively takes up ammonium and phosphate and hence, minimizes nutrient loads causing eutrophication. Its capacity to develop in the mixotrophic conditions also increases its nutrient uptake capacity since it has the capacity to use inorganic carbon as well as organic nutrients emitted by bacterial collaborators. Oxygen produced in the process of photosynthesis facilitates aerobic metabolism of bacteria, thus eliminating the requirement of external aeration (Lu et al., 2011). Besides being effective in treatment, *C. vulgaris* forms high biomass that contains proteins, lipids, and carbohydrates. This biomass can be used as a biofertilizer, animal feed supplement, or bioenergy feedstock, which supports its use in the valorization of wastewater as well as in the system of a circular bioeconomy (Ahorsu et al., 2018).

4.2 *Pseudomonas putida* Bacterial Partner

Pseudomonas putida is a metabolically adaptable bacterium, which is known to have the ability to degrade a broad spectrum of organic pollutants that are usually present in wastewater (Weimer et al., 2020). It has strong enzymatic mechanisms, thus facilitating the degradation of hydrocarbons, aromatic compounds, and other organic contaminants, contributing to massive decreases in chemical oxygen demand. There is a high level of compatibility of the species with the growth of microalgae, especially *Chlorella vulgaris*. *P. putida* feeds on organic exudates produced by algae as it produces carbon dioxide, vitamins and growth-stimulating metabolites to stimulate algal growth. This symbiosis stabilizes the co-culture and inhibitory buildup of organic compounds within it. Additionally, *P. putida* helps in creating resilience to the system by surviving the environmental stresses like nutrient changes and oxidation (Moreno & Rojo, 2013). The capacity of its biofilm formation, as well as microbial signaling, promotes community stability and functional redundancy, and therefore, it is a useful bacterial partner in algal-bacteria wastewater treatment systems.

4.3 Co-Culture System Dynamics of Interaction

The dynamics of interaction between algal and bacterial co-culture systems are controlled by the growth kinetics, resource availability, and environmental conditions. Mutualistic interactions require equal population ratios to ensure that there is no competitive dominance (Johnson & Amarasekare, 2013). When there is optimal condition, algal photosynthesis provides oxygen and organic carbon and bacterial respiration ensures that there is regeneration and degradation of nutrients and organic matter, respectively. In cases where carbon, nitrogen and phosphorus cycling between mates are synchronized, the interactions tend to be mutualistic. Nevertheless, the mechanism of competitive interactions can occur in nutrient limitation or unfavorable light requirements and can disrupt the system (Craine & Dybzinski, 2013). To achieve equilibrium, it is therefore important to design the reactor effectively and control its operation. Biofilm and microbial aggregates formation further increases the effectiveness of interaction by promoting close physical interaction and exchange of metabolites. The dynamics enhance the stability of treatments, the efficiency of the removal of nutrients, and environmental perturbation resistance.

5. EXPERIMENTAL DESIGN AND LABORATORY-SCALE TEST

The data acquired and tested on a laboratory-scale will be of significant insights into the performance and stability of algal-bacterial consortia under controlled conditions. Systematic manipulation of parameters like reactor configuration, light availability, aeration, and nutrient loading is systematically done to determine the efficiency of pollutant removal, biomass productivity, and resilience of the system (Pholchan et al., 2010). These findings provide baseline information on the upscaling of the algae-bacteria wastewater treatment systems.

5.1 Reactor Design and Operating Environment

Photobioreactors that are used in algal-bacterial co-cultures are typically configured to maximize light exposure, mixing, and exchange of gases (Scognamiglio et al., 2021). The intensity of light and photoperiod have a direct impact on photosynthetic activity and oxygen generation, whereas controlled aeration provides a sufficient supply of carbon dioxide and prevents the biomass from settling. The hydraulic retention time (HRT) is a very important parameter of operation, which defines the time of contact between microorganisms and wastewater constituents (Estrada-Arriaga & Mijaylova, 2011). Short HRTs are more suitable for quick treatment but can restrict the biomass accumulation, and long HRTs promote the removal of nutrients and increase biomass. Nutrient loading rates are set to match the growth of microbes to avoid inhibitory effects. These parameters should be properly controlled in order to have the stability in co-culture performance and reproducible experimental results.

5.2 Pollutant Removal Performance

The performance of algal-bacterial consortia in the removal of pollutants is observed to be better than those of monoculture systems. The use of *Chlorella vulgaris* as a co-culture with *Pseudomonas putida* has demonstrated up to 90 percent loss in chemical oxygen demand, which is an efficient degradation of organic matter (Mujtaba et al., 2015). Combined assimilation algal and bacterial transformation pathways increase ammonia removal whereas algal uptake and bacteria mobilization increases phosphate removal (Li et al., 2023a). Such synergistic processes lead to much greater nutrient removal efficiencies as compared to the level of efficiencies in algal-only and bacterial-only systems. A comparative study indicates the security and resilience of consortia operating in fluctuating operating environments, which suggests their appropriateness in sustainable wastewater treatment systems. Figure 2 compares the removal efficiency of chemical oxygen demand (COD), nitrogen, and phosphorus in microalgal monoculture, bacterial monoculture, and algal–bacterial consortium systems, highlighting the superior performance of synergistic consortia.

Figure 2. illustrates the enhanced pollutant removal efficiency achieved by algal–bacterial consortia compared with monoculture systems

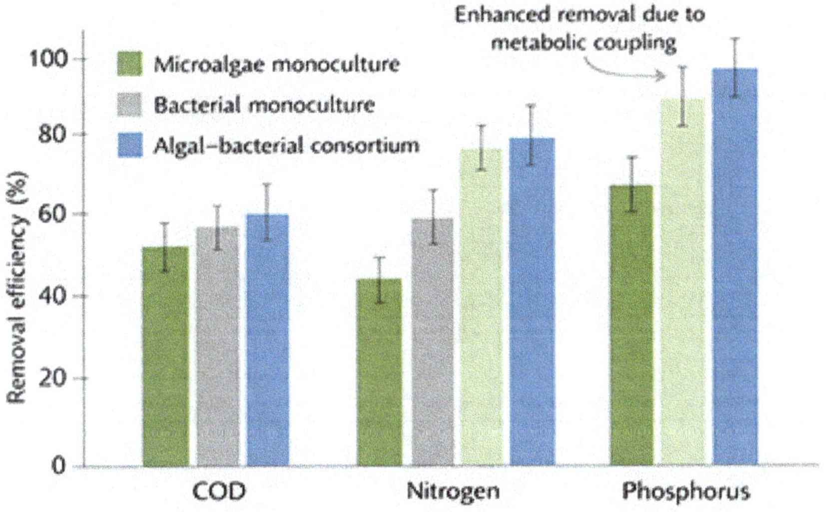

5.3 Yield and Stability of Biomass

The performance of algal-bacterial systems is determined by the biomass yield, which directly determines the potential of recovering resources. Metabolic cooperation and increased nutrient availability are the two factors that usually increase the concentrations of biomass in co-culture systems. The resulting biomass has desirable settling and harvesting properties, especially when biofilms or aggregates are developed. Enhanced settleability minimizes the downstream processing expenses, as well as enables the biomass collection. Biochemical examination shows that there are high protein, lipids, and carbohydrates in the biomass, thus it can be used in bioenergy generation, biofertilizers, and bioproduct production (Fernández-Reiriz et al., 1989). Stability of the system during long working periods also contributes to its practical application.

Table 2. Resource recovery pathways and valorization potential of algal–bacterial biomass in wastewater treatment systems

Valorization pathway	Biomass component	Representative organisms	End-use application	Sustainability benefit
Biofertilizer (Gonçalves et al., 2023)	N, P, organic matter, micronutrients	*Chlorella vulgaris*, *Pseudomonas putida*, *Bacillus* spp.	Soil amendment, crop nutrition	Reduced chemical fertilizer use
Bioenergy (biogas) (Bohutskyi & Bouwer, 2013)	Carbohydrates, proteins	Algal–bacterial mixed biomass	Anaerobic digestion, biogas production	Renewable energy generation
Biodiesel (Liu et al., 2024)	Lipids	*Chlorella vulgaris*, *Scenedesmus obliquus*	Biodiesel production	Fossil fuel substitution
Bioproducts (Dolganyuk et al., 2020)	Pigments, proteins	*Chlorella*, *Scenedesmus* spp.	Feed additives, biochemicals	Value-added products
Carbon sequestration (Singh & Ahluwalia, 2013)	Fixed algal carbon	Photosynthetic microalgae	Carbon capture and storage	Climate change mitigation

6. OVERVIEW OF INSIGHTS OF OMICS OF FUNCTIONING OF THE CONSORTIUM

The results of the omics in the form of molecular-level insights into algal-bacterial consortia in wastewater treatment systems are achieved. With high-throughput metagenomics and metabolomics, it is possible to identify good taxa including *Chlorella vulgaris*, *Scenedesmus obliquus*, *Pseudomonas putida*, and *Bacillus subtilis*, and the functional genes and exchange of metabolic processes. These tools explain metabolic processes of nutrient transformation, microbial co-operation and adaptive responses that provide stability to systems, removal efficiency of pollutants and biomass productivity (Tegegne & Savidge, 2025).

6.1 Metagenomic Analysis

Metagenomics can be used to identify both the taxonomic and functional capacity of algal-bacterial consortia in wastewater (Gan et al., 2024). Photosynthetic, carbon concentration, and ammonium assimilation genes of microalgae are often found through sequencing, especially in *Chlorella vulgaris* and *Chlamydomonas reinhardtii*. *Pseudomonas*, *Bacillus*, *Nitrosomonas* and *Nitrobacter* are the genera that usually dominate the bacterial communities and contain the genes that are related to the process of degradation of organic matter, nitrification, denitrification and metabolism of phosphorus. Functional clusters of genes involve nitrogen cycling (e.g., amoA, nirS, nirK) and phosphorus solubilization (e.g., phoA, ppx) (Ramzan et al., 2025). Categorize complementary products in nutrient-processing functional roles within a consortium. The genes associated with biofilm formation, oxidative stress tolerance and nutrient transport explain the resilience of algalbacterial systems in changing wastewater conditions as well. The insights facilitate rational consortium design and optimization of the wastewater treatment performance.

6.2 Metabolomic Profiling

The direct evidence of metabolic exchange and nutritional complementation is given by metabolomic profiling between microalgae and bacteria. Glucose, acetate, lactate, amino acids, and organic acids are common metabolites that can be found in the case of algal-bacterial systems. *Chlorella vulgaris* algal species emit carbohydrates and low-molecular-weight organic compounds, which act as carbon sources to heterotrophic bacteria such as *Pseudomonas putida* (Shi et al., 2022). On the other hand, bacterial metabolism exudes carbon dioxide, ammonium, phosphate, and vitamins (e.g., B-group vitamins), which encourage the growth of algae. The occurrence of signaling molecules such as quorum-sensing-related metabolites shows that there is coordinated control over the activity of microbes. The adaptive metabolic responses to changes in metabolite profiles under different light or nutrient conditions indicate changes in the stability of the consortium. In general, the metabolomic studies verify that bidirectional metabolite exchange is a focal process of enhanced pollutant elimination and biomass enrichment.

6.3 Control Mechanisms of Regulatory and Symbiotic Control

Symbiotic and regulatory control has been identified as important in ensuring functional stability in algal-bacterial consortia (Zhao et al., 2025). In bacteria, quorum sensing controls biofilm development, enzyme release, and stress response and enhances interactions among bacterial partners like Pseudomonas putida. These signal pathways also affect the rate of degradation of organic matter and the efficiency of the transformation of nutrients. In turn, microalgae react to the environmental stimuli by regulating photosynthetic activities and releasing metabolites, which control the growth of bacteria. The responses to the conditions of nutrient limitation and oxidative stress control mechanisms help both algae and bacteria to survive the changing conditions of wastewater. Adaptive gene expression and metabolic flexibility enable consortia to be stable in terms of its performance even during the period of operation. These regulatory processes individually maintain a long-lasting synergy, effective nutrient elimination, and resilience of the system.

7. THE VALORIZATION OF WASTEWATER AND RECOVERY OF RESOURCES

Algal-bacterial wastewater treatment systems are not limited to the removal of pollutants, but they also promote a recovery of useful resources. The biomass produced in consortia of Pseudomonas putida and Chlorella vulgaris has carbon, nitrogen, phosphorus, and bioactive substances that are useful in agriculture and energy production (Mujtaba et al., 2017). These systems can minimize waste creation by combining wastewater purification and biomass conversion, increase resource usage efficiency, and align with the idea of a circular bioeconomy, which helps to achieve sustainable and climate-smart wastewater management.

7.1 Biofertilizer Potential

The wastewater-based algal-bacteria biomass contains high levels of nutrients, organic matter, and useful microorganisms, hence a potential biofertilizer. Microalgae like Chlorella vulgaris and related bacteria generate biomass that is full of nitrogen, phosphorus, potassium, micronutrients, and growth-promoting agents (Hernandez et al., 2009). This biomass enhances the nutrient availability, the activity of soil microbes and the content of organic matter when introduced to soils (Akhter, et al., 2025). In contrast to synthetic fertilizers, nutrients in the algal-bacterial biomass are discharged slowly, and the losses due to leaching are minimal, and it is more efficient in utilizing nutrients. The beneficial bacterial genera, including *Bacillus* and *Pseudomonas*, found in the biomass can improve the nutrient cycling and inhibit soil-borne pathogens (Ding et al., 2024). It has been found that by using algal-based biofertilizers, the growth of crops like wheat, maize and vegetables has improved and harvest increased. The reuse of wastewater-based nutrients in agriculture will involve nutrient looping and less reliance on chemical fertilisers.

7.2 Bioenergy and Biofuel Applications

Biomass of algal-bacterial is another good feedstock that can be used in the production of bioenergy and biofuels. Some microalgae, like *Chlorella vulgaris* and *Scenedesmus obliquus*, are able to build lipids that can be used to produce biodiesel during conditions of nutrient stress. Moreover, carbohydrates and proteins in the biomass may be transformed into bioethanol or biohydrogen. The anaerobic digestion of remaining algalbacterial biomass can result in the production of methane-rich biogas which can serve as a source of renewable energy that can replace energy needs of the operational wastewater treatment plant (Marín et al., 2022). Research has demonstrated that bio-digestion of algal biomass in wastewater sludge improves biogas output and stability of the process. The combined recovery of wastewater treatment and bioenergy enhances the sustainability and economic viability of the system in general.

7.3 Inclusion in the Circular Bioeconomy Models

Circular bioeconomic systems of algal-bacterial wastewater treatment are a good example of waste stream transformation into marketable products (Wang et al., 2023). The nutrients in the wastewater are transformed into reusable biomass that can be reused as biofertilizers, bioenergy feedstock, or as raw material to make bioproducts, including pigments and biopolymers. The process of algal photosynthesis of carbon can be used to mitigate greenhouse gases, and nutrient recycling can help to minimize environmental pollution. The reduction of carbon associated with the treatment of wastewater and the recovery of resources combined with carbon management makes such systems capable of producing carbon-negative and resource-efficient wastewater management (Lu et al., 2018). Inclusion in the circular bioeconomy models will facilitate sustainable production processes, decrease the use of fossil-based inputs, and lead to environmental and economic sustainability in the long term.

8. CONCEPTUAL FRAMEWORK: ALGAL-BACTERIAL SYSTEM OF WASTEWATER BIOREMEDIATION SYNERGETICALLY

The synergistic algal-bacterial system conceptual framework comprises wastewater treatment, nutrient recovery, and carbon sequestration as a result of biological interactions in which the living organisms are synchronized. Wastewater is a potential source of organic carbon, nitrogen, phosphorus and micronutrients that form the appropriate growth medium of microalgae-bacteria co-culture platforms. Oxygenic photosynthesis is carried out by microalgae in this system, yielding oxygen and biomass and consuming inorganic carbon, ammonium, and phosphate (Grobbelaar et al., 1988). The emitted oxygen helps in the aerobic respiration of bacteria and effective breakdown of organic material, leading to significant decreases in chemical oxygen demand. Metabolism of bacteria on their part replenishes carbon dioxide and mineral nutrients, which maintains the growth of algae and completes the carbon, nitrogen, and phosphorus cycles. The interconnected biogeochemical cycling of carbon, nitrogen, and phosphorus in algal–bacterial consortia is schematically presented in Figure 3.

Figure 3. Conceptual framework of synergistic algal–bacterial wastewater bioremediation and resource recovery

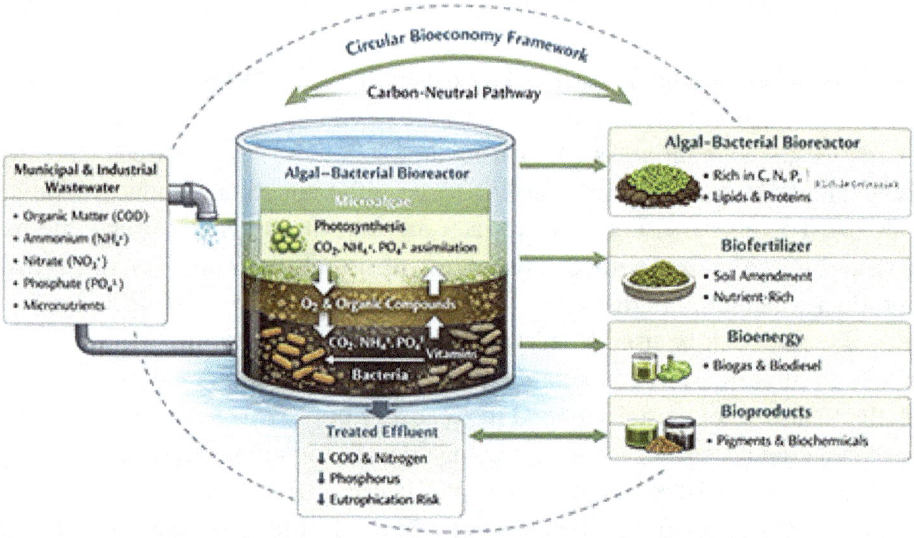

System stability is based on nutrient exchange and feedback whereby ammonium and phosphate assimilation by microalgae is countered by bacterial mineralization and organic matter degradation and these processes produce a self-regulating microbial equilibrium (Zhang et al., 2024). In addition to pollutant removal, the framework includes resource recovery pathways, with the valorization of the algal-bacterial biomass as biofertilizers, bioenergy feedstock, or bio-products being the main one to both assist with carbon sequestration. This system aligns with the principles of sustainability and closed-loop nutrient recycling because of reduced energy inputs, especially reduced aeration demand, and its disposition to closed-loop recycling. In general, the conceptual framework highlights that the efficiency and resilience of treatment are due to metabolic coupling and interactions at the ecosystems, and not the independent functions of microbes.

9. FINDINGS

The experimental assessment of algal-bacterial consortia showed a stable high performance of wastewater treatment in terms of various parameters. The fact that co-culture systems displayed a greater efficiency to remove the pollutants is a significant finding that showed that the chemical oxygen demand had reduced significantly, and this is attributable to the fact that organic matter had been effectively degraded. Combined algal uptake of ammonium and bacterial transformation was used to remove nitrogen whereas concomitant algal uptake and microbial mobilization helped in removing phosphorus (Li et al., 2023b). The algal bacteria consortia had shown to be more effective in nutrient removal and more stable in adverse environmental conditions than monoculture systems. The co-cultures were able to have balanced population processes, and system collapse that occurred frequently during single-species treatments was avoided. The analyses of microbial communities demonstrated high levels of functional synergy, as demonstrated by metagenomic evidence of complementary metabolic pathways that are present during carbon fixation, organic matter degradation processes, nitrogen cycling processes, and phosphorus metabolism. Nutritional complementation was further proved by metabolomic profiling by exchanging organic acids, sugars, amino acids, and signaling molecules that were used to coordinate the activity of microbes. The consortia was much more productive in biomass production because of enhanced nutrient supply and metabolic collaboration (Makut et al., 2019). The biomass collected was shown to have a good biochemical makeup, with a good content of proteins, lipids, and carbohydrates, which improved the downstream valorization of the biomass. Taken together, the available findings can attest to the fact that synergistic algal-bacterial systems are more efficient than traditional monoculture methods in the sense of combining the removal of pollutants, biomass production, and the stability of the system in one biological platform.

10. DISCUSSION

The enhanced performance of algal-bacterial consortia could be explained by mechanistic synergy based on the metabolic coupling and ecological cooperation. Co-cultures are more successful than the monocultures since the photosynthetic generation of oxygen by microalgae maintains the respiration of bacteria, and the mineralization by bacteria restores carbon dioxide and nutrients required by algae. Such a two-way process reduces resource constraints and contributes to the increased efficiency of treatment. The environmental parameters are very important in controlling such interactions, the availability of light controls the activity of photosynthesis, aeration of soil controls the exchange of gases, and the balance of nutrient loading of the microbe determines the balance of the growth of microbes. These parameters must be optimally set to ensure the stability of the system and the best performance. Algal-bacterial systems require less energy compared to traditional wastewater treatment systems, measured by lower energy consumption through less aeration, less chemical use to eliminate nutrients, and less sludge production (Oruganti et al., 2022). Nonetheless, its scalability is still an issue, especially concerning reactor design, light penetration, and harvesting of biomass on large scales. Light and temperature changes over the seasons can also affect the functionality of the system, which requires the strategy of adaptive operation. Nevertheless, as the discussion highlights, algalbacterial consortia is a potentially useful substitute to conventional treatment systems, with its ability to combine both environmental performance and resource recovery. The further development of technology and optimization of systems is the primary solution to the conversion of the success in labs into a reality.

Despite promising laboratory-scale performance, large-scale implementation faces challenges related to reactor design, land requirements, light penetration, and biomass harvesting costs. Economic feasibility depends on balancing operational savings from reduced aeration with costs associated with system setup and maintenance. Integration with existing wastewater infrastructure and valorization of biomass into marketable products such as biofertilizers and bioenergy can significantly improve economic viability. Techno-economic assessments are therefore essential for commercialization.

10.1 Comparison with Conventional Wastewater Treatment Systems

Conventional wastewater treatment systems such as activated sludge and physicochemical processes are effective but energy-intensive and generate large amounts of sludge. In contrast, algal–bacterial consortia reduce aeration demand, enable nutrient recovery, and produce valuable biomass. While traditional systems focus primarily on pollutant removal, algal–bacterial systems integrate treatment with resource recovery, making them more suitable for sustainable and circular wastewater management. However, conventional systems remain more established and scalable, highlighting the need for hybrid approaches.

11. RECOMMENDATIONS

In order to improve the viable application of algal-bacterial wastewater treatment systems, a number of recommendations are suggested. The optimization of system design must focus on designs that give the highest light distribution, gas exchange, and microbial contact with sufficient hydraulic retention time to balance between treatment effectiveness and biomass generation. The selection of the species in question must be directed towards the selection of strong microalgal and bacterial isolates that will allow withstanding the existing variability of wastewater compositions and environmental conditions. Consortia that are stress resistant and complementary in the metabolic functions have higher chances of being long-term stable. It is suggested to be incorporated into the current wastewater treatment plants by using hybrid types of treatment incorporating algal-bacterial systems as well as standard processes to facilitate a gradual adoption and minimize the risk. Decentralized wastewater systems provide more avenues of implementation, especially where there are few centralized treatment facilities that might be central to treatment (Massoud et al., 2009). Making a policy stance, policy support in terms of regulations, and offering financial incentives are necessary to encourage the implementation of the sustainable wastewater technologies. Commercialization can be spurred by standardized guide lines on how biomass can reuse it and incentives on carbon reduction and nutrient recovery. On the whole, it is necessary that researchers, engineers, policymakers, and stakeholders conjoin their efforts to ensure that algal-bacterial systems are not only available on experimental platforms but are also implemented in wastewater treatment systems on a large scale.

12. CONCLUSION

The synergistic algal-bacterial systems are a revolutionary concept in the treatment of wastewater through the combination of the elimination of pollutants, recovery of resources and carbon sequestration into a single biological system. These systems are able to remove organic matter, nitrogen, and phosphorus efficiently through metabolic coupling to produce valuable biomass. The symbiotic relations between microalgae and bacteria are more stable, less energy-demanding, and less chemical-intensive than the traditional treatment technologies. These systems have environmental as well as economic importance because they enable the transformation of wastewater, which is regarded as a problem of waste disposal, into a resource stream, which would facilitate the production of biofertilizers, bioenergy, and the implementation of a circular bioeconomy. Algal-bacterial consortia will be a climate-neutral approach to wastewater treatment, as they can be used in the future to take over conventional treatment methods and achieve the vision of a carbon-neutral and sustainable state of the environment. The results and conclusions that are being provided in this chapter reveal the possibility and benefits of such systems, but also emphasizes the fact that further innovation and optimization is required. Altogether, algal-bacterial wastewater management facilities have a great potential of positive influence on environmental protection and sustainable water management.

13. FUTURE STUDIES

The prospective studies concerning the algal-bacterial wastewater treatment systems must focus on the long-term field-based research to confirm the laboratory results in the actual wastewater environment. These studies are necessary to evaluate the stability of operations, seasonal changes, and efficiency of treatment in practice. Further omics-based studies, such as integrating metagenomics, transcriptomics, and metabolomics, will be used to learn more about the systems biology of algal-bacterial interactions and uncover important regulatory networks. Genetic and metabolic engineering strategies present the prospects of improving the efficiency of nutrient uptake, biomass output and stress robustness of both bacterial and algal partners (Choi et al., 2019). Also, life cycle analysis and techno-economic analysis are essential in order to measure environmental impacts, economic viability, and carbon footprints of large-scale implementation. Such evaluations will be helpful in decision-making and policy formulation because they will determine the pathways of trade-offs and optimality. The incorporation into renewable energy systems and digital monitoring tools should also be studied in the future with the aim of enhancing process control and scalability. Together, further interdisciplinary studies will be instrumental to develop algal-bacterial systems beyond conceptual experimental approaches to popular wastewater treatment technologies.

REFERENCES

Ahorsu, R., Medina, F., & Constantí, M. (2018). Significance and Challenges of Biomass as a Suitable Feedstock for Bioenergy and Biochemical Production: A Review. *Energies, 11*(12), 3366. DOI: 10.3390/en11123366

Akhter, Md. S., Miah, M. S., Rahman, H., Rahman, S. M. M., Karim, A. J. M. S., Ratul, S. I., & Rahman, M. A. (2025). Agricultural Applications of Chalcogenide-Based Materials: Toward Sustainable, Smart, and Nutrient-Efficient Farming. In K. Kannan & V. Tari (Eds.), Chalcogenide-Based Materials for Optoelectronics, Energy, and Sustainability (pp. 409–438). IGI Global Scientific Publishing. DOI: 10.4018/979-8-3373-3962-7.ch016

Bohutskyi, P., & Bouwer, E. (2013). Biogas Production from Algae and Cyanobacteria Through Anaerobic Digestion: A Review, Analysis, and Research Needs. In Lee, J. W. (Ed.), *Advanced Biofuels and Bioproducts* (pp. 873–975). Springer., DOI: 10.1007/978-1-4614-3348-4_36

Chaudhary, R., Dikshit, A. K., & Tong, Y. W. (2018). Carbon-dioxide biofixation and phycoremediation of municipal wastewater using Chlorella vulgaris and Scenedesmus obliquus. *Environmental Science and Pollution Research International, 25*(21), 20399–20406. DOI: 10.1007/s11356-017-9575-3 PMID: 28656576

Chaudhary, R., Tong, Y. W., & Dikshit, A. K. (2018). CO2-assisted removal of nutrients from municipal wastewater by microalgae Chlorella vulgaris and Scenedesmus obliquus. *International Journal of Environmental Science and Technology, 15*(10), 2183–2192. DOI: 10.1007/s13762-017-1571-0

Choi, K. R., Jang, W. D., Yang, D., Cho, J. S., Park, D., & Lee, S. Y. (2019). Systems Metabolic Engineering Strategies: Integrating Systems and Synthetic Biology with Metabolic Engineering. *Trends in Biotechnology, 37*(8), 817–837. DOI: 10.1016/j.tibtech.2019.01.003 PMID: 30737009

Craine, J. M., & Dybzinski, R. (2013). Mechanisms of plant competition for nutrients, water and light. *Functional Ecology, 27*(4), 833–840. DOI: 10.1111/1365-2435.12081

Ding, H., Luo, C., Li, Y., Li, Q., & Dong, Y. (2024). Impact of *Bacillus subtilis* and *Pseudomonas fluorescens* beneficial bacterial agents on soil-borne diseases, growth, and economics of continuous cropping of flue-cured tobacco. *Crop Protection (Guildford, Surrey), 177*, 106556. DOI: 10.1016/j.cropro.2023.106556

Djaoudi, K., Van Wambeke, F., Barani, A., Bhairy, N., Chevaillier, S., Desboeufs, K., Nunige, S., Labiadh, M., Henry des Tureaux, T., Lefèvre, D., Nouara, A., Panagiotopoulos, C., Tedetti, M., & Pulido-Villena, E. (2020). Potential bioavailability of organic matter from atmospheric particles to marine heterotrophic bacteria. *Biogeosciences*, *17*(24), 6271–6285. DOI: 10.5194/bg-17-6271-2020

do Carmo, T. S., Moreira, F. S., Cabral, B. V., Dantas, R. C. C., de Resende, M. M., Cardoso, V. L., & Ribeiro, E. J. (2019). Phosphorus Recovery from Phosphate Rocks Using Phosphate-Solubilizing Bacteria. *Geomicrobiology Journal*, *36*(3), 195–203. DOI: 10.1080/01490451.2018.1534901

Dolganyuk, V., Belova, D., Babich, O., Prosekov, A., Ivanova, S., Katserov, D., Patyukov, N., & Sukhikh, S. (2020). Microalgae: A Promising Source of Valuable Bioproducts. *Biomolecules*, *10*(8), 1153. DOI: 10.3390/biom10081153 PMID: 32781745

Estrada-Arriaga, E. B., & Mijaylova, P. N. (2011). Influence of operational parameters (sludge retention time and hydraulic residence time) on the removal of estrogens by membrane bioreactor. *Environmental Science and Pollution Research International*, *18*(7), 1121–1128. DOI: 10.1007/s11356-011-0461-0 PMID: 21331588

Fernández-Reiriz, M. J., Perez-Camacho, A., Ferreiro, M. J., Blanco, J., Planas, M., Campos, M. J., & Labarta, U. (1989). Biomass production and variation in the biochemical profile (total protein, carbohydrates, RNA, lipids and fatty acids) of seven species of marine microalgae. *Aquaculture (Amsterdam, Netherlands)*, *83*(1), 17–37. DOI: 10.1016/0044-8486(89)90057-4

Gan, Y., Ji, X., & Yang, R. (2024). Metagenomic profiling of antibiotic resistance genes/bacteria removal in urban water: Algal-bacterial consortium treatment system. *Bioresource Technology*, *404*, 130905. DOI: 10.1016/j.biortech.2024.130905 PMID: 38801952

Gonçalves, J., Freitas, J., Fernandes, I., & Silva, P. (2023). Microalgae as Biofertilizers: A Sustainable Way to Improve Soil Fertility and Plant Growth. *Sustainability (Basel)*, *15*(16), 12413. DOI: 10.3390/su151612413

Grobbelaar, J. U., Soeder, C. J., Groeneweg, J., Stengel, E., & Hartig, P. (1988). Rates of biogenic oxygen production in mass cultures of microalgae, absorption of atmospheric oxygen and oxygen availability for wastewater treatment. *Water Research*, *22*(11), 1459–1464. DOI: 10.1016/0043-1354(88)90105-4

Han, D., Hu, Z., Li, D., & Tang, R. (2022). Nitrogen Removal of Water and Sediment in Grass Carp Aquaculture Ponds by Mixed Nitrifying and Denitrifying Bacteria and Its Effects on Bacterial Community. *Water (Basel)*, *14*(12), 1855. DOI: 10.3390/w14121855

Han, J., Zhang, L., Wang, S., Yang, G., Zhao, L., & Pan, K. (2016). Co-culturing bacteria and microalgae in organic carbon containing medium. *Journal of Biological Research (Thessaloniki)*, *23*(1), 8. DOI: 10.1186/s40709-016-0047-6 PMID: 27119075

Hernandez, J.-P., de-Bashan, L. E., Rodriguez, D. J., Rodriguez, Y., & Bashan, Y. (2009). Growth promotion of the freshwater microalga *Chlorella vulgaris* by the nitrogen-fixing, plant growth-promoting bacterium *Bacillus pumilus* from arid zone soils. *European Journal of Soil Biology* [PGPB]. *European Journal of Soil Biology*, *45*(1), 88–93. DOI: 10.1016/j.ejsobi.2008.08.004

James, S. N., & Vijayanandan, A. (2023). Recent advances in simultaneous nitrification and denitrification for nitrogen and micropollutant removal: A review. *Biodegradation*, *34*(2), 103–123. DOI: 10.1007/s10532-023-10015-8 PMID: 36899211

Johnson, C. A., & Amarasekare, P. (2013). Competition for benefits can promote the persistence of mutualistic interactions. *Journal of Theoretical Biology*, *328*, 54–64. DOI: 10.1016/j.jtbi.2013.03.016 PMID: 23542049

Khalili, A., Najafpour, G. D., Amini, G., & Samkhaniyani, F. (2015). Influence of nutrients and LED light intensities on biomass production of microalgae Chlorella vulgaris. *Biotechnology and Bioprocess Engineering; BBE*, *20*(2), 284–290. DOI: 10.1007/s12257-013-0845-8

Li, Q., Xu, Y., Liang, C., Peng, L., & Zhou, Y. (2023a). Nitrogen removal by algal-bacterial consortium during mainstream wastewater treatment: Transformation mechanisms and potential N2O mitigation. *Water Research*, *235*, 119890. DOI: 10.1016/j.watres.2023.119890 PMID: 36958220

Li, Q., Xu, Y., Liang, C., Peng, L., & Zhou, Y. (2023b). Nitrogen removal by algal-bacterial consortium during mainstream wastewater treatment: Transformation mechanisms and potential N2O mitigation. *Water Research*, *235*, 119890. DOI: 10.1016/j.watres.2023.119890 PMID: 36958220

Liu, J., Booma Devi, P., Chinnathambi, A., & Ali Alharbi, S. (2024). Mitigating fossil fuel deficiency and environmental impacts: Performance analysis of *Scenedesmus obliquus* microalgae biodiesel in a diesel engine. *Fuel*, *364*, 131033. DOI: 10.1016/j.fuel.2024.131033

Lock, G. A., & Hoyer, W. A. (1974). Carbon-Oxygen (C/O) Log: Use and Interpretation. *Journal of Petroleum Technology, 26*(09), 1044–1054. DOI: 10.2118/4639-PA

Lu, H., Zhang, G., Wan, T., & Lu, Y. (2011). Influences of light and oxygen conditions on photosynthetic bacteria macromolecule degradation: Different metabolic pathways. *Bioresource Technology, 102*(20), 9503–9508. DOI: 10.1016/j.biortech.2011.07.114 PMID: 21871794

Lu, L., Guest, J. S., Peters, C. A., Zhu, X., Rau, G. H., & Ren, Z. J. (2018). Wastewater treatment for carbon capture and utilization. *Nature Sustainability, 1*(12), 750–758. DOI: 10.1038/s41893-018-0187-9

Luo, Y., Le-Clech, P., & Henderson, R. K. (2020). Characterisation of microalgae-based monocultures and mixed cultures for biomass production and wastewater treatment. *Algal Research, 49*, 101963. DOI: 10.1016/j.algal.2020.101963

Makut, B. B., Das, D., & Goswami, G. (2019). Production of microbial biomass feedstock via co-cultivation of microalgae-bacteria consortium coupled with effective wastewater treatment: A sustainable approach. *Algal Research, 37*, 228–239. DOI: 10.1016/j.algal.2018.11.020

Marín, D., Méndez, L., Suero, I., Díaz, I., Blanco, S., Fdz-Polanco, M., & Muñoz, R. (2022). Anaerobic digestion of food waste coupled with biogas upgrading in an outdoors algal-bacterial photobioreactor at pilot scale. *Fuel, 324*, 124554. DOI: 10.1016/j.fuel.2022.124554

Masojídek, J., Torzillo, G., & Koblížek, M. (2013). Photosynthesis in Microalgae. In *Handbook of Microalgal Culture* (pp. 21–36). John Wiley & Sons, Ltd., DOI: 10.1002/9781118567166.ch2

Massoud, M. A., Tarhini, A., & Nasr, J. A. (2009). Decentralized approaches to wastewater treatment and management: Applicability in developing countries. *Journal of Environmental Management, 90*(1), 652–659. DOI: 10.1016/j.jenvman.2008.07.001 PMID: 18701206

Moreno, R., & Rojo, F. (2013). The contribution of proteomics to the unveiling of the survival strategies used by Pseudomonas putida in changing and hostile environments. *Proteomics, 13*(18–19), 2822–2830. DOI: 10.1002/pmic.201200503 PMID: 23625785

Mujtaba, G., Rizwan, M., & Lee, K. (2015). Simultaneous removal of inorganic nutrients and organic carbon by symbiotic co-culture of Chlorella vulgaris and Pseudomonas putida. *Biotechnology and Bioprocess Engineering; BBE, 20*(6), 1114–1122. DOI: 10.1007/s12257-015-0421-5

Mujtaba, G., Rizwan, M., & Lee, K. (2017). Removal of nutrients and COD from wastewater using symbiotic co-culture of bacterium *Pseudomonas putida* and immobilized microalga *Chlorella vulgaris. Journal of Industrial and Engineering Chemistry, 49*, 145–151. DOI: 10.1016/j.jiec.2017.01.021

Oruganti, R. K., Katam, K., Show, P. L., Gadhamshetty, V., Upadhyayula, V. K. K., & Bhattacharyya, D. (2022). A comprehensive review on the use of algal-bacterial systems for wastewater treatment with emphasis on nutrient and micropollutant removal. *Bioengineered, 13*(4), 10412–10453. DOI: 10.1080/21655979.2022.2056823 PMID: 35441582

Peters, R. W., Ku, Y., Boardman, G. D., Walker, T. J., Hoffman, C. P., Chang, T.-K., & Parks, J. T. (1984). Wastewater Treatment-Physical and Chemical Methods. *Journal - Water Pollution Control Federation, 56*(6), 553–568.

Pholchan, M. K., Baptista, J. de C., Davenport, R. J., & Curtis, T. P. (2010). Systematic study of the effect of operating variables on reactor performance and microbial diversity in laboratory-scale activated sludge reactors. *Water Research, 44*(5), 1341–1352. DOI: 10.1016/j.watres.2009.11.005 PMID: 19945137

Qin, C., Liu, H., Liu, L., Smith, S., Sedlak, D. L., & Gu, A. Z. (2015). Bioavailability and characterization of dissolved organic nitrogen and dissolved organic phosphorus in wastewater effluents. *The Science of the Total Environment, 511*, 47–53. DOI: 10.1016/j.scitotenv.2014.11.005 PMID: 25527968

Rajta, A., Bhatia, R., Setia, H., & Pathania, P. (2020). Role of heterotrophic aerobic denitrifying bacteria in nitrate removal from wastewater. *Journal of Applied Microbiology, 128*(5), 1261–1278. DOI: 10.1111/jam.14476 PMID: 31587489

Ramzan, M. N., Shen, D., Wei, Y., Emmanuel, A., Nicholaus, R., Yang, W., & Zheng, Z. (2025). Nitrogen and phosphorus-related functional genes enhance nutrient removal in the integrated aquaculture wastewater bioremediation system in the presence of photosynthetic bacteria. *Aquaculture International, 33*(2), 131. DOI: 10.1007/s10499-024-01809-5

Safitri, R., Priadie, B., Miranti, M., & Astuti, A. W. (2015). Ability of bacterial consortium Bacillus coagulans Bacilus licheniformis Bacillus pumilus Bacillus subtilis Nitrosomonas sp and Pseudomonas putida in bioremediation of waste water in cisirung waste water treatment plant. *AgroLife Scientific Journal, 4*(1), 146–152. https://www.agrolifejournal.usamv.ro/index.php/agrolife/article/view/402. DOI: 10.17930/AGL2015122

Saravanan, A., Kumar, P. S., Varjani, S., Jeevanantham, S., Yaashikaa, P. R., Thamarai, P., Abirami, B., & George, C. S. (2021). A review on algal-bacterial symbiotic system for effective treatment of wastewater. *Chemosphere, 271*, 129540. DOI: 10.1016/j.chemosphere.2021.129540 PMID: 33434824

Scognamiglio, V., Giardi, M. T., Zappi, D., Touloupakis, E., & Antonacci, A. (2021). Photoautotrophs–Bacteria Co-Cultures: Advances, Challenges and Applications. *Materials (Basel), 14*(11), 3027. DOI: 10.3390/ma14113027 PMID: 34199583

Shen, Y. (2014a). Carbon dioxide bio-fixation and wastewater treatment via algae photochemical synthesis for biofuels production. *RSC Advances, 4*(91), 49672–49722. DOI: 10.1039/C4RA06441K

Shen, Y. (2014b). Carbon dioxide bio-fixation and wastewater treatment via algae photochemical synthesis for biofuels production. *RSC Advances, 4*(91), 49672–49722. DOI: 10.1039/C4RA06441K

Shen, Y., Gao, J., & Li, L. (2017). Municipal wastewater treatment via co-immobilized microalgal-bacterial symbiosis: Microorganism growth and nutrients removal. *Bioresource Technology, 243*, 905–913. DOI: 10.1016/j.biortech.2017.07.041 PMID: 28738545

Shi, Z., Du, H., Wang, C., & Xu, H. (2022). Quantifying the bioaccumulation of Pb to Chlorella vulgaris in the presence of dissolved organic matters with different molecular weights. *Environmental Science and Pollution Research International, 29*(47), 70921–70932. DOI: 10.1007/s11356-022-19699-6 PMID: 35593980

Singh, S. P., & Singh, P. (2014). Effect of CO2 concentration on algal growth: A review. *Renewable & Sustainable Energy Reviews, 38*, 172–179. DOI: 10.1016/j.rser.2014.05.043

Singh, U. B., & Ahluwalia, A. S. (2013). Microalgae: A promising tool for carbon sequestration. *Mitigation and Adaptation Strategies for Global Change, 18*(1), 73–95. DOI: 10.1007/s11027-012-9393-3

Tegegne, H. A., & Savidge, T. C. (2025). Gut microbiome metagenomics in clinical practice: Bridging the gap between research and precision medicine. *Gut Microbes, 17*(1), 2569739. DOI: 10.1080/19490976.2025.2569739 PMID: 41137523

Viswanaathan, S., Perumal, P. K., & Sundaram, S. (2022). Integrated Approach for Carbon Sequestration and Wastewater Treatment Using Algal–Bacterial Consortia: Opportunities and Challenges. *Sustainability (Basel), 14*(3), 1075. DOI: 10.3390/su14031075

Wang, J., Deng, Y., Chen, W., Song, H., Zan, F., Wu, X., & Ren, Y. (2023). Revealing the role of algae in algae enhanced bacteria consortia for municipal wastewater treatment: Performance, characteristics, and microbial pathways. *Journal of Water Process Engineering*, *53*, 103640. DOI: 10.1016/j.jwpe.2023.103640

Weimer, A., Kohlstedt, M., Volke, D. C., Nikel, P. I., & Wittmann, C. (2020). Industrial biotechnology of Pseudomonas putida: Advances and prospects. *Applied Microbiology and Biotechnology*, *104*(18), 7745–7766. DOI: 10.1007/s00253-020-10811-9 PMID: 32789744

Yu, T., Wang, S., Yang, H., Sun, Y., Chen, Z., Xu, G., & Zhang, C. (2024). Carbon Capture and Resource Utilization by Algal–Bacterial Consortium in Wastewater Treatment: A Mini-Review. *Water (Basel)*, *16*(15), 2197. DOI: 10.3390/w16152197

Zhang, L., Ali, A., Su, J., Huang, T., & Wang, Z. (2024a). Ammonium nitrogen and phosphorus removal by bacterial-algal symbiotic dynamic sponge bioremediation system in micropolluted water: Operational mechanism and transformation pathways. *The Science of the Total Environment*, *947*, 174636. DOI: 10.1016/j.scitotenv.2024.174636 PMID: 38992368

Zhang, L., Ali, A., Su, J., Huang, T., & Wang, Z. (2024b). Ammonium nitrogen and phosphorus removal by bacterial-algal symbiotic dynamic sponge bioremediation system in micropolluted water: Operational mechanism and transformation pathways. *The Science of the Total Environment*, *947*, 174636. DOI: 10.1016/j.scitotenv.2024.174636 PMID: 38992368

Zhao, W., Tian, K., Zhang, L., Tang, Y., Chen, R., Zheng, X., & Zhao, M. (2025). Harnessing an Algae–Bacteria Symbiosis System: Innovative Strategies for Enhancing Complex Wastewater Matrices Treatment. *Sustainability (Basel)*, *17*(15), 7104. DOI: 10.3390/su17157104

Chapter 5
Microalgal Wastewater Valorization: Mechanisms, Emerging Contaminant Removal, and Carbon Sequestration

Nilotpal Das
http://orcid.org/0009-0005-0083-4093
Visva-Bharati University, India

Atin Kumar
http://orcid.org/0000-0002-1653-2146
Uttaranchal University, India

Sharad Sachan
Lovely Professional University, India

ABSTRACT

Emerging contaminants refer to natural and human-made and diverse types of chemicals that are not regularly monitored within environments and are by-products of human activity. The wastewater usually has such pollutants in large amounts with industrial, agro industrial as well as agricultural sources with heavy metals, oils and dangerous compounds being discharged. Microalgae have demonstrated great potential in terms of biological assay to remove such contaminants in wastewater. This review identifies the use of microalgae to remove the new pollutants with a

DOI: 10.4018/979-8-2600-3080-6.ch005

demonstration of the important mechanisms which include absorption and biodegradation. It also talks about the contemporary issues and outlooks of microalgae-based wastewater treatments. The review also highlights the effectiveness of the phycoremediation in enabling sustainable wastewater treatment and limiting the emergent contaminants in accordance with the UN sustainable development goals (SDGs).

1. INTRODUCTION

As a result of the releases by houses, commerce, and agriculture, chemicals of emergent concern (CECs), which are largely anthropogenic, are becoming more widely discovered in various aquatic environments. These chemicals are so-called emerging pollutants (EPs), and they are often present in the range of micrograms to milligrams per litre (Wanda et al., 2017). EPs contain a very large variety of organic and inorganic materials. Though the inorganic contaminants are mostly made up of heavy metals, the organic contaminants generally include hormones, endocrine-disrupting substances, personal care products, and pharmaceutical residues. These pollutants create severe ecological imbalances as they have a high biological and chemical oxygen demand (BOD and COD), and therefore their presence in aquatic habitats causes a serious imbalance (Kundu et al., 2024).

The treatment of wastewater is normally conducted by a combination of physical, biological, and chemical treatment processes in primary, secondary, and tertiary processes. The goal of the first therapy is to get rid of readily soluble substances that could otherwise make later phases of treatment more difficult. Secondary treatment oxidizes nutrients into nitrate and orthophosphate and breaks down dissolved organic debris using biological processes. But the resultant effluent frequently contains high amounts of inorganic phosphate and nitrogen, which worsen eutrophication and add heavy metals and toxic organic compounds to aquatic systems (Gondi et al., 2022; Rathod, 2014; Kundu et al., 2024; Shitu et al., 2024). In order to create safe and clean effluent for discharge, tertiary treatment—which is thought to be the most advanced stage—tries to further reduce nitrates, phosphates, and remaining organic matter. In this step of denitrification, nitrate is reduced to nitrite, which is then converted to nitrogen gas and discharged into the environment (Molazadeh et al., 2019).

Despite improvements, a number of nations' wastewater treatment plants continue to operate below ideal levels because of poor design, a lack of technical know-how, and budgetary limitations (Farazaki & Gikas, 2019). By using activated sludge processes, conventional treatment systems mainly aim to reduce BOD and remove suspended particles. Consequently, they have no significant use in removing micropollutants and inorganic nutrients. The fact that traditional biodegradation methods often cannot effectively decompose organic and inorganic substances makes wastewaters with complex mixtures of heavy metals, xenobiotics, and surplus nutrients a problem (Gondi et al., 2022; Wollmann et al., 2019; Kundu et al., 2024). This inefficiency may pose a significant threat to aquatic life and have harmful consequences on the ecosystem, including depletion of oxygen and high toxicity of effluents (Umamaheswari & Shanthakumar, 2016).

Alongside that, the untreated persistence of nutrients in effluents may complicate the process of disinfection, increase the demand for chlorine, and increase the risks to human health and aquatic life. Therefore, the necessity of more advanced treatment approaches that will remove the pollutants and nutrients before the wastewater release is increasing (Chai et al., 2021). Microalgae absorb the inorganic minerals in an astounding capacity using nitrogen and phosphate to make their proteins and take up some of the heavy metals as micronutrients that aid in their growth (Kundu et al., 2024). Therefore, microalgae can effectively eliminate pollutants of wastewater such as nitrogen, phosphorus, and others through the application of microalgae as bioremediation agents. Microalgae also assist in reducing the bacteria and pathogens in wastewater, besides maintaining the dissolved oxygen concentration (Rathod, 2014).

Because microalgal therapy may treat carbon, nitrogen, and phosphorus all at once in a single step, it is thought to be a more effective and sustainable option than traditional approaches. By transforming carbon dioxide into useful bio-based molecules and fuel products, it also aids in carbon sequestration by lowering greenhouse gas emissions (Klinthong et al., 2015). Operational costs can be reduced by using the collected microalgal biomass to produce high-value bio-products such as health supplements, biohydrogen, bioalcohols, and biohydrocarbons (Koyande et al., 2019; Perez-Garcia & Bashan, 2015). Both eukaryotic and prokaryotic (cyanobacterial) species are often used in both experimental and practical settings, and microalgae have been widely used in wastewater treatment systems.

Enough amounts of vital nutrients, like phosphorus and nitrogen, which are non-renewable resources necessary for biomass production and algal growth, are found in wastewater itself. In addition to increasing algal productivity and carbon neutrality, this nutrient availability lessens the need for synthetic and freshwater nutrient inputs that are normally needed in traditional biological treatment systems, which lowers operating costs and has a smaller negative impact on the environment (Delrue et al., 2016).

2. PHYTOREMEDIATION OF WASTEWATER

Microalgae are made of eukaryotic cells, which are minute photosynthetic organisms that utilize sunlight to produce organic matter, as in the case of higher plants (Kundu et al., 2024). Their cellular structure consists of a nucleus, cytoplasm, cell wall, plasma membrane, and other organelles, such as plastids, which contain a pigment known as chlorophyll that is necessary in photosynthesis. Microalgae lack a vascular system like higher plants do to transport nutrients. This simplifies the intake process of nutrients and enhances productivity in aquatic ecosystems because every microalgal cell is photoautotrophic, i.e., it takes in nutrients directly itself (Emparan et al., 2019).

Due to their unique metabolic and photosynthetic characteristics, microalgae-based wastewater treatment is effective in the removal of organic and inorganic impurities, such as phosphates, nitrates, heavy metals, carbon compounds, and toxic substances, chemical oxygen demand (COD), biochemical oxygen demand (BOD), and other dissolved pollutants (Borowitzka, 1998). To synthesize biomass and emit oxygen, microalgae require carbon dioxide (CO_2) from exhaust gases or respiration of microbes, nutrients from wastewater, and photons in their chloroplasts. This oxygen does not undergo secondary pollution, and the remaining organic substances of wastewater are decomposed by aerobic bacteria (Kundu et al., 2024).

2.1 Nutrient Removal

Microalgae can be used to extract nutrients in wastewater, particularly phosphate and nitrogen. Phosphorus is required in the metabolism of energy, nucleic acids, and phospholipids, and nitrogen helps in protein synthesis; thus, these nutrients are essential in algae growth (Kundu et al., 2024). Studies of microalgal strains, including *Oocystis* sp., *Chlamydomonas* sp., *and Chlorella* sp., have indicated that during the treatment processes, the levels of NO_3 -N and PO_4 -P reduce significantly (Rasoul-Amini et al., 2014). In most cases, algal absorption takes hours to a few days to remove nutrients, depending on the organism and the environmental conditions (Kaloudas et al., 2021; Kundu et al., 2024).

2.1.1 Phosphorus Removal

Protein, lipid, and nucleic acid inorganic phosphorus are important in the development and transfer of energy in microalgal cells in wastewater. Microalgae absorb phosphorus through their plasma membrane, and phosphorus is phosphorylated to form organic chemicals, such as adenosine diphosphate (ADP) and subsequently adenosine triphosphate (ATP). The process is driven by the photosynthetic light processes, the mitochondrial electron transport, or the respiratory oxidation, which is energy-dependent (Chai et al., 2021; Emparan et al., 2019).

2.1.2 Nitrogen Removal

Nitrogen, which is usually present in wastewater in the form of ammonium (NH_4^+), nitrite (NO_2^-), and nitrate (NO_3^-), is obtained through fertilizers, sewage release, and decomposing organic substances. Microalgae absorb inorganic nitrogen through the process of the ingestion of NO_3^-, which is reduced to NO_2^-, followed by the reduction of NH_4^+. The reduced ammonium is subsequently reduced to amino acids such as glutamine by the assistance of ATP and NADPH produced during the photosynthesis process using the glutamine synthetase-glutamate synthase (GS-GOGAT) pathway (Chai et al., 2021; Emparan et al., 2019; Kaloudas et al., 2021; Kundu et al., 2024).

2.2 Xenobiotic Compound Removal

2.2.1 PAH Removal

Also, microalgae can degrade synthetic organic substances that are not degraded naturally. These include petroleum hydrocarbons, insecticides, chlorophenols, and polycyclic aromatic hydrocarbons (PAHs). PAHs are toxic, long-lasting, and carcinogenic (they are formed in cases when organic material is not fully burnt), which is why they need to be removed with the highest level of care (Torres et al., 2008; Kundu et al., 2024). Certain microalgae can also survive in environments that have PAHs and enzymatically transform them to less toxic intermediates. Selenastrum capricornutum acquired removal efficiencies of 88-98% towards pyrene and fluoranthene in a 7-day incubation interval (Lei et al., 2007; Olmos-Espejel et al., 2012). Monoaromatic hydrocarbons, common in the oily wastewater of oil-polluted environments, can substitute carbon sources for microalgae in nutrient-rich environments. Growth inhibition was not observed at gasoline concentrations as high as 50% in Tetrapurtum chui research, which demonstrates that the species is tolerant and capable of hydrocarbon decomposition (Semple et al., 1999; Kundu et al., 2024).

2.2.2 Heavy Metal Removal

Elements having atomic densities more than 4000 kg/m^3 are classified as heavy metals (HMs) (Vardhan et al., 2019). They are significant environmental contaminants due to their toxicity, persistence, and bioaccumulation. HMs may be extremely harmful even at low doses, endangering both human and animal health (Edelstein & Ben-Hur, 2018). They can be generically categorized as radionuclides (like U, Ra, Am, and Th), poisonous metals (like Cu, Cr, As, Zn, Ni, Ag, Sn, Co, and Pb), and valuable metals (like Au, Pd, Pt, and Ru). Although there are a number of physicochemical techniques for removing heavy metals, these conventional methods are frequently expensive and produce secondary waste. As a result, microalgae-based bioaccumulation has become a viable and affordable substitute (Vardhan et al., 2019).

In general, there are two stages of metal uptake: (1) quick surface adsorption or ion exchange, and (2) delayed chemisorption, including intracellular accumulation and active metabolism. The majority of metals attach to cell walls, according to cellular investigations, whereas lesser portions build up inside the cell (Omar, 2002). According to Shen et al. (2013), *C. vulgaris* can bioconvert hazardous Cr (VI) into the less toxic Cr (III), mostly through the use of secondary alcohols and functional groups like $-NH_2$ and -COOH. Magro et al. (2000) reported that when *Spirulina platensis* was grown under ideal light, air, and temperature conditions, it removed 60.92% of Cr (VI). Similar to this, *S. quadricauda's* wide surface area allowed it to achieve a 97% removal efficiency, reducing 30 mg/L of Ni and Zn to 0.9 mg/L in 5 minutes (Chong et al., 2000).

Because the inorganic forms of arsenic (As) [As (III) and As (V)] are especially poisonous, pollution with this element is also a serious hazard. Skin, cardiovascular, neurological, and metabolic issues are brought on by prolonged exposure (Mestrot et al., 2013). Fascinatingly, through the process of biomethylation, the sea algae *Ostreococcus tauri* may biotransform inorganic arsenic into organic forms that are less harmful, allowing it to be incorporated into natural biogeochemical cycles (Zhang et al., 2013).

2.3 Removal of Dyes

Microalgae have been effectively utilized in the decolorization and removal of textile effluent due to their huge surface area and binding affinity. The cell wall of microalgae plays a crucial role in the process of removing dye, and complexation, biosorption, electrostatic attraction, as well as bioconversion are the processes involved in this process (Andrade et al., 2018). The dye molecules initially adsorb onto the surface of the algal cells before they diffuse into the solid part of the algal biopolymers. Extracellular polymers also enhance dye molecule adsorption on polymer surfaces, but using functional groups of different types (Saha et al., 2023).

Among the most successful microalgal species that have proved to be a very effective biosorbent with regard to reactive dyes degradation is *Spirogyra* biomass. Similarly, simple dyes can also be removed using biosorption in *Caulerpa lentillifera* and *Caulerpa scalpelliformis*. Much research has been done on the capacity of *Chlorella vulgaris* to remove reactive dyes such as Remazol black B (Aksu and Tezer, 2005). Also, more complicated dyes may be transformed by microalgae into less toxic, simpler molecules. Mono-azo dye, as an example, was transformed into aniline after 10 days of culture in the textile wastewater with Supranol Red 3BW, eliminating 63-69 per cent of the colour.

2.4 Removal of Pathogens

Nitrogen assimilation generates OH^- ions, which further raise alkalinity and prevent pathogenic survival. Photosynthetic CO_2 assimilation raises the medium's pH. Bacterial respiration and microalgae-induced oxygenation raise DO levels, which can occasionally surpass 0.5 mg/L, which is harmful to fecal bacteria such as *Clostridium perfringens*, *Enterococci*, and *E. coli* (Ansa et al., 2011). Additionally, certain microalgae generate bioactive substances that directly prevent infections. For example, in high pH circumstances, *C. vulgaris* generates long-chain fatty acids, while Synechocystis sp. secretes microcystin-LR, both of which are poisonous to pathogenic bacteria (Mohamed, 2008).

2.5 Removal of Pharmaceuticals

Unused pharmaceuticals also create significant health and environmental risks when they enter the aquatic environments as a result of industrial effluents, hospital effluents, domestic waste, and agricultural effluents. Modern wastewater treatment plants (WWTPs) have a removal rate of 10-100% in accordance with the specific pollutant and treatment method (Kruglova et al., 2016). The effect of the molecular complexity of the medicinal substances on the biodegradability is quite high. As an illustration, in plant-microalgae hybrid systems, *Chlorella sorokiniana* has been shown to completely degrade ibuprofen, in addition to paracetamol, with a higher decolonization efficiency with ibuprofen compared to ketoprofen (Bai and Acharya, 2017). Microalgae exude a meshwork of extracellular polymeric matter (EPS), and hence, medications are more readily adsorbed and concentrated. Carboxyl, hydroxyl, and phosphoryl functional groups tend to make the algal surfaces negatively charged; they can therefore readily absorb positively charged medicinal compounds (Hom-Diaz et al., 2015). Regardless of the specific type of drug and the microalgal type, research always demonstrates that bioadsorption is the major path of pharmaceutical removal (Bai and Acharya, 2017). An example is *Scenedesmus dimorphus*, which has been demonstrated to biodegrade approximately 85 percent of 17alpha-estradiol within seven days (Zhang et al., 2014; Kundu et al., 2024).

Figure 1. Microalgae wastewater treatment process

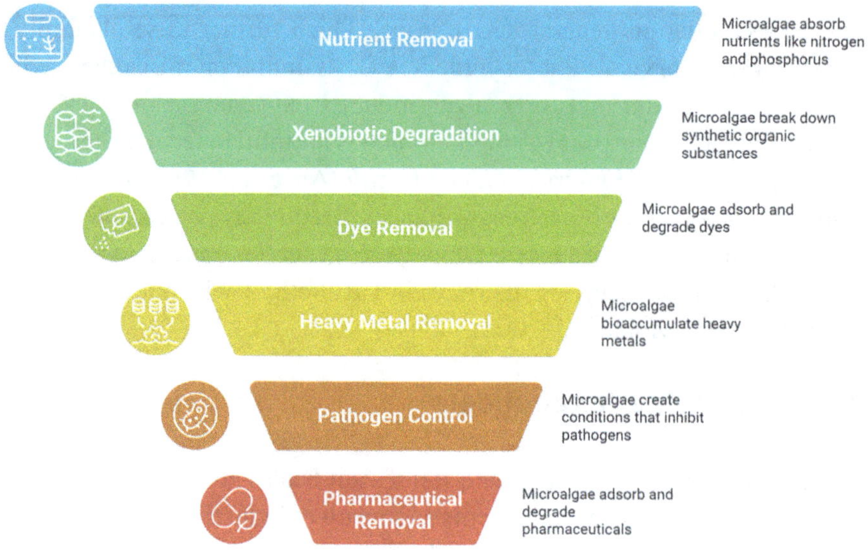

2.6 Carbon Dioxide Fixation

The good approach to fix the CO_2 biologically consists of the use of industrial effluents that are often 10-15 % CO_2 and could be employed as a source of carbon to be used by the microalgae (Ji et al., 2013). Microalgae offer a viable solution to curtailing the atmospheric CO_2 level because they have the ability to absorb and convert CO_2 to biomass much faster than land plants (Farhadian et al., 2008). Having in mind that more than half of the microalgal cells consist of carbon, 1.83 tonnes of CO_2 are required to form one tonne of biomass. It reduces the pre-treatment costs, but utilisation of flue gases directly off industrial chimneys to provide CO_2 to algal cultures requires strains that can survive in challenging conditions such as elevated CO_2 levels and other contaminants such as NOx and SOx (Kumar et al., 2010; Kundu et al., 2024).

The loss of dissolved oxygen (DO) due to high biochemical oxygen demand (BOD) in water bodies also causes fish death and the development of anaerobic environments. A major objective of wastewater treatment is the reduction of BOD. Microalgae help in reducing the levels of BOD through the production of oxygen by the process of photosynthesis and the breakdown of organic pollutants, e.g., phenolic compounds (Kaloudas et al., 2021). Experimental studies indicate that *Spirulina sp.*, *Chlorella pyrenoidosa*, and *Chlorella kessleri* are effective in the extraction of the phenolic chemicals in wastewater (Zhang C. et al., 2020). Nevertheless, the supply of carbon sources is often constraining the quantity of phenol that can be degraded by microalgae. Microalgae selectively absorb easier carbon compounds, as most wastewaters are already high in organic carbon, reducing the absorption of phenol (Kaloudas et al., 2021).

3. MECHANISMS OF MICROALGAL REMOVAL OF EMERGING POLLUTANTS

To cleanse wastewater, microalgae apply a diversity of biochemical and physicochemical operations to eliminate the developing contaminants. Bio-uptake: the contaminants penetrate the live cells through the cell wall of the plants, and attach to intracellular protein or organelles. (Kundu et al., 2024). Bioaccumulation: The pollutants accumulate in the algae cells that are not living through surface binding and active diffusion. Biosorption: Contaminants bind themselves onto the extracellular polymeric substances (EPS) or the algal cell wall by way of ion exchange as well as electrostatic reactions. Biodegradation: It is a process through which complex organic contaminants are decomposed to less hazardous, simple forms by enzymes. Photodegradation: The disintegration of pollutants includes direct exposure to the direct sun or ultraviolet rays to break down the pollutants into non-toxic compounds. (Kundu et al., 2024)

3.1 Biosorption

The process of attaching the contaminants to the surface of microorganisms is called biosorption, and algae can be used as bio-sorbents to a great extent. Proteins, lipids, and polysaccharides are some of the functional groups that the algal cell surface can present adsorption sites (El Naggar et al., 2019; Kundu et al., 2024). Two types of biosorption exist: chemisorption, also known as chemical biosorption, and physisorption, also known as physical biosorption. Although in chemisorption, the pollutant and the sorbent surface are chemically combined, in physisorption, the solute molecules bind to the binding sites through weak forces. The advantages of bio-sorption are that the method is cheap, simple, does not produce sludge, and can remediate large volumes of wastewater containing low pollutant concentrations (Lo et al., 2014; Olawale, 2021; Kundu et al., 2024).

3.2 Bioaccumulation and Bio-Uptake

Bioaccumulation is the process through which the algal cells absorb and store the pollutants present in the surrounding environment in the cytoplasm. One of the species that is able to absorb and retain pollutants to grow is Spirogyra. Indicatively, as explained by Rezania et al. (2016), *Nannochloris sp.* uptakes 11 percent of sulfamethoxazole, 11 percent of trimethoprim, 27 percent of triclosan, and 13 percent of carbamazepine (Kundu et al., 2024). Bioaccumulation and bio-adsorption are different processes that occur simultaneously; it is not easy to measure either of the two. Before internalization, a first adsorption phase precedes the bioaccumulation, and the contaminants are first adsorbed onto the cell wall. It was shown that certain drugs, including triclosan and sulfamethoxazole, accumulate in the cells of algae, leading to the surplus of reactive oxygen species (ROS) and the eventual death of the cells (Bai and Acharya, 2017; Kundu et al., 2024).

3.3 Biodegradation

Usually, the procedure is divided into three stages:

- **Phase I (Transformation):** Hydrolysis, oxidation, or reduction processes are catalyzed by enzymes such as cytochrome P450, which add hydroxyl groups to transform lipophilic compounds into hydrophilic forms.
- **Phase II (Conjugation):** To shield cells from oxidative stress, the electrophilic chemicals that are produced combine with glutathione to create conjugates.
- **Phase III (Enzymatic Degradation):** The compounds are further broken down by enzymes such as laccase, dehydrogenase, carboxylase, and decarboxylase.

Molecular structure plays a major role in the biodegradability of pollutants; unsaturated and linear compounds break down more quickly than cyclic or very complex ones. Up to 95% biodegradation efficiency has been shown by species including Scenedesmus obliquus, Chlorella pyrenoidosa, and Navicula sp. Levofloxacin is also efficiently broken down by C. vulgaris. Around algal cells, extracellular polymeric substances (EPSs) create hygroscopic matrices that facilitate enzymatic breakdown, adsorption, and accumulation (Xiong et al., 2016; Posadas et al., 2014; Kundu et al., 2024).

Biodegradability is greatly impacted by wastewater's C: N:P ratio, which is best at 100:18:2 (Xiong et al., 2016). In contrast to biosorption or bioaccumulation, biodegradation lowers the toxicity of contaminants and makes it possible to use the resultant algal biomass to produce biofuel and bioproducts. Biodegradation can take place through metabolic degradation, in which pollutants act as the main source of carbon, or co-metabolism, in which enzymes break down contaminants without utilizing them as a carbon source (Norvill et al., 2016).

3.4 Photodegradation

In photodegradation (breakdown of pollutants by light), the light energy may often dissolve even those pollutants that resist normal bioremediation methods. The two major processes involved are photooxidative degradation and photolysis. Photooxidation occurs with pollutants and oxidants generated by exposure to light, like hydroxyl radicals. The pollutants absorb photons directly during photolysis and undergo a change in their chemical composition, which contributes to faster deterioration. Releasing reactive oxygen species as well as participating in redox cycling, dissolved organic matter (DOM), like hemicellulose, fulvic acids, and humic acids, enhances photodegradation.

The pollutants that are weakly absorbed may be photosensitized by DOM and Extracellular Organic Matter (EOM), accelerating their change (Gondi et al., 2022). Photodegradation can occur in two ways, namely, direct and indirect. In order to maximize light exposure using turbulent mixing, algae systems utilize photobioreactors and algae scrubbers. Nevertheless, not every pollutant is photodegradable, so it makes the process selective (Reymann et al., 2020).

3.5 Hydrolysis

Some newly discovered contaminants in wastewater can be deactivated by hydrolysis, a chemical process that breaks down molecules by reacting with water. Temperature, pH, concentration, and pollutant structure all affect process efficiency. However, some pharmaceutical substances, such as lactam antibiotics, can be efficiently broken down, while others, like sulphonamides and fluoroquinolones, are resistant to hydrolysis (Banu et al., 2020). Algal pond pH variations affect the rate of hydrolysis, which rises with warmth. While neutral-pH pollutants continue to degrade at night when pH stabilizes, contaminants that hydrolyze more easily at higher pH levels often break down more quickly during the day when algal photosynthesis raises pond pH (Zhang S. et al., 2020; Gondi et al., 2022; Kundu et al., 2024).

Figure 2. Microalgal Pollutant Removal Process

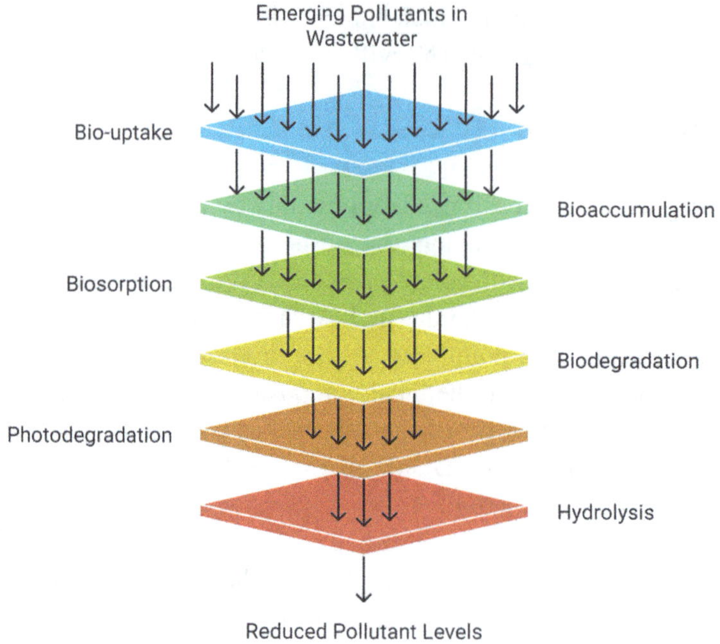

4. FACTORS AFFECTING WASTEWATER TREATMENT BY MICROALGAE

Hydraulic retention time (HRT), temperature, light intensity and duration, adsorbent size, pH, redox potential, and other physicochemical parameters all affect how effective microalgal bioremediation is. Together, these factors dictate the pace and scope of pollution removal. Temperature, oxidation-reduction potential, and pH are examples of environmental factors that impact algal growth, metabolism, and pollutant-absorbing capability. Photosynthetic activity is directly affected by the length and intensity of light, but the contact time and accessible surface area for pollutant removal are determined by particle size and HRT (Kundu et al., 2024).

4.1 pH

The development of microalgae and the effectiveness of pollution removal are significantly influenced by pH. According to Ummalyma et al. (2016), it has an impact on the functional groups' ionization state on algal cell surfaces. Algal surfaces get positively charged at low pH values, which reduces adsorption effectiveness; at higher pH values, the surface gets negatively charged, which increases adsorption (Chen et al., 2016). However, because of decreased RuBisCO enzyme activity and lower CO_2 solubility, a very high pH (>9) prevents algal development (Sutherland et al., 2015).

4.2 Temperature

High temperatures seen in industrial effluent can cause thermal pollution in aquatic environments. The majority of microalgae can withstand a wide range of temperatures (15°C to 35°C), allowing them to be used in the treatment of thermally fluctuating wastewater, even if the ideal temperatures vary by species (Chen et al., 2020). The productivity of biomass drastically decreases over this level. *Spirulina platensis*, for example, grows best around 35°C, whereas *Scenedesmus sp.* can tolerate a range of temperatures. For the bioremediation of emerging pollutants, thermotolerant microalgae that can grow in wastewater at 30 to 40°C are therefore perfect (Cheah et al., 2015; Kundu et al., 2024).

4.3 Light Intensity

Being phototrophic, microalgae use light energy for both photosynthesis and nutritional absorption. The blue (400–500 nm) and red (600–700 nm) portions of the spectrum are ideal for photosynthetic activity because they promote the absorption of nitrogen and phosphate (Whitton et al., 2015). With less energy usage, light-emitting diode (LED) technology offers effective, narrowband light sources that encourage algae growth (Ibrahim et al., 2014). For instance, green light boosted Chlorella vulgaris productivity, whereas red light raised Spirulina platensis growth by 38% (Wang et al., 2007; Kundu et al., 2024).

4.4 Hydraulic Retention Time (HRT)

The duration of pollutant–algae interaction is influenced by hydraulic retention time, which controls how long wastewater stays in the treatment system. While too lengthy HRT might decrease efficiency because of nutrient deficiency or overgrowth, short HRT may result in insufficient pollutant removal. To maximize removal efficiency, optimal HRT strikes a balance between contaminant concentration and algal biomass density. Up to a certain point, removal rates are improved by increasing both algal biomass and retention time; after that, performance starts to deteriorate, according to studies (De-Bashan & Bashan, 2010; Kundu et al., 2024).

Figure 3. Unveiling Factors in Microalgal Wastewater Treatment

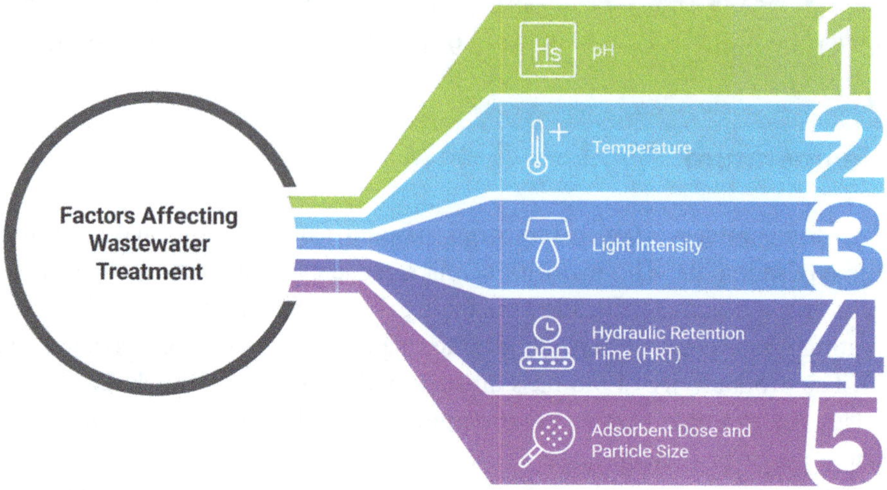

4.5 Dose of Adsorbent and Particle Size

The dosage and size of the adsorbent used influence the pollutant absorption ability of microalgae (Kundu et al., 2024). This interaction is also influenced by the toxicity of the pollutant, though increased permeability of the pollutant particles favors increased absorption into the algal cell wall. The increased area of the binding sites enhances the availability of the binding sites and, therefore, leads to the absorption of smaller particles such as substances of nanoscale size. Such factors as surface area, electrostatic capability, and the presence of functional groups are also important in adsorption-mediated bioremediation (Hlongwane et al., 2019). Long hydraulic retention times are also likely to contribute to the enhancement of pollutant capture, which is the prolonged contact of the pollutants and the algal biomass (Kundu et al., 2024).

5. TRANSFORMATION AND FATE OF CONTAMINANTS INSIDE MICROALGAL CELLS

The fate of pollutants in the algal cells remains a significant issue, though bioaccumulation is an excellent technique for eliminating pollutants. The safe disposal of the harmful algal biomass after bioaccumulation is under discussion (Kundu et al.,

2024). Biodegradation or biotransformation is the enzymatic or metabolic degradation of complex chemicals into simple molecules that may occur both in and outside of the cell. Unlike bioaccumulation or biosorption, the process reduces the toxicity of the contaminants in the algal cells and the bulk medium. Moreover, the algal biomass can be used to produce value-added goods based on the resultant product.

There are two major mechanisms through which biodegradation takes place:

Metabolic disturbance, where the pollutants supply the microalgae with a source of carbon. Co-metabolism is a mechanism whereby the pollutants in the bulk media are decomposed or broken down by the enzymes without depending on the pollutants as a carbon source (Tiwari et al., 2017). Microalgal biodegradation involves enzymatic activity, which has shown a high level of effectiveness in destroying medications and personal care products (Nguyen et al., 2021; Kundu et al., 2024). Activation of enzymes by the concentration of emerging contaminants (ECs) in the medium might be monitored repeatedly; threshold quantities are necessary to trigger biodegradation pathways (Xiong et al., 2018). The exact role of enzymes in biodegradation remains debatable, but, at the same time, the process is complex (Sutherland and Ralph, 2019).

Not every pollutant is readily biodegradable; meanwhile, certain pollutants, such as carbamazepine, in high-rate algal ponds, are not readily biodegraded or even biodegraded by photolysis. To attain a good cleanup process, microalgae should be pre-acclimated to non-toxic EC levels (Kundu et al., 2024). Long-term exposure to ECs can increase microalgal tolerance by activating the enzymatic processes that decrease the toxicity of pollutants. An example is Chlorella pyrenoidosa, which was better eliminated after pre-exposure to the antibiotic cefradine (Norvill et al., 2016; Kundu et al., 2024).

6. ADVANTAGES OF MICROALGAL BIOREMEDIATION SYSTEMS FOR EMERGING POLLUTANTS

Examples of traditional methods of wastewater treatment are flocculation, chemical precipitation, activated carbon adsorption, reverse osmosis, UV disinfection, ultrafiltration, electrocoagulation, and ion exchange, which are often costly and energy-consuming (Kundu et al., 2024). Microalgal-based bioremediation is an effective and viable alternative due to the following advantages: Cultivation of microalgae enables the prevention of eutrophication and the extraction of useful components of wastewater like carbon, phosphorus, and nitrogen, among others. The conventional treatment involves phosphorus precipitation, CO_2 oxidation to CO_2, and release of nitrogen as ambient N_2 (Nagarajan et al., 2020).

Emerging Pollutants Tolerance

The tolerance and efficiency of such strains as Scenedesmus, Chlamydomonas, and Chlorella to pollution have been successfully studied (Wang et al., 2016). Microalgae absorb atmospheric CO_2 through photosynthesis and convert it to organic compounds. This process reduces the amount of greenhouse gases and is a natural carbon sink that reduces global warming (Kannah et al., 2021).

Terrain Usage

Although it fosters the restoration of ecosystems through the cultivation of microalgae on dry or non-farmable land, competition between the traditional crops is reduced, and the land area is optimized (Farooq et al., 2015). Value-Added Bioproduct Production: Biofuels, lipids, carbohydrates, pigments, enzymes, biofertilizers, and bioplastics are other biomaterials that can be produced using microalgal biomass. The research into protein and oil composition helps to find proper applications (Banu et al., 2020). The growth rate of microalgae is very high and is 10 to 50 times higher than that of terrestrial plants. Pittman et al. (2011) note that wastewater treatment using several species has been successful, such as *Chlorella, Chlamydomonas, Botryococcus, Scenedesmus, Arthrospira,* and *Phormidium.*

7. LIMITATIONS AND CHALLENGES OF MICROALGAE-BASED WASTEWATER TREATMENT

Although microalgae have a lot of promise, there are several restrictions and difficulties:

Incomplete Elimination of Emerging Contaminants: Conventional WWTPs have removal rates as low as 50% for certain ECs, making them ineffective at eliminating hormones, medications, antibiotic resistance genes (ARGs), and microplastics (Kundu et al., 2024). Although effective, advanced techniques (such as ozonation, reverse osmosis, and AOPs) are expensive, energy-intensive, and may result in hazardous byproducts like bromates.

Scalability and Space Requirements: In low- and middle-income nations, microalgal systems' scalability is constrained by their need for expansive growing spaces and regulated environmental conditions. The percentage of wastewater treated in these areas is decreased by operational and financial limitations.

Operational and Environmental Factors

Turbidity and high particle concentration in wastewater can restrict light penetration, which hinders photosynthesis. To improve light availability, pre-treatment techniques like coagulation, adsorption, or sedimentation would be required (Amenorfenyo et al., 2019).

Cost Issues

Harvesting biomass and downstream processing are still costly even when growing is straightforward. Operating expenses are further increased by enclosed photobioreactors that need artificial light and sterilizing agents (Umamaheswari & Shanthakumar, 2016).

Selection of Species

The ideal species of microalgae should be capable of surviving a variety of environmental changes, exchange of metabolites under stress, and be able to survive competition or nutritional limitation. Facultative species are uncommon, and only a few of them can develop in a heterotrophic or mixotrophic form (Amenorfenyo et al., 2019).

Climatic and Altitude Limitations

High elevations can offer a challenge in growing the microalgae, and this may affect the biomass production. This limitation can be reduced by selecting extremophile strains, although they need to evaluate how effective they are in bioremediation (Gondi et al., 2022).

8. THE USE OF MICROALGAE TO MEET SUSTAINABLE DEVELOPMENT GOALS (SDGS)

The microalgal strains have demonstrated much potential in the removal of emerging contaminants (ECs) in wastewater to help minimize health risks and improve health in general. Microalgae addition to the already-established wastewater treatment plants (WWTPs) could be beneficial by complementing the conventional techniques, such as membrane bioreactors (MBRs), activated carbon adsorption, or advanced oxidation processes (AOPs). This mixed approach is able to fill the efficiency gap in EC removal. An example of nature-based solutions to eliminate

EC, which uses less energy, is the microalgae cultivation and artificial wetlands. In addition, other useful byproducts are generated by such systems, such as fertilizer and biofuels, which facilitate a circular economy (Dutta et al., 2023; Kundu et al., 2024).

The use of real-time monitoring systems also enhances the efficiency of the treatment process by means of prioritizing the top-priority pollutants and taking targeted actions at their centre (Kundu et al., 2024). Business, government, and academia should collaborate to increase the pilot technologies to ensure that they are viable under various environmental conditions. Modular photobioreactors or algal ponds, otherwise known as decentralized technologies, are scalable and relatively inexpensive in low- and middle-income regions. Microalgal bioremediation should be supported with support of public awareness and policy to be successfully adopted. Indicatively, such programs as promoting proper drug disposal and more restrictive regulations on EC discharges contribute to reducing pollution loading. Resistance gene and antibiotic-residue targeting technologies are also able to mitigate the threat of antibiotic resistance to world health.

9. RECOMMENDATIONS AND FUTURE PERSPECTIVES

Although the possibilities of microalgae being used on an industrial scale have been researched in-depth, there are still several challenges:

- **Scaling Up:** Microalgae can often be subjected to unfavourable conditions when the pilot scale is being scaled up to an industrial scale, which reduces the yields of bioproducts. In order to integrate robust algae strains with cutting-edge bioprocess engineering techniques and guarantee economic and environmental viability, more research is required. In addition to increasing bioproduct yields, genetic and metabolic engineering may improve stress tolerance and the effectiveness of pollution removal.
- **Biomass Harvesting:** It might be difficult to separate microalgal biomass from treated wastewater, particularly in suspended culture. Hydraulic retention durations can be shortened and harvesting made easier by immobilizing algae on media or supports.
- **Cost-Effectiveness:** Not much research has thoroughly examined how much microalgae-based therapies are more affordable than traditional techniques. It is important to establish design and operating principles to enhance resilience and flexibility across a variety of wastewater types.
- **Integration with Biological Processes:** To fully comprehend how native bacterial populations and microalgae interact in wastewater treatment systems, more study is needed.

- **Nutrient Management:** For the best algal development, pretreatment or influent composition control is essential. Blends of various wastewater sources can be optimized to provide a microalgae nutrition medium that is well-balanced.
- **Considerations related to geography:** Climate, temperature, and sunlight availability can all affect cultivation methods. Photobioreactor-based culture might be advantageous for producing high-value algal products in colder climates.
- **Monitoring and Management:** To treat microalgal wastewater effectively, pH, temperature, dissolved oxygen, BOD, and algal cell health must all be closely monitored. The advancement of technology for remote control and online monitoring will make widespread deployment easier.

10. CONCLUSION

Microalgae have shown great promise in wastewater bioremediation, efficiently eliminating heavy metals, nutrients, and new pollutants. They are appropriate for a variety of wastewater streams due to their quick growth, strong nitrogen absorption, and resilience to harsh environments. Ammonium, phosphate, heavy metals, dyes, pesticides, and medicines are among the organic, inorganic, and synthetic toxins that may be removed via phytoremediation, which uses algae to remove biological and chemical pollutants. Without additional feeding, algal growth may be sustained by high nitrogen concentrations in wastewater from homes, businesses, and farms. Although live cells often show better efficiency, algae biomass, whether alive or dead, may biosorbent heavy metals. Algal strains with improved phytoremediation capacities are being developed using molecular techniques. Further research is required to evaluate the effectiveness of various algal species alone or combined to treat a gamut of pollutants to fully use the potential and utility of large-scale algal micro-treatment systems. There is also a need to conduct further studies on the concept of incorporating the technologies of microalgae with the traditional wastewater treatment plant, with the aim of enhancing the overall treatment performance and sustainability. Moreover, there is a need to improve the quality of the algal bioproducts and to optimize the methods of efficient and cost-effective large-scale production. All things considered, microalgae-based wastewater treatment is an environmentally responsible, economically feasible, and sustainable method of recovering nutrients, eliminating contaminants, and producing bioresources, all of which support sustainable development and environmental preservation.

REFERENCES

Aksu, Z., & Tezer, S. (2005). Biosorption of reactive dyes on the green alga *Chlorella vulgaris*. *Process Biochemistry (Barking, London, England)*, *40*(4), 1347–1361. DOI: 10.1016/j.procbio.2004.06.007

Amenorfenyo, D. K., Huang, X., Zhang, Y., Zeng, Q., Zhang, N., Ren, J., & Huang, Q. (2019). Microalgae brewery wastewater treatment: Potentials, benefits, and the challenges. *International Journal of Environmental Research and Public Health*, *16*(11), 1910. DOI: 10.3390/ijerph16111910 PMID: 31151156

Andrade, C., Andrade, L., Dias, M., Nascimento, C., & Mendes, M. (2018). Microalgae for bioremediation of textile wastewater: An overview. *MOJ Food Processing & Technology*, *6*(6), 432–433. DOI: 10.15406/mojfpt.2018.06.00200

Ansa, E., Lubberding, H. J., Ampofo, J., & Gijzen, H. (2011). The role of algae in the removal of *Escherichia coli* in a tropical eutrophic lake. *Ecological Engineering*, *37*(3), 317–324. DOI: 10.1016/j.ecoleng.2010.11.023

Bai, X., & Acharya, K. (2017). Algae-mediated removal of selected pharmaceutical and personal care products (PPCPs) from Lake Mead water. *The Science of the Total Environment*, *581–582*, 734–740. DOI: 10.1016/j.scitotenv.2016.12.192 PMID: 28089530

Banu, J. R., Kannah, R. Y., Kavitha, S., Ashikvivek, A., Bhosale, R. R., & Kumar, G. (2020). Cost-effective biomethanation via surfactant-coupled ultrasonic liquefaction of mixed microalgal biomass harvested from an open raceway pond. *Bioresource Technology*, *304*, 123021. DOI: 10.1016/j.biortech.2020.123021 PMID: 32086031

Borowitzka, M. A. (1998). Limits to growth. In *Wastewater treatment with algae* (pp. 203–226). DOI: 10.1007/978-3-662-10863-5_12

Chai, W. S., Tan, W. G., Munawaroh, H. S. H., Gupta, V. K., Ho, S.-H., & Show, P. L. (2021). Multifaceted roles of microalgae in the application of wastewater biotreatment: A review. *Environmental Pollution*, *269*, 116236. DOI: 10.1016/j.envpol.2020.116236 PMID: 33333449

Cheah, W. Y., Show, P. L., Chang, J.-S., Ling, T. C., & Juan, J. C. (2015). Biosequestration of atmospheric CO2 and flue gas-containing CO2 by microalgae. *Bioresource Technology*, *184*, 190–201. DOI: 10.1016/j.biortech.2014.11.026 PMID: 25497054

Chen, R., Liu, Y., & Liao, W. (2016). Using an environmentally friendly process combining electrocoagulation and algal cultivation to treat high-strength wastewater. *Algal Research*, *16*, 330–337. DOI: 10.1016/j.algal.2016.03.032

Chen, Z., Shao, S., He, Y., Luo, Q., Zheng, M., Zheng, M., Chen, B., & Wang, M. (2020). Nutrient removal from piggery wastewater coupled to lipid production by a newly isolated self-flocculating microalga *Desmodesmus* sp. PW1. *Bioresource Technology*, *302*, 122806. DOI: 10.1016/j.biortech.2020.122806 PMID: 31982846

Chong, A., Wong, Y., & Tam, N. (2000). Performance of different microalgal species in removing nickel and zinc from industrial wastewater. *Chemosphere*, *41*(2), 251–257. DOI: 10.1016/S0045-6535(99)00418-X PMID: 10819208

De-Bashan, L. E., & Bashan, Y. (2010). Immobilized microalgae for removing pollutants: Review of practical aspects. *Bioresource Technology*, *101*(6), 1611–1627. DOI: 10.1016/j.biortech.2009.09.043 PMID: 19931451

Delrue, F., Álvarez-Díaz, P. D., Fon-Sing, S., Fleury, G., & Sassi, J.-F. (2016). The environmental biorefinery: Using microalgae to remediate wastewater, a win-win paradigm. *Energies*, *9*(3), 132. DOI: 10.3390/en9030132

Dutta, N., Kundu, P., Lee, J. T. E., & Bhattacharya, S. (2023). Implementation and optimization of algal biomass in value-added products recovery: A step towards algae-based green economy. *Hydrobiology*, *2*(2), 326–346. DOI: 10.3390/hydrobiology2020021

Edelstein, M., & Ben-Hur, M. (2018). Heavy metals and metalloids: Sources, risks and strategies to reduce their accumulation in horticultural crops. *Scientia Horticulturae*, *234*, 431–444. DOI: 10.1016/j.scienta.2017.12.039

Emparan, Q., Harun, R., & Danquah, M. (2019). Role of phycoremediation for nutrient removal from wastewaters: A review. *Applied Ecology and Environmental Research*, *17*(1), 889–915. DOI: 10.15666/aeer/1701_889915

Farazaki, M., & Gikas, P. (2019). Nitrification-denitrification of municipal wastewater without recirculation, using encapsulated microorganisms. *Journal of Environmental Management*, *242*, 258–265. DOI: 10.1016/j.jenvman.2019.04.054 PMID: 31048231

Farhadian, M., Vachelard, C., Duchez, D., & Larroche, C. (2008). In situ bioremediation of monoaromatic pollutants in groundwater: A review. *Bioresource Technology*, *99*(14), 5296–5308. DOI: 10.1016/j.biortech.2007.10.025 PMID: 18054222

Farooq, W., Suh, W. I., Park, M. S., & Yang, J.-W. (2015). Water use and its recycling in microalgae cultivation for biofuel application. *Bioresource Technology*, *184*, 73–81. DOI: 10.1016/j.biortech.2014.10.140 PMID: 25465788

Gondi, R., Kavitha, S., Kannah, R. Y., Karthikeyan, O. P., Kumar, G., & Tyagi, V. K.. (2022). Algal-based system for removal of emerging pollutants from wastewater: A review. *Bioresource Technology*, *344*, 126245. DOI: 10.1016/j.biortech.2021.126245 PMID: 34743994

Hlongwane, G. N., Sekoai, P. T., Meyyappan, M., & Moothi, K. (2019). Simultaneous removal of pollutants from water using nanoparticles: A shift from single pollutant control to multiple pollutant control. *The Science of the Total Environment*, *656*, 808–833. DOI: 10.1016/j.scitotenv.2018.11.257 PMID: 30530150

Hom-Diaz, A., Llorca, M., Rodríguez-Mozaz, S., Vicent, T., Barceló, D., & Blánquez, P. (2015). Microalgae cultivation on wastewater digestate: B-estradiol and 17α-ethynylestradiol degradation and transformation products identification. *Journal of Environmental Management*, *155*, 106–113. DOI: 10.1016/j.jenvman.2015.03.003 PMID: 25785785

Ibrahim, M. A., MacAdam, J., Autin, O., & Jefferson, B. (2014). Evaluating the impact of LED bulb development on the economic viability of ultraviolet technology for disinfection. *Environmental Technology*, *35*(4), 400–406. DOI: 10.1080/09593330.2013.829858 PMID: 24600880

Ji, M.-K., Abou-Shanab, R. A., Kim, S.-H., Salama, E.-S., Lee, S.-H., Kabra, A. N., Lee, Y.-S., Hong, S., & Jeon, B.-H. (2013). Cultivation of microalgae species in tertiary municipal wastewater supplemented with CO2 for nutrient removal and biomass production. *Ecological Engineering*, *58*, 142–148. DOI: 10.1016/j.ecoleng.2013.06.020

Kaloudas, D., Pavlova, N., & Penchovsky, R. (2021). Phytoremediation of wastewater by microalgae: A review. *Environmental Chemistry Letters*, *19*(4), 2905–2920. DOI: 10.1007/s10311-021-01203-0

Kannah, R. Y., Kavitha, S., Karthikeyan, O. P., Rene, E. R., Kumar, G., & Banu, J. R. (2021). A review on anaerobic digestion of energy and cost-effective microalgae pretreatment for biogas production. *Bioresource Technology*, *332*, 125055. DOI: 10.1016/j.biortech.2021.125055 PMID: 33813179

Klinthong, W., Yang, Y.-H., Huang, C.-H., & Tan, C.-S. (2015). A review: Microalgae and their applications in CO2 capture and renewable energy. *Aerosol and Air Quality Research*, *15*(2), 712–742. DOI: 10.4209/aaqr.2014.11.0299

Koyande, A. K., Chew, K. W., Rambabu, K., Tao, Y., Chu, D.-T., & Show, P.-L. (2019). Microalgae: A potential alternative to health supplementation for humans. *Food Science and Human Wellness*, *8*(1), 16–24. DOI: 10.1016/j.fshw.2019.03.001

Kruglova, A., Kråkström, M., Riska, M., Mikola, A., Rantanen, P., Vahala, R., & Kronberg, L. (2016). Comparative study of emerging micropollutants removal by aerobic activated sludge of large laboratory-scale membrane bioreactors and sequencing batch reactors under low-temperature conditions. *Bioresource Technology, 214*, 81–88. DOI: 10.1016/j.biortech.2016.04.037 PMID: 27128192

Kumar, A., Ergas, S., Yuan, X., Sahu, A., Zhang, Q., Dewulf, J., Malcata, F. X., & van Langenhove, H. (2010). Enhanced CO2 fixation and biofuel production via microalgae: Recent developments and future directions. *Trends in Biotechnology, 28*(7), 371–380. DOI: 10.1016/j.tibtech.2010.04.004 PMID: 20541270

Kundu, P., Dutta, N., & Bhattacharya, S. (2024). Application of microalgae in wastewater treatment with special reference to emerging contaminants: A step towards sustainability. *Frontiers in Analytical Science, 4*, 1513153. DOI: 10.3389/frans.2024.1513153

Lei, A.-P., Hu, Z.-L., Wong, Y.-S., & Tam, N. F.-Y. (2007). Removal of fluoranthene and pyrene by different microalgal species. *Bioresource Technology, 98*(2), 273–280. DOI: 10.1016/j.biortech.2006.01.012 PMID: 16517155

Mestrot, A., Xie, W.-Y., Xue, X., & Zhu, Y.-G. (2013). Arsenic volatilization in model anaerobic biogas digesters. *Applied Geochemistry, 33*, 294–297. DOI: 10.1016/j.apgeochem.2013.02.023

Mohamed, Z. A. (2008). Polysaccharides as a protective response against microcystin-induced oxidative stress in *Chlorella vulgaris* and *Scenedesmus quadricauda* and their possible significance in the aquatic ecosystem. *Ecotoxicology (London, England), 17*(6), 504–516. DOI: 10.1007/s10646-008-0204-2 PMID: 18389369

Molazadeh, M., Ahmadzadeh, H., Pourianfar, H. R., Lyon, S., & Rampelotto, P. H. (2019). The use of microalgae for coupling wastewater treatment with CO2 biofixation. *Frontiers in Bioengineering and Biotechnology, 7*, 42. DOI: 10.3389/fbioe.2019.00042 PMID: 30941348

Nagarajan, D., Lee, D.-J., Chen, C.-Y., & Chang, J.-S. (2020). Resource recovery from wastewaters using microalgae-based approaches: A circular bioeconomy perspective. *Bioresource Technology, 302*, 122817. DOI: 10.1016/j.biortech.2020.122817 PMID: 32007309

Nguyen, H. T., Yoon, Y., Ngo, H. H., & Jang, A. (2021). The application of microalgae in removing organic micropollutants in wastewater. *Critical Reviews in Environmental Science and Technology, 51*(12), 1187–1220. DOI: 10.1080/10643389.2020.1753633

Norvill, Z. N., Shilton, A., & Guieysse, B. (2016). Emerging contaminant degradation and removal in algal wastewater treatment ponds: Identifying the research gaps. *Journal of Hazardous Materials*, *313*, 291–309. DOI: 10.1016/j.jhazmat.2016.03.085 PMID: 27135171

Olawale, S. A. (2021). Biosorption of heavy metals from aqueous solutions: An insight and review. *Archives of Industrial Engineering*, *3*, 1–31.

Olmos-Espejel, J. J., de Llasera, M. P. G., & Velasco-Cruz, M. (2012). Extraction and analysis of polycyclic aromatic hydrocarbons and benzo[a]pyrene metabolites in microalgae cultures by off-line/on-line methodology based on matrix solid-phase dispersion, solid-phase extraction, and high-performance liquid chromatography. *Journal of Chromatography. A*, *1262*, 138–147. DOI: 10.1016/j.chroma.2012.09.015 PMID: 23031438

Omar, H. H. (2002). Bioremoval of zinc ions by *Scenedesmus obliquus* and *Scenedesmus quadricauda* and its effect on growth and metabolism. *International Biodeterioration & Biodegradation*, *50*(2), 95–100. DOI: 10.1016/S0964-8305(02)00048-3

Pérez-García, O., & Bashan, Y. (2015). Microalgal heterotrophic and mixotrophic culturing for bio-refining: From metabolic routes to techno-economics. In *Algal Biorefineries: Volume 2 Production, Refining, Design* (pp. 61–131). DOI: 10.1007/978-3-319-20200-6_3

Pittman, J. K., Dean, A. P., & Osundeko, O. (2011). The potential of sustainable algal biofuel production using wastewater resources. *Bioresource Technology*, *102*(1), 17–25. DOI: 10.1016/j.biortech.2010.06.035 PMID: 20594826

Posadas, E., Bochon, S., Coca, M., García-González, M., García-Encina, P., & Muñoz, R. (2014). Microalgae-based agro-industrial wastewater treatment: A preliminary screening of biodegradability. *Journal of Applied Phycology*, *26*(6), 2335–2345. DOI: 10.1007/s10811-014-0263-0

Rasoul-Amini, S., Montazeri-Najafabady, N., Shaker, S., Safari, A., Kazemi, A., Mousavi, P., Mobasher, M. A., & Ghasemi, Y. (2014). Removal of nitrogen and phosphorus from wastewater using microalgae free cells in a bath culture system. *Biocatalysis and Agricultural Biotechnology*, *3*(2), 126–131. DOI: 10.1016/j.bcab.2013.09.003

Rathod, H. (2014). *Algae-based wastewater treatment* [Master's seminar report, University of Roorkee, India].

Reymann, T., Kerner, M., & Kümmerer, K. (2020). Assessment of the biotic and abiotic elimination processes of five micropollutants during cultivation of the green microalgae *Acutodesmus obliquus*. *Bioresource Technology Reports*, *11*, 100512. DOI: 10.1016/j.biteb.2020.100512

Rezania, S., Taib, S. M., Din, M. F. M., Dahalan, F. A., & Kamyab, H. (2016). Comprehensive review on phytotechnology: Heavy metals removal by diverse aquatic plant species from wastewater. *Journal of Hazardous Materials*, *318*, 587–599. DOI: 10.1016/j.jhazmat.2016.07.053 PMID: 27474848

Saha, I., Datta, S., & Biswas, D. (2023). Mechanistic approach to comparative biosorption of dyes by the extracellular polymer of *Ochrobactrum pseudintermedium* C1 utilizing waste mineral lubricating oil. *Bioremediation Journal*, *28*(3), 266–285. DOI: 10.1080/10889868.2023.2269220

Semple, K. T., Cain, R. B., & Schmidt, S. (1999). Biodegradation of aromatic compounds by microalgae. *FEMS Microbiology Letters*, *170*(2), 291–300. DOI: 10.1111/j.1574-6968.1999.tb13386.x

Shen, Q.-H., Zhi, T.-T., Cheng, L.-H., Xu, X.-H., & Chen, H.-L. (2013). Hexavalent chromium detoxification by nonliving *Chlorella vulgaris* cultivated under tuned conditions. *Chemical Engineering Journal*, *228*, 993–1002. DOI: 10.1016/j.cej.2013.05.074

Shitu, A., Tadda, M. A., Zhao, J., Danhassan, U. A., Ye, Z., Liu, D., Chen, W., & Zhu, S. (2024). Review of recent advances in utilising aquaculture wastewater for algae cultivation and microalgae-based bioproduct recovery. *Environmental Geochemistry and Health*, *46*(12), 485. DOI: 10.1007/s10653-024-02286-8 PMID: 39508916

Sutherland, D. L., Howard-Williams, C., Turnbull, M. H., Broady, P. A., & Craggs, R. J. (2015). Enhancing microalgal photosynthesis and productivity in wastewater treatment high-rate algal ponds for biofuel production. *Bioresource Technology*, *184*, 222–229. DOI: 10.1016/j.biortech.2014.10.074 PMID: 25453429

Sutherland, D. L., & Ralph, P. J. (2019). Microalgal bioremediation of emerging contaminants: Opportunities and challenges. *Water Research*, *164*, 114921. DOI: 10.1016/j.watres.2019.114921 PMID: 31382151

Tiwari, B., Sellamuthu, B., Ouarda, Y., Drogui, P., Tyagi, R. D., & Buelna, G. (2017). Review on fate and mechanism of removal of pharmaceutical pollutants from wastewater using biological approach. *Bioresource Technology*, *224*, 1–12. DOI: 10.1016/j.biortech.2016.11.042 PMID: 27889353

Torres, M. A., Barros, M. P., Campos, S. C., Pinto, E., Rajamani, S., Sayre, R. T., & Colepicolo, P. (2008). Biochemical biomarkers in algae and marine pollution: A review. *Ecotoxicology and Environmental Safety*, *71*(1), 1–15. DOI: 10.1016/j.ecoenv.2008.05.009 PMID: 18599121

Umamaheswari, J., & Shanthakumar, S. (2016). Efficacy of microalgae for industrial wastewater treatment: A review on operating conditions, treatment efficiency, and biomass productivity. *Reviews in Environmental Science and Biotechnology*, *15*(2), 265–284. DOI: 10.1007/s11157-016-9397-7

Ummalyma, S. B., Mathew, A. K., Pandey, A., & Sukumaran, R. K. (2016). Harvesting of microalgal biomass: Efficient method for flocculation through pH modulation. *Bioresource Technology*, *213*, 216–221. DOI: 10.1016/j.biortech.2016.03.114 PMID: 27036330

Vardhan, K. H., Kumar, P. S., & Panda, R. C. (2019). A review on heavy metal pollution, toxicity, and remedial measures: Current trends and future perspectives. *Journal of Molecular Liquids*, *290*, 111197. DOI: 10.1016/j.molliq.2019.111197

Wanda, E. M., Nyoni, H., Mamba, B. B., & Msagati, T. A. (2017). Occurrence of emerging micropollutants in water systems in Gauteng, Mpumalanga, and North West provinces, South Africa. *International Journal of Environmental Research and Public Health*, *14*(1), 79. DOI: 10.3390/ijerph14010079 PMID: 28098799

Wang, C.-Y., Fu, C.-C., & Liu, Y.-C. (2007). Effects of using light-emitting diodes on the cultivation of *Spirulina platensis*. *Biochemical Engineering Journal*, *37*(1), 21–25. DOI: 10.1016/j.bej.2007.03.004

Wang, Y., Ho, S.-H., Cheng, C.-L., Guo, W.-Q., Nagarajan, D., Ren, N.-Q., Lee, D.-J., & Chang, J.-S. (2016). Perspectives on the feasibility of using microalgae for industrial wastewater treatment. *Bioresource Technology*, *222*, 485–497. DOI: 10.1016/j.biortech.2016.09.106 PMID: 27765375

Whitton, R., Ometto, F., Pidou, M., Jarvis, P., Villa, R., & Jefferson, B. (2015). Microalgae for municipal wastewater nutrient remediation: Mechanisms, reactors, and outlook for tertiary treatment. *Environmental Technology Reviews*, *4*(1), 133–148. DOI: 10.1080/21622515.2015.1105308

Wollmann, F., Dietze, S., Ackermann, J. U., Bley, T., Walther, T., Steingroewer, J., & Krujatz, F. (2019). Microalgae wastewater treatment: Biological and technological approaches. *Engineering in Life Sciences*, *19*(12), 860–871. DOI: 10.1002/elsc.201900071 PMID: 32624978

Xiong, J.-Q., Kurade, M. B., Abou-Shanab, R. A., Ji, M.-K., Choi, J., Kim, J. O., & Jeon, B.-H. (2016). Biodegradation of carbamazepine using freshwater microalgae *Chlamydomonas mexicana* and *Scenedesmus obliquus* and the determination of its metabolic fate. *Bioresource Technology, 205*, 183–190. DOI: 10.1016/j.biortech.2016.01.038 PMID: 26826958

Xiong, J.-Q., Kurade, M. B., & Jeon, B.-H. (2018). Can microalgae remove pharmaceutical contaminants from water? *Trends in Biotechnology, 36*(1), 30–44. DOI: 10.1016/j.tibtech.2017.09.003 PMID: 28993012

Zhang, C., Wang, X., Ma, Z., Luan, Z., Wang, Y., Wang, Z., & Wang, L. (2020a). Removal of phenolic substances from wastewater by algae: A review. *Environmental Chemistry Letters, 18*(2), 377–392. DOI: 10.1007/s10311-019-00953-2

Zhang, S., Xiao, J., Wang, G., & Chen, G. (2020b). Enzymatic hydrolysis of lignin by ligninolytic enzymes and analysis of the hydrolyzed lignin products. *Bioresource Technology, 304*, 122975. DOI: 10.1016/j.biortech.2020.122975 PMID: 32086036

Zhang, S.-Y., Sun, G.-X., Yin, X.-X., Rensing, C., & Zhu, Y.-G. (2013). Biomethylation and volatilization of arsenic by the marine microalgae *Ostreococcus tauri*. *Chemosphere, 93*(1), 47–53. DOI: 10.1016/j.chemosphere.2013.04.063 PMID: 23726009

Zhang, Y., Habteselassie, M. Y., Resurreccion, E. P., Mantripragada, V., Peng, S., Bauer, S., & Colosi, L. M. (2014). Evaluating the removal of steroid estrogens by a model alga as a possible sustainability benefit of hypothetical integrated algae cultivation and wastewater treatment systems. *ACS Sustainable Chemistry & Engineering, 2*(11), 2544–2553. DOI: 10.1021/sc5004538

Chapter 6

Microalgae for the Removal of Emerging Contaminants From Wastewater

K. Kesavardhini
http://orcid.org/0009-0005-8235-042X
Department of Microbiology, Sacred Heart College, Tirupattur, India

A. Sabina Banu
Department of Microbiology, Sacred Heart College, Tirupattur, India

P. Saranraj
http://orcid.org/0000-0002-5052-6532
Department of Microbiology, Sacred Heart College, Tirupattur, India

K. Gayathri
http://orcid.org/0009-0007-3786-2009
Department of Microbiology, Sacred Heart College, Tirupattur, India

S. Dayana
Department of Microbiology, Sacred Heart College, Tirupattur, India

B. Lokeshwari
http://orcid.org/0009-0005-6612-5134
Department of Microbiology, Sacred Heart College, Tirupattur, India

ABSTRACT

The aquatic environment and human health are seriously threatened by the increasing amount of Emerging Contaminants (ECs) in sewage. Because of their adaptable metabolic as well as capacity to adapt to the environment, microalgae have been given a lot of focus, as a viable and sustainable solution of eliminating these contaminants. Microalgal systems have extremely broad capabilities of eliminating a variety of ECs, including steroids, insecticides, hygiene items, medications, and heavy metals, with the mechanisms of bioaccumulation, biosorption and biodegradation, as well as photodegradation. Besides the elimination of contaminants, microalgae have been found to help in the recovery of nutrients, sequestration of carbon dioxide

DOI: 10.4018/979-8-2600-3080-6.ch006

and biomass to be used in value-added products. Microalgae-based processes are environmentally friendly and efficient in terms of energy usage compared to the conventional treatment technologies. Nevertheless, there are process optimization issues, high scale deployment and harvesting of biomass.

1. INTRODUCTION

Microalgae have proven to be effective biological agents in Sewage treatment, particularly for the elimination of Emerging Contaminants (ECs). Conventional wastewater treatment methods, including activated sludge processes, trickling filters, and stabilization ponds, are often inefficient at removing ECs. Traditional wastewater treatment methods inadequately eradicate pollutants, including Pharmaceuticals, Personal Grooming Products, Endocrine-Disrupting Chemical Substances, and Microplastics. ECs pose serious risks to human health, including endocrine disruption, antibiotic resistance propagation, and carcinogenic effects, while in aquatic environments they cause toxicity, bioaccumulation in food webs, and disruption of ecosystem functioning. Microalgal technologies have been reported to receive attention because they have the capacity to assimilate, degrade, and adsorb such contaminants through various mechanisms like Biosorption, Bioaccumulation, and Biodegradation (Khan et al., 2021; Luo et al., 2022). In addition, microalgae present a greener and cost-effective option compared to physicochemical processes. Due to their photosynthetic capability, microalgae utilize sunlight and CO_2, making the treatment process both cost-efficient and carbon neutral. The elevated ratio of surface area to volume of microalgal cells also ensures effective interaction with contaminants (Khan et al., 2021).

Moreover, microalgae produce bioactive compounds that can be synthesized into biofuels, Fertilizer, or Bioplastics, with additional economic and sustainability advantages (Rashid et al., 2023). Various species like *Chlorella, Scenedesmus*, and *Spirulina* have shown impressive efficiencies in removing ECs. Recent progress in molecular biology, such as CRISPR-Cas9 gene editing and overexpression of degradation-related enzymes, has further augmented their capacities for degradation. Still, complexity in wastewater matrices and heterogeneity in ECs call for specific approaches for maximum efficiency. Synergistic outcomes have been demonstrated through combining microalgae with other biological and physicochemical treatment processes. Such systems can be fine-tuned to function under varying environmental as well as operational conditions, making the process even more versatile. Thus, microalgae-based treatment continues to evolve, bridging the gap between laboratory research and practical application. Their sustainability, eco-compatibility, and multifunctionality establish them as a cornerstone of future wastewater treatment strategies (Rashid et al., 2023)

Bioactive chemicals that can be converted to fuels, fertilizers, or plastics are synthesized by microalgae, which add more economic significance (Rashid et al., 2023). Microalgae like *Chlorella, Scenedesmus*, and *Spirulina* exhibit remarkable EC degrading abilities. The use of innovations in molecular biology, such as the CRISPR-Cas9 technique, has increased the efficiency of their degradation mechanisms. But the diversity of ECs and wastewater complicates matters, so specific strategies need to be applied. A synergistic effect can be created when combined with other treatments.

The chemical versatility of microalgae has made them able to cope with various ECs found in sewage, effluents from industries, and farms. They produce oxidoreductase enzymes such as laccases, peroxidases, and cytochrome P450 while having cell walls containing functional groups, which help to promote biosorption (Koupaie et al., 2021). The adaptive strategies exhibited by microalgae make them suitable for functioning under variable conditions (Luo et al., 2022). In addition, microalgae engage in mutualistic biofilm/bacteria cultures, which aid in EC decomposition and biomass production better than monocultures (Gonçalves et al., 2017). Omic technologies (such as transcriptomics and proteomics) have been used to study how gene expressions change in response to ECs for strain selection (Zhang et al., 2020). However, EC removal rates still differ and require field testing.

Other than the removal of ECs, there is potential for nutrient recovery (as nitrogen and phosphorous assimilators) and oxygenation of the water column as well as preventing secondaries from entering the environment. Some algae species (*Chlorella vulgaris, Scenedesmus obliquus*) have natural flocculation ability, facilitating the harvesting process (Villar-Navarro et al., 2020; Arbib et al., 2014; Olguín, 2012). Closed systems may be more expensive, but they offer more control for valuable ECs. A combination system, such as using membrane bioreactors, constructed wetlands, or activated sludge processes can increase efficiency and lower costs.

2. APPROACHES TO REMOVE EMERGING CONTAMINANTS

In marine and aquatic ecosystems, emerging contaminants of concern (CECs) are present today at microgram per liter to nanogram per liter concentrations. Prominent CECs include residues from pharmaceuticals, personal care products, pesticides, industrial chemicals, flame retardants, and microplastics. Major pharmaceutical manufacturers release their water waste into sewerage systems connected to city plants that purify sewage. In broadly, these do not have the choice of disposing of pharmaceutical residues. Accordingly, there has been an effort to experiment with various techniques for the removal of CEC from wastewater. These are Activated Carbon Adsorption, Ozonelayer, Membrane Operation, Biological Treatment, and Advanced Oxidation Process (AOP). A novel method using microalgae has recently been valued. To determine the most efficient treatment approach for a specific wastewater, one must consider process nature, cost, benefits, and limitations. Traditional treatments such as the ozonation active carbon absorption, membrane filtering, chlorination, and UV disinfection were not considered for the reduction of pollutants of rising significance in wastewater. Biological treatments - Conventional Activating Sludge (CAS), Artificial Wetland, Membrane Bioreactor (MBR), Biofiltration, and Enzymatic Processes are some of them (Salimi et al., 2017; Rizzo et al., 2018). Besides pharmaceuticals, other CECs like bisphenol A, triclosan, perfluorinated compounds, and microplastics are increasingly detected in wastewater, necessitating advanced treatment.

2.1 Activated Carbon Adsorption

The process of activated carbon adsorption involves the passage of wastewater through activated carbon particles, either granular or powdered form, whereby CECs attach to the large surface area of the carbon via van der Waals forces and hydrophobic bonds. Activated carbon adsorption is an efficient treatment option that removes a variety of organic pollutants, such as pharmaceuticals and pesticides, with no secondary hazardous substances generated. Nonetheless, the adsorptive capability is limited and needs to be replenished by regeneration or disposal of used carbon, both of which are costly. Natural organic matter present in the wastewater will also affect adsorption efficacy (Luo et al., 2022).

2.2 Ozonation

The ozonation technique makes use of bubbling of ozone gas (O_3) into the wastewater to react directly with pollutants, or it converts into hydroxyl radicals (-OH) that react with any compound present non-specifically. This method successfully oxidizes many drugs, endocrine disruptors, and pesticides in a matter of minutes. There is no waste stream formation during ozonation, and it enhances the biodegradability of the effluent. The main drawback of this method is the presence of toxic by-products like bromate in case bromides are available in water and aldehydes (Koupaie et al., 2021).

2.3 Membrane Filtration

The method of membrane filtration such as nanofiltration and reverse osmosis employs semi-permeable membranes, where pores have dimensions between 0.1-10 nm and can reject CECs by sieving out based on molecular size and charge. Membrane filtration is highly efficient (>90%) and does not require any chemicals. High water quality of permeate (effluent) makes this technology applicable for reuse. The downside of using membrane filtration technologies is that they are expensive to construct and operate, require a high pressure (5-30 bars), and experience fouling — buildup of organic matter, bacteria, and salts on membranes (Olguín, 2012).

2.4 Advanced Oxidation Processes

The Advanced Oxidation Processes (AOPs) involve the use of combinations of ultraviolet radiation, hydrogen peroxide, ozone, and photocatalysis to create reactive hydroxyl radicals (•OH, oxidation potential = 2.8 V) that will indiscriminately attack and mineralize even the most difficult-to-breakdown CECs into CO_2 and H_2O. AOPs have been found to be very effective in the treatment of industrial wastewaters that contain persistent compounds. Nevertheless, such processes are associated with a great number of chemical reagents and energy inputs in addition to the cost-intensive requirement of maintaining special equipment. Moreover, there is always a chance of interference from radical scavengers (carbonates, chlorides), as well as formation of toxic byproducts (Rizzo et al., 2018).

2.5 Biological Treatments

The biological techniques such as Conventional Activated Sludge (CAS), constructed wetlands, Membrane Bioreactors (MBRs), biofiltration, and enzymatic techniques use microbial metabolic activities to metabolize the CECs. In CAS, heterotrophic bacteria and protozoans in the aerobic tanks metabolize organic substances, partially removing some CECs (ibuprofen, caffeine) via co-metabolism or direct catabolism. The constructed wetlands imitate nature, where the plant absorption, microbial degradation, and sorption processes eliminate the CECs using little energy. MBRs incorporate the biological treatment process with membrane filtration, keeping the slow-growing bacteria and offering more removal of CECs than CAS systems. Enzymatic processes use isolated enzymes (laccases, peroxidases) to oxidize CECs specifically. Biological processes are economically viable, environmentally friendly, and generate minimal chemical wastes. Nevertheless, removal efficiencies are very inconsistent (0-95%, depending on the CEC and method used), some CECs (carbamazepine, perfluorinated compounds) resist biodegradation, and effectiveness is influenced by temperature, pH, and hydraulic retention time (Salimi et al., 2017; Rizzo et al., 2018).

Each technique has its own strengths and weaknesses. The choice of the technique depends on several factors such as the composition of the wastewater, the cost of the process, practicality of operations, and the purpose of the purification. Microalgae-based purification is a new technique that is an effective and economical process. However, no one technique can work effectively in all cases, and a combination of different techniques is usually preferred for better outcomes (Wang et al., 2021).

3. TREATMENT OF WASTEWATER USING MICROALGAE-BASED SYSTEMS

Microalgae can grow autotrophically, mixotrophically, or heterotrophically. The fact that microalgae are recognized as the world's primary oxygen source has also seen this characteristic exploited for enhancing wastewater treatment using bacteria. The oxygen-requiring process of bacterial breakdown of organic matter in sewage is normally supplied by mechanical equipment like blowers and diffusers in traditional sewage systems. For the activated sludge system to be more efficient, the dissolved oxygen level is to be maintained above 2 mg/L. Thus, the highest amount of energy that would be needed for conventional wastewater treatment employing activated sludge technology is 0.5 kWh/m^3, which is basically for aeration. After the decomposition of the organic matter into mineral form substances (e.g., NH_4^+, PO_4^{3}) and units (e.g., kWh/m^3, mg/L), CO_2 and N_2 because of the nitrification and denitrification procedures are released onto the atmosphere. Phosphate complexes are activated or dissolved by water. In addition to using phosphorus, carbon, and ammonia-containing compounds released into the medium by bacteria and recycling them as valuable biomass, the application of microalgae in this process stimulates the production of oxygen *via* photosynthesis. Having an excellent balance between the amounts of organic material in the load, how fast it is decomposed with the aid of bacteria, and how fast microalgae produce oxygen depending on light intensity will ensure maximum process efficiency. The efficacy of microalgae-derived processes is based on the balance between bacterial and microalgal performance (Rizzo et al., 2018).

The quantity of organic materials added to the process has the maximum influence on the population of bacteria. It affects the rate of oxygen consumption and chemical nutrients dissolved into the growth medium. The rates should be equivalent to the rates with which microalgae can do it, i.e., remove the inorganic nutrients supplemented by bacteria and supply oxygen consumed by bacteria at an equivalent pace. Microalgae concentration and light availability are determined by geometric configuration of the reactors and strain-induced optical characteristics, which in turn are dependent on inorganic compound concentration used. The density of biomass from microalgae should be considerably greater compared to that bacterial biomass to produce a steady biological matrix because microalgae grow much slower than bacteria (1 day to several hours). Microalgal biomass concentration in such a system is five times higher than bacterial biomass density. Microalgae biomass growth is the primary reason for core competence of the yield of biomass. A yield up to one kilogram of biomass from microalgae per cubic meter of wastewater results from the normal concentration of sewage water at 10 mg/L phosphorus, 60 mg/L ammonium, and 500 mg/L COD. Besides, net positive energy utilization is achieved considering biomass combustion heats (20 MJ/kg), as energy used in generating biomass in less than 5 MW for open raceway reactors. Microalgae wastewater is not treated with pure monoalgal cultures. Therefore, maintaining pristine cultures is unequivocally unfeasible on the industrial level, despite having much literature regarding successful treatment of wastewater using most of the microalgae strains (Urbina-Suarez et al., 2021).

The use of pure cultures is not preferable in large-scale applications. Naturally occurring Bacterial and Microalgal consortia have been demonstrated to be significantly more reliable and stable than pure cultures when employed in large-scale treatment systems. Operating conditions, environmental conditions, and wastewater composition are the key parameters affecting the species during wastewater treatment. Hydraulic retention time, dilution rate, water depth, and pH control are the most critical parameters influencing the process's execution. Wastewater cannot be carbonated in terms of pH, and thus the efficiency of microalgae cultures is negatively affected under ideal conditions. Hence, in the absence of surplus CO_2, the cultures are limited by carbon. The primary consequence of this event is not a decrease in biomass production but a rise in pH, causing a lack of bacteria and consequently destabilizing the interaction between Microalgae and Bacteria (Olguín, 2012).

Contradictory results have been reported on this issue, with observed variability largely dependent on environmental conditions. In freshwater environments, cellular responses differ markedly from those observed under atmospheric exposure. The effect of supplementary CO_2 supply increases as key Environment-related variables such as temperature, amount of light, and access to nutrients approach optimal conditions. Even though it has been proposed by some authors to decrease the water depth to 10–20 cm, it is better to run wastewater treatment reactors at 30–40 cm depth. Though every proposal is aimed at a different target, they are not mutually exclusive. Hence, it would be logical to utilize deeper water in a bid to treat as much sewage as is feasible. Although the rate of photosynthesis is low, it is adequate to provide the oxygen necessary for aerobic bacterial decomposition of organic matter. Conversely, shallow depth utilization of water is recommended for maximum nutrient recovery and for the large-scale assimilation of ammonium as useful biomass from microalgae. The alga's dominant development rate strain under the reactor's present culture conditions must be used to calculate the optimum dilution rate or hydraulic retention time. Since microalgae can grow at a certain maximum rate of about one per day, the maximum dilution rate is between 0.4 and 0.5 per day. The efficient dilution rates of open raceway reactors, nonetheless, would typically fall between 0.2 and 0.3 per day, which is equivalent to an average holding time of 5 to 7 days, since external cultural environments are not optimal (Arbib et al., 2014).

The problem of harvesting remains one of the key barriers to scaling up wastewater treatment through microalgae cultivation. The microalgal cells are usually characterized by a small size (3-30 μm), a small bulk density (about 1.05-1.10 g/cm^3), and a constant negative surface charge, preventing them from settling in the culture broth. Harvesting techniques are divided into centrifugation, filtration, chemical flocculation, bio flocculation, and sedimentation. Centrifugation works well; however, it is expensive in terms of energy consumption (1-8 kWh per m^3 of treated water) and makes up 20-30% of all production expenses (Wang et al., 2021). Membrane fouling is a common problem of filtration because of membrane clogging in cases of high biomass concentrations. The application of chemicals such as aluminum sulfate and ferric chloride is less costly; however, metals get accumulated in biomass, making it unsuitable for feeding purposes. Though DAF is promising, appropriate pressure control and chemical dosages must be exercised. Moreover, the effectiveness of harvesting varies depending on the type of microalgae being cultivated, stage of growth, and composition of the effluent (Rashid et al., 2023). Filamentous or auto-flocculating microalgae like certain species of *Scenedesmus* and *Chlorella* will precipitate more rapidly compared to unicellular auto-flocculating species. Harvesting methods influence subsequent operations as well; biofuel production requires drying to more than 80 percent dryness whereas fertilizers can accept higher levels of moisture content. Further studies continue to examine innovative ways for effective harvesting at a low cost, which include the use of magnetic nanoparticles, electro-coagulation, and photobioreactors with *in-situ* harvesting (Matamoros et al., 2015). The development of economically viable harvesting methods is crucial for microalgae-based wastewater treatment to take place on a commercial scale, until then the prospects are limited.

4. MICROALGAL REMOVAL OF CECS

The ability of specific microalgae species to eliminate CECs was investigated, and the choice may influence the process effectiveness. Resistance of certain Microalgal species is exposed even at a higher concentration compared with the concentrations of pollutants in wastewater, with greater survival upon exposure to conditions of the toxic effect of CECs. Microalga *Chlorella vulgaris* resisted and biodegraded wastewater pollutants with high organophosphorus compound concentrations effectively. The same was discovered in *C. vulgaris* when it was exposed to diazinon. The microalga was challenged with contamination by some Contaminants of Emerging Concern, especially hormones. A mechanism for adapting to a contaminated environment is necessary and leads to enhanced removal efficiency. Using an adaptation technique, Escapa et al. (2016) assessed three different species of phytoplankton (*C. sorokiniana, C. vulgaris*, and *S. obliquus*) in wastewater for their susceptibility to diclofenac. The findings confirm elevated Microalgal density and development rate as well as reduced diclofenac susceptibility (high elimination ability of as much as 79%). According to Xiong et al. (2019), *S. obliquus* used a chemical adaptation process to remove varying doses of the sulfamethazine and sulfamethoxazole, with notable removal efficiencies for greater amounts (31.4–62.3% around 27.7–46.8% more than 0.025–0.25 mg/L of sulfamethoxazole and 0.025–0.25 mg/L of sulfamethazine).

Microalgae have an impact on other mechanisms of degradation, such as Volatilization and degradation by sunlight Ge et al. (2009) recognized hydroxyl radicals in Microalgal extracellular polymers which promoted CECs by activation *via* photodegradation. Examples include diclofenac, ciprofloxacin, Tetracycline (TC), 17α-ethynylestradiol (EE2), and 17β-estradiol (E2) pollutants, that photodegrade upon exposure to microalgae. Volatilization of CECs is phase transfer to the air phase of organic volatiles molecules from the fluid phase, dependent on the physicochemical property of wastewater and operating conditions (i.e., Temperature, Aeration, and Agitation). Microalgal cultures with high rates of aeration, light, and fluctuating operation temperatures (Cold and Hot Seasons) can cause significant improvement in the elimination of newly emerging issue volatile pollutants like the fragrances galaxolide and musky with tonalide (Matamoros et al., 2015).

Biosorption refers to the passive adsorption of CECs by functional groups on the surface of microalgae cell walls (carboxyl, hydroxyl, amino), without metabolic energy. Bioaccumulation refers to the active process of contamination transport into microalgae cells requiring metabolic energy. Intracellular biodegradation follows up, where the microalgae degrade contaminants into more mineralized substances using enzyme systems such as cytochrome P450 and glutathione-S-transferases. Contaminant fate includes complete mineralization to carbon dioxide and water, transformation into conjugated metabolites, or uptake into biomass. Scientific limitations include: (i) limited knowledge of CECs degradation mechanisms; (ii) possible generation of transformation products that are toxic; (iii) variable efficiency resulting from changing environmental factors; (iv) the presence of interfering substances within the wastewater matrix; and (v) the absence of standard procedures to compare removal efficiencies (Gentili & Fick, 2017).

4.1. Planktonic Cultures

Microalgae are cultivated separately or mixed in open or closed systems. Microalgae are increasingly being found suitable for sewage treatment that includes primary as well as secondary processing facilities and can be employed for the removal of CECs. Open systems may also be referred to as stabilization ponds or High-Rate Algal Ponds (HRAPs). Easy construction and utilization and low running cost are the primary benefits of open platforms. Open systems are prone to microbial contamination, removal of CO_2 from the atmosphere, and large areas of land are needed. Temperature and light dependency slow down development. Shriwastav et al. (2017) have shown that a 10°C rise in temperature enhanced removal effectiveness by CECs of *Chlamydomonas reinhardtii* and *C. vulgaris*. Rise in temperature also reduces solubility of CO_2, decreasing the possible utilization as a source of carbon. It has been observed that the clearance rate also has tremendous variability, the most significant factor differing with the reactor types and biological therapy.

According to Matamoros et al. (2016), stabilisation lakes are optimally effective for the elimination of certain Emerging Concern Contaminants (CECs) such as Guaranine, Calprofen, Acetaminophen, Hedione, Benzophenone-3, and Furosemide compared to the sewage purification methods and across all seasons. Gentili and Fick (2017) demonstrated the effective elimination of substantial amounts of pharmaceutical agents (i.e., Tenormin Bisoprolol, Lopressor, Bupropion, Atracurium, Clarithromycin, Diltiazem, and Terbutaline) from a stabilization lagoon, observing a relationship between light intensity and removal efficiency. Advances in conventional stabilization ponds resulted in more efficient design, i.e. High-Rate Algal Ponds (HRAPs). These technologies are as effective as or even more effective than traditional stabilization ponds in decontaminating many types of CECs with less aeration and land space for deployment (Gonçalves et al., 2017). Compared to immobilized or biofilm cultures, planktonic systems are simpler to operate but suffer from lower biomass density, higher harvesting costs, and greater susceptibility to washout.

Figure 1 depicts a microalgae-based system for the removal of Contaminants of Emerging Concern (CECs). It illustrates how microalgae uptake, degrade, or convert contaminants using absorption and metabolism mechanisms. The technology demonstrates a renewable and environmentally friendly method of treating sewage and environmental remediation.

Figure 1. Microalgae- based system for removal of Contaminants of Emerging Concern (CECs)

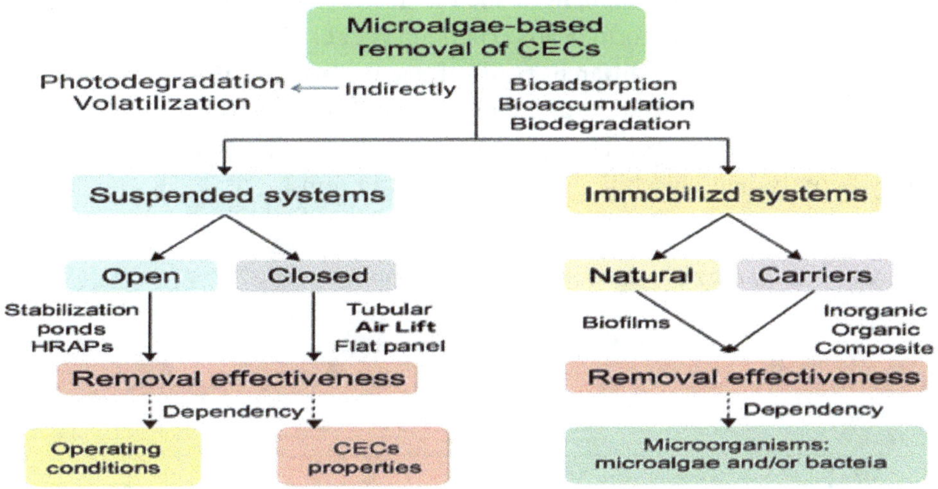

4.2. Immobilized Cultures

Cell immobilization is a term that is used to describe limitation of movement by adsorption or entrapment onto an insoluble solids transporter that is impervious in liquid, either inorganic or organic in nature. Covalent bonding, surfaces adsorption, deposition, entrapment in semi-permeable membranes, and entrapment in fibers or porous polymers are the major immobilization techniques. Natural and normal activity of microbe capacity to absorb on to a surface lead to cell immobilization as a biofilm. Immobilized systems possess several merits, such as possessing high microalgae activity and concentration, enhanced removal efficiency of Contaminants of Emerging Concern (CECs), easy harvesting process, and enhanced stress tolerance, including toxicity against harmful contaminants. Immobilized microalgae eliminated nitrogen, phosphate, and additional heavy metals by wastewater effectively. In recent decades, immobilized microalgae have been employed for the elimination of effluent Contaminants of Emergent Concern (CECS). According to Solé and Matamoros (2016), immobilized Nutrition (nitrogen and phosphate) and several Endocrine Interfering Chemicals (EDCs) including bisphenol A or as F, 17α-ethinylestradiol, 4-octylphenol, and 2, 4-dichlorophenol were effectively eliminated *via* wastes by phytoplankton in alginate shells. However, diffusion resistance through the immobilization matrix can limit mass transfer of nutrients and contaminants, reducing removal rates at high biomass densities. On a large scale, immobilized reactors face challenges such as matrix degradation, clogging, and difficulty in scaling up uniform beads (Hamouda et al., 2019)

Figure 2 illustrates the different types of carriers used in immobilization systems. It shows natural, synthetic, and inorganic carriers employed to support cells, enzymes, or microorganisms. The figure highlights how carrier selection influences stability, activity, and reusability in immobilized systems.

Figure 2. Types of carriers used in Immobilization systems

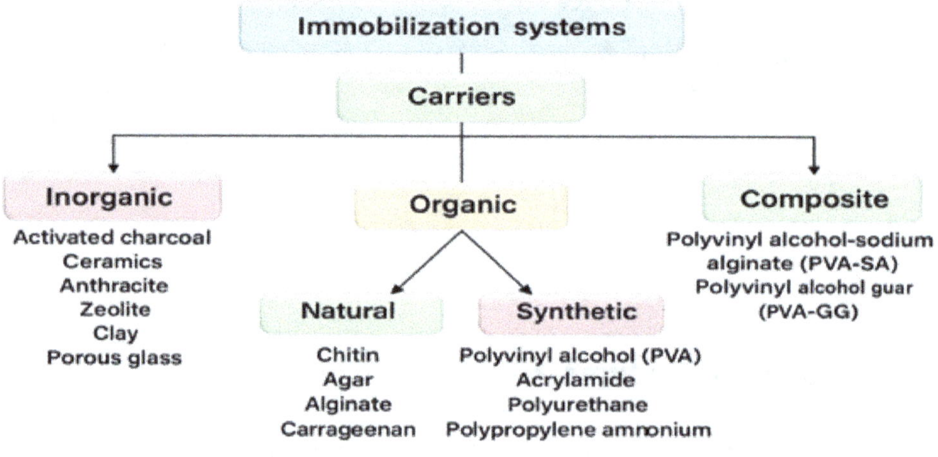

Enhancement in their use of Microphytes biofilm-based Sewage treatment processes has evolved achieved by the different reactor geometries, following the benefits of microalgae immobilization. They can be constructed horizontally, vertically, or as a cylindrical rotating structure with many layers, the justification for increasing productivity and unit land area efficiency Microalgae biofilms have been pilot-scaled in tertiary treatment of sewage facilities employing reactors with various geometries. Cylindrical reactor with internal biofilm, movable vertical submerged microbe films reactor, bed microbial films reactor, there are three types of algae biofilm reactors: Horizontal, Rotating, and Parallel Plate Membrane Photobioreactor are the most used reactor types. Zhang et al. (2018) utilized a multi layered algal-biofilm reactor for removing wastewater CECs. Biofilm reactors are also easy to install and operate and are very effective in removal and are an ideal substitute for traditional sludge activated and membrane bioreactors (Zhang et al., 2018).

4.3. Mixed Biofilm Cultures

As opposed to monospecies, microbial consortia are more stable and resistant to abrupt changes in the environment, invasion by competitive organisms and predators. By respiration and emission of carbon dioxide, bacteria can transform the environment and supply the microalgae with important nutrition and growth determinants, i.e., Phytohormones and Chelators, which can stimulate and enhance the growth of microalgae (Krzeminski et al., 2019). Several studies have been conducted and have proven that the excretion of growth-promoting substances by bacteria, i.e. microalgae Extracellular Polymeric Substances (EPS), promotes microalgae growth and vice versa. Microbial degradation of CECs may also provide substrates required for bacterial growth (Krzeminski et al., 2019).

Figure 3 illustrates the formation and disintegration processes of a microalgae bacteria biofilm. It shows the stages of microbial attachment, biofilm maturation, and structural stabilization through extracellular polymeric substances. The figure also highlights biofilm disruption and detachment under environmental or operational changes.

Figure 3. Formation and disintegration process of Microalgae-Bacteria Biofilm

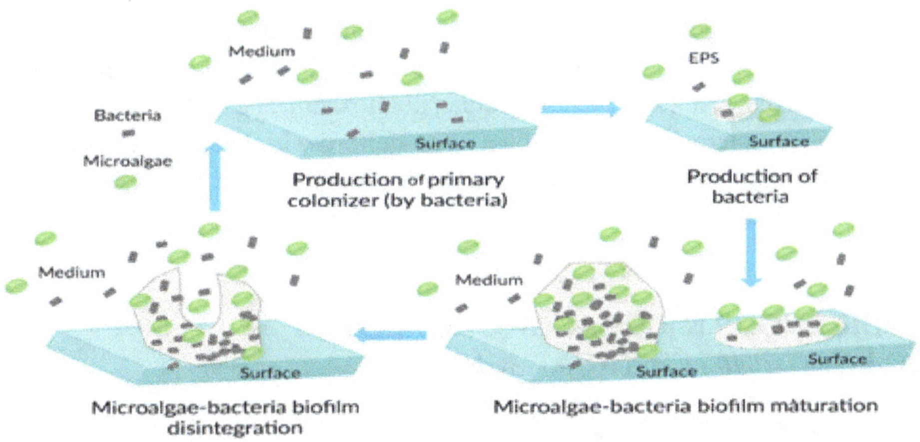

Mixed species of biofilms were employed to remediate different hydrocarbon contaminants from industrial wastewater in a variety of biofilm reactors, such as fluidized-bed biofilm reactors, fixed column biofilm reactors, and rotating perforation tube biofilm reactors. *Pseudomonas sp.* was used in a biofilm reactor with revolving perforated tubes to eliminate 2,4-dichlorophenol, according to Kargi and Eker (2005). The results of de-chlorinating several compounds with chlorinated aromatics utilizing fluidized bed biofilm reactors injected with *Pseudomonas* and *Rhodococcus* species were reported by Puhakka et al. (1995). The use a fixed bed biofilm reactor with a mixed culture of immobilized microorganisms, Oh and Tuovinen (1994) achieved both whole and partial extraction of two herbicides, 2-(2-Methyl-4-Chlorophenoxy) Propionic Acid (MCPP) and 2,4-dichlorophenoxyacetic acid. To the best of our knowledge, Contaminants of Emergent Concerns (CECs) have not been removed using microalgae-bacteria biofilms. Carbon and nutrients were previously removed from wastewater using suspended mixed cultures of bacteria and microalgae. The bacterial genera *Achromobacter, Alcaligenes faecalis, Hydrogenophaga, Pseudomonas, Bacillus, Flavobacterium, Rhizobium*, and *Muricauda* were identified for this purpose. Bacteria can neutralize the toxic or inhibitory action of high levels of pollutants in wastewater and lower removal efficiency of microalgae. For decontamination of black oil, Safonova et al. (1999) investigated a series of microalgae species in combination with alcanotrophic bacteria.

Bacteria were beneficial to black oil-treated microalgae, with a degradation rate ranging from 36% to 61%. Microalgae and bacteria were used in assessing the elimination of the insecticide diclofop-methyl; Therapeutic intervention utilizing bacteria and microalgae gave 36% enhancement of removal efficiency. The clearance efficiencies of ibuprofen were increased from 15% to 60% to 90% and 40% by the microalgae-bacteria association for caffeine. In addition, elimination of significant Contaminants of Emerging Concern (CECs) like atrazine, bromacil, Linear Alkylbenzene Sulphonate (LAS), phenol, phenanthrene, sulfadiazine, sulfamethazine, sulfamethoxazole, bisphenol A, tetracycline, norfloxacin, and sulfadimidine highly improved when microalgae were grown in co-culture with bacteria. The outcome confirms that bacterial-microalgae cultures efficiently remove CECs from wastewater (Zhang et al., 2020).

5. INDIGENOUS MICROALGAE EMPLOYED FOR WASTEWATER TREATMENT

5.1. Microalgae for the remediation of Nitrogen and Phosphorus

Human beings emit colossal amounts of both phosphorus and nitrogen need to be reclaimed through sewage since they are contaminants in water, and fertilizers that contribute to the algae blooms of aquatic systems. Because biomass from algae has a greater capacity to absorb nutrients than soil, it is a suitable method for recovering minerals from sewage by algae species. Various studies on the treatment of phosphorus and nitrogen fertilizer by micro algae have been done. Li's study (2020) indicates

Spirulina can achieve overall elimination effectiveness for both overall phosphorus and ammonium nitrogen are present in foul-smelling, black water. According to Su and Jacobsen (2021), the elimination effectiveness of *Scenedesmus obliquus* and *Chlorella vulgaris* was 54% of phosphorus and nitrogen in food factory wastewater, respectively. Shi Jing et al. (2014) also designed dual-layer Photobiotic reactor prototypes, which are in resistance against suspended particulates loss, bacteria, or algae biomass loss during leaching. The outcomes of the experiments show that nitrogen and phosphorus content in wastewater may be decreased by 70–99%. Mennaa et al. (2015) treated municipal wastewater with seven species of microalgae and obtained more than 80% and 87% removal efficiency for the content of total dissolved nitrogen and phosphorus, respectively. Microalgae remove diluted swine effluent efficiently, and studies observed 100% ammonia or phosphate elimination in seven and eleven days, respectively. Wang et al. (2021) carried out a comparative study between dual microalgae treatment of fishing effluent using monotypic wastewater treatment and two distinct species of micro algae (*Isochrysis* and *Thalassiosira pseudonana*). The findings confirm that optimal inoculation volume and ratio facilitate the treatment of sewage through dual microalgae to obtain appreciable removal efficiency. The advantage of using indigenous strains lies in their pre-acclimation to local environmental conditions (temperature, pH, light, and native microbiota), which enhances treatment stability and reduces the need for external inoculation. The link to circular economy is that the harvested nitrogen- and phosphorus-rich biomass can be valorized as biofertilizer, closing nutrient loops (Krzeminski et al., 2019).

5.2. Reduction of Biochemical and Chemical Oxygen Demand (BOD and COD) by Algal Cells

Microalgae possess inherent purification effects on some wastewater. Soto et al. (2021) utilized *Scenedesmus abundans* and *Chlorella minutissima* to reduce both BOD and COD by 56% and 70%, respectively (Soto et al., 2021). To overcome the limitation of microalgae growth in certain food processing wastewaters, Chenlu (2019) tried blending different food wastewaters with varying acidity, alkalinity, and organic matter contents. Under optimized conditions, Chlorella eliminated 72.75% of total nitrogen, 65.97% of ammonia nitrogen, 94.3% of total phosphorus, and 62.75% of Chemical Oxygen Demand (COD) (Chenlu, 2019). The removal efficiency of various microalgal species from identical wastewater types cannot be directly compared without standardized conditions. Chandra et al. (2021) utilized numerous species of microalgae for the remediation of dairy effluent. The results show that whereas *Nostoc muscorum* and *Spirulina sp.* exhibit negligible decreases in BOD and COD, with efficiencies of roughly 38% and 24%, respectively, *Chlorella minutissima* and *Scenedesmus abundans* reduced BOD and COD by 70% and 56%, respectively (Chandra et al., 2021).

Metabolic plasticity of microalgae plays an important role in BOD and COD treatment. Microalgae may function in three different trophic forms: (i) phototrophic – with only light and CO_2; (ii) heterotrophic – using organic matter in darkness; and (iii) mixotrophic – using light and organic matter. Regarding BOD and COD treatment, only heterotrophic and mixotrophic metabolisms are relevant since they allow microalgae to take up organic contaminants as nutrients, for example, sugars, organic acids, and hydrocarbons. Besides, oxygen formed during the process of photosynthesis facilitates aerobic bacteria, breaking down organic material, hence indirectly decreasing BOD and COD. Microalgae can decrease BOD and COD in wastewater in two different ways (Vassalle et al., 2020). First, there is a direct uptake of organic materials by the microalgae while growing under mixotrophy or heterotrophy conditions. Second, the oxygen produced during photosynthesis increases the performance of aerobic bacteria that decompose organic materials, thus decreasing oxygen consumption in water (Matamoros et al., 2016)

5.3. Microalgae for the Remediation of Contaminants of Emerging Concern

Developing apprehension Contaminants of Emerging Concern (CECs) are defined as recognized chemicals or chemical substances found in minute quantities within the environment with a current or potential threat to environmental, human, and animal health. Some research shows that specific microalgal species possess improved survivability because of their inhibitory nature against CECs. Microalgae eliminate CECs by Biosorption, Bioaccumulation, and Biodegradation. The microalgae-supported biodegradation of CECs has been showing promising potential in the biodegradation of diazinon (pesticide), sulfamethazine, ciprofloxacin, etc. Mulla et al. (2021) were conducting a study on the impact on the breakdown of many kinds of aquatic microalgae produce contaminants called Diethyl Phthalate (DEP) and Dibutyl Phthalate (DBP). The outcomes showed that microalgae within the marine ecosystem would biodegrade phthalate pollutants in the marine ecosystem, and the best species to degrade phthalates was *Cydrotheca closterium* (Mulla et al., 2021).

Sulfamethylpyrimidine was deemed a significant emerging micropollutant due to its environmental stability. Chen et al. (2020) identified that sulfamethazine was removed by *Chlorella vulgaris* and *Scenedesmus obliquus* by means of biodegradation processes such as ring cleavage, the oxygenation, hydrogenating, and methylated processes. The outcomes documented 52% and 43.3% removals within a span of only Fifteen days. Vassalle et al. (2020) attained an 84.91% reduction of oestradiol in an upward flow anaerobic wastewater bed reactor and a 95% reduction of oestrone within a collection of rapidly growing microscopic algal pools. The removal of microalgae Contaminants of Emerging Concern (CECs) is not universally significant. Microalgae (*Chlorella sp.*) can degrade caffeine and ibuprofen, but not carbamazepine and tris(2-chloroethyl) phosphate, which exhibit resistance to biodegradation (Yan et al., 2021).

Transformation mechanisms of CECs after microalgae treatment include the following: i) decomposition of compounds to CO_2, water, and salts; ii) biotransformation to intermediates; or iii) conjugation with cellular macromolecules such as glutathione, carbohydrates, or proteins. TPs may have different physicochemical and toxicity profiles compared to the parental compound. Thus, sulfamethazine metabolism by *Chlorella vulgaris* results in the formation of desamino- and hydroxysulfamethazine, which display decreased bactericidal effects but remain ecotoxic (Escapa et al., 2016). Analogously, phthalate esters, e.g., DEP and DBP, undergo sequential degradation to monoethyl phthalate, further converted to phthalic acid, and then enter the TCA pathway. Determination of TPs is accomplished using sensitive techniques such as LC-MS/MS or GC-MS. Notably, some TPs can be more persistent/toxic than the original CEC, as evidenced by photodegradation byproducts of diclofenac and tetracycline. Therefore, comprehensive risk assessment of microalgae-based treatment must include identification and toxicity evaluation of transformation products, not merely removal efficiency of the parent compound. Long-term studies are needed to confirm that TPs do not accumulate in biomass or effluent, ensuring environmental safety.

5.4. Microalgae for Heavy Metal Remediation

This worldwide environment is seriously threatened by toxic metal contamination. Metals that are heavy including arsenic, lead, cadmium, and mercurial are poisonous and dangerous that cause serious risks to human health and environmental biodiversity. Microalgae can be used to reduce heavy metal contamination, as previous research (Yan et al., 2021) has shown. Four algae species were used by Meichen (2020) to remove iron and manganese Heavy metals extracted from shallow groundwater in a mining region. *Microcystis aeruginosa* constituted the most effective at removing these metals from the water. By using algal adsorption and cation exchange, *M. aeruginosa* makes it easier to remove heavy metals from groundwater. By releasing extracellular polysaccharides to stop extracellular adsorption, secreting cell-surface organic ligands like Trehalose, Phytochelatins, Glutamic Acid, or altering the DNA production of proteins and transports that regulate consumption, microalgae reduce the toxicity of heavy metals. Tritiated and monounsaturated fatty acids predominate under lead stress. Microalgae can thus be utilized for the bioremediation of lead-contaminated wastewater and simultaneously for biofuel production (Nanda et al., 2021).

Figure 4 illustrates the microalgae-mediated wastewater treatment process. It shows how microalgae uptake nutrients, degrade organic pollutants, and facilitate removing pollutants through wastewater. The image highlights a green and environmentally responsible strategy for water purification and resource recovery.

Figure 4. Microalgae-mediated wastewater treatment process

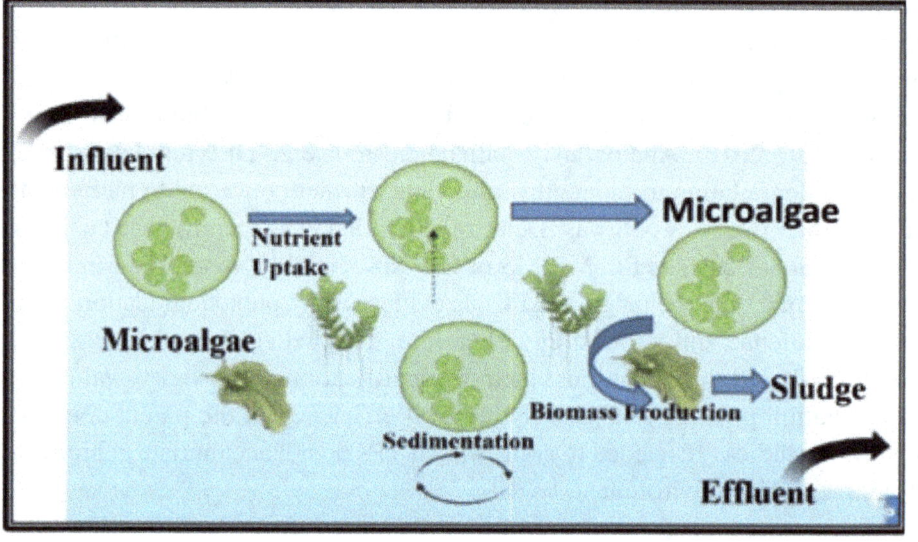

Microalgae can eliminate the heavy metal chromium. Hamouda et al. (2019) utilized *Chlorella sorokiniana* for the elimination of chromium. The amount of 100 ppm Cr (VI) in this study demonstrates a chromium removal efficiency of 99.6793% following three days of exposure. Hedayatkhah et al. (2018) investigated the efficacy of two microalgae as bioremediation agent for dichromate. The most significant decrease in dichromate and lipid formation was observed in both cultures at the peak light intensity of the 55 μmol m^{-2} s^{-1}. Microalgae have diverse capacities of converting acid-reducible Hg^{2+} towards elemental Hg^0 and metacinnabar. Furthermore, microalgae demonstrate remarkable effectiveness in the removal of arsenic from wastewater. Singh et al. (2016) quantified the absorbed arsenic concentration in microalgae as 760 μg/g. *Hydrodictyon reticulatum* and *Oscillatoria sp.* from toxic-contaminated sites exhibited potential for arsenic biological remediation. Researchers employed four strains of Arsenic that may be eliminated from replicated soft-water samples using microalgae. These results demonstrate that *Chlorella minutissima* and *Scenedesmus sp.* IITRIND2 exhibits tolerance to arsenic concentrations of up to 500 mg/L and can absorb arsenic at a rate of up to 161 μg per gram of biomass. Besides direct arsenic recovery from wastewater, microalgae can enhance plant resistance, such as in rice, by inhibiting and modulating uptake by antioxidant enzymes, therefore fostering the tolerance of plants to arsenic (Upadhyay et al., 2016).

The importance of metal speciation for micro-algal remediation cannot be emphasized enough. Metal speciation denotes the different types of chemical species that the metal forms, which could either be its oxidation state, its complex form, or even its organometallic form. This affects bioavailability and toxicity and removal efficiency. In chromium, for instance, hexavalent chromium, or Cr (VI), is highly soluble, mobile, and carcinogenic, while Cr (III) is not toxic, insoluble, and precipitates easily and adsorbs readily on cell walls (Gentili & Fick, 2017). In microalgae like *Chlorella sorokiniana*, reduction of hexavalent chromium into trivalent chromium happens via chromate reductases, and then the metal ions are either sequestered or effluxed inside cells, making their removal rates quite high, as indicated in Table 1. With mercury, microalgae can reduce Hg^{2+} to elemental mercury (volatilization) or metacinnabar (HgS) depending on the sulfides available. In arsenic, the order of toxicity is As (III) > As(V) > Organic arsenic. The microalgae *Chlorella minutissima* and Scenedesmus sp. can absorb both As (III) and As(V); however, the mechanisms vary in terms of the carriers used. Arsenate is absorbed using phosphate transporters, while arsenite is transported using aquaglyceroporins. After absorption, the arsenate is reduced to arsenite, which can then be chelated by phytochelatins and stored in vacuoles or converted into dimethylarsinic acid, which is much less toxic. Not considering the species may result in an overestimation or underestimation of the potential for bioremediation. Thus, in evaluating microalgae's potential for removing heavy metals, it is imperative to consider which chemical species were used since the efficiency of removal will vary greatly between Cr(VI) and Cr (III), Hg^{2+} and methylmercury, or As(III) and As(V).

Table 1 summarizes the ability of various algae species to remove toxic metals from contaminated water. It compares the efficiency of each species in adsorbing or bio-accumulating heavy metals. The table demonstrates the possibilities of technologies centered around microalgae for eco-friendly water purification and heavy metal remediation.

Table 1. Effectiveness of different algae in the elimination of toxic metals

Microalgae species	Metal	Initial pH	Initial metal concentration	Contact time (day)	Removal (%)	References
Chlorella sorokiniana	Cr (VI)	7.0	100 ppm	3	99.68	(Ren et al., 2015)
Microcystis aerugunosa	Fe	9.0	350 μg/L	4	54.13	(Qiao et al., 2020)
Anabaena flosaquae	Mn	9.0	150 μg/L	6	72.71	(Mujtaba et al., 2018)
Desmodesmus sp. CHX1	Cu(II)	6.0	0.41 g/L	4	88.35	(Rosli et al., 2020)
Scenedesmus quadricauda	Ni(II)	6.6	5 mg/L	10 min	66.00	(Moreno Osorio et al., 2019)
Neochloris pseudoalveolaris	Cr(III)	6.6	5 mg/L	10 min	80.60	(Akao et al., 2021)

6. WASTEWATER TREATMENT WITH A MICROALGAE-SLUDGE SYMBIOTIC SYSTEM

Microalgae can take various organic compounds as substrates for carbon and energy to produce lipids. Anaerobic waste microalgae co-culture technique possesses high Substrate transformation and effectiveness of power transmission. The two main solubility fermented byproducts are butyrate and acetic excluding co-culture with most of these compounds. Mujtaba et al. (2018) elaborated that the use of the starch from sweet potatoes wastewater as a substrate yielded the highest rates of COD removal and efficiencies of energy conversion of TN and TP as 88.7%, 80.5%, 34.2%, and 80.1%, in that order. Starch wastewater constitutes an efficient substrate for fermentation which is anaerobic with high levels of Starch, Nitrogen, Phosphorus, and other nutrients. Anaerobic sludge and algae co-culture is a novel approach for soluble compound assimilation to enhance cellular growth and lipid synthesis and thus save energy (Chen et al., 2021).

Besides anaerobic sludge activated sludge process is most important in industrial treatment of wastewater as well as for the well-being of the human beings and hydrological integrity, but the activation of sludge will release three greenhouse gases directly or indirectly, i.e., CO_2, N_2O, and CH_4. The simultaneous co-cultivation of microalgae and sludge can solve this issue. Microalgae emit oxygen during photosynthesis, thus improving the content of dissolved oxygen and lowering N_2O and CH_4 emission. In addition, microalgae possess nitrogen assimilation capability. Experiments (Ji et al., 2019) confirm that COD, TN, and TP removal rate of wastewater treatment by microalgae sludge co-culture could exceed 95% at various light intensities. Mujtaba's et al. (2018) results agree since the elements carbon, phosphorus, and ammonia are all concurrently eliminated within a brief time frame (only two days) through the Co-cultivation of immobilized *Chlorella vulgaris* with activated sludge suspended. Side by side with this, COD removal was not influenced largely by the inoculation rate but at an inoculation rate of 0.5, microalgal biomass was maximal up to 0.76 g/L which shows the need for a proper inoculum volume (Wang et al., 2022).

The dynamics of microbial growth and gas exchange in the microalgae-sludge relationship are complex and must be delicately managed. Although microalgae and activated sludge microorganisms have a symbiotic relationship, there is competition among them over nutrients, light, and other requirements. Sludge bacteria could compete with microalgae over organic carbon, especially under dark or low-light conditions when heterotrophs would flourish, limiting the growth of microalgae and their oxygen generation. Under highly intense light conditions, microalgae would thrive and may even inhibit the activities of the bacteria through oxygen supersaturation or high pH values (up to 10-11 because of CO_2 absorption) (Oh & Tuovinen, 1994). As far as gas exchange is concerned, an equilibrium must exist between the two types of organisms. However, during the nighttime phase, microalgae undergo respiration where they consume O_2 and generate CO_2, causing oxygen depletion and potentially increasing emissions of greenhouse gases such as CO_2, N_2O, and CH_4 due to poor mixing. Additionally, there is high competition for nutrients such as nitrogen and phosphorus (Shi et al., 2014). Although microalgae can absorb both NH_4^+ and NO_3^-, the process of oxidation of NH_4^+ to NO_3^- by nitrifying bacteria requires O_2, while reduction of NO_3^- to N_2 by denitrifying bacteria requires organic carbon. Poor allocation of nutrients leads to incomplete nitrogen removal or accumulation of toxic intermediaries such as nitrite that adversely affect microalgae and aquatic organisms. Predatory protozoans and rotifers that occur in sludge can feed on microalgal cells and reduce their biomass and pollutant removal efficiency. Consequently, it is imperative to optimize inoculation rates, hydraulic retention time, photoperiods, and mixing intensity to reduce competition and promote synergy (Khan et al., 2021).

7. CARRIER-ASSISTED MICROALGAL WASTEWATER TREATMENT

The microalgae remained suspended within the culture medium owing to their diminutive diameter (3–30 μm), poor bulk density, and consistent surface charge for the whole duration. Transportation is a significant issue, and the financial load associated with harvesting operations is exorbitant; microalgae immobilization technology can mitigate this challenge. The process entails creating a thin layer of algae on a solid base and immobilizing it. A recent study revealed that the microalgae biofilms of *Chlorella vulgaris* and the *Scenedesmus vacuolans* can completely the microscopic algae continued to float in the solution used for developing cells owing to their diminutive diameter (3–30 μm), poor bulk density, and consistent surface charge (Wahid et al., 2013). The researchers demonstrated that the treatment of water with microalgae biofilm from Kaduna River water yielded elevated levels of O_2 and H_2O_2, with these rates exhibiting a strong correlation to decomposition of contaminants. The GC–MS analysis of the Microalgal biofilm identified several phytochemical substances as responsible for pollutants breakdown. Additionally, the pretreatment surface characteristics of the biofilm exhibited irregularities, thus demonstrating its ability to depollute water by the process of biosorption. Akao et al. (2021) submerged the control polyether sponge in the same initial sewage liquid sample to facilitate the development of Biofilm on the cushion.

Following a seven-day period, their treatment enhances the removal effectiveness of BOD, nitrogen from ammonia, and phosphates by 80%, 64%, and 95%, respectively, through a reduction in the sponge's hydrophobicity. Chen et al. (2021) using polyurethane sponge as a substrate for microalgae cultivation to purify swine effluents and the analytical assessment indicated that the wastewater outflow from pig farms adhered to local regulations, since the COD and BOD levels were below 600 mg/L and 80 mg/L, respectively. Ji et al. (2019) employed Potassium carbonate (K_2CO_3) was used to activate roasted shellfish powder (calcined mussel shell powder, CMSP), followed by surface treatment of the resulting porous material K-CMSP with a bit of L-arginine to impart a positive charge, which was then utilized as an adsorption carrier for the electrostatic immobilization of microalgae. At 25 °C, the clearance rates for Nitrogen and phosphorus levels were 88.63% and 95.0%, respectively. Among a microalgae- bacteria consortia using foam carriers the levels of general nitrogen substances, ammonia nitrogen emissions, and nitrite ammonia decreased by 29.15%, 51.28%, and 33.48%, accordingly relative to the non-bio-fortified treatment while fish production increased by 13.3% (Wang et al., 2022). The application of the nano biotic hybrid method for wastewater treatment is the result of recent developments in membrane carriers and microalgae.

Wahid et al. (2013) reported that the multilayer graphene-algae composite had a higher performance in nitrate removal than the normal microalgae cells. Microalgae, zinc oxide nanoparticles, and wastewater treatment are utilized. ZnO nanoparticles possess excellent adsorption capabilities owing to elevated surface energy, thus enabling them to form stable hybrid materials with adjacent cells. Low biocompatibility, being in expensive, and lesser toxicity made ZnO-NPs among the most widely used Nanoparticles composed of metal oxide. While ZnO-NPs demonstrate significant cytotoxicity towards microalgae, it is confirmed that a reduced concentration (10 mg/L) of the ZnO-NPs is non-toxic to cells of Microalgal and, in fact, enhances cell proliferation. MAS04-ZnO nanoparticle utilization raises removal of nutrients from wastewater after treatment with the use of zinc oxide nanoparticles and microalgae. Microalgae suspended in water are immobilized with Nano-mixing and membrane carrier technologies. The advantage significantly improves sewage removal efficiency and enables microalgae biomass recovery from treated wastewater, making the cost of recovery less expensive. Apart from that, research has confirmed immobilized micro algae in wastewater treatment as being highly resistant to toxins (Formosa-Dague et al., 2016).

Microalgal-carrier material interactions occur through physical and chemical forces responsible for the attachment of microalgae to the carrier material followed by biofilm development. The first step involves attachment mediated by electrostatic forces, van der Waals forces, and hydrophobicity. Most microalgal cells are negatively charged because of carboxyl, hydroxyl, and phosphate functionalities on the cell wall. This indicates that carriers bearing either positive charges or hydrophobic properties will ensure better and more rapid attachment. According to Ji et al. (2019), treatment of calcined mussel shell powder (K-CMSP) with L-arginine creates positive charges, leading to efficient electrostatic binding of negatively charged microalgae. Once attached reversibly, the microalgae release extracellular polymeric substances, including mainly polysaccharides, proteins, and humic-like compounds, securing irreversible attachment. The EPS also assists in cell-cell communication, matrix stabilization, and shear stress tolerance (Gonçalves et al., 2017). In contrast to surface attachment, entrapment refers to the physical encapsulation of microalgae within the pores of carriers, including polyurethane sponges, polyether foams, or alginate beads. Entrapped microalgae cells are resistant to grazing, toxic shock, and hydraulic washing, yet may face issues related to light transmission and mass diffusion, particularly when carriers are thick or densely filled. The decision about whether to apply attachment or entrapment is guided by the objective of the process; thus, biofilm attachment is used for applications where it is important to have high oxygen productivity and contact with pollutants, while entrapment can be applied for applications requiring long-term residence and protection of sensitive strains (Koupaie et al., 2021). Porosity, pore diameter, surface texture, and hydrophobicity are some of the design characteristics of carriers that require attention. Ideally, carriers should combine high porosity with sufficient light transmission and mass transfer capacity. In the case of macroporous carriers, i.e., those with pore size exceeding 100 μm, microalgae can colonize the external surface of the carriers as well as their interior, which enhances the biomass density without limiting light transmission (Nanda et al., 2021).

8. WASTEWATER TREATMENT BY THE CO-CULTIVATION OF MICROALGAE AND MICROORGANISMS

8.1. Wastewater Treatment Using a Microalgae–Fungus Co-Culture

The co-culture of filamentous fungi with microalgae is a new effective microalgae harvesting method because filamentous fungi possess the ability to trap microalgae by mycelial contact. Biomass of fungal-microalgae consortium was several-fold higher than monoculture, and as discovered in the study. Extracellular fungal enzymes possess the capacity to decompose hard natural substances to soluble nutrients and carbon dioxide and therefore facilitate the ingestion of Microalgal cells. Unlike oxygen, microalgae cells produce Oxygen to enable fungal respiration. As stated by Weixing Cao et al. (2017), the highest efficiencies of *Chlorella vulgaris* and *Ganoderma lucidum* were 78.09%, 86.24%, and 86.74%, respectively. Fungal hyphae co-culture with microalgae is intricate *via* Proteins Interface Connection, Electrical Neutralization, and Extracellular Polysaccharide Adhesion enabling attachment of the fungal hyphae to microalgae, which eventually leads to co-spheroidization of the two organisms. Wang et al. (2021) demonstrated that symbiotic relationships of the fungal microalgae yielded significantly higher total extracellular polysaccharides compared to the control microalgae and fungi. It is crucial for mitigating heavy metal toxicity and defending against metal-induced stress. Symbiotic interaction of microalgae-fungi is that enhances growth in biomass by giving the microalgae high specific surface area, high mass transfer rates, and good mechanical strength. Large cell particles of microalgae facilitate higher filtration harvesting and therefore lower harvesting cost. Microalgae to fungi concentration regulates Microalgal biomass yield. Gultom et al. (2014) demonstrated that the optimal microalgae harvest efficiency, above 90%, was achieved with a 1:300 *Aspergillus niger* towards *Chlorella vulgaris* ratio.

Fungal microalgae pelletization under varying pH conditions remains significant in co-cultivation processes. Fungal-myceal palletisation refers to the aggregation of fungi into compact spherical formations which encapsulate microalgae. This process relies greatly on pH conditions as fungi pellets form optimally when pH ranges between 4.0 and 5.5. At such an environment, fungi produce long branches with high surface charge capability that attracts the negatively charged cells of microalgae through electrostatic interactions (Solé & Matamoros, 2016). Besides, acidity also ensures that bacteria growth is minimized in the system as well as increased functionality of extracellular enzymes of fungi. In case where pH rises above 7.0, fungi mycelia become clumped and scattered, making the formation of pellets impossible. Also, higher pH levels tend to interfere with the growth of fungi as most fungi are acidophilic or neutrophilic. Hence, pH should be kept at 4.0 to 5.5 (Rawat et al., 2016).

The specific enzymes associated with fungal-microalgae interactions are the enzymes secreted by fungi as well as the stress reactions of the microalgae. In fungi, there are extracellular hydrolytic enzymes produced by the fungi which act on complex organic compounds to convert them to simpler compounds to ensure availability of nutrients to microalgae (Matamoros et al., 2016). The enzymes include: (i) cellulases (endoglucanases, exoglucanases, and β-glucosidases)—acts on cellulose from either plant-based or algal wastewater; (ii) proteases—hydrolyses proteins to amino acids and peptides; (iii) amylases—which acts on starches and converts them to fermentable sugars; (iv) lipases—convert lipids to fatty acids and glycerol; and (v) chitinases—involved in modification of fungal cell walls during pellet formation. In microalgae, when there are metals present, the stress reaction involves induction of antioxidant enzymes like SOD, CAT, and GPx. Moreover, microalgae can produce enzymes such as laccase and peroxidase, which play an important role in the biodegradation of refractory organic contaminants. The combined action of enzymes from fungi and microalgae increases the decomposition of pollutants, and thus, improves the efficacy of sewage treatment (Gonçalves et al., 2017).

8.2. Wastewater Treatment Using a Microalgae–Yeast Co-Culture

Yeast has wide ranges of applications in Medicine, Brewing, and Baking. Sedimentation and dehydration properties of yeast are also good; therefore, biomass separation becomes economical and efficient. Lim et al. (2003) proved in their experiments that yeast can remove 70% COD from industrial wastewater within 6.3 hours. Yeast was also very effective alone for wastewater removal in previous work. Some researchers (Walls et al., 2019) used a microalgae-yeast co-culture as recently as wasted for wastewater treatment and found that the co-culture reduced the nutrients at a very high rate, i.e., 96% nitrates, completely Total Ammonia Nitrogen (TAN) and 93% orthophosphate. These scientists found that Microalgae and the biomass of yeast yield were very low even during pH fluctuation. Co-cultivation of microalgae and yeast has a benefit of utilization of oxygen through consumption by utilization through microalgae production, whereas yeast uses it as an available carbon source and produces carbon dioxide, which is helpful in the photosynthetic processes of microalgae. Yeast and microalgae culture protocol is very important. Zhizhan (2020) categorized microalgae into a stable form mode and co-cultured using *Rhodosporidium sp.* for the treatment of prawn culture effluent in a non-fixed mode. The outcome showed that the suspended yeast and *Chlorella sp.* co-culture system gave the highest nutrient removal efficiency, with efficiencies of 85.47%

for NH^{4+}, 100% for NO^{2-}, 96.09% for NO^{3-}, and 93.38% for PO_4^{3-}, respectively. The microalgae-yeast consortium removes nutrients efficiently, but both microalgae and yeast growth are seriously repressed by pH change. Limited data are available on wastewater microalgae-yeast treatment process and on it again, more studies on other parameters should be conducted (Wang et al., 2020).

Crabtree effect represents an essential physiological aspect related to yeast metabolism and affecting significantly the operation of algae-yeast co-cultures. The Crabtree effect denotes the tendency of specific yeast species such as *Saccharomyces cerevisiae, Rhodosporidium sp.,* some *Candida*, etc., to conduct anaerobic fermentation and produce ethanol despite the presence of oxygen whenever the concentrations of glucose and other fermentable sugars are higher than certain limit (above 5-10 mM). At higher organic carbon concentration levels, yeast stops the oxidative phosphorylation process and resorts to fermentation to convert sugars into ethanol despite the availability of oxygen (Arbib et al., 2014). There is a range of negative consequences associated with the Crabtree effect concerning microalgae-yeast co-cultivation including: (i) ethanol toxicity for microalgae and negative effect on photosynthetic capacity and growth; (ii) reduced biomass generation per mole of carbon used; (iii) decreased carbon dioxide production needed for photosynthesis compared to complete oxidation; and (iv) reduced oxygen consumption by microorganisms at night. The Crabtree effect may be avoided in co-cultures by limiting organic carbon supplementation to less than the critical value or choosing yeast strains that are not Crabtree positive. Fed-batch or continuous cultivation processes with low glucose levels may also ensure that aerobic fermentation does not occur (Upadhyay et al., 2016).

The role of yeast in nutrient cycling in microalgae-yeast symbiosis goes beyond carbon metabolism. Nutrient cycling by yeast takes place in various ways: (i) Assimilation of ammonium – yeast actively absorbs ammonium ions (NH_4^+), using it to form amino acids and nucleic acids. As a result, ammonia concentration becomes lower and does not pose risks to microalgae survival; (ii) Solubilization of phosphate – yeasts produce organic acids, such as citric, gluconic, and oxalic acid, which dissolve phosphate precipitates and provide microalgae with free orthophosphate ions (PO_4^{3-}); (iii) Synthesis of vitamins – yeast produces thiamine, biotin, and cobalamin (B group vitamins), which act as growth factors for most microalgae lacking complete pathways for their biosynthesis; (iv) Supply of CO_2 – yeast breathes down organic carbon to carbon dioxide, used by algae in photosynthesis; (v) Trace metal sequestration – Yeast excretes siderophores and metallothioneins, sequestering iron, zinc, and other trace metals to allow for their absorption by microalgae without causing metal toxicity; and (vi) Breakdown of complex organic molecules – Yeast excretes enzymes like invertase, amylase, and proteases, breaking down polysaccharides, proteins, and lipids into simple monomers (glucose, amino acids, fatty acids), which could be metabolized by yeast as well as microalgae. The above-described physiological characteristics of yeast allow it to become an important partner in the wastewater remediation process involving microalgae, with the only condition being that pH and carbon loading need to be monitored to avoid the Crabtree effect (Walls et al., 2019).

8.3. Wastewater Treatment Using a Microalgae–Bacteria Co-Culture

Co-culturing bacterium and phytoplankton is an emerging field that has the possibility to solve some of the problems associated with bacteria and antifungal methods of treatment in wastewater treatment has been observed. Microalgae-bacteria consortia have been proved by researchers to even tolerate 80 mg/L concentrations of Chlortetracycline, whereas microalgae pure cultures tolerate up to only 60 mg/L concentrations. The collaborative relationship between algae and bacteria will improve the effectiveness of wastewater treatment. Wang et al. (2022) illustrated that the incorporation of bacteria-microalgae systems, in contrast to conventional ponds, resulted in reductions of 17.29% in Total Nitrogen (TN), 29.69% in Total Ammonia Nitrogen (TAN), and 15.92% in nitrite. Qu et al. (2021) examined the treatment of swine sewage water utilizing a bacterial algae symbiotic system including *Chlorella sp.* MA1 and *Coelastrella sp.* KE4, which shown an enhancement in the elimination the removal efficiency of NH_4^+–N increased from 12.79% to 99.52%, while that of PO_4^{3-}–P increased from 35.66% to 96.06%. Table 2 presents the effectiveness of co-cultivating microalgae with microorganisms for wastewater treatment. It highlights how their synergistic interactions enhance nutrient removal, organic pollutant degradation, and overall water quality. The table emphasizes the possibility for combined biological structures for environmentally friendly treatment of wastewater.

Metabolic interaction between microalgae and bacteria is the key to the success of symbiotic associations in treating wastewaters. The classic mutualist interactions include: (i) oxygen exchange — microalgae synthesize O_2 through photosynthesis while aerobic bacteria require it for respiration and decomposition of organics; (ii) CO_2 exchange — bacteria breathe inorganic carbon from organics to CO_2, which is fixed by the microalgae through photosynthesis; (iii) nutrients cycling — bacteria convert organics to NH_4^+ and PO_4^{3-} that can be used by the microalgae; on the other hand, microalgae provide DOC including sugars, amino acids, and organic acids needed by the bacteria; (iv) vitamin exchange — many microalgae are dependent on extrinsic sources of the B group vitamins (B_1, B_7, B_{12}) provided by specific bacteria like *Mesorhizobium* and *Rhodobacteraceae;* (v) phytohormones production — bacteria produce auxins, gibberellins, and cytokinins that help microalgae grow;

and (vi) siderophore-based iron uptake — bacteria use siderophores to chelate Fe^{3+} thus making iron accessible to microalgae.

However, there are several practical constraints for using microalgae-bacteria consortia in applications that should be considered. First, maintaining a constant microbial community structure within the consortium is difficult since environmental factors such as temperature, pH, light intensity, and organic load may alter the balance between the two partners. Second, when scaled up from bench top tests to pilot plants or larger scale systems, hydrodynamic stresses, light gradients, and mixing issues will make proximity of the two partners impossible. Third, natural wastewater comprises a complex matrix with numerous microorganisms, including protozoans and rotifers, as well as pathogens, all of which might compete with or even prey upon the beneficial microbial community members. Fourth, extracting microalgal and bacterial biomass from the consortium requires a cumbersome process since it involves not only small microalgae but also filamentous bacteria. Finally, there may be regulatory and biosafety issues when the bacterial member is a pathogen or an antibiotic-resistant organism, making it unsuitable for outdoor application (Zhang et al., 2018).

Another set of adverse impacts that bacteria have on the microalgae-bacteria process is notable. This is mainly caused by increased population levels due to excess organic carbon, and the problems include: (i) anoxia – oxygen is depleted during the nighttime as the result of bacterial respiration but not photosynthesis. In addition, there may be denitrification and methane formation in these regions; (ii) light limitation – excess biomass will increase the turbidity of the water column, reducing its transparency for light; (iii) nitrogen/phosphorus limitation – heterotrophs may compete with microalgae for ammonium and phosphate ions in organic rich waters; (iv) algicidal effect – some species of bacteria release algicidal substances or enzymes to lyse microalgal cells. (v) toxin secretion — certain bacteria may produce cyanotoxins, lipopolysaccharides, or secondary metabolites that can hinder photosynthesis and growth of microalgae; (vi) alteration in pH — the production of organic acids or CO_2 by bacteria can decrease pH, whereas overconsumption of CO_2 by microalgae may increase pH beyond 10, either of which can negatively impact sensitive members of the consortium; and (vii) biofilm clogging — excessive proliferation of bacteria in attached growth systems can lead to pore clogging on carrier materials or membranes, thereby lowering mass transfer efficiency (Zhang et al., 2020). Hence, the successful utilization of microalgae-bacteria consortia necessitates stringent control of operational factors such as organic loading rate, hydraulic retention time, light intensity, pH, and intermittent harvesting of biomass (Zhizhan, 2020).

Table 2. Wastewater remediation via co-cultivation with microalgae and microorganisms

Microalgae species	Micro Organisms	Species	Wastewater	TN (%)	TP (%)	COD (%)	References
Chlorella vulgaris	Fungi	*Ganoderma lucidum*	Biogas slurry	86.24	86.74	78.09	(Wang et al., 2021)
Chlorella sp.	Fungi	*Penicillium sp.*	Byproduct of the Hydrothermal carbonization	32.91	95.30	64.38	(Su and Jacobsen, 2021)
Chlorella sp.	Fungi	*Chlorella sp.*	Molasses wastewater	67.09	88.39	70.68	(Shi et al., 2014)
Scenedesmus sp.	Yeast	*Barnettomyza californica*	Municipal wastewater	96.00(nitrate)	93.00	97.00	(Mennaa et al., 2015)
Chlorella vulgaris	Yeast	*Rhodosporidium sp.*	Shrimp culture wastewater	85.47(NH^{4+}-N)	93.38	100	(Michelon et al., 2022)
Chlorella vulgaris	Bacteria	*Bacillus sp.* AT07-1	Marine culture wastewater	70.70 (NH^{4+}-N)	90.00	98.10	(Soto et al., 2021)

9. FUTURE PERSPECTIVES OF MICROALGAE FOR CONTAMINANT REMOVAL IN WASTEWATER TREATMENT

The future of algal-based sewage treatment is set to witness revolutionary progress, evolving from an obscure biological treatment technology to a crucial element of sustainable water circularity (Wang et al., 2020). The potential is hinged on the development of comprehensive microalgal biorefinery plants that not only treat sewage but also capture CO_2 emissions and produce a wide range of bioresources for biofuel, bioplastic, and biochemical production. One of the emerging areas is genetic and metabolic engineering, which will see the creation of "designer" strains of algae engineered for efficient pollutant assimilation and detoxification, as well as resistance to complex contaminants such as pharmaceutical compounds and heavy metals (Chenlu, 2019). The other promising area is engineering microbial communities by pairing selected species of algae and bacteria to achieve optimal degradation of various pollutants. Innovative processes will involve hybrid and intensified technologies that integrate micro-algae with more advanced oxidative or membrane processes for reliable sewage treatment with potable water as the product. Moreover, smart monitoring and AI-based control will facilitate real-time adjustment of growth conditions for efficiency and resilience. The broad paradigm shift will encompass not only decontamination but also resource recovery, where nutrients such as nitrogen, phosphorus, and even rare earth metals will be captured and utilized efficiently. For successful implementation of innovations, research should address the issue of scaling by developing modular approaches, conducting thorough techno-economic analyses, and establishing a favorable political environment. In conclusion, the next stage of microalgae treatment systems will include the realization of a closed-loop system addressing water purification, resource scarcity, and greenhouse gases mitigation (Chandra et al., 2021).

10. CONCLUSION

A successful and environmentally friendly method for eliminating Emerging Contaminants (ECs) through sewage is phytoplankton. Their capacity to eliminate steroid medications, nutrition, and individual hygiene items, and heavy metals highlights their multifunctional treatment potential. Microalgal removal mechanisms include bioaccumulation, biosorption, biodegradation, and photodegradation, often enhanced by algal–bacterial interactions. Compared to conventional treatment methods, microalgae- these solutions are economical, beneficial to the environment, and capable of carbon dioxide sequestration. The production of value-added biomass further improves the economic feasibility of this approach. However, factors such as light availability, contaminant concentration, algal species selection, and operational conditions significantly influence removal efficiency. There are still challenges with garbage gathering, large-scale agriculture, and procedure optimization Integration of microalgal systems with existing wastewater treatment technologies can improve overall performance. Additional research should concentrate on reactors design and gene enhancement, and long-term ecotoxicological assessments. Overall, microalgae represent a viable and eco-efficient strategy for addressing emerging contaminants in wastewater treatment.

REFERENCES

Akao, P. K., Singh, B., Kaur, P., Sor, A., Avni, A., Dhir, A., Verma, S., Kapoor, S., Phutela, U. G., Satpute, S., Sharma, S., Avisar, D., Sandha, K. S., & Mamane, H. (2021). Coupled microalgal–bacterial biofilm for enhanced wastewater treatment without energy investment. *Journal of Water Process Engineering, 41*, 102029. DOI: 10.1016/j.jwpe.2021.102029

Arbib, Z., Ruiz, J., Álvarez-Díaz, P. A., Garrido-Pérez, C., & Perales, J. A. (2014). Capability of different microalgae species for phytoremediation processes: Wastewater tertiary treatment, CO_2 bio-fixation and low cost biofuels production. *Water Research, 49*, 465–474. DOI: 10.1016/j.watres.2013.10.036

Cao, W., Wang, X., Sun, S., Hu, C., & Zhao, Y. (2017). Simultaneously upgrading biogas and purifying biogas slurry using cocultivation of *Chlorella vulgaris* and three different fungi under various mixed light wavelength and photoperiods. *Bioresource Technology, 241*, 701–709. DOI: 10.1016/j.biortech.2017.05.194

Chandra, R., Pradhan, S., Patel, A., & Ghosh, U. K. (2021). An approach for dairy wastewater remediation using mixture of microalgae and biodiesel production for sustainable transportation. *Journal of Environmental Management, 297*, 113210. DOI: 10.1016/j.jenvman.2021.113210

Chen, C. Y., Kuo, E. W., Nagarajan, D., Dong, C. D., Lee, D. J., Varjani, S., Lam, S. S., & Chang, J. S. (2021). Semi-batch cultivation of *Chlorella sorokiniana* AK-1 with dual carriers for the effective treatment of full strength piggery wastewater treatment. *Bioresource Technology, 326*, 124773. DOI: 10.1016/j.biortech.2021.124773

Chen, Q., Zhang, L., Han, Y., Fang, J., & Wang, H. (2020). Degradation and metabolic pathways of sulfamethazine and enrofloxacin in *Chlorella vulgaris* and *Scenedesmus obliquus* treatment systems. *Environmental Science and Pollution Research International, 27*(22), 28198–28208. DOI: 10.1007/s11356-020-09008-4

) Chenlu, S. (2019). *Efficient Microalgae Purification and Resource Utilization of Mixed Food Wastewater* [Master's thesis, Tianjin University].

Escapa, C., Coimbra, R. N., Paniagua, S., García, A. I., & Otero, M. (2016). Comparative assessment of diclofenac removal from water by different microalgae strains. *Algal Research, 18*, 127–134. DOI: 10.1016/j.algal.2016.06.008

Formosa-Dague, C., Speziale, P., Foster, T. J., Geoghegan, J. A., & Dufrêne, Y. F. (2016). Zinc-dependent mechanical properties of *Staphylococcus aureus* biofilm-forming surface protein SasG. *Proceedings of the National Academy of Sciences of the United States of America, 113*(2), 410–415. DOI: 10.1073/pnas.1519265113

Ge, L., Deng, H., Wu, F., & Deng, N. (2009). Microalgae-promoted photodegradation of two endocrine disrupters in aqueous solutions. *Journal of Chemical Technology and Biotechnology*, *84*(3), 331–336. DOI: 10.1002/jctb.2043

Gentili, F. G., & Fick, J. (2017). Algal cultivation in urban wastewater: An efficient way to reduce pharmaceutical pollutants. *Journal of Applied Phycology*, *29*(1), 255–262. DOI: 10.1007/s10811-016-0950-0

Gonçalves, A. L., Pires, J. C. M., & Simões, M. (2017). A review on the use of microalgal consortia for wastewater treatment. *Algal Research*, *24*, 403–415. DOI: 10.1016/j.algal.2016.11.008

Gultom, S., Zamalloa, C., & Hu, B. (2014). Microalgae harvest through fungal pelletization—Co-culture of *Chlorella vulgaris* and *Aspergillus niger*. *Energies*, *7*(7), 4417–4429. DOI: 10.3390/en7074417

Hamouda, A. S., Mahmoud, H. M., El-Belely, E. F., Labena, A., & Husien, S. (2019). Absorption of hexavalent chromium by green micro algae *Chlorella sorokiniana*: Live planktonic cells. *Water Practice & Technology*, *14*(3), 515–529. DOI: 10.2166/wpt.2019.034

Hedayatkhah, A., Cretoiu, M. S., Emtiazi, G., Stal, L. J., & Bolhuis, H. (2018). Bioremediation of chromium contaminated water by diatoms with concomitant lipid accumulation for biofuel production. *Journal of Environmental Management*, *227*, 313–320. DOI: 10.1016/j.jenvman.2018.09.011

Ji, L., Song, W., Wei, D., Jiang, D., Cai, L., Wang, Y., Guo, J., & Zhang, H. (2019). Modified mussel shell powder for microalgae immobilization to remove N and P from eutrophic wastewater. *Bioresource Technology*, *284*, 36–42. DOI: 10.1016/j.biortech.2019.03.112

Kargi, F., & Eker, S. (2005). Removal of 2,4-dichlorophenol and toxicity from synthetic wastewater in a rotating perforated tube biofilm reactor. *Process Biochemistry (Barking, London, England)*, *40*(6), 2105–2111. DOI: 10.1016/j.procbio.2004.07.013

Khan, N. A., Nancharaiah, Y. V., & Venkata Mohan, S. (2021). Microalgae-based treatment of emerging contaminants in wastewater. *Bioresource Technology*, *337*, 125429. DOI: 10.1016/j.biortech.2021.125429

Koupaie, E. H., Eskicioglu, C., & Azizi, S. (2021). Role of microalgae in removal of emerging contaminants from wastewater: Mechanisms and applications. *Chemosphere*, *263*, 128305. DOI: 10.1016/j.chemosphere.2020.128305

Krzeminski, P., Tomei, M. C., Karaolia, P., Langenhoff, A., Almeida, C. M. R., Felis, E., Gritten, F., Andersen, H. R., Fernandes, T., Manaia, C. M., Rizzo, L., & Fatta-Kassinos, D. (2019). Performance of secondary wastewater treatment methods for the removal of contaminants of emerging concern implicated in crop uptake and antibiotic resistance spread: A review. *The Science of the Total Environment*, *648*, 1052–1081. DOI: 10.1016/j.scitotenv.2018.08.130

Lim, J., Kim, T., & Hwang, S. (2003). Treatment of fish-processing wastewater by co-culture of *Candida rugopelliculosa* and *Brachionus plicatilis. Water Research*, *37*(9), 2228–2232. DOI: 10.1016/S0043-1354(02)00586-9

Luo, H., Cheng, Y., He, D., Liu, Y., Wang, L., & Lin, H. (2022). Recent advances in microalgae-based technologies for emerging contaminants removal: A review. *The Science of the Total Environment*, *836*, 155643. DOI: 10.1016/j.scitotenv.2022.155643

Matamoros, V., Gutiérrez, R., Ferrer, I., García, J., & Bayona, J. M. (2015). Capability of microalgae-based wastewater treatment systems to remove emerging organic contaminants: A pilot-scale study. *Journal of Hazardous Materials*, *288*, 34–42. DOI: 10.1016/j.jhazmat.2015.02.002

Matamoros, V., Rodríguez, Y., & Albaigés, J. (2016). A comparative assessment of intensive and extensive wastewater treatment technologies for removing emerging contaminants in small communities. *Water Research*, *88*, 777–785. DOI: 10.1016/j.watres.2015.10.058

Matamoros, V., Uggetti, E., García, J., & Bayona, J. M. (2016). Assessment of the mechanisms involved in the removal of emerging contaminants by microalgae from wastewater: A laboratory scale study. *Journal of Hazardous Materials*, *301*, 197–205. DOI: 10.1016/j.jhazmat.2015.08.050

Meichen, W. (2020). *Experimental study on removal of iron and manganese in groundwater by microalgae* [Master's thesis, Anhui University of Technology].

Mennaa, F. Z., Arbib, Z., & Perales, J. A. (2015). Urban wastewater treatment by seven species of microalgae and an algal bloom: Biomass production, N and P removal kinetics and harvestability. *Water Research*, *83*, 42–51. DOI: 10.1016/j.watres.2015.06.007

Michelon, W., Zuchi, I. D. P., Reis, J. G., Matthiensen, A., Viancelli, A., da Cruz, A. C. C., Silva, I. T., Fongaro, G., & Soares, H. M. (2022). Virucidal activity of microalgae extracts harvested during phycoremediation of swine wastewater. *Environmental Science and Pollution Research International*, *29*(19), 13427–13436. DOI: 10.1007/s11356-021-17912-6

Moreno Osorio, J. H., Pinto, G., Pollio, A., Frunzo, L., Lens, P. N. L., & Esposito, G. (2019). Startup of a nutrient removal system using *Scenedesmus vacuolatus* and *Chlorella vulgaris* biofilms. *Bioresources and Bioprocessing*, *6*(1), 1–16. DOI: 10.1186/s40643-019-0259-3

Mujtaba, G., Rizwan, M., Kim, G., & Lee, K. (2018). Removal of nutrients and COD through co-culturing activated sludge and immobilized *Chlorella vulgaris. Chemical Engineering Journal*, *343*, 155–162. DOI: 10.1016/j.cej.2018.03.007

Mulla, S. I., Bagewadi, Z. K., Faniband, Z., Bilal, M., Chae, J. C., Bankole, P. O., Saratale, G. D., Bhargava, R. N., & Gurumurthy, D. M. (2021). Various strategies applied for the removal of emerging micropollutant sulfamethazine: A systematic review. *Environmental Science and Pollution Research International*, *28*(28), 56–72. DOI: 10.1007/s11356-021-14259-w

Nanda, M., Jaiswal, K. K., Kumar, V. M., Vlaskin, S., Gautam, P., Bahuguna, V., & Chauhan, P. K. (2021). Micro-pollutant Pb(II) mitigation and lipid induction in oleaginous microalgae *Chlorella sorokiniana* UUIND6. *Environmental Technology & Innovation*, *23*, 101613. DOI: 10.1016/j.eti.2021.101613

Oh, K. H., & Tuovinen, O. H. (1994). Biodegradation of the phenoxy herbicides MCPP and 2,4-D in fixed-film column reactors. *International Biodeterioration & Biodegradation*, *33*(1), 93–99. DOI: 10.1016/0964-8305(94)90057-4

Olguín, E. J. (2012). Dual purpose microalgae-bacteria-based systems that treat wastewater and produce biodiesel and chemical products within a biorefinery. *Biotechnology Advances*, *30*(5), 1031–1046. DOI: 10.1016/j.biotechadv.2012.05.001

Puhakka, J. A., Melin, E. S., Järvinen, K. T., Koro, P. M., Rintala, J. A., Hartikainen, P., Shieh, W. K., & Ferguson, J. F. (1995). Fluidized-bed biofilms for chlorophenol mineralization. *Water Science and Technology*, *31*(1), 227–235. DOI: 10.2166/wst.1995.0051

Qiao, S., Hou, C., Wang, X., & Zhou, J. (2020). Minimizing greenhouse gas emission from wastewater treatment process by integrating activated sludge and microalgae processes. *The Science of the Total Environment*, *732*, 139032. DOI: 10.1016/j.scitotenv.2020.139032

Qu, W., Zhang, C., Chen, X., & Ho, S. H. (2021). New concept in swine wastewater treatment: Development of a self-sustaining synergetic microalgae-bacteria symbiosis (ABS) system to achieve environmental sustainability. *Journal of Hazardous Materials*, *418*, 126264. DOI: 10.1016/j.jhazmat.2021.126264

Rashid, N., Bharathiraja, B., Ali, M., & Yun, J. (2023). Microalgae in wastewater treatment: Current status and future prospects. *Journal of Environmental Management, 326*, 116673. DOI: 10.1016/j.jenvman.2022.116673

Rawat, I., Kumar, R. R., Mutanda, T., & Bux, F. (2016). Dual role of microalgae: Phycoremediation of domestic wastewater and biomass production for sustainable biofuels production. *Applied Energy, 88*(10), 3411–3424. DOI: 10.1016/j.apenergy.2010.11.025

Ren, H. Y., Liu, B. F., Kong, F., Zhao, L., & Ren, N. (2015). Hydrogen and lipid production from starch wastewater by co-culture of anaerobic sludge and oleaginous microalgae with simultaneous COD, nitrogen and phosphorus removal. *Water Research, 85*, 404–412. DOI: 10.1016/j.watres.2015.08.057

Rizzo, L., Malato, S., Antakyali, D., Beretsou, V. G., Dolic, M. B., Gernjak, W., Heath, E., Ivancev-Tumbas, I., Karaolia, P., Ribeiro, A. R. L., Mascolo, G., McArdell, C. S., Schaar, H., Silva, A. M. T., & Fatta-Kassinos, D. (2018). Consolidated vs new advanced treatment methods for the removal of contaminants of emerging concern from urban wastewater. *The Science of the Total Environment, 655*, 986–1008. DOI: 10.1016/j.scitotenv.2018.11.265

Rosli, S. S., Amalina Kadir, W. N. A., Wong, C. Y., Han, F. Y., Lim, J. W., Lam, M. K., Yusup, S., Kiatkittipong, W., Kiatkittipong, K., & Usman, A. (2020). Insight review of attached microalgae growth focusing on support material packed in photobioreactor for sustainable biodiesel production and wastewater bioremediation. *Renewable & Sustainable Energy Reviews, 134*, 110358. DOI: 10.1016/j.rser.2020.110306

Safonova, E. T., Dmitrieva, I. A., & Kvitko, K. V. (1999). The interaction of algae with alkanotrophic bacteria in black oil decomposition. *Resources, Conservation and Recycling, 27*(1-2), 193–201. DOI: 10.1016/S0921-3449(99)00014-2

Salimi, M., Esrafili, A., Gholami, M., Jafari, A. J., Kalantary, R. R., Farzadkia, M., Kermani, M., & Sobhi, H. R. (2017). Contaminants of emerging concern: A review of new approach in AOP technologies. *Environmental Monitoring and Assessment, 189*(8), 414. DOI: 10.1007/s10661-017-6097-x

Shi, J., Podola, B., & Melkonian, M. (2014). Application of a prototype-scale twin-layer photobioreactor for effective N and P removal from different process stages of municipal wastewater by immobilized microalgae. *Bioresource Technology, 154*, 260–266. DOI: 10.1016/j.biortech.2013.11.100

Shriwastav, A., Thomas, J., & Bose, P. (2017). A comprehensive mechanistic model for simulating algal growth dynamics in photobioreactors. *Bioresource Technology*, *233*, 7–14. DOI: 10.1016/j.biortech.2017.02.080

Singh, N. K., Raghubanshi, A. S., Upadhyay, A. K., & Rai, U. N. (2016). Arsenic and other heavy metal accumulation in plants and algae growing naturally in contaminated area of West Bengal, India. *Ecotoxicology and Environmental Safety*, *130*, 224–233. DOI: 10.1016/j.ecoenv.2016.04.024

Solé, A., & Matamoros, V. (2016). Removal of endocrine disrupting compounds from wastewater by microalgae co-immobilized in alginate beads. *Chemosphere*, *164*, 516–523. DOI: 10.1016/j.chemosphere.2016.08.047

Soto, M. F., Diaz, C. A., Zapata, A. M., & Higuita, J. C. (2021). BOD and COD removal in vinasses from sugarcane alcoholic distillation by *Chlorella vulgaris*: Environmental evaluation. *Biochemical Engineering Journal*, *176*, 108209. DOI: 10.1016/j.bej.2021.108191

Su, Y., & Jacobsen, C. (2021). Treatment of clean in place (CIP) wastewater using microalgae: Nutrient upcycling and value-added byproducts production. *The Science of the Total Environment*, *785*, 147277. DOI: 10.1016/j.scitotenv.2021.147337

Upadhyay, A. K., Singh, N. K., Singh, R., & Rai, U. N. (2016). Amelioration of arsenic toxicity in rice: Comparative effect of inoculation of *Chlorella vulgaris* and *Nannochloropsis* sp. on growth, biochemical changes and arsenic uptake. *Ecotoxicology and Environmental Safety*, *124*, 68–73. DOI: 10.1016/j.ecoenv.2015.10.002

Urbina-Suarez, N. A., Machuca-Martínez, F., & Barajas-Solano, A. F. (2021). Advanced oxidation processes and biotechnological alternatives for the treatment of tannery wastewater. *Molecules (Basel, Switzerland)*, *26*(11), 3222. DOI: 10.3390/molecules26113222

Vassalle, L., García-Galán, M. J., Aquino, S. F., Afonso, R., Ferrer, I., Passos, F., & R Mota, C. (2020). Can high rate algal ponds be used as post-treatment of UASB reactors to remove micropollutants? *Chemosphere*, *248*, 126003. DOI: 10.1016/j.chemosphere.2020.125969

Villar-Navarro, E., García-Galán, M. J., Pérez, S., & Rodríguez-Mozaz, S. (2020). Microalgae-based processes for the removal of pharmaceuticals from wastewater: A review. *Environmental Pollution*, *265*, 115114. DOI: 10.1016/j.envpol.2020.115114

Wahid, M. H., Eroglu, E., Chen, X., Smith, S. M., & Raston, C. L. (2013). Functional multi-layer graphene-algae hybrid material formed using vortex fluidics. *Green Chemistry*, *15*(3), 650–655. DOI: 10.1039/c2gc36892g

Walls, L. E., Velasquez-Orta, S. B., Romero-Frasca, E., Leary, P., Yáñez Noguez, I., & Orta Ledesma, M. T. (2019). Non-sterile heterotrophic cultivation of native wastewater yeast and microalgae for integrated municipal wastewater treatment and bioethanol production. *Biochemical Engineering Journal, 151*, 107319. DOI: 10.1016/j.bej.2019.107319

Wang, C., Jiang, C., Gao, T., Peng, X., Ma, S., Sun, Q., Xia, B., Xie, X., Bai, Z., Xu, S., & Zhuang, X. (2022). Improvement of fish production and water quality in a recirculating aquaculture pond enhanced with bacteria-microalgae association. *Aquaculture (Amsterdam, Netherlands), 547*, 737440. DOI: 10.1016/j.aquaculture.2021.737420

Wang, H., Qi, M., Bo, Y., Zhou, C., Yan, X., Wang, G., & Cheng, P. (2021). Treatment of fishery wastewater by co-culture of *Thalassiosira pseudonana* with *Isochrysis galbana* and evaluation of their active components. *Algal Research, 60*, 102541. DOI: 10.1016/j.algal.2021.102498

Wang, J., Chen, R., Fan, L., Cui, L., Zhang, Y., Cheng, J., Wu, X., Zeng, W., Tian, Q., & Shen, L. (2021). Construction of fungi-microalgae symbiotic system and adsorption study of heavy metal ions. *Separation and Purification Technology, 268*, 118689. DOI: 10.1016/j.seppur.2021.118689

Wang, J., Liu, W., & Yang, J. (2020). Microalgae-yeast co-culture for wastewater treatment: Challenges and perspectives. *Bioresource Technology, 315*, 123789. DOI: 10.1016/j.biortech.2020.123789

Xiong, J. Q., Govindwar, S., Kurade, M. B., Paeng, K. J., Roh, H. S., Khan, M. A., & Jeon, B. H. (2019). Toxicity of sulfamethazine and sulfamethoxazole and their removal by a green microalga, *Scenedesmus obliquus. Chemosphere, 218*, 551–558. DOI: 10.1016/j.chemosphere.2018.11.146

Yan, C., Qu, Z., Wang, J., Cao, L., & Han, Q. (2021). Microalgal bioremediation of heavy metal pollution in water: Recent advances, challenges, and prospects. *Chemosphere, 286*, 131870. DOI: 10.1016/j.chemosphere.2021.131870

Zhang, B., Wei, L., Yuan, G., Zhang, Z., Shi, W., Cui, F., Lens, P. N. L., & Tay, J. H. (2020). Microalgal-bacterial consortia: From interspecies interactions to biotechnological applications. *Renewable & Sustainable Energy Reviews, 118*, 109563. DOI: 10.1016/j.rser.2019.109563

Zhang, Y., Roostaei, J., Asadi, A., Doosti, Z., Ewing, M., Guan, J., Uddin, T., & van Wagnen, J. (2018). *Stratified multilayer algal-biofilm reclamation technology. USA Science and Engineering Festival.* Wayne State University.

Zhizhan, L. (2020). *Application of microalgae yeast co-culture system in shrimp culture wastewater treatment*. Guangzhou University.

Chapter 7
Microalgal Adsorption of Emerging Pollutants: An Analytical and Mechanistic Perspective

Eshraq Aljamal
http://orcid.org/0009-0002-1521-7918
Irbid National University, Jordan

ABSTRACT

Microalgal adsorption offers a sustainable approach to removing emerging contaminants from wastewater systems. This chapter provides an analytical evaluation of microalgal systems, focusing on key physicochemical factors governing surface interactions, including functional group chemistry, surface charge behavior, and solution-dependent speciation. The discussion integrates equilibrium, kinetic, and thermodynamic modeling with analytical techniques such as Fourier-transform infrared spectroscopy and zeta potential analysis to support mechanistic interpretation. Distinctions between surface adsorption and intracellular bioaccumulation are clarified to improve conceptual accuracy. The chapter also examines the limitations of classical modeling approaches in complex wastewater systems and highlights the influence of dissolved organic matter and competing ions. Finally, microalgal treatment systems are discussed within sustainability frameworks, emphasizing scalability, reproducibility, and environmental integration.

DOI: 10.4018/979-8-2600-3080-6.ch007

1. INTRODUCTION

1.1 Emerging Pollutants in Modern Wastewater Systems

Wastewater systems have evolved into highly complex chemical environments characterized by the coexistence of diverse classes of contaminants collectively referred to as emerging pollutants. These include pharmaceuticals and personal care products (PPCPs), endocrine-disrupting compounds (EDCs), synthetic dyes, trace metals, and microplastics, many of which remain insufficiently regulated despite increasing evidence of their ecological and human health risks (aus der Beek et al., 2016; Daughton & Ternes, 1999; Lapworth et al., 2012).

A defining feature of emerging pollutants is not only their occurrence at trace concentrations but also their pseudo-persistent nature, resulting from continuous discharge and incomplete removal in conventional treatment systems (Deblonde et al., 2011). Importantly, these contaminants do not exist as isolated species; rather, they coexist and interact within multi-component wastewater matrices. Such interactions significantly influence their physicochemical behavior, including speciation, mobility, and bioavailability (Fu & Wang, 2011; Richardson & Kimura, 2020).

For example, pharmaceutical compounds such as antibiotics can form complexes with dissolved metal ions (e.g., Cu^{2+} or Fe^{3+}), altering their transport behavior and resistance to removal processes. Similarly, microplastics can act as vectors for hydrophobic organic pollutants and metal ions, enhancing their environmental mobility and ecological impact (Fu & Wang, 2011; Koelmans et al., 2016). Furthermore, dissolved organic matter (DOM) adds an additional layer of complexity by exerting a dual influence: it can enhance pollutant transport through complexation processes, while also inhibiting adsorption by competing with contaminants for available active binding sites (La Farre et al., 2008; Luo et al., 2014).

1.2 Limitations of Conventional Treatment Technologies

Conventional wastewater treatment plants are primarily designed to remove suspended solids, biodegradable organic matter, and nutrients through physical separation and biological degradation processes. While effective for these conventional targets, such systems often exhibit limited efficiency in removing emerging pollutants, particularly those that are chemically stable, poorly biodegradable, or present at trace concentrations (Luo et al., 2014; Zhou et al., 2025). To address these limitations, advanced treatment technologies such as activated carbon adsorption, membrane filtration, and advanced oxidation processes have been widely implemented. Although these approaches can enhance pollutant removal, their application is frequently constrained by high energy demand, operational complexity, and the generation of secondary waste streams (Crini & Lichtfouse, 2019; Rodriguez-Narvaez et al., 2017).

Furthermore, many of these technologies are optimized under controlled laboratory conditions and may not perform consistently in real wastewater systems characterized by fluctuating composition, high ionic strength, and significant levels of dissolved organic matter. For instance, advanced oxidation processes that demonstrate high efficiency in laboratory conditions may exhibit reduced performance in actual wastewater due to interference from organic matter, pH variability, and competing contaminants (Wen et al., 2023).

These limitations underscore the need for alternative treatment strategies that are not only effective but also adaptable, sustainable, and resilient under realistic environmental conditions.

1.3 Microalgae as Multifunctional Agents for Sustainable Remediation

Within this context, microalgae have emerged as promising and sustainable remediation agents capable of integrating pollutant removal with resource recovery. As photosynthetic microorganisms, microalgae assimilate nutrients, capture carbon dioxide, and generate biomass while interacting with a wide range of contaminants present in wastewater systems (Rawat et al., 2011; L. Wang et al., 2010).

Microalgal systems operate through multiple removal mechanisms, primarily surface-mediated biosorption and intracellular bioaccumulation. From a physicochemical perspective, microalgal cell walls are enriched with reactive functional groups, such as carboxyl, hydroxyl, phosphate, and amine moieties that facilitate interactions with metal ions, dyes, and charged organic compounds (Volesky, 2007; J. Wang & Chen, 2009). These interactions are governed by electrostatic attraction, ion exchange, and surface complexation mechanisms, which are strongly influenced by environmental parameters such as pH, ionic strength, and the presence of competing species.

In addition to their pollutant removal capabilities, microalgae offer significant sustainability advantages. The biomass generated during treatment can be valorized into biofuels, biofertilizers, or other value-added products, aligning microalgal systems with circular bioeconomy principles (Abdel-Raouf et al., 2012; Trivedi et al., 2015)

However, the performance of microalgal systems in real wastewater environments remains highly variable due to system complexity and environmental fluctuations. Factors such as temperature variation, light intensity, nutrient availability, and microbial competition can significantly influence both microalgal growth and pollutant removal efficiency (Velásquez-Orta et al., 2024)

1.4 Knowledge Gaps and the Need for an Analytical–Mechanistic Approach

Despite the growing body of literature on microalgal pollutant removal, many studies remain predominantly performance-oriented, focusing on removal efficiencies without sufficient mechanistic interpretation. Adsorption processes are often described using empirical models such as Langmuir and Freundlich isotherms, which assume idealized conditions and homogeneous surface properties. However, these assumptions rarely hold in complex wastewater systems (Monteiro et al., 2012).

As systems transition from controlled laboratory conditions to real-world environments, the presence of dissolved organic matter, competing ions, and multi-contaminant interactions can significantly alter adsorption behavior, leading to discrepancies between predicted and observed performance.

From an analytical chemistry perspective, addressing these limitations requires the integration of adsorption studies with advanced characterization techniques. Analytical tools such as Fourier-transform infrared (FTIR) spectroscopy, Brunauer–Emmett–Teller (BET) surface analysis, zeta potential measurements, and electron microscopy provide critical insights into surface functionality, charge behavior, and adsorption mechanisms (Foo & Hameed, 2010; Volesky, 2007; J. Wang & Chen, 2009). Without such analytical validation, biosorption studies remain largely descriptive rather than predictive (Michalak et al., 2013).

1.5. Objectives and Scope of the Chapter

This chapter provides an analytically grounded framework for understanding microalgal adsorption of emerging pollutants, integrating physicochemical principles, analytical characterization, and the complexity of environmental systems.

Specifically, it aims to: (1) examine adsorption mechanisms for metals, dyes, and particulate pollutants, (2) highlight the role of advanced analytical techniques, (3) evaluate limitations of conventional models in multi-component systems, (4) compare microalgal systems with conventional adsorbents, and (5) assess sustainability aspects, including life cycle analysis and circular bioeconomy considerations.

2. BACKGROUND

Emerging pollutants represent a persistent and complex challenge in aquatic environments due to their continuous input from domestic, industrial, and agricultural sources and their incomplete removal in conventional treatment systems (aus der Beek et al., 2016; Lapworth et al., 2012). Despite occurring at trace levels, their pseudo-persistence and cumulative exposure raise concerns about long-term ecological and human health effects (Luo et al., 2014; Richardson & Kimura, 2020). Their behavior in multi-component systems is governed by processes such as speciation, complexation, and competitive interactions, which are strongly influenced by environmental conditions, often limiting the effectiveness of purely biological treatment approaches (Rawat et al., 2011). Microalgae have emerged as promising remediation agents by integrating metabolic activity with surface-mediated physicochemical interactions, enabling the removal of diverse contaminants under environmentally sustainable conditions (Abdel-Raouf et al., 2012; Markou et al., 2014). In addition, microalgal systems support circular bioeconomy concepts through biomass valorization into bioenergy and other value-added products (Trivedi et al., 2015). However, significant variability in reported removal efficiencies reflects the lack of robust mechanistic frameworks capable of predicting system behavior under realistic conditions, highlighting a key barrier to scalable application.

2.1. Application of Microalgae in Wastewater Treatment

Microalgae have been widely applied in wastewater treatment, initially for nutrient removal and oxygen production, and more recently for the removal of diverse contaminants, including heavy metals, dyes, and pharmaceutical residues (Abdel-Raouf et al., 2012; De-Bashan & Bashan, 2010; Rawat et al., 2011). Unlike conventional biological systems that rely primarily on metabolic degradation, microalgal systems combine biological and physicochemical mechanisms, including surface adsorption and intracellular uptake (Monteiro et al., 2012; J. Wang & Chen, 2009). However, removal efficiency is strongly influenced by species characteristics, cultivation conditions, wastewater composition, and environmental factors such as pH, ionic strength, and temperature. In complex matrices, dissolved organic matter (DOM) and competing ions can significantly alter adsorption behavior and reduce performance (He & Chen, 2014; Markou et al., 2014). These limitations highlight the need for mechanistic approaches capable of predicting system behavior under realistic conditions.

2.2. Classification of Emerging Pollutants

Emerging pollutants comprise a chemically diverse group of contaminants that are increasingly detected in wastewater effluents and aquatic environments, despite not being routinely regulated (aus der Beek et al., 2016; Lapworth et al., 2012). While classification is often presented descriptively, its primary significance lies in its mechanistic implications for pollutant behavior and treatment interactions. Key physicochemical properties, including ionic charge, molecular size, hydrophobicity, and structural complexity, govern the interaction of pollutants with microalgal biomass and ultimately determine removal pathways. These relationships are particularly critical in adsorption-dominated systems, where surface interactions are highly sensitive to both molecular structure and environmental conditions (Luo et al., 2014).

2.2.1. Heavy Metals

Heavy metals are among the most extensively studied contaminants in microalgal systems due to their persistence, non-biodegradability, and toxicity even at low concentrations (Volesky, 2007). Their interaction with microalgal biomass is primarily mediated through binding to functional groups such as carboxylate, phosphate, and hydroxyl moieties located on the cell wall (Wang & Chen, 2009).

These interactions are predominantly adsorption-driven and are strongly influenced by solution chemistry, including pH, ionic strength, and metal speciation. In multi-component systems, competitive interactions among metal ions and complexation with dissolved ligands can significantly affect adsorption capacity and selectivity, highlighting the limitations of single-solute experimental models (Fu & Wang, 2011; He & Chen, 2014; Monteiro et al., 2012).

2.2.2. Synthetic Dyes

Synthetic dyes are structurally complex organic compounds widely used in industrial applications and are known for their resistance to biodegradation (Yagub et al., 2014). In microalgal systems, dye removal is primarily governed by surface-mediated interactions, including electrostatic attraction, π–π interactions, and hydrophobic effects.

The adsorption behavior of dyes depends strongly on their molecular structure, charge characteristics, and environmental conditions. In complex wastewater systems, competition with other organic compounds and interactions with dissolved organic matter can significantly influence removal efficiency and adsorption dynamics (De-Bashan & Bashan, 2010).

2.2.3. Pharmaceuticals and Personal Care Products (PPCPs)

PPCPs represent a highly diverse group of biologically active compounds designed to exert specific physiological effects, many of which exhibit resistance to conventional wastewater treatment processes (aus der Beek et al., 2016). Their widespread occurrence in aquatic environments has raised concerns regarding chronic exposure and long-term ecological impacts (Richardson & Kimura, 2020). The interaction of PPCPs with microalgal biomass is highly compound-specific and governed by factors such as ionization state, hydrophobicity, and molecular structure. These compounds may undergo surface adsorption, limited intracellular uptake, or partial transformation, depending on environmental conditions and microalgal physiology (Hejna et al., 2022; Rawat et al., 2011).

2.2.4. Microplastics and Nanoplastics

Microplastics and nanoplastics differ fundamentally from dissolved contaminants due to their particulate nature and colloidal behavior. Their interaction with microalgal systems is governed primarily by physical processes such as attachment, aggregation, and entrapment rather than classical molecular adsorption (Bhattacharya et al., 2010).

In addition, these particles can act as vectors for co-contaminants, influencing the transport and bioavailability of associated pollutants in aquatic systems (Koelmans et al., 2016; Rochman et al., 2013). Such interactions are dominated by colloidal-scale forces, including electrostatic interactions, van der Waals forces, and steric effects, which limit the applicability of conventional adsorption models and necessitate alternative analytical approaches.

2.3. Adsorption vs. Bioaccumulation in Microalgal Systems

Adsorption and bioaccumulation are widely recognized as the two primary mechanisms governing pollutant removal in microalgal systems (Volesky, 2007). Adsorption involves the rapid association of contaminants with functional groups located on the external cell surface and is largely independent of metabolic activity. In contrast, bioaccumulation refers to intracellular uptake processes that depend on cellular metabolism and physiological state.

For many inorganic contaminants and synthetic dyes, adsorption is considered the dominant removal pathway, particularly under short-term conditions where surface interactions prevail (Monteiro et al., 2012). Because adsorption is governed primarily by physicochemical parameters, it is often more predictable and reproducible than metabolically driven processes.

By contrast, bioaccumulation is highly sensitive to environmental conditions, cellular viability, and metabolic activity, making it more variable and less predictable across systems (De-Bashan & Bashan, 2010). In complex wastewater environments, stress conditions and chemical heterogeneity may suppress metabolic activity, thereby shifting removal mechanisms toward adsorption-dominated pathways.

Distinguishing between these mechanisms is therefore essential for interpreting adsorption models and assessing the scalability of microalgal treatment systems.

This distinction is not merely conceptual but has direct implications for process design and scalability, as adsorption-based systems can operate with non-living biomass, whereas bioaccumulation requires viable and metabolically active cultures under controlled environmental conditions.

2.4. Overview of Adsorption Studies Using Microalgae

A significant portion of the literature on microalgal pollutant removal focuses on adsorption-based mechanisms, typically analyzed using equilibrium, kinetic, and thermodynamic models (Volesky, 2007; Wang & Chen, 2009). These models are widely used to estimate adsorption capacity, infer surface properties, and identify rate-controlling steps.

However, most studies are conducted under simplified laboratory conditions using single-solute systems, which do not adequately capture the complexity of real wastewater matrices. In multi-component systems, adsorption behavior is strongly influenced by competition for binding sites, interactions with dissolved organic matter, and variations in ionic strength.

Consequently, classical models such as Langmuir and Freundlich isotherms often fail to accurately predict adsorption behavior in real systems unless modified to account for heterogeneous and competitive interactions.

Moreover, the mechanistic validity of adsorption models depends heavily on complementary analytical characterization. Without integrating surface analysis techniques, such as spectroscopy and microscopy, adsorption modeling remains largely descriptive and offers limited predictive capability.

These limitations highlight the need for integrative approaches that combine mechanistic understanding, advanced analytical tools, and validation under realistic environmental conditions. Such integration is essential for bridging the gap between laboratory-scale studies and practical implementation, as well as for informing reactor design, process optimization, and regulatory acceptance.

Although equilibrium and kinetic models provide valuable insights into adsorption mechanisms under controlled laboratory conditions, their direct applicability to real wastewater systems remains inherently limited. Most classical models, including Langmuir and Freundlich isotherms, are developed based on single-solute systems in simplified aqueous matrices. In contrast, real wastewater represents a highly complex, multi-component environment characterized by dissolved organic matter (DOM), competing cations and anions, variable ionic strength, and fluctuating physicochemical conditions. These factors introduce competitive adsorption, site blocking, and complexation, significantly altering adsorption behavior and reducing model predictability. As a result, model parameters obtained from synthetic systems often fail to accurately describe pollutant removal in real effluents, necessitating the use of modified or multi-component isotherm approaches and emphasizing the need for experimental validation under realistic conditions

3. PHYSICOCHEMICAL BASIS OF MICROALGAL ADSORPTION

The adsorption behavior of microalgal biomass toward emerging pollutants is governed by a dynamic interplay between surface chemistry, solution conditions, and pollutant speciation. Unlike conventional inert adsorbents, microalgae possess chemically heterogeneous and biologically derived surfaces that enable multiple interaction pathways—including electrostatic attraction, ion exchange, surface complexation, and hydrogen bonding—to operate concurrently (Volesky, 2007; J. Wang & Chen, 2009).

This intrinsic heterogeneity implies that adsorption cannot be interpreted as a single-process phenomenon, but rather as the cumulative outcome of coupled physicochemical interactions. Consequently, adsorption performance is highly sensitive to both intrinsic biomass properties and external environmental conditions.

A mechanistic understanding of these interactions is essential for interpreting experimental results, enabling predictive modeling, and supporting the rational design of adsorption systems under realistic wastewater conditions.

3.1 Microalgal Cell Wall Composition and Functional Binding Sites

Microalgal cell walls consist of a structurally complex matrix composed primarily of polysaccharides, proteins, and lipids, enriched with reactive functional groups such as carboxyl (–COOH), hydroxyl (–OH), phosphate ($-PO_4^{3-}$), sulfate ($-SO_4^{2-}$), and amine ($-NH_2$) moieties (Davis et al., 2003; Romera et al., 2006).

These functional groups serve as the primary binding sites governing adsorption capacity and selectivity. However, their contribution is not fixed; rather, it is dynamically influenced by solution chemistry, pollutant properties, and environmental conditions.

Carboxyl and phosphate groups typically dominate the adsorption of metal cations due to their strong affinity for positively charged species, particularly under circumneutral conditions where deprotonation results in a negatively charged surface (Gadd, 2009; Michalak et al., 2013). In contrast, hydroxyl and amine groups are more relevant for interactions with organic contaminants through hydrogen bonding and dipole-driven interactions (Gupta & Rastogi, 2009).

Importantly, adsorption capacity depends not only on the presence of these functional groups but also on their accessibility, spatial distribution, and chemical reactivity. These characteristics vary across microalgal species and cultivation conditions, explaining the variability reported in adsorption performance (Monteiro et al., 2012). The diversity of functional groups and their corresponding interaction pathways introduces significant surface heterogeneity, which directly governs adsorption selectivity and mechanism. These relationships are summarized in Table 1.

Table 1. Major surface functional groups of microalgae and their dominant adsorption mechanisms

Functional group	Primary surface origin	Charge behavior (pH dependent)	Dominant interaction mechanisms	Representative target pollutants	Key references
–COOH (Carboxyl)	Cell wall polysaccharides, uronic acids	Deprotonates above pKa → negatively charged surface; protonated at low pH	Electrostatic attraction, ion exchange, surface complexation	Pb^{2+}, Cd^{2+}, Cu^{2+}, Zn^{2+}	(Davis et al., 2003; Volesky, 2007)
$–PO_4^{3-}$ (Phosphate)	Phospholipids, phosphorylated biomolecules	Strongly negative under circumneutral to alkaline conditions; partial protonation at low pH	Surface complexation, ligand exchange	Cu^{2+}, Zn^{2+}, Ni^{2+}	(Gadd, 2009; Michalak et al., 2013)
–OH (Hydroxyl)	Structural polysaccharides, cellulose-like components	Weakly ionizable; participates mainly in neutral form	Hydrogen bonding, van der Waals interactions	Synthetic dyes, PPCPs	(Gupta & Rastogi, 2009)
$–NH_2$ (Amine)	Cell wall proteins, peptides	Protonated at low pH → positively charged; neutral at higher pH	Dipole–dipole interactions, coordination bonding	Pharmaceuticals, anionic dyes	(Fomina & Gadd, 2014)
$–SO_4^{2-}$ (Sulfate)	Sulfated polysaccharides (species-dependent)	Negatively charged across broad pH range	Electrostatic attraction, weak complexation	Divalent metal ions	(Romera et al., 2006)

Furthermore, protonation–deprotonation equilibria control the charge state of these functional groups, making parameters such as the point of zero charge (pHpzc) critical descriptors of adsorption behavior. In multicomponent systems, competitive binding further modulates adsorption selectivity, reinforcing the interpretation of adsorption as a dynamic and competitive process.

3.2 Surface Charge, Zeta Potential, and pH Effects

Surface charge is a key determinant of adsorption selectivity and efficiency in microalgal systems and is commonly evaluated through zeta potential measurements (Hunter, 1981).

Most microalgal species exhibit negative surface charge under neutral to mildly alkaline conditions due to the deprotonation of acidic functional groups (Ozkan & Berberoglu, 2013). This enhances adsorption of cationic species while limiting interactions with anionic pollutants.

The influence of pH arises from its dual role in controlling both surface chemistry and pollutant speciation. At low pH, protonation reduces binding-site availability and introduces competition between protons and metal ions. As pH increases, enhanced deprotonation strengthens electrostatic attraction and coordination interactions until precipitation or competing processes become significant (Sheng et al., 2004; Volesky & Holan, 1995).

For organic contaminants, pH influences molecular ionization and intermolecular interactions, while for particulate pollutants such as microplastics, it affects aggregation and colloidal stability. Figure 1 shows that adsorption depends on the interplay of surface charge, functional groups, and pathways, emphasizing system-level interactions over isolated equilibrium processes.

Figure 1. shows adsorption depends on solution conditions affecting surface charge and binding sites; combining surface data with modeling is crucial for accurately interpreting microalgal systems.

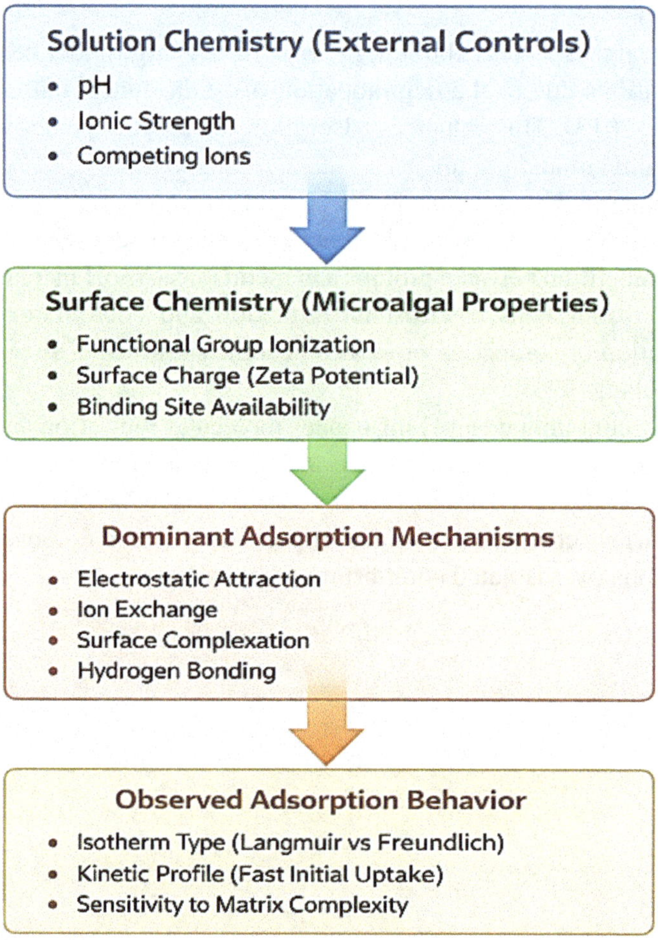

3.3 Multimodal Adsorption Mechanisms

Adsorption in microalgal systems is inherently multimodal, with pollutant uptake resulting from the simultaneous contribution of multiple physicochemical mechanisms (Fomina & Gadd, 2014).

Electrostatic attraction governs interactions between oppositely charged species, while ion exchange involves the displacement of loosely bound ions associated with surface functional groups. Surface complexation represents a more specific interaction involving coordination between metal ions and electron-donating ligands.

In contrast, interactions with organic pollutants are typically governed by hydrogen bonding and weak intermolecular forces. For particulate contaminants such as microplastics, physical processes, including attachment, aggregation, and entrapment become dominant.

The coexistence of these mechanisms explains the commonly observed adsorption behavior characterized by a rapid initial uptake phase followed by slower diffusion-controlled stages.

This complexity highlights the importance of integrating adsorption modeling with analytical characterization techniques (e.g., FTIR, XPS) to provide mechanistic validation.

3.4 Quantitative Interpretation Using Isotherm and Kinetic Models

Equilibrium and kinetic models are widely used to describe adsorption behavior and extract quantitative parameters. The Langmuir isotherm assumes monolayer adsorption on homogeneous surfaces, whereas the Freundlich model accounts for surface heterogeneity and multilayer adsorption (Foo & Hameed, 2010). Kinetic models such as pseudo-first-order and pseudo-second-order equations are commonly applied. Although pseudo-second-order models often provide better fits, this does not necessarily indicate a specific adsorption mechanism (Ho & McKay, 1999). Model parameters should therefore be interpreted cautiously and supported by physicochemical evidence.

3.5 Effects of Ionic Strength and Competitive Interactions

In realistic wastewater systems, adsorption occurs in complex, multicomponent environments where ionic strength, competing ions, and dissolved organic matter significantly influence performance.

Increased ionic strength reduces adsorption efficiency through electrostatic screening, while competing ions-particularly multivalent cations, can occupy binding sites and reduce adsorption selectivity (Kratochvil & Volesky, 1998).

Dissolved organic matter further modifies adsorption behavior by altering pollutant speciation or blocking active sites. The combined effects of solution chemistry parameters on adsorption behavior are summarized in Table 2.

Table 2. Influence of solution chemistry on microalgal adsorption behavior

Parameter	Effect on surface chemistry	Resulting impact on adsorption performance
pH	Protonation / deprotonation of functional groups; alteration of pollutant speciation	Enhanced metal uptake at neutral to mildly alkaline pH; reduced adsorption under acidic conditions
Ionic strength	Electrostatic charge screening	Decreased electrostatic attraction and reduced adsorption capacity
Competing ions	Competition for active binding sites	Reduced selectivity and uptake efficiency

3.6 Integrated Surface Chemistry-Adsorption Framework

The relationships between microalgal surface chemistry, solution conditions, and adsorption behavior form an integrated mechanistic framework that underpins biosorption performance.

Rather than treating adsorption as a set of independent variables, this framework emphasizes the coupling between functional group chemistry, surface charge, pollutant speciation, and environmental conditions.

Together, Table 1, Figure 1, and Table 2 illustrate that adsorption is governed by interdependent processes rather than isolated parameters. This integrated perspective is essential for advancing microalgal adsorption toward predictive, scalable, and application-oriented systems.

4. ANALYTICAL CHARACTERIZATION OF ADSORPTION MECHANISMS

Analytical characterization is essential for validating adsorption mechanisms in microalgal systems and distinguishing between biosorption, bioaccumulation, and physicochemical interactions. Due to cell wall heterogeneity and pollutant diversity, reliance on a single technique is insufficient.

A multi-technique approach is therefore employed, integrating methods such as FTIR, SEM, BET, and zeta potential to provide complementary insights into functional groups, surface morphology, and charge-dependent interactions. However, each technique offers only partial information, necessitating combined interpretation to achieve reliable mechanistic understanding (Davis et al., 2003; He & Chen, 2014).

The principal analytical techniques and their mechanistic contributions are summarized in Table 3.

Table 3. Analytical techniques used for mechanistic interpretation of microalgal adsorption

Analytical technique	Information obtained	Mechanism supported	Strengths	Limitations
FTIR	Functional group shifts	Surface complexation, H-bonding, ion exchange	Chemical specificity	Qualitative / semi-quantitative
SEM (schematic)	Surface morphology changes	Surface adsorption, particle adherence	Visual conceptualization	Not quantitative
BET	Surface area & porosity change	Physical adsorption, site occupation	Quantitative textural data	Limited sensitivity for biomass
Zeta potential	Surface charge variation	Electrostatic attraction, ion exchange	Direct electrostatic evidence	pH & ionic strength sensitive

4.1 Fourier Transform Infrared Spectroscopy (FTIR)

FTIR spectroscopy is widely used to identify functional groups involved in microalgal adsorption. Microalgal cell walls contain diverse reactive moieties—including carboxyl, hydroxyl, phosphate, sulfate, and amine groups—that act as binding sites for pollutant interaction (Davis et al., 2003; J. Wang & Chen, 2009)

Mechanistic interpretation is based on comparative spectral analysis before and after adsorption. Shifts in characteristic absorption bands particularly in carboxyl (≈1700–1600 cm^{-1}) and hydroxyl (≈3600–3200 cm^{-1}) regions are commonly associated with surface complexation, hydrogen bonding, and ion exchange (Romera et al., 2007).

The relationship between functional groups and adsorption behavior is summarized in Table 4.

Table 4. Typical FTIR functional groups involved in microalgal adsorption

Functional group	FTIR region (cm^{-1})	Role in adsorption	Typical pollutants
–COOH	1700–1600	Metal complexation	Heavy metals
–OH	3600–3200	Hydrogen bonding	Pharmaceuticals, dyes
$-NH_2$	1650–1550	Electrostatic interaction	Organic pollutants
$-PO_4^{3-}$	1250–1000	Ion exchange	Metal cations

The distribution and functional role of these groups are illustrated in Figure 2.

Figure 2. shows carboxyl, hydroxyl, and amine groups on microalgae act as binding sites, enabling adsorption of positively charged pollutants via electrostatic attraction and surface complexation.

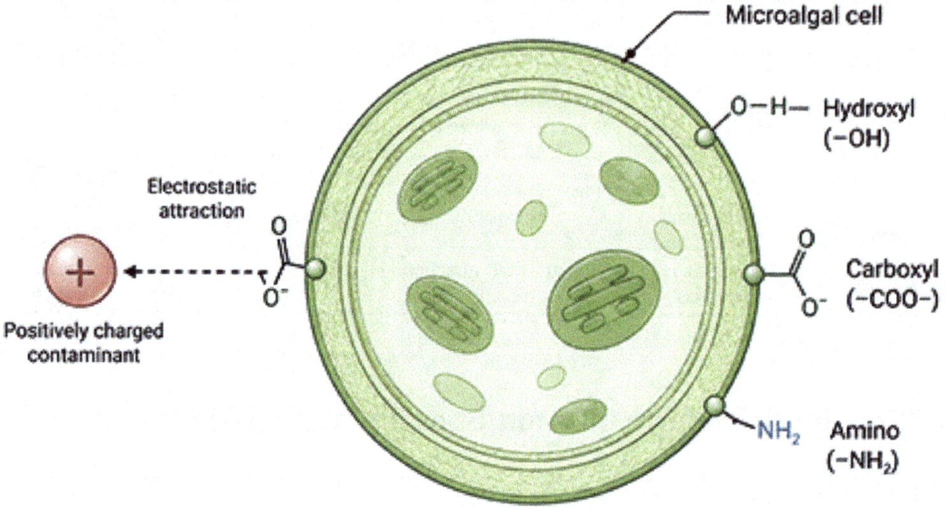

Although FTIR does not provide quantitative adsorption capacity, it offers critical chemical evidence confirming the surface-mediated nature of adsorption processes.

4.2 SEM-Based Morphological Interpretation

SEM provides qualitative insight into surface morphology and structural changes following adsorption. Typical observations include surface roughening, particle adherence, aggregation, and partial masking of surface features, particularly in systems involving dyes or particulate contaminants (Crini & Lichtfouse, 2019; Davis et al., 2003).

However, SEM does not directly identify chemical interactions and should therefore be interpreted as supportive rather than definitive mechanistic evidence.

In this chapter, schematic representations are used to illustrate generalized morphological trends and enhance conceptual clarity, particularly in comparative analyses.

4.3 BET Surface Area and Porosity Analysis

BET analysis provides quantitative information on surface area, pore volume, and pore size distribution. Although microalgal biomass lacks the well-defined porosity of engineered adsorbents, its heterogeneous surface matrix still influences adsorption behavior (Volesky, 2007). Changes in BET parameters after adsorption may indicate surface coverage or partial obstruction of accessible sites. However, such variations must be interpreted cautiously, as they may also arise from structural rearrangements during sample preparation.

Importantly, adsorption in microalgal systems is often governed by surface functional-group chemistry rather than classical pore-filling mechanisms. Therefore, BET analysis should be interpreted within an integrated analytical framework rather than as a standalone mechanistic indicator.

The conceptual relationship between surface structure and adsorption is illustrated in Figure 3.

Figure 3. adsorption alters surface morphology and pore occupation, with particle adherence and textural changes matching BET-based interpretations.

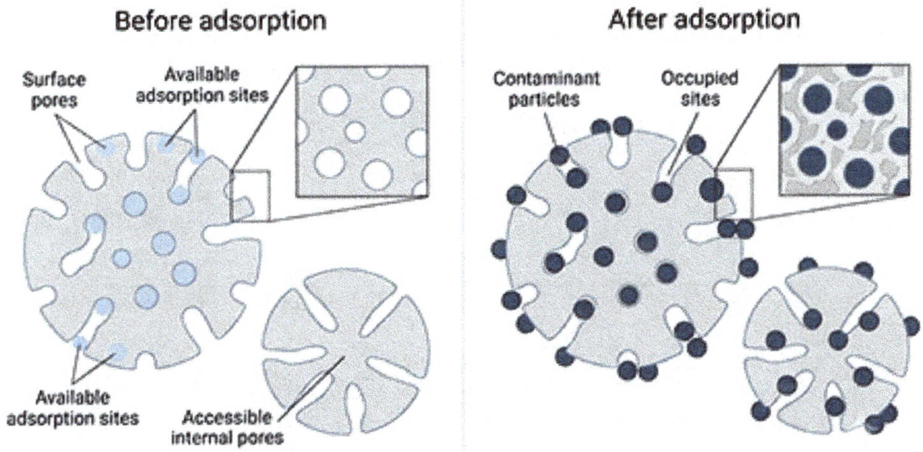

4.4 Zeta Potential and Surface Charge Behavior

Zeta potential measurements provide insight into electrostatic interactions governing adsorption processes. Surface charge in microalgal systems is strongly pH-dependent due to the protonation and deprotonation of functional groups.

Most microalgae exhibit a negative surface charge under neutral conditions, facilitating the adsorption of cationic species. Changes in zeta potential after adsorption provide indirect evidence of mechanisms such as electrostatic attraction, ion exchange, and charge neutralization.

The pH-dependent variation of surface charge is illustrated in Figure 4.

Figure 4. adsorption alters zeta potential and surface charge as pH changes, highlighting the role of pollutant binding and electrostatic interactions.

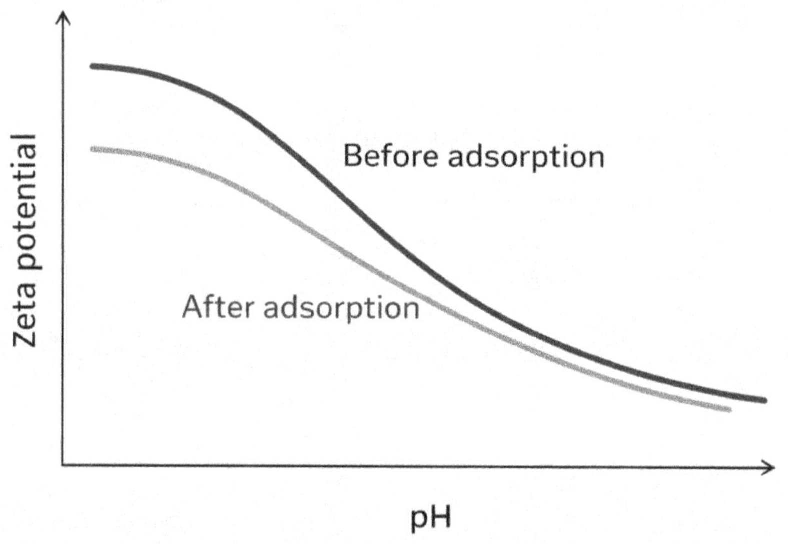

However, zeta potential shifts must be interpreted cautiously, as they may also reflect variations in ionic strength or electrical double-layer effects rather than specific chemical interactions.

A simplified interpretation framework is presented in Table 5.

Table 5. Interpretation of zeta potential shifts after adsorption

Observation	Interpretation	Dominant mechanism
Reduced negative charge	Metal ion binding	Electrostatic attraction
Charge neutralization	Surface coating	Physical adsorption
Shift in isoelectric point	Chemical surface modification	Surface complexation

4.5 Integrated Analytical Validation of Adsorption Mechanisms

No single analytical technique can independently confirm adsorption mechanisms in microalgal systems. Robust mechanistic validation emerges from the integration of complementary analytical evidence. FTIR confirms functional-group participation, SEM provides morphological context, BET contributes structural insights, and zeta potential analysis elucidates electrostatic behavior. Together, these techniques enable differentiation between adsorption, bioaccumulation, and non-specific interactions.

This integrated analytical framework is conceptually illustrated in Figures 5–7, demonstrating how spectroscopic, structural, and modeling approaches converge to support mechanistic interpretation. This multi-analytical approach provides a robust and transferable basis for mechanism-driven optimization of microalgal adsorption systems targeting emerging pollutants.

Figure 5. Conceptual illustration of pH-dependent microalgal surface charge governing adsorption behavior.

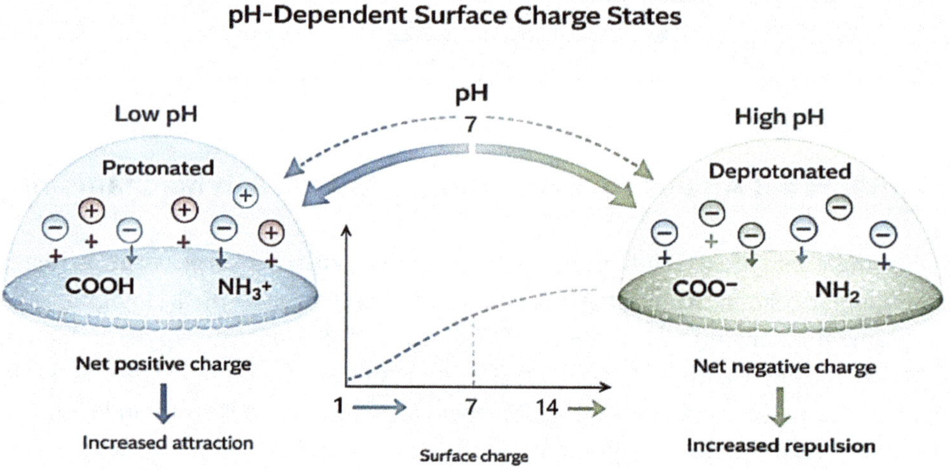

Figure 6. Integrated analytical framework illustrating the complementary roles of FTIR spectroscopy, SEM-based morphology, BET surface analysis, and zeta potential measurements in validating adsorption mechanisms.

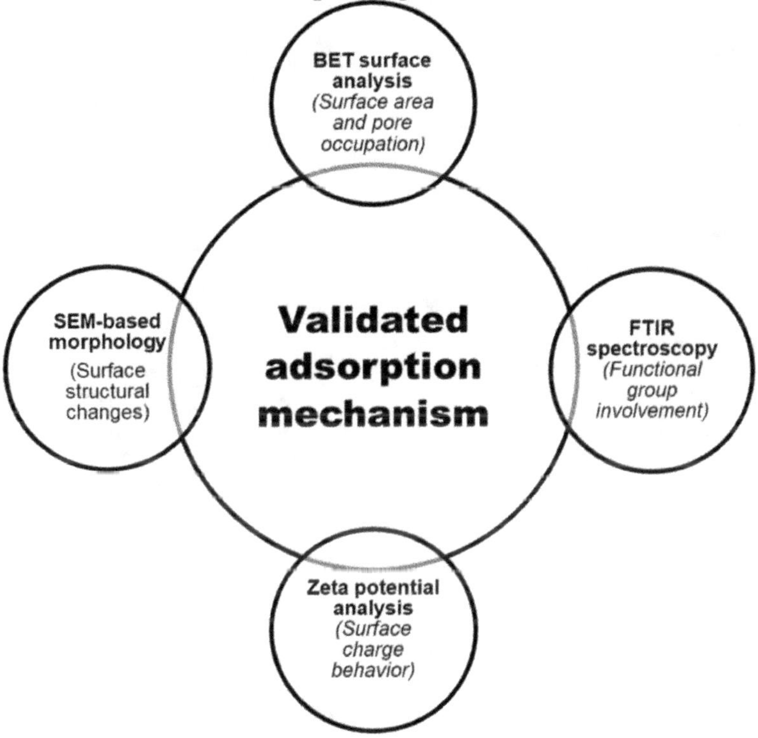

Figure 7. The equilibrium, kinetic, and thermodynamic models together provide a comprehensive understanding of microalgal adsorption, reflecting general literature trends rather than specific experimental results.

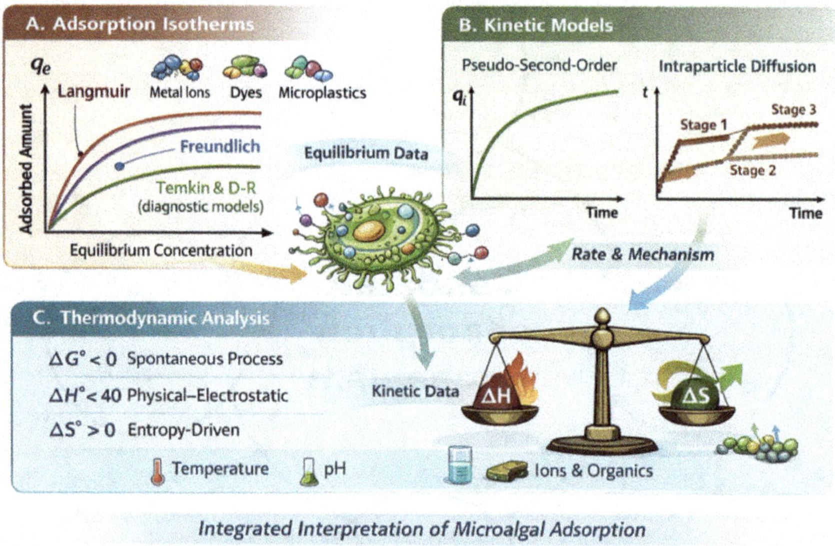

5. QUANTITATIVE ADSORPTION MODELING AND INTERPRETATION

Quantitative adsorption modeling provides the primary analytical framework for interpreting microalgal adsorption systems across diverse pollutant classes and aquatic matrices. Isotherm, kinetic, and thermodynamic models enable systematic comparison of adsorption capacity, rate-controlling mechanisms, and environmental influences such as pH, ionic strength, temperature, and water-matrix composition. Due to the biological origin and chemical heterogeneity of microalgal surfaces, model-derived parameters should be interpreted as system-dependent descriptors rather than intrinsic material constants, unless experimental conditions are rigorously standardized.

5.1 Adsorption Isotherm Models

Adsorption isotherms describe the equilibrium relationship between adsorption capacity (q_E, mg g^{-1}) and equilibrium concentration (C_E, mg L^{-1}) at constant temperature. In microalgal systems, isotherm analysis is widely used to compare adsorption performance, evaluate surface heterogeneity, and assess pollutant affinity for functional groups (J. Wang & Chen, 2009). However, mechanistic interpretation of "best-fit" models requires caution, as similar statistical fits may arise from different physicochemical processes operating simultaneously.

Equilibrium adsorption data are commonly interpreted using established isotherm models, including Langmuir, Freundlich, Temkin, and Dubinin–Radushkevich formulations (Langmuir (1918), Freundlich (1906), Temkin and Pyzhev (1940), and Dubinin and Radushkevich (1947)). However, the mechanistic interpretation of a statistically "best-fit" isotherm requires caution, as comparable fits may arise from different or concurrent physicochemical pathways operating on chemically heterogeneous and biologically responsive microalgal surfaces. Table 6 shows the Common adsorption isotherm models applied to microalgal systems.

Langmuir Isotherm

$$\frac{C_e}{q_e} = \frac{1}{q_{max}\ K_L} + \frac{C_e}{q_{max}}$$

The Langmuir model assumes monolayer adsorption on energetically uniform sites. In microalgal systems, it is often interpreted as indicative of dominant site-specific interactions involving functional groups such as carboxylate and phosphate moieties.

However, apparent agreement does not confirm surface homogeneity, but rather reflects the dominance of similar binding energies within the studied concentration range.

Freundlich Isotherm

$$\mathbf{Log\ q_e = Log\ K_F + \frac{1}{n}\ Log\ C_e}$$

The Freundlich model accounts for heterogeneous adsorption and is frequently reported in microalgal systems. Its applicability reflects the chemical diversity of cell-wall functional groups and their variable affinities for pollutants.

Because the model lacks a saturation limit, extrapolation beyond experimental conditions should be avoided.

Temkin Isotherm

$$q_e = \frac{RT}{b} Ln\ A_T + \frac{RT}{b} Ln\ C_e$$

Where B = RT/b

The Temkin model incorporates adsorbate–adsorbent interactions and assumes a linear decrease in adsorption energy with surface coverage. It is particularly relevant under high concentration conditions where electrostatic screening becomes significant.

Dubinin–Radushkevich (D–R) Isotherm

$$\mathbf{Ln}\ (\mathbf{q}_e) = \mathbf{Ln}\ (\mathbf{q}_{max}) - \beta\, \varepsilon^2$$

Where the Polanyi potential ε = RT ln(1 + 1/Ce).

In this formulation, q_e is the equilibrium adsorption capacity (mg g^{-1}), C_e is the equilibrium solute concentration (mg L^{-1}), q_{max} is the theoretical maximum adsorption capacity, and β is a constant related to adsorption energy. The mean adsorption energy E (kJ mol^{-1}) is commonly calculated from β using:

$$E = 1/\sqrt{(2\beta)}$$

The D–R model provides a comparative estimate of adsorption energy. However, its application in biological systems should be considered qualitative, as microalgal adsorption involves multiple concurrent mechanisms.

Table 6. Common adsorption isotherm models applied to microalgal systems

Model	Governing equation	Key parameters	Mechanistic interpretation in microalgae
Langmuir	$C_e/q_e = 1/(q_{max} K_L) + C_e/q_{max}$	q , K_L	Dominant site-specific surface binding
Freundlich	Log q_e =Log K_F + (1/n) Log C_e	K_F, 1/n	Surface heterogeneity and non-uniform energies
Temkin	q = (RT/b) ln AT + (RT/b) ln C .	B, A_T	Decreasing adsorption energy with coverage
D–R	Ln (q_e) = Ln (q_{max}) - β ε^2	β, E	Qualitative energy classification

The most widely applied equilibrium models in microalgal adsorption studies include the Langmuir, Freundlich, Temkin, and Dubinin–Radushkevich isotherms (Crini & Badot, 2008; Foo & Hameed, 2010).

Collectively, these models describe complementary aspects of adsorption behavior rather than distinct mechanisms. Their primary value lies in comparative interpretation and benchmarking rather than mechanistic confirmation, as summarized in Table 7.

Table 7. Standardized notation used for adsorption modeling in microalgal systems

Category	Symbol	Definition	Units
Equilibrium variables	(q_e)	Adsorption capacity at equilibrium	mg g^{-1}
	(qt)	Adsorption capacity at time *t*	mg g^{-1}
	(Ce)	Equilibrium solute concentration	mg L^{-1}
	(C_0)	Initial solute concentration	mg L^{-1}
Isotherm parameters	(q_{max})	Maximum adsorption capacity (Langmuir, D–R)	mg g^{-1}
	(K_L)	Langmuir affinity constant	L mg^{-1}
	(K_F)	Freundlich adsorption constant	(mg g^{-1})(L $mg^{-1})^{1/N}$
	(1/n)	Freundlich heterogeneity factor	—
	(A_T)	Temkin equilibrium binding constant	L g^{-1}
	(b)	Temkin constant related to adsorption energy	J mol^{-1}
	(β)	D–R activity coefficient	mol^2 kJ^{-2}
	(E)	Mean adsorption energy (D–R)	kJ mol^{-1}
	(ε)	Polanyi potential	kJ mol^{-1}
Kinetic parameters	(k_1)	Pseudo-first-order rate constant	min^{-1}
	(k_2)	Pseudo-second-order rate constant	g mg^{-1} min^{-1}
	k_{id}	Intraparticle diffusion rate constant	mg g^{-1} $min^{-1\ 2}$
	(C)	Boundary layer thickness constant	mg g^{-1}
Thermodynamic parameters	(ΔG°)	Standard Gibbs free energy change	kJ mol^{-1}
	(ΔH°)	Standard enthalpy change	kJ mol^{-1}
	(ΔS°)	Standard entropy change	J mol^{-1} K^{-1}
	(R)	Universal gas constant	8.314 J mol^{-1} K^{-1}
	(T)	Absolute temperature	K

Classical isotherm models like Langmuir and Freundlich are based on idealized, single-solute systems, limiting their predictive power in real wastewater with multiple contaminants and competing ions. These models often require modification, and their parameters should be validated under realistic conditions for reliable interpretation.

5.2 Adsorption Kinetic Models

Kinetic models describe the rate of adsorption and provide insight into rate-limiting steps, including external mass transfer, surface interaction, and diffusion processes.

Pseudo-First-Order (PFO) Model

$$\mathbf{Ln}\,(\mathbf{q}_e - \mathbf{q}_t) = \mathbf{ln}\,\mathbf{q}_e - (\mathbf{k}_1 \mathbf{t})$$

Primarily describes early-stage adsorption governed by external mass transfer

Pseudo-Second-Order (PSO) Model

$$\mathbf{t}/\mathbf{q}_t = 1/(\mathbf{k}_2 * \mathbf{q}_e^2) + \mathbf{t}/\mathbf{q}_e$$

Frequently provides the best statistical fit in microalgal systems. However, it should not be interpreted as definitive evidence of chemisorption.

Elovich model:

$$\mathbf{q}_t = \left(\frac{1}{\beta}\right)\mathbf{ln}(\alpha\beta) + \left(\frac{1}{\beta}\right)\mathbf{ln}\,\mathbf{t}$$

Describes adsorption on heterogeneous surfaces with variable activation energies.

Intraparticle Diffusion Model:

Multi-linear behavior indicates multiple rate-controlling steps, including diffusion and surface interaction.

The model is expressed as:

$$\mathbf{q}_t = \mathbf{K}_{id}\,\mathbf{t}^{1/2} + \mathbf{C}$$

Where q_t and q_e (mg g^{-1}) represent the adsorption capacities at time t and at equilibrium, respectively k_{id} (mg g^{-1} $min^{-1/2}$) is the Intraparticle diffusion rate constant; and C (mg g^{-1}) is the intercept associated with boundary layer thickness and external mass-transfer resistance.

Figure 8. sequential adsorption in microalgae, from bulk-solution transfer to surface binding and intraparticle diffusion, providing a generalized framework for understanding the mechanistic stages.

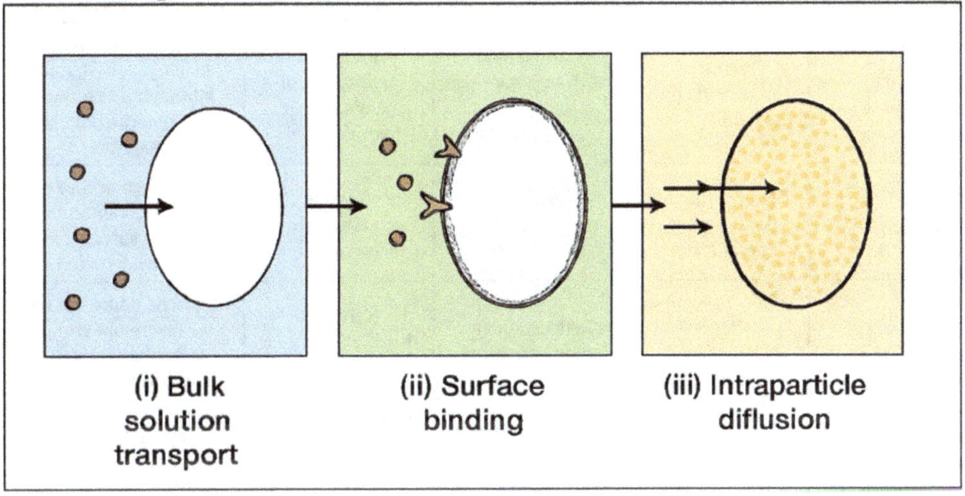

5.3 Thermodynamic Parameters and Nature of Adsorption

Thermodynamic parameters (ΔG°, ΔH°, ΔS°) provide insight into adsorption feasibility, heat effects, and interfacial disorder. The standard Gibbs free energy change is calculated as:

$$\Delta G^\circ = -RT \ln K$$

The enthalpy and entropy changes are obtained from the Van't Hoff equation:

$$\ln K = \left(\frac{\Delta S^\circ}{R}\right) - \left(\frac{\Delta H^\circ}{RT}\right)$$

Negative ΔG° values indicate spontaneous adsorption, while positive ΔH° values suggest endothermic behavior.

In microalgal systems, adsorption is typically governed by physico-chemical interactions such as electrostatic attraction, ion exchange, and surface complexation rather than covalent bonding.

Adsorption behavior is strongly influenced by pH, ionic strength, competing ions, and dissolved organic matter. Therefore, thermodynamic parameters should be interpreted comparatively within defined environmental conditions.

Table 8 synthesizes methodological trends reported in recent microalgal adsorption studies published between 2020 and 2025, highlighting the progressive transition from empirical isotherm fitting toward mechanism-informed model selection.

Table 8. Evolution of isotherm modeling strategies in microalgal adsorption studies (2020–2025)

Aspect	Early Phase (~2020)	Transitional Phase (2021–2023)	Recent Phase (2024–2025)	Mechanistic Implication
Dominant isotherm model	Frequent preference for Freundlich fitting	Comparable Langmuir–Freundlich performance	Increased application of Langmuir and hybrid models	Transition from heterogeneity-based interpretation toward site-specific binding analysis
Langmuir capacity (q) reporting	Broad variability with limited normalization	Improved parameter consistency	Greater reproducibility across studies	Improved control of pH, biomass conditioning, and surface characterization
Freundlich exponent (1/n)	Wide dispersion reflecting heterogeneous surfaces	Reduced inter-study variability	Narrower distribution across matrices	Enhanced understanding of functional group contributions and competitive ion effects
Temkin model application	Rarely applied	Applied in selected concentration-dependent systems	Increasingly used in interaction-sensitive conditions	Recognition of adsorbate–adsorbent energetic interactions
Dubinin–Radushkevich (D–R) usage	Limited and largely qualitative	Increasing diagnostic application	More frequent estimation of mean adsorption energy	Emphasis on differentiating physico-chemical adsorption mechanisms
Reporting standards	Inconsistent units and regression approaches	Improved methodological transparency	Standardized regression practices and error analysis	Field maturation toward quantitative rigor and reproducibility

Between 2020 and 2025, microalgal adsorption studies demonstrate a methodological transition from empirical curve fitting toward mechanism-informed isotherm selection. Recent investigations increasingly integrate equilibrium modeling with complementary surface characterization techniques (e.g., zeta potential, FTIR, XPS), enabling improved discrimination between heterogeneity-driven adsorption and site-specific binding behavior.

Recent kinetic analyses (Table 9) increasingly integrate diffusion modeling and multi-step interpretation, reflecting the growing recognition that pseudo-second-order fitting alone does not establish adsorption mechanism (Ho & McKay, 1999).

Table 9. Evolution of kinetic modeling strategies in microalgal adsorption studies (2020–2025)

Kinetic Aspect	Early Phase (~2020)	Transitional Phase (2021–2023)	Recent Phase (2024–2025)	Mechanistic Implication
Best-fit model selection	Pseudo-second-order frequently reported without mechanistic justification	Pseudo-second-order predominates with improved fitting practices	Pseudo-second-order combined with diffusion modeling	Increasing recognition of coupled surface reaction and diffusion contributions
Pseudo-first-order relevance	Applied mainly to initial adsorption stage	Limited mechanistic significance	Primarily descriptive for early-time behavior	Represents external mass transfer control
Intraparticle diffusion analysis	Often omitted	Included but weakly interpreted	Multilinear behavior systematically analyzed	Evidence for multi-stage adsorption mechanisms
Identification of rate-limiting step	Not clearly distinguished	External diffusion frequently proposed	Combined surface binding and diffusion control	Integrated kinetic interpretation reflecting biomass heterogeneity
Temperature-dependent kinetics	Rarely evaluated	Occasionally incorporated	Systematically assessed in relation to thermodynamics	Increasing linkage between kinetic and energetic parameters

The evolution of kinetic modeling indicates a transition from single-model empirical fitting toward integrated multi-step interpretations. Recent investigations increasingly recognize that microalgal adsorption involves coupled external mass transfer, surface binding, and diffusion processes consistent with the structural and chemical heterogeneity of biological surfaces.

The thermodynamic trends in Table 10 align with established biosorption mechanisms, showing adsorption is mainly entropy-driven and controlled by physicochemical interactions. This supports widely reported interpretations for adsorption in similar systems (Crini & Badot, 2008; J. Wang & Chen, 2009).

Table 10. Reported thermodynamic trends in microalgal adsorption systems (2020–2025)

Parameter	Typical Reported Range	Observed Trend	Mechanistic Interpretation
ΔG°	Negative (typically −5 to −25 kJ mol^{-1})	Consistently negative across systems	Spontaneous adsorption under investigated conditions
ΔH°	Low to moderate magnitude (≈ ±5–40 kJ mol^{-1})	Increasing reporting of positive values	Predominantly endothermic behavior in many systems
ΔS°	Frequently positive	Consistently positive in recent studies	Increased disorder due to desolvation and interfacial restructuring
Temperature effect	Initially underreported	Increasingly examined	Adsorption efficiency often improves with increasing temperature
Mechanistic classification	Previously ambiguous	More consistently described as physico-chemical	Dominance of electrostatic interactions and surface complexation rather than covalent bonding

Overall Integration: Equilibrium, kinetic, and thermodynamic analyses together reveal adsorption capacity, rate, and feasibility. Integrating these with surface characterization builds a quantitative basis for mechanistic understanding and cross-system comparisons.

6. THERMODYNAMIC INTERPRETATION OF MICROALGAL ADSORPTION SYSTEMS

This section provides a critical interpretation of thermodynamic parameters beyond their mathematical formulation, focusing on their mechanistic and practical limitations in microalgal systems. Thermodynamic analysis is commonly used to assess adsorption feasibility; however, in biosorption systems, "equilibrium" is more accurately described as pseudo-equilibrium, reflecting steady-state conditions rather than true thermodynamic equilibrium.

Standard parameters (ΔG°, ΔH°, ΔS°), typically derived from isotherm models, do not always represent true equilibrium due to system heterogeneity and partial irreversibility (Volesky, 2007). Consequently, their interpretation must consider kinetic limitations and multi-site interactions.

Reliable thermodynamic analysis requires careful experimental design, including sufficient equilibration time, temperature-dependent studies, and validation using complementary analytical techniques. While a negative ΔG° indicates spontaneity and ΔH° reflects energy changes, these trends must be interpreted cautiously in light of surface heterogeneity and diffusion effects (Davis et al., 2003; J. Wang &

Chen, 2009). Table 11 summarizes the relationships between surface properties, adsorption mechanisms, and energy profiles.

Table 11. System-Level Comparison of Surface Generation Pathways, Adsorption Mechanisms, and Production Energy Profiles

Criterion	Microalgal Biomass	Activated Carbon	Biochar	Zeolites
Surface generation pathway	Biological synthesis under ambient conditions	Thermochemical activation	Pyrolytic conversion	Mineral crystallization
Dominant mechanism	Electrostatic interaction, biosorption, aggregation	Pore filling, physical adsorption	Mixed surface–pore effects	Ion exchange
Typical performance metric	q_{AX}; removal %; aggregation indices	q_{AX}	q_{AX}	Exchange capacity
Embedded production energy profile	Low-temperature growth	High-temperature activation	Moderate thermal input	Moderate
Regeneration profile	Limited or valorized post-use	Thermal/chemical regeneration	Thermal	Chemical

Table 11 shows that material selection should consider thermodynamic pathways, adsorption mechanisms, and regeneration needs, not just adsorption capacity, for effective evaluation.

Entropy changes (ΔS°) in microalgal adsorption are often associated with variations in hydration shells, ion displacement, and structural organization at the biomass–solution interface. These effects differ from classical assumptions of purely configurational entropy and highlight the importance of coupling thermodynamic analysis with surface characterization techniques. For instance, FTIR spectroscopy can reveal changes in functional groups following adsorption, while zeta potential measurements provide direct evidence of surface charge shifts and ion exchange processes. Correlating these analytical findings with thermodynamic parameters enables a more robust linkage between molecular-level interactions and macroscopic adsorption behavior. Overall, thermodynamic parameters should be regarded as supportive indicators rather than definitive descriptors of adsorption mechanisms in heterogeneous biosorption systems.

It is important to emphasize that thermodynamic parameters reported in microalgal biosorption studies should be interpreted as *apparent or conditional values* rather than true equilibrium constants. This distinction arises from the inherently heterogeneous, partially irreversible, and kinetically constrained nature of biosorption systems. Consequently, parameters such as ΔG°, ΔH°, and ΔS° reflect system-specific interactions under pseudo-equilibrium conditions, rather than strict thermodynamic equilibrium. Therefore, their interpretation must be supported by complementary kinetic and surface characterization analyses to ensure mechanistic validity.

Equilibrium constants from isotherm models are often used to estimate thermodynamic parameters; however, in biosorption systems they may not represent true equilibrium due to surface heterogeneity and irreversible interactions. Therefore, these values should be interpreted with caution.

7. INTEGRATION OF MICROALGAL ADSORPTION INTO CIRCULAR TREATMENT SYSTEMS

Beyond laboratory-scale performance, the practical significance of microalgal adsorption lies in its integration within sustainable wastewater treatment and circular bioeconomy frameworks. Unlike conventional adsorbents, microalgae offer simultaneous pollutant removal, carbon capture, and biomass production, positioning them as multifunctional remediation platforms (Abdel-Raouf et al., 2012; Rawat et al., 2011). Figure 9 highlights microalgal adsorption as part of an integrated treatment chain, from wastewater input and pollutant removal to biomass harvesting and valorization within defined system boundaries.

These constraints highlight the critical gap between laboratory-based adsorption modeling and real wastewater performance.

Figure 9. Outlines microalgal adsorption within integrated treatment, showing wastewater input, CO_2 assimilation, pollutant removal, and biomass valorization. System-specific carbon fluxes and regeneration depend on operational design.

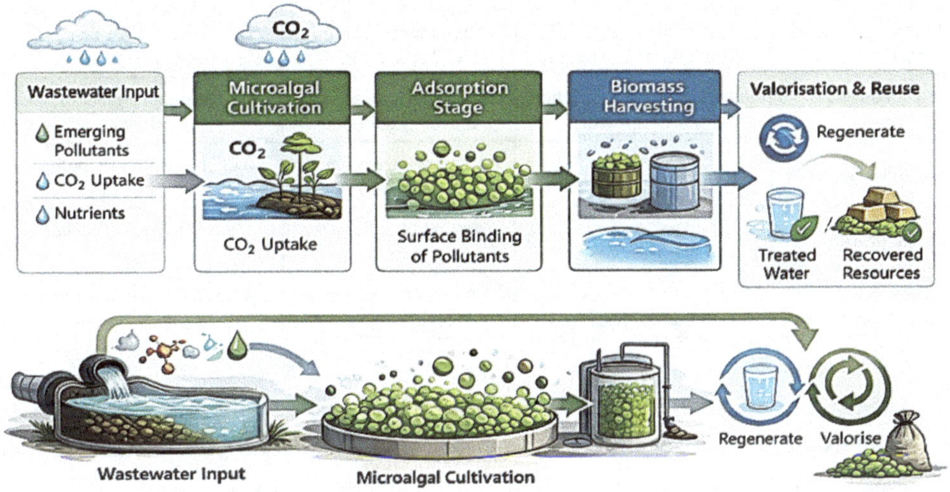

Integrating microalgal adsorption in treatment systems allows simultaneous environmental cleanup and resource recovery, such as biofuels and biofertilizers. Performance depends on light, nutrients, and retention time, which affect growth and pollutant removal. Optimized control and adaptive management ensure efficient, stable operation even under changing conditions.

Cultivating microalgae in wastewater can offset nutrient needs and lower environmental impacts. In contrast, producing activated carbon requires energy-intensive processes and generates greenhouse gas emissions (Crini & Lichtfouse, 2019).

Real-world applications show microalgal adsorption works in industrial and urban systems, supporting both wastewater treatment and biomass generation. Case studies confirm that mechanistic insights and system optimization lead to scalable, cost-effective technologies.

Because life-cycle assessments (LCAs) vary in scope, this chapter focuses on comparative trends instead of specific values. Clear definitions of comparison units are crucial for accurate sustainability assessment.

Table 12 summarizes life-cycle differences between microalgal and activated carbon systems, covering energy use, greenhouse gas emissions, chemical inputs, and disposal pathways.

Table 12. Comparative Life-Cycle Indicators for Microalgal and Activated Carbon Adsorption Systems

Impact Category	Microalgal Adsorption Systems	Activated Carbon Systems	Interpretative Trend
Production energy demand	Low to moderate (cultivation-dependent)	High (carbonization and activation)	Lower embodied energy in biological systems
Greenhouse gas profile	Potential CO_2 assimilation during growth	Net CO_2 emission during production	System-dependent climate implications
Chemical inputs	Minimal external reagents	Acids, bases, activation agents	Reduced secondary pollution burden
Waste profile	Biomass (potentially valorisable)	Spent carbon requiring regeneration	Circular vs. linear material flow
Regeneration requirement	Often single-use with valorization pathways	Thermal or chemical regeneration	Additional energy burden in conventional systems

Adsorption capacity alone does not define environmental suitability; sustainability depends on production pathways, energy demand, and end-of-life management. Therefore, adsorption efficiency and environmental impact must be evaluated concurrently. A major challenge lies in managing contaminant-laden microalgal biomass, as its valorization for biofuels or fertilizers may risk reintroducing pollutants if not properly controlled. This creates a critical paradox: the same biomass used for pollutant removal can become a secondary source of contamination. Consequently, safe downstream strategies such as controlled disposal, metal recovery, or restricted reuse are essential to mitigate environmental risks.

Microalgal adsorption also enables integration into circular bioeconomy frameworks, offering resource recovery pathways not readily achievable with spent activated carbon, which often requires energy-intensive regeneration or disposal. Figure 10 illustrates the integration of pollutant removal, biomass generation, and downstream valorization. However, practical implementation remains limited by challenges related to process efficiency, biomass handling, and system stability under real wastewater conditions.

Figure 10. Microalgal adsorption links pollutant removal, biomass production, and valorization in a circular bioeconomy, but actual efficiency depends on operational and regulatory factors.

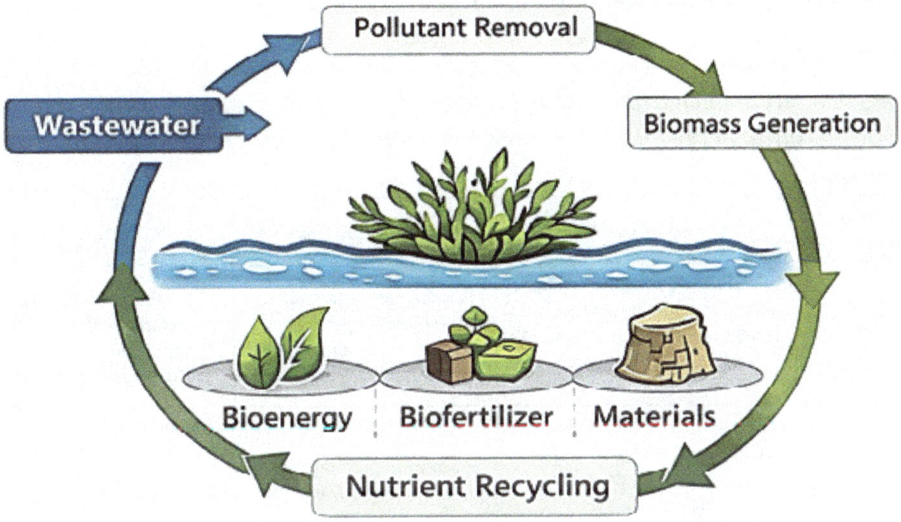

8. CONSTRAINTS IN SCALING MICROALGAL ADSORPTION TECHNOLOGIES

The transition of microalgal adsorption from laboratory-scale studies to real-world applications is limited by a combination of technical, environmental, economic, and regulatory challenges.

A key limitation arises from the complexity of real wastewater matrices, which differ substantially from synthetic systems commonly used in experimental studies. Real wastewater contains dissolved organic matter (DOM), competing ions, and fluctuating physicochemical conditions that significantly influence adsorption behavior.

DOM may exert both enhancing and inhibitory effects depending on its composition and concentration. For example, humic substances can introduce additional binding pathways or form ternary complexes with metals and biomass, whereas high DOM concentrations may block active sites or compete with target pollutants for adsorption.

Similarly, multi-component systems introduce competitive adsorption effects, with pollutants exhibiting selective binding hierarchies (e.g., Pb > Cu > Cd > Zn > Ni), resulting in reduced adsorption efficiency compared to single-solute systems.

These barriers rarely operate independently; rather, they reinforce one another across scales of analysis and implementation (Table 13).

Table 13. Interconnected analytical, biological, technical-economic, and regulatory constraints influencing the scalability and standardization of microalgal adsorption systems.

Challenge Domain	Core Limitation	Scientific Consequence	Scale-Up Implication
Analytical	Overreliance on isotherm/kinetic modeling without mechanistic validation; lack of harmonized and validated analytical protocols	Limited reproducibility; context-dependent adsorption capacities; weak cross-study comparability	Reduced predictive reliability and difficulty establishing transferable performance benchmarks
Biological	Species-dependent surface chemistry; growth-condition sensitivity; physiological heterogeneity	Variability in functional-group availability and binding affinity	Limited standardization and difficulty developing generalized predictive models
Technical–Economic	Energy-intensive biomass harvesting; instability under dynamic wastewater conditions; incomplete techno-economic assessments	Elevated operational costs; uncertain long-term sustainability	Unclear cost-performance competitiveness relative to conventional treatment technologies
Regulatory	Inconsistent regulation of emerging contaminants; unclear biomass management guidelines	Weak policy drivers for innovation; regulatory uncertainty	Delayed industrial adoption and limited integration into existing treatment frameworks

Figure 11. Conceptual representation of the interdependent analytical, biological, technical–economic, and regulatory barriers constraining the large-scale implementation of microalgal adsorption technologies.

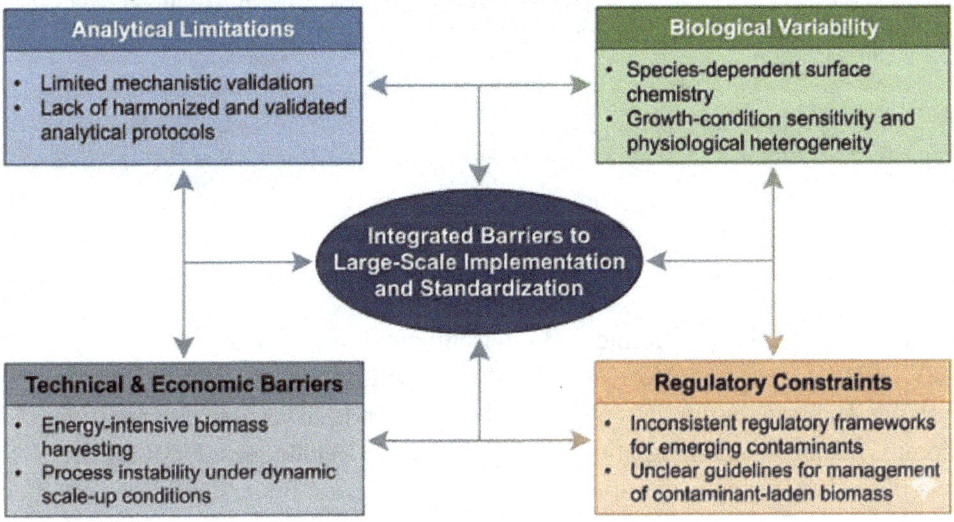

From a technical perspective, challenges include biomass harvesting, contamination risks, and sensitivity to environmental fluctuations such as light and temperature. Economically, high operational costs, particularly for biomass separation and processing, remain a major barrier to large-scale deployment. Life Cycle Assessment (LCA) studies indicate that the environmental performance of microalgal systems depends strongly on energy sources and process optimization, with the integration of renewable energy significantly improving carbon outcomes. Regulatory uncertainty further complicates implementation, particularly regarding biomass handling, reuse, and compliance with discharge standards. Figure 11 illustrates the interrelated analytical, technical–economic, and regulatory constraints limiting large-scale application.

9. BRIDGING ANALYTICAL RESEARCH AND PRACTICAL IMPLEMENTATION

A key limitation in current research is the disconnect between mechanistic laboratory studies and scalable wastewater applications, as adsorption models often fail to predict behavior under complex conditions. Bridging this gap requires standardized frameworks that incorporate real wastewater matrices, multi-component systems, and integrated analytical–modeling approaches. The transition toward implementation-ready systems is conceptually illustrated in Figure 12.

Figure 12. Conceptual progression from analytical uncertainty to implementation readiness in microalgal adsorption research.

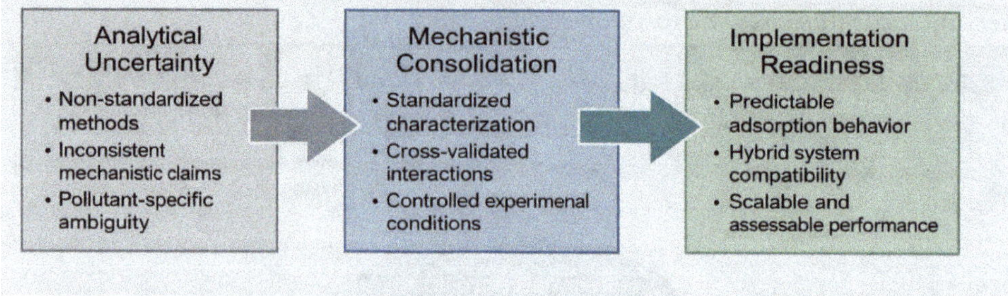

Integrated techno-economic and environmental assessments combining LCA, NPV, and IRR are essential for evaluating the competitiveness of microalgal adsorption systems. These analyses must consider energy demand, harvesting, regeneration, and biomass management. Integrating advanced analytical techniques (FTIR, SEM, BET, zeta potential) during scale-up ensures that mechanistic understanding is retained under real conditions. Reactor-scale factors, including retention time, mixing, light distribution, and biomass recovery, must be optimized for effective system design.

Transitioning from empirical evaluation to predictive, mechanistically validated approaches is critical for bridging the gap between laboratory studies and real wastewater performance. Industrial applications support this transition, with raceway ponds achieving >23 g/m^{2}d productivity and facultative systems reaching 80–95% BOD removal at lower energy demand. These outcomes highlight the importance of integrating mechanistic insight with engineering optimization for successful large-scale implementation.

9.1. Industrial Case Studies and Techno-Economic Validation

The practical viability of microalgal adsorption technologies can be demonstrated through documented industrial-scale and pilot-scale applications. The following case studies illustrate how mechanistic principles established in laboratory research translate into operational systems treating complex, real-world wastewater matrices.

Case Study 1: Pilot-Scale Cultivation in Industrial Munition Wastewater (Abraham et al., 2023).

A pilot-scale system cultivated algae in industrial wastewater, achieving 23.9 g/m^{2}d biomass productivity. Biomass analysis showed 22% oil content under low nitrogen and 350 scc/gVS BMP, confirming strong bioenergy potential.

The study validates the chapter's mechanistic framework; showing resilient microbial consortia can thrive in toxic, nitrogen-rich wastewater. Environmental factors affected growth and pollutant removal, while biomass valorization supports a circular bioeconomy.

Integrating wastewater treatment with bioenergy production enhances economic viability by offsetting costs and making the process competitive with conventional methods.

Case Study 2: Treated Spirulina platensis efficiently removed toxic metals from industrial effluents, achieving 95% Al, 87% Ni, and 63% Cu removal under optimal conditions (4.8 g/L biomass, pH 6). Zeta potential confirmed electrostatic and ion-exchange mechanisms, with data fitting Langmuir and second-order kinetic models. These results experimentally validate the chapter's mechanistic framework and highlight the importance of surface charge and functional groups. Acidic pre-treatment enhanced removal to meet discharge standards, supporting its cost-effectiveness for regulatory compliance (Almomani & Bohsale, 2021).

10. FUTURE RESEARCH DIRECTIONS

Future research should focus on integrating mechanistic modeling with real wastewater validation to improve the predictability and scalability of microalgal systems under variable environmental conditions.

11. CONCLUSION

Microalgal adsorption is a promising and sustainable approach for removing emerging pollutants. Its effectiveness arises from the integration of biological and physicochemical mechanisms. However, the complexity of real wastewater systems requires mechanistic approaches supported by analytical validation to enable reliable and scalable applications.

REFERENCES

Abdel-Raouf, N., Al-Homaidan, A. A., & Ibraheem, I. B. M. (2012). Microalgae and wastewater treatment. *Saudi Journal of Biological Sciences, 19*(3), 257–275. DOI: 10.1016/j.sjbs.2012.04.005 PMID: 24936135

Abraham, J., Abimbola, T., Braida, W. J., Terracciano, A., Su, T. L., Christodoulatos, C., Koutsospyros, A., RoyChowdhury, A., Smolinski, B., & Lawal, A. (2023). On-Site Pilot-Scale Microalgae Cultivation Using Industrial Wastewater for Bioenergy Production: A Case Study towards Circular Bioeconomy. *Bioengineering (Basel, Switzerland), 10*(12), 1339. DOI: 10.3390/bioengineering10121339 PMID: 38135931

Almomani, F., & Bohsale, R. R. (2021). Bio-sorption of toxic metals from industrial wastewater by algae strains Spirulina platensis and Chlorella vulgaris: Application of isotherm, kinetic models and process optimization. *The Science of the Total Environment, 755*, 142654. DOI: 10.1016/j.scitotenv.2020.142654 PMID: 33082037

Armbruster, M. H., & Austin, J. B. (1938). The Adsorption of Gases on Plane Surfaces of Mica. *Journal of the American Chemical Society, 60*(2), 467–475. DOI: 10.1021/ja01269a066

aus der Beek, T., Weber, F. A., Bergmann, A., Hickmann, S., Ebert, I., Hein, A., & Küster, A. (2016). Pharmaceuticals in the environment-Global occurrences and perspectives. *Environmental Toxicology and Chemistry, 35*(4), 823–835. DOI: 10.1002/etc.3339 PMID: 26666847

Bhattacharya, P., Lin, S., Turner, J. P., & Ke, P. C. (2010). Physical adsorption of charged plastic nanoparticles affects algal photosynthesis. *The Journal of Physical Chemistry. C, Nanomaterials and Interfaces, 114*(39), 16556–16561. DOI: 10.1021/jp1054759

Crini, G., & Badot, P. M. (2008). Application of chitosan, a natural aminopolysaccharide, for dye removal from aqueous solutions by adsorption processes using batch studies: A review of recent literature. *Progress in Polymer Science, 33*(4), 399–447. DOI: 10.1016/j.progpolymsci.2007.11.001

Crini, G., & Lichtfouse, E. (2019). Advantages and disadvantages of techniques used for wastewater treatment. *Environmental Chemistry Letters, 17*(1), 145–155. DOI: 10.1007/s10311-018-0785-9

Daughton, C. G., & Ternes, T. A. (1999). Pharmaceuticals and personal care products in the environment: Agents of subtle change? *Environmental Health Perspectives, 107*(SUPPL. 6), 907–938. DOI: 10.1289/ehp.99107s6907 PMID: 10592150

Davis, T. A., Volesky, B., & Mucci, A. (2003). A review of the biochemistry of heavy metal biosorption by brown algae. *Water Research*, *37*(18), 4311–4330. DOI: 10.1016/S0043-1354(03)00293-8 PMID: 14511701

De-Bashan, L. E., & Bashan, Y. (2010). Immobilized microalgae for removing pollutants: Review of practical aspects. *Bioresource Technology*, *101*(6), 1611–1627. DOI: 10.1016/j.biortech.2009.09.043 PMID: 19931451

Deblonde, T., Cossu-Leguille, C., & Hartemann, P. (2011). Emerging pollutants in wastewater: A review of the literature. *International Journal of Hygiene and Environmental Health*, *214*(6), 442–448. DOI: 10.1016/j.ijheh.2011.08.002 PMID: 21885335

Dubinin, M. M. (1947). The equation of the characteristic curve of activated charcoal. *Doklady Akademii Nauk SSSR*, *55*, 327–329.

Freundlich, H. (1906/1907), "Über die Adsorption in Lösungen," Zeitschrift für Physikalische Chemie.

Fomina, M., & Gadd, G. M. (2014). Biosorption: Current perspectives on concept, definition and application. *Bioresource Technology*, *160*, 3–14. DOI: 10.1016/j.biortech.2013.12.102 PMID: 24468322

Foo, K. Y., & Hameed, B. H. (2010). Insights into the modeling of adsorption isotherm systems. *Chemical Engineering Journal*, *156*(1), 2–10. DOI: 10.1016/j.cej.2009.09.013

Fu, F., & Wang, Q. (2011). Removal of heavy metal ions from wastewaters: A review. *Journal of Environmental Management*, *92*(3), 407–418. DOI: 10.1016/j.jenvman.2010.11.011 PMID: 21138785

Gadd, G. M. (2009). Biosorption: Critical review of scientific rationale, environmental importance and significance for pollution treatment. *Journal of Chemical Technology and Biotechnology*, *84*(1), 13–28. DOI: 10.1002/jctb.1999

Gupta, V. K., & Rastogi, A. (2009). Biosorption of hexavalent chromium by raw and acid-treated green alga Oedogonium hatei from aqueous solutions. *Journal of Hazardous Materials*, *163*(1), 396–402. DOI: 10.1016/j.jhazmat.2008.06.104 PMID: 18691812

He, J., & Chen, J. P. (2014). A comprehensive review on biosorption of heavy metals by algal biomass: Materials, performances, chemistry, and modeling simulation tools. *Bioresource Technology*, *160*, 67–78. DOI: 10.1016/j.biortech.2014.01.068 PMID: 24630371

Hejna, M., Kapuścińska, D., & Aksmann, A. (2022). Pharmaceuticals in the Aquatic Environment: A Review on Eco-Toxicology and the Remediation Potential of Algae. *International Journal of Environmental Research and Public Health*, *19*(13), 7717. DOI: 10.3390/ijerph19137717 PMID: 35805373

Ho, Y. S., & McKay, G. (1999). Pseudo-second order model for sorption processes. *Process Biochemistry (Barking, London, England)*, *34*(5), 451–465. DOI: 10.1016/S0032-9592(98)00112-5

Hunter, R. J. (1981). *Zeta potential in colloid science: Principles and applications*. Academic Press.

Koelmans, A. A., Bakir, A., Burton, G. A., & Janssen, C. R. (2016). Microplastic as a Vector for Chemicals in the Aquatic Environment: Critical Review and Model-Supported Reinterpretation of Empirical Studies. *Environmental Science & Technology*, *50*(7), 3315–3326. DOI: 10.1021/acs.est.5b06069 PMID: 26946978

Kratochvil, D., & Volesky, B. (1998). Advances in the biosorption of heavy metals. *Trends in Biotechnology*, *16*(7), 291–300. DOI: 10.1016/S0167-7799(98)01218-9

Langmuir, I. (1918). The adsorption of gases on plane surfaces of glass, mica and platinum. *Journal of the American Chemical Society*, *40*(9), 1361–1403.

La Farre, M., Pérez, S., Kantiani, L., & Barceló, D. (2008). Fate and toxicity of emerging pollutants, their metabolites and transformation products in the aquatic environment. *Trends in Analytical Chemistry*, *27*(11), 991–1007. DOI: 10.1016/j.trac.2008.09.010

Lapworth, D. J., Baran, N., Stuart, M. E., & Ward, R. S. (2012). Emerging organic contaminants in groundwater: A review of sources, fate and occurrence. *Environmental Pollution*, *163*, 287–303. DOI: 10.1016/j.envpol.2011.12.034 PMID: 22306910

Luo, Y., Guo, W., Ngo, H. H., Nghiem, L. D., Hai, F. I., Zhang, J., Liang, S., & Wang, X. C. (2014). A review on the occurrence of micropollutants in the aquatic environment and their fate and removal during wastewater treatment. *The Science of the Total Environment*, *473–474*, 619–641. DOI: 10.1016/j.scitotenv.2013.12.065 PMID: 24394371

Markou, G., Vandamme, D., & Muylaert, K. (2014). Microalgal and cyanobacterial cultivation: The supply of nutrients. *Water Research*, *65*, 186–202. DOI: 10.1016/j.watres.2014.07.025 PMID: 25113948

Michalak, I., Chojnacka, K., & Witek-Krowiak, A. (2013). State of the art for the biosorption process - A review. *Applied Biochemistry and Biotechnology*, *170*(6), 1389–1416. DOI: 10.1007/s12010-013-0269-0 PMID: 23666641

Monteiro, C. M., Castro, P. M. L., & Malcata, F. X. (2012). Metal uptake by microalgae: Underlying mechanisms and practical applications. *Biotechnology Progress*, *28*(2), 299–311. DOI: 10.1002/btpr.1504 PMID: 22228490

Ozkan, A., & Berberoglu, H. (2013). Physico-chemical surface properties of microalgae. *Colloids and Surfaces. B, Biointerfaces*, *112*, 287–293. DOI: 10.1016/j.colsurfb.2013.08.001 PMID: 24001448

Rawat, I., Ranjith Kumar, R., Mutanda, T., & Bux, F. (2011). Dual role of microalgae: Phycoremediation of domestic wastewater and biomass production for sustainable biofuels production. *Applied Energy*, *88*(10), 3411–3424. DOI: 10.1016/j.apenergy.2010.11.025

Richardson, S. D., & Kimura, S. Y. (2020). Water analysis: Emerging contaminants and current issues. *Analytical Chemistry*, *92*(1), 473–505. DOI: 10.1021/acs.analchem.9b05269 PMID: 31825597

Rochman, C. M., Hoh, E., Kurobe, T., & Teh, S. J. (2013). Ingested plastic transfers hazardous chemicals to fish and induces hepatic stress. *Scientific Reports*, *3*(1), 3263. DOI: 10.1038/srep03263 PMID: 24263561

Rodriguez-Narvaez, O. M., Peralta-Hernandez, J. M., Goonetilleke, A., & Bandala, E. R. (2017). Treatment technologies for emerging contaminants in water: A review. *Chemical Engineering Journal*, *323*, 361–380. DOI: 10.1016/j.cej.2017.04.106

Romera, E., González, F., Ballester, A., Blázquez, M. L., & Muñoz, J. A. (2006). Biosorption with algae: A statistical review. *Critical Reviews in Biotechnology*, *26*(4), 223–235. DOI: 10.1080/07388550600972153 PMID: 17095433

Romera, E., González, F., Ballester, A., Blázquez, M. L., & Muñoz, J. A. (2007). Comparative study of biosorption of heavy metals using different types of algae. *Bioresource Technology*, *98*(17), 3344–3353. DOI: 10.1016/j.biortech.2006.09.026 PMID: 17624771

Sheng, P. X., Ting, Y. P., Chen, J. P., & Hong, L. (2004). Sorption of lead, copper, cadmium, zinc, and nickel by marine algal biomass: Characterization of biosorptive capacity and investigation of mechanisms. *Journal of Colloid and Interface Science*, *275*(1), 131–141. DOI: 10.1016/j.jcis.2004.01.036 PMID: 15158390

Temkin, M. I., & Pyzhev, V. (1940), . Kinetics of ammonia synthesis on promoted iron catalyst. *Acta Phys. Chim. USSR*, *12*, 327–356.

Trivedi, J., Aila, M., Bangwal, D. P., Kaul, S., & Garg, M. O. (2015). Algae based biorefinery - How to make sense? *Renewable & Sustainable Energy Reviews*, *47*, 295–307. DOI: 10.1016/j.rser.2015.03.052

Velásquez-Orta, S. B., Yáñez-Noguez, I., Ramírez, I. M., & Ledesma, M. T. O. (2024). Pilot-scale microalgae cultivation and wastewater treatment using high-rate ponds: A meta-analysis. *Environmental Science and Pollution Research International, 31*(34), 46994–47021. DOI: 10.1007/s11356-024-34000-7 PMID: 38985422

Volesky, B. (2007). Biosorption and me. *Water Research, 41*(18), 4017–4029. DOI: 10.1016/j.watres.2007.05.062 PMID: 17632204

Volesky, B., & Holan, Z. R. (1995). Biosorption of Heavy Metals. *Biotechnology Progress, 11*(3), 235–250. DOI: 10.1021/bp00033a001 PMID: 7619394

Wang, J., & Chen, C. (2009). Biosorbents for heavy metals removal and their future. *Biotechnology Advances, 27*(2), 195–226. DOI: 10.1016/j.biotechadv.2008.11.002 PMID: 19103274

Wang, L., Min, M., Li, Y., Chen, P., Chen, Y., Liu, Y., Wang, Y., & Ruan, R. (2010). Cultivation of green algae Chlorella sp. in different wastewaters from municipal wastewater treatment plant. *Applied Biochemistry and Biotechnology, 162*(4), 1174–1186. DOI: 10.1007/s12010-009-8866-7 PMID: 19937154

Wen, L., Yang, F., Li, X., Liu, S., Lin, Y., Hu, E., Gao, L., & Li, M. (2023). Composition of dissolved organic matter (DOM) in wastewater treatment plants influent affects the efficiency of carbon and nitrogen removal. *The Science of the Total Environment, 857*, 159541. DOI: 10.1016/j.scitotenv.2022.159541 PMID: 36265625

Yagub, M. T., Sen, T. K., Afroze, S., & Ang, H. M. (2014). Dye and its removal from aqueous solution by adsorption: A review. *Advances in Colloid and Interface Science, 209*, 172–184. DOI: 10.1016/j.cis.2014.04.002 PMID: 24780401

Zhou, Y., Zhou, Y., Yao, L., Luo, X., Kong, Q., Leung, K. S.-Y., & Yang, X. (2025). Insights into DOM-Enhanced Periodate Degradation of Emerging Contaminants: The Organic Fenton-like Reactions. *Environmental Science & Technology, 59*(40), 21682–21690. DOI: 10.1021/acs.est.5c09780 PMID: 41036757

ADDITIONAL READING

Ali, I., & Gupta, V. K. (2006). Advances in water treatment by adsorption technology. *Nature Protocols, 1*(6), 2661–2667. DOI: 10.1038/nprot.2006.370 PMID: 17406522

Amrutha Jeppu, G., Girish, C. R., Prabhu, B., & Mayer, K. (2023). Multi-component adsorption isotherms: Review and modeling studies. *Environmental Processes, 10*(2), 38. DOI: 10.1007/s40710-023-00631-0

Ordóñez, J. I., Cortés, S., Maluenda, P., & Soto, I. (2023). Biosorption of heavy metals with algae: Critical review of its application in real effluents. *Sustainability (Basel), 15*(6), 5521. DOI: 10.3390/su15065521

Volesky, B. (2001). Detoxification of metal-bearing effluents: Biosorption for the next century. *Hydrometallurgy, 59*(2–3), 203–216. DOI: 10.1016/S0304-386X(00)00160-2

KEY TERMS AND DEFINITIONS

Bioaccumulation: An active process involving the uptake and accumulation of contaminants within living cells through metabolic pathways.

Biosorption: A passive physicochemical process in which contaminants are adsorbed onto the surface of biological materials without metabolic involvement.

Circular Bioeconomy: A sustainability framework that promotes resource recovery, reuse, and waste minimization in environmental systems.

Dissolved Organic Matter: A complex mixture of organic compounds in water that influences adsorption through competition and complexation.

Isotherm Model: A mathematical representation describing the relationship between adsorbed and dissolved contaminant concentrations at equilibrium.

Pseudo-Equilibrium: A steady-state condition in adsorption systems that does not represent true thermodynamic equilibrium.

Surface Complexation: The formation of chemical bonds between dissolved species and functional groups on a solid surface.

Zeta Potential: A measure of surface charge that affects particle stability and electrostatic interactions in aqueous systems.

Chapter 8
Heavy Metal Remediation in Wastewater Using Microalgae

B. Lokeshwari
http://orcid.org/0009-0005-6612-5134
Department of Microbiology, Sacred Heart College, Tirupattur, India

P. Saranraj
http://orcid.org/0000-0002-5052-6532
Department of Microbiology, Sacred Heart College, Tirupattur, India

Ashish Kumar Nayak
http://orcid.org/0000-0002-3211-7065
Prasad V. Potluri Siddhartha Institute of Technology, Vijayawada, India

Alexander Machado Cardoso
http://orcid.org/0000-0003-2974-0232
Rio de Janeiro State University, Brazil

K. Kesavardhini
http://orcid.org/0009-0005-8235-042X
Department of Microbiology, Sacred Heart College, Tirupattur, India

K. Gayathri
http://orcid.org/0009-0007-3786-2009
Department of Microbiology, Sacred Heart College, Tirupattur, India

ABSTRACT

The rapid industrialization and urbanization cause the release of wastewater that is contaminated with heavy metals, dyes, pharmaceuticals, and pathogenic microorganisms, which poses serious hazards to the environment and human health. Traditional physicochemical treatment processes tend to be expensive, energy-intensive, and generate secondary pollutants, prompting a search for sustainable alternatives. The enzymatic transformation, chelation, and cell wall binding are the major remediation pathways used by fungi in the process of detoxification of heavy metals. Although they have great potential, such challenges as harvesting biomass, metal toxicity levels, scale potential, and economic viability continue to be the main

DOI: 10.4018/979-8-2600-3080-6.ch008

limitations. Microalgae and fungi represent a promising and eco-friendly approach for treating toxic wastewater and bioremediating heavy metals, contributing to the development of circular and sustainable environmental management strategies.

1. INTRODUCTION

Heavy metal contamination of wastewater is one of the most severe and multifaceted environmental problems resulting from rapid industrialization, urbanization, mining, and the escalation of agricultural activities. Non-biodegradable metals such as cadmium, lead, mercury...tend to accumulate in water bodies, posing serious health risks to humans and causing environmental instability (Delanka-Pedige et al., 2020). The traditional physicochemical treatment processes, such as chemical precipitation, ion exchange, membrane filtration, and adsorption, tend to be expensive, energy consuming, and produce secondary pollutants, such as toxic sludge. Consequently, biological methods for wastewater treatment have garnered increasing interest due to their sustainability, cost-effectiveness, and reduced environmental impact. Microalgae have emerged as one of the most promising biological agents for heavy metal remediation due to high growth rates, high surface-volume ratio, metabolic plasticity, and outstanding metal stress tolerance (Ahmed et al., 2022).

Microalgae are microorganisms that are photosynthetic and can use inorganic nutrients as well as carbon dioxide and eliminate pollutants in wastewater. Their application in wastewater treatment extends beyond metal removal to include nutrient recovery, reduction of organic loading, and oxygenation of water bodies (Brar et al., 2022). Notably, microalgae-based remediation aligns with the principles of a circular bioeconomy, as the resulting biomass can be utilized for the production of biofuels, fertilizers, animal feed, and bioactive compounds. The processes through which microalgae respond to heavy metals are biosorption to cell walls, biosorption into cells, intracellular sorbent, and biotransformation into less toxic compounds. This combination of multidimensional interactions makes microalgae particularly effective for treating complex wastewater streams containing mixtures of metals and other pollutants (Deshaware et al., 2021).

The adoption of microalgal systems in wastewater treatment represents a paradigm shift, transforming the process from one of simple pollutant removal to one of resource recovery. The microalgal technologies also have the added benefit of environmental remediation in addition to biomass production in contrast to traditional treatment systems which mostly concentrate on the removal of contaminants (Vazirzadeh et al., 2022). This chapter presents an overall review of microalgae-based heavy metal remediation in wastewater, such as algal classification, metal uptake mechanism, application in industrial and domestic effluent treatment, the role of fungi in metal detoxification, recent technological achievements, and the main challenges that need to be overcome to use microalgae when removing heavy metals in wastewater (Dudhagara et al., 2023).

Sustainable and eco-friendly alternatives to chemical precipitation, ion exchange, and membrane filtration are needed due to their high operational costs, energy consumption, and secondary pollutant production. Due to their rapid growth rate, high metal-binding capacity, and metabolic adaptability, microalgae can remove contaminants through biosorption, bioaccumulation, and biotransformation (Deshaware et al., 2021). Microalgae also improve water quality by removing nutrients, sequestering carbon dioxide, and producing oxygen and biomass for biofuels, fertilizers, and other bioproducts (Vazirzadeh et al., 2022). By combining multiple detoxifying pathways and enhancing system resilience in complex wastewater, microalgae and other microbes, especially fungus, improve remediation efficiency. Genetic engineering, nanobiotechnology, omics-based technologies, and upgraded bioreactor systems have improved microalgae-based remediation performance and scalability. Thus, microalgae-driven bioremediation is becoming a potential, cost-effective, and environmentally sustainable solution to wastewater heavy metal pollution.

2. CLASSIFICATION OF ALGAE

Algae are a heterogeneous group of photoautotrophic organisms, ranging from unicellular microalgae to multicellular macroalgae. Depending on their size and complexity, algae can be divided into microalgae and macroalgae with microalgae being the major subject of wastewater bioremediation because of their microscopic dimensions, rapid development, and adaptation to designed systems. Microalgae are spread over a number of divisions taxonomically such as Chlorophyta (green algae), Bacillariophyta (diatoms), Cyanobacteria (blue-green algae), Euglenophyta and Chrysophyta. These groups are characterized by distinct morphological, physiological, and biochemical features that determine their capacity for metal absorption and tolerance (El-Naggar et al., 2021).

Some of the most widely studied microalgae that have been used in the treatment of wastewater include the genera *Chlorella* (green microalgae). These organisms possess cell walls rich in functional groups, such as carboxyl, hydroxyl, phosphate, and amine groups, which are effective in binding metal ions (Gondi et al., 2021). Diatoms, with their silica-based frustules, possess distinctive adsorption characteristics that make them effective for removing metals such as cadmium and lead. Although cyanobacteria are prokaryotic, they are highly metal tolerant and have extracellular polymeric substances that increase the biosorption. The taxonomic variety of algal organisms offers high mobility of remediation strategies to the particular wastewater mixtures (Gao, 2022).

Ecological preferences and metabolic characteristics of algae classification are also applicable to bioremediation. There are microalgae that grow in freshwater and microalgae that are adapted to marine or hypersaline, and thus, varied streams of wastewater can be treated. Moreover, mixotrophic and heterotrophic microalgae are also able to use organic sources of carbon and hence are suitable to wastewater that is highly organic. These algal classification is essential for selecting strains with optimal metal removal capabilities, environmental resistance, and suitability for wastewater treatment systems (Hasanin et al., 2022). Figure 1 presents the classification of microalgae employed in heavy metal bioremediation, grouping them into major taxonomic categories such as green algae (Chlorophyceae), diatoms (Bacillariophyceae), and cyanobacteria based on their structural and physiological traits that influence their efficiency in biosorption, bioaccumulation, and detoxification processes.

Figure 1. Classification of Microalgae Used for Heavy Metal Bioremediation

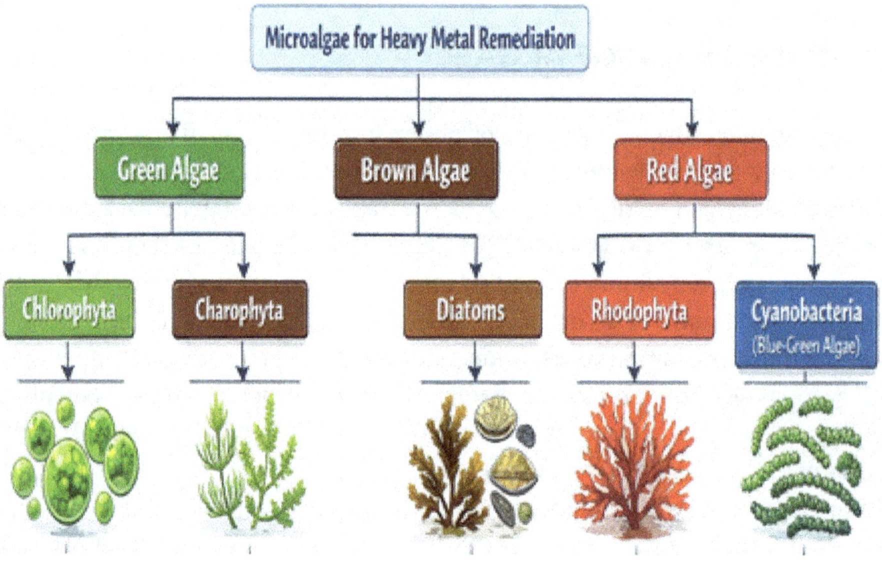

3. MICROALGAE FOR WASTEWATER TREATMENT

The concept of microalgae usage as the part of the efficient wastewater treatment systems has become more significant because of the presence of various functions of this organism in terms of contaminants elimination and recovery of the ecosystem (Mojiri et al., 2022). Microalgae in wastewater conditions absorb nutrients like nitrogen and phosphorus, lower chemical oxygen demand, and promote the extraction of the hard metal by passive and active methods. Their ability to photosynthesize leads to the production of oxygen which helps sustain aerobic bacterial communities and increases the overall performance of the treatment. These properties render microalgae especially desirable to be adopted in secondary and tertiary treatment processes of wastewater (Husain, 2022).

Microalgae efficiency in wastewater treatment is determined by a number of factors, such as the species choice, the composition of wastewater, light, temperature, pH, and the hydraulic retention time. High-rate algal ponds, photobioreactors, and algal-bacterial consortia have been developed as a type of algal-based system to maximize these parameters. Under these systems, there is a synergistic interaction of microalgae with bacteria whereby algae provide oxygen and bacteria decompose organic matter to give a self-sustaining and balanced treatment process. Such symbiotic relationship increases the process of pollutant removal and reduces the energy inputs (Ji et al., 2019).

Microalgal wastewater treatment has important environmental and economic advantages in addition to the elimination of pollutants. The algal biomass collected can be transformed into value added products, hence compensating the cost of operation. Besides, the microalgae help in capturing carbon dioxide in the atmosphere and thereby reduce greenhouse gas emissions that are related to traditional treatment methods. With the growing shift in wastewater treatment plants towards sustainability-based initiatives, microalgae-based systems will be vital in the shift towards more environmentally friendly and resource efficient infrastructure (Kapahi and Sachdeva, 2019).

4. HEAVY METAL BIOREMEDIATION USING MICROALGAE

The ability of microalgae to eliminate heavy metals in the wastewater lies in their cellular organization and plasticity in metabolism. The microalgae remove metals via biosorption, bioaccumulation, and biotransformation. Biosorption is a fast, metabolism-free process that entails the attachment of metal ions on the functional groups present on the algal cell surface. This is especially efficient when the metals are in low concentrations and plays a major role in the removal of initial metals.

Bioaccumulation on the other hand is an energy-dependent process in which metals are carried into the cell and retained in intracellular compartments (Zhan et al., 2022).

Microalgae have different levels of tolerance to heavy metals mediated by synthesis of metal binding peptides including phytochelatins and metallothioneins. These molecules chelate metal ions and help in their accumulation in vacuoles, in turn lowering the cytotoxic effects. There are also a number of microalgae that can convert metals to less toxic or less bioavailable forms via redox reaction. As an example, some species have the ability of reducing hexavalent chromium to trivalent chromium which is less mobile and less toxic. These detoxification systems improve stability of metal removal in the long run (Ajiboye et al., 2021).

Environmental factors, and the nature of wastewater, have an effect on the effectiveness of heavy metal bioremediation with the help of microalgae. The pH, competing ions, metal specification, and the physiological state of the algal activities are factors that are important in determining the removal performance. In spite of these complexities, most laboratory and pilot-scale experiments have shown high metals removal efficiency such as cadmium, lead, copper, zinc, and mercury. These results highlight the possibility of microalgae as strong and multifunctional remover of heavy metal in a wide range of wastewater scenarios (Chu et al., 2019). Table 1 summarizes how microalgae remove heavy metals through key mechanisms such as biosorption, bioaccumulation, and biotransformation. These processes involve surface binding of metal ions, intracellular uptake, and enzymatic conversion into less toxic forms, making microalgae effective agents for wastewater treatment.

Table 1. Mechanisms of Heavy Metal Removal by Microalgae

Mechanisms	Process	Process Type	Key Features	Examples
Biosorption	Attachment of heavy metal ions onto functional groups present on the surface of microalgal cells	Passive (metabolism-independent)	Rapid process Does not require energy Effective at low metal concentrations Involves functional groups like hydroxyl, amino	Plays a major role in initial metal removal from wastewater
Bioaccumulation	Uptake and accumulation of heavy metals inside the microalgal cells	Active (energy-dependent)	Requires metabolic activity Metals transported into intracellular compartments Enables long-term storage	Metals stored in vacuoles and cytoplasm
Biotransformation	Conversion of heavy metals into less toxic or less bioavailable forms	Enzymatic / metabolic	Involves redox reactions Reduces toxicity and mobility of metals Enhances detoxification efficiency	Example: $Cr^{6+} \rightarrow Cr^{3+}$ (less toxic form)
Metal-binding Peptides	Synthesis of peptides that bind and detoxify heavy metals	Biochemical defense mechanism	Includes phytochelatins and metallothioneins Chelate metal ions Facilitate sequestration into vacuoles	Reduces cytotoxic effects and increases tolerance
Environmental Influence	External factors affecting metal removal efficiency	Conditional	pH Competing ions Metal speciation Physiological state of algae	Determines efficiency and stability of bioremediation process
Removal Efficiency	Combined effect of all mechanisms in removing metals	Integrated process	Effective for multiple metals Works in lab and pilot-scale systems Adaptable to different wastewater conditions	Efficient removal of Cd, Pb, Cu, Zn, Hg

5. DETOXIFICATION OF HEAVY METALS BY FUNGI

Fungi are another significant group of microorganisms involved in the detoxification of heavy metals in addition to the microalgae. The fungi species have large mycelial networks and cell walls that are laden with chitin, glucans and proteins that offer them many binding sites to metal ions. Fungi metal removal mechanisms are biosorption, bioaccumulation, enzyme transformation, and precipitation. These are processes that usually supplement algal-mediated remediation especially in complicated wastewater systems (Kaparapu et al., 2018). Some filamentous fungi are able to immobilize metals by the generation of insoluble metal complexes or by changing metal speciation.

Additionally, fungi enzymes contribute to the decomposition of organic pollutants, and thus the toxicity of the wastewater is minimized. Treatment can be more effective with incorporation of fungal systems with microalgal processes because of covering a larger range of contaminants (Pradhan et al., 2019). Combination of microalgae and fungi in the treatment of wastewater acts as a bioremediation approach. The combination of the photosynthetic potential of algae and the high metal-binding and destructive abilities of fungi has been achieved in algal-fungal consortia. These integrated systems are more resilient, have better removal efficiencies, and adaptability to varying wastewater conditions, and they would be good solutions in future bioremediation technologies (Rashidi and Trindade, 2018).

6. APPLICATIONS OF MICROALGAE IN HEAVY METAL BIOREMEDIATION

There have been numerous microalgae uses in heavy metal bioremediation, including both industrial and domestic wastewater treatment. They can be adapted to various environmental factors and wastewater mixtures and hence implemented in any sector flexibly. Microalgae are used in treating effluents in mining, electroplating, textile, and pharmaceutical industries since their effluents usually have high levels of metals and other toxic substances. Microalgae play a significant role in the treatment of domestic wastewater such as nutrient removal, reduction of pathogens and general enhancement of water quality (Saavedra et al., 2018).

The use of microalgae in bioremediation of heavy metals is facilitated by technology in growing, reactor or reactor development and biomass harvesting. This has facilitated the upscaling of algal systems of laboratory experiments and pilots as well as full scale. With the ever tightening of regulatory requirements on wastewater discharge, the technologies based on microalgae provide the possible and the sustainable alternative to traditional ones (Sedlakova-Kadukova et al., 2019). Figure 2 illustrates the diverse applications of microalgae in treating industrial and domestic effluents by removing nutrients, heavy metals, organic pollutants, and reducing Chemical Oxygen Demand (COD) through bioaccumulation and biodegradation mechanisms for their dual role in wastewater purification and biomass generation for value-added products such as biofuels, biofertilizers, and bioproducts, promoting a circular and sustainable treatment approach.

Figure 2. Applications of Microalgae in Industrial and Domestic Effluent Treatment

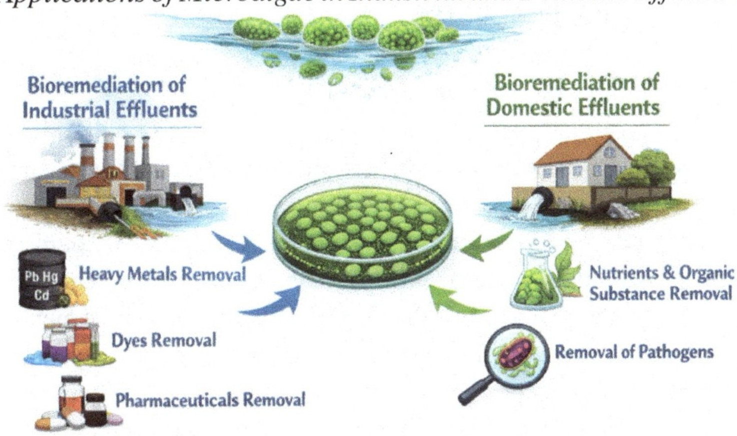

6.1. Bioremediation of Industrial Effluents

Through industrialization, there has been high release of complex and hazardous effluents into both water and soil. These effluents are usually rich in heavy metals, synthetic dyes, pharmaceutical residues and other recalcitrant organic pollutants which are very dangerous to the ecosystem and human health (Torres, 2020). Traditional forms of treatment that include physicochemical treatment like chemical precipitation, adsorption, and advanced oxidation process can be very costly, energy-consuming, and might produce byproducts. Bioremediation on the other hand provides a more environmentally friendly and less expensive alternative by utilizing the metabolic power of the microorganisms, plants, and their enzymes to get rid of the contaminants. Industrial effluent bioremediation depends on natural or artificial biological systems that convert the pollutants to less toxic forms, immobilize them or totally mineralizes them into harmless end products like carbon dioxide, water and biomass (Solisio et al., 2019).

Microorganisms such as bacteria, fungi, and algae, and actinomycotes are at the center of industrial effluent bioremediation because of their metabolic versatility and extreme environmental tolerance. These organisms are able to use pollutants either as a source of carbon or energy or co-metabolize them with the other potential substrates. Recent progress in molecular biology and systems biotechnology has further improved the performance of bioremediation by the discovery of functional genes, metabolic pathways and regulatory networks governing the pollutant degradation process. Interest in the use of microbial consortia (instead of single strains) has been awakened and synergetic interaction of microorganisms typically leads to enhanced degradation performance and stability under changing environmental conditions (Dixon and Wilken, 2018).

6.1.1. Heavy Metals Removal

Common industrial effluents include heavy metals like cadmium, lead, mercury, chromium, arsenic and nickel as a result of mining, electroplating, battery production, textile processing and metallurgical industries. Contrary to organic pollutants, the heavy metals are non-biodegradable, and accumulate in the environment thus biomagnification by the food chains. Prolonged contact with heavy metals may lead to acute health conditions, such as nervous system disorders, kidney complications, and cancer. The main concept of bioremediation plans to eliminate heavy metals is not based on degradation but on immobilization, sequestration or transformation (Sathya et al., 2023).

Biosorption of microbes is among the most popular processes of removing heavy metals. It entails passive adsorption of metal ions to functional groups e.g. carboxyl, hydroxyl, phosphate and amino groups on microbial cell walls. The process is flexible and cost-effective since both living and dead biomass can be employed in the process of biosorption. The biosorption capacity of fungal biomass, especially filamentous ones, has been high owing to the high surface area and the high number of binding sites. Algal biomass has also been widely investigated in terms of its capacity to adsorb a number of metal ions concurrently (Ye et al., 2023).

Besides biosorption, metal ions can be actively recycled in cells using microorganisms by the process of bioaccumulation. After internalization, the metals can be trapped by the metallothioneins, phytochelatins or polyphosphate granules, hence lowering their bioavailability and toxicity. Some bacteria can enzymatically convert metals to less toxic or less mobile forms, e.g. the reduction of hexavalent chromium to trivalent chromium or the methylation and demethylation of mercury. These microbial changes are very important in the purification of industrial effluents and in averting the propagation of metal contamination (Bhuyan, 2023).

6.1.2. Dyes Removal

Synthetic dyes have a very wide application in textile, leather, paper, cosmetic, and food industries which leads to release of colored effluents to water bodies. Dyes are highly coloured even in low concentration, lower light penetration and disrupt photosynthesis in aquatic ecosystems. Most dyes especially azo dyes cannot be treated by traditional wastewater treatment regimens and they might release toxic, mutagenic, or even carcinogenic by-products when partially degraded. Thus, the process of extracting dyes out of industrial effluents is a key environmental problem (Briffa et al., 2020).

The major processes involved in the bioremediation of dye contaminated effluents are microbial degradation, enzyme transformation and adsorption. The bacteria that have the capacity to generate azo reductases are able to cut the azo bonds either in anaerobic or microaerophilic environment, which results in decolorization. The further degradation and mineralization of aromatic amines that occur during the initial reduction step is possible through subsequent aerobic treatment. Fungi, in particular, white-rot fungi have shown exceptional dye-degrading capabilities because of their extracellular ligninolytic enzyme systems, such as laccases, lignin peroxidases and manganese peroxidases. These enzymes have limited substrate specificity thus can oxidize a broad variety of structurally diverse dyes (Hosseinzadeh et al., 2022).

Dye removal has potential to the use of algal-based bioremediation as a combination of biosorption and biodegradation. Algae have the ability to absorb dyes on their cells and in certain instances to metabolise them by enzymatic processes. Microbial and algal systems in bioreactors have provided a deeper capacity of dye removal and at the same time lower nutrient loads during wastewater treatment. In general, biological dye remediation gives a sustainable method that reduces the secondary pollution and contributes to the innovation of the environmentally friendly industrial wastewater treatment methods (Abdelfattah et al., 2023).

6.1.3. Pharmaceuticals Removal

Due to the wide usage, persistence, and low-level biological activity, pharmaceutical compounds have become the contaminants of concern in the industrial effluents. The effluents discharged by pharmaceutical manufacturing units, hospitals, and health facilities normally include antibiotics, analgesics, anti-inflammatory medicine, hormones, and psychiatric drugs. These drugs may have a negative impact on aquatic life, interfere with endocrine functions, and increase resistance to antimicrobials. Traditional wastewater treatment facilities do not usually treat pharmaceutical residues fully and therefore alternative treatment approaches should be employed (Nagarajan et al., 2022).

Bioremediation provides a good option to eliminate the pharmaceutical contaminant by microbial degradation and transformation. Some bacteria and fungi have enzymatic systems that can break down more complex pharmaceutical molecules into relatively simple, less toxic metabolites. Monooxygenases, dioxylases, dehydrogenases, and hydrolases are enzymes that are important in oxidative and reductive changes of pharmaceuticals. Compounds that do not act as primary sources of either carbon or energy are of particular interest, because degradation processes of the pharmaceuticals in the presence of the primary growth substrates are referred to as co-metabolic processes (Demertzioglou et al., 2022).

There are improved bioremediation techniques, such as biofilm reactor, membrane bioreactor and constructed wetland, which have shown improved pharmaceutical removal efficiencies. These systems offer good environments of microbial attachment, growth, and metabolism. Development of bioremediation can be enhanced with the other treatment methods, and the overall efficacy of industrial effluent treatment will be increased, minimizing the environmental impact of pharmaceutical production and consumption (Yang et al., 2022).

6.2. Bioremediation of Domestic Effluents

Municipal wastewater or sewage is domestic effluents that are emitted by households, institutions, and commercial premises. Such effluents are normally organic, nutrient, pathogen, suspended solids and trace chemicals. When released without proper treatment or treatment with insufficient treatment, domestic effluents may cause water pollution and eutrophication as well as transmission of waterborne diseases. Bioremediation is the core of domestic wastewater treatment which is based on naturally available microbial communities to stabilize organic matter and eliminate contaminants (Wei et al., 2022).

Activated sludge systems, trickling filters, anaerobic digesters or constructed wetlands are some of the biological treatment processes common in the treatment of domestic effluents. They are systems that make use of the metabolic processes of the varied microbial populations to transform organic pollutants into stable end products. Bioremediation-based domestic wastewater treatment is economical compared to physicochemical methods, is scalable and environmentally friendly. Recurrent research and technology development have resulted in the designing of better reactors and operational measures that result in better efficacy of treatment and recovery of resources (Lalucat et al., 2022).

6.2.1. Nutrients and Organic Substance Removal

Nitrogen and phosphorus excess in domestic effluents are other significant contributors of eutrophication in the receiving water bodies. Sewage pollutants that are organic in nature elevate the Biochemical Oxygen Demand (BOD) and Chemical Oxygen Demand (COD), which results in the depletion of oxygen and the harmful effects on aquatic organisms. The objective of bioremediation processes is to reduce the amount of organic load and nutrients to tolerable levels to either discharge or reuse effluents (Morseletto et al., 2022).

Decomposition of organic matter by microbes is done either by aerobic or anaerobic processes. The heterotrophic bacteria found in aerobic systems oxidize organic materials to produce carbon dioxide and water resulting in the production of biomass. Anaerobic digestion is a set of microbially mediated processes, which constitute hydrolysis, acidogenesis, acetogenesis, and methanogenesis that lead to generation of biogas. Anaerobic treatment is also economical to high-strength domestic wastewater because it will enable recovery of energy and low production of sludge (Chen et al., 2020).

Removal of biological nutrient depends on the specialized groups of microbes. At aerobic conditions, ammonia is transformed to nitrate by nitrifying bacteria, whereas at anoxic conditions, nitrate is reduced to form nitrogen gas by the denitrifying bacteria: a complete elimination of nitrogen in wastewater. The removal of phosphorus is done by the activity of polyphosphate-accumulating organisms which accumulate phosphorus in their cell. Optimal mechanisms of these biological processes are important towards meeting high standards of effluent discharges and ensuring the safety of aquatic life (Wang et al., 2022).

6.2.2. Removal of Pathogens

The effluents that are produced at the domestic scenes contain a large variety of pathogenic microorganisms such as bacteria, viruses, protozoa and helminths that bring about serious health hazards to the population. Untreated or poorly treated wastewater may be containing pathogens that may cause waterborne diseases like cholera, dysentery, typhoid, and hepatitis. Thus, efficient destructions of pathogens are a crucial goal of domestic treatment of the waterways (Hena et al., 2021).

Bioremediation helps in reducing pathogens by natural die-off, predation, competition and unfavourable environment in the biological treatment systems. In biofilm-based and activated sludge processes, the pathogens are lowered because of the long retention periods, antagonism of microbes, and adsorption onto the biomass and then removed together with the sludge. Further pathogen inactivation occurs by anaerobic digestion that increases temperatures, duration of exposure, and metabolic by-products like volatile fatty acids (Mo et al., 2022).

Waste stabilization ponds and constructed wetlands are especially good in the removal of pathogens and they associated physical filtration, sedimentation, ultraviolet radiation, and biological interactions. These are natural solutions that offer low-cost and sustainable solutions to domestic effluents, particularly in developing areas. Altogether, the pathogen elimination method, based on bioremediation, is crucial to the protection of human health and facilitation of the safe usage or discharge of treated domestic wastewater (Chan et al., 2022).

Bioremedation is a dynamic and environmental friendly method of treating industrial effluents as well as domestic effluents. Using the strategic application of microorganisms and biological systems, one can treat a broad spectrum of contaminants, such as heavy metals, dyes, pharmaceuticals, nutrients, organic matter, and pathogens. Further efficiency and reliability in bioremediation will be achieved through continued research, which will assist in protecting the environment and sustainable water management in the world (Hena et al., 2020).

7. RECENT TECHNOLOGIES IN HEAVY METAL BIOREMEDIATION

Molecular Biology, Nanotechnology, Systems Biology, and Process Engineering Molecular biology, nanotechnology, systems biology, and process engineering have played a major role in the recent developments in heavy metal bioremediation. Standard bioremediation methods, which mainly depended on naturally present microorganisms and plants, were usually afflicted with slowness of kinetics, lack of specificity of metals, and lack of adverse environmental response. The modern technologies are supposed to address these shortcomings by increasing the uptake of metals, transformation efficiency and the controllability of the system (Luo et al., 2021). The most significant technological advancement of these is the Genetically Engineered Microorganisms (GEMs). Microbes can be engineered to overexpress metallothioneins, phytochelatins, and metal transporters in recombinant DNA techniques to achieve high levels of intracellular toxic metal uptake of cadmium, mercury, lead and arsenic. Moreover, metabolic pathway engineering can be used to make microorganisms convert metals into less toxic or easily extractable biosites, including reduction of hexavalent chromium to trivalent chromium by microbes. Such developments provide the possibility of direct and expedited remediation and still relevant to the ecology when used under controlled circumstances (Lin et al., 2022).

The other technology that has transformed heavy metal bioremediation is in the use of nanobiotechnology, more so the use of biogenic nanoparticles that are produced by the microorganisms, algae and fungi. The synthesis of metal nanoparticles by microbes does not only detoxify the dissolved metal ions, but it also allows the recovery of metal nanoparticles as valuable nanomaterials (Arora et al., 2018). An example is that bacteria that can reduce silver, gold, and selenium ions can produce stable nanoparticles with industrial uses efficiently integrating remediation and resource recovery. In order to improve the adsorption capacity and easy separation of treated effluents with the aid of external magnetic fields, functionalized nanoparticles have been developed, such as magnetic nanoparticles coated with biopolymers or microbial biomass. Such nanomaterials are characterized by high surface area, tunable reactivity, and high metal affinity and are very efficient at low metal concentrations. Notably, bio-synthesis of nanoparticles reduces the risks of secondary pollution of chemically synthesized nanomaterials (Balzano et al., 2020).

With the advent of omics-based technologies (genomics, transcriptomics, proteomics, and metabolomics) has been the revolution in how heavy metal bioremediation processes are understood and optimized. High-throughput sequencing permits identification of the community of microorganisms resistant to metals and functional genes that facilitate the transportation, sequestration, and conversion of metals (Carvalho et al., 2020). The metagenomic techniques are also useful when one aims to find out more about unculturable microorganisms in polluted environments, they have demonstrated complex microbial consortia that synergize during bioremediation. Transcriptomic and proteomic studies also reveal the mechanisms and regulatory networks of stress responses to exposure to metal. These lessons can be used to rationally design microbial consortia and to select robust strains that can be used to operate under variable environmental conditions. Combined with bioinformatics and machine learning algorithms, omics data can predict the bioremediation performance, and the process of moving experimental work on the laboratory scale to the field is promoted (Danouche et al., 2020).

Recent developments in bioreactor design and intensification of the process have been an important factor in recent high-tech bioremediation bioreactors. Current bioreactors use immobilized cells, biofilms, and alginate- or chitosan-entrusted biomass to enhance the stability and reuse of microbes. Fixed-bed, fluidized-bed and membrane bioreactors have better mass transfer, minimal biomass washout, and greater metal removal efficiencies (Hamida et al., 2020). By combining the biosorption and bioaccumulation processes in the hybrid treatment systems, it is possible to remove various types of metals in complex effluents at the same time. Moreover, electro-biotechnology, e.g., microbial electrochemical systems, has provided new opportunities in reclaiming and detoxification of metals. These systems are used to enable microorganisms to contact electrodes to promote redox reactions to allow the reduction, precipitation, or recovery of metals and produce electrical energy or value-added products (Ibuot et al., 2020).

Digital technologies and artificial intelligence (AI) are slowly coming into the heavy metal bioremediation protocols to streamline process manipulation and decision-making. Environmental monitoring, omics, and reactor performance can be predicted using AI-driven models to analyze large amounts of data to determine metal removal efficiency and system stability in different conditions (Tsednee et al., 2019). Bioremediation systems that have digital twins enable virtual experimentation of operational conditions, thereby lowering the cost and risks of experiments. The monitoring systems by remote sensing and Internet of Things (IoT) also contribute to real-time monitoring of contamination levels and the progress of remediation. Collectively, these technologies are a paradigm shift to intelligent and adaptive and sustainable bioremediation systems that should overcome the complex and large scale heavy metal contamination (Wang et al., 2021).

8. CHALLENGES AND LIMITATIONS OF HEAVY METAL BIOREMEDIATION

The heavy metal bioremediation has some inherent challenges that restrict the large-scale application despite its great technological progress. Toxicity of heavy metals to the biological agents themselves is one of the major limitations. Metals are capable of enzyme inhibition, cell membrane disruption, and the formation of oxidative stress at high concentrations, which decreases microbial viability and remediation efficiency (Yu et al., 2019). Even though there are metal-resistant strains, the performance of these strains tend to reduce when there is mixed-metal contamination, or even when the environment changes. In addition, the metabolic load of the metal detoxification process may retard the growth of microorganisms, and thus, it may take more time to eliminate microorganisms than using physico-chemical treatments (Zhang et al., 2020).

Another significant constraint to the heavy metal bioremediation is environmental variability. PH, temperature, salinity, availability of oxygen, and presence of competitive ions are some factors that have a great impact on the speciation and bioavailability of metals (Boretti and Rosa, 2019). As an example, metals which are attached to organic matter or are in the insoluble complex form may not be taken up by microbes, which will decrease the effectiveness of the remediation. In a natural ecosystem, unevenness in soil or sediment organization also makes the process of uniform treatment complicated. Such environmental restrictions complicate extrapolation of laboratory-scale findings to the field-scale application and in most cases produces varying or unpredictable results (Zhang and Liu, 2021).

Another issue that is related to bioremediation processes is the risk of secondary pollution. Although biotransformation can lessen the toxicity of metals, it does not remove the metals. Mishandling of biomass containing metals will result in re-enterprises of the contaminants into the environment (Wu and Vaneeckhaute, 2021). Moreover, ecological and regulatory issues associated with the use of genetically engineered microorganisms are horizontal gene transfer and unintended ecological interactions, as well as issues of public acceptance. The application of GEMs in non-regulated settings is usually restricted by strict biosafety rules and ethical aspects that reduce the potential benefits (Zhang and Liu, 2022).

The large-scale use of heavy metal bioremediation is also hampered by economic and operational problems. The rearing and culture of microorganisms, feeding, and support of reactor technology may boost the operation fees, especially in the less developed countries (Zhang and Liu, 2021). Bioremediation can take long treatment periods as compared to the traditional chemical methods, which is not appealing to industries that want to be guided with the environment as soon as possible. Also, the absence of the common procedures related to the performance assessment and long-term observation makes the regulatory approval and technology transfer more difficult (Ahmed et al., 2022).

The interactions may modify metal bioavailability and microbial activity, which may change the effectiveness of the entire remediation process (Imwene et al., 2022). Majority of the existing literature is concentrated on single-metal systems and there is a gap in the knowledge about real-life, multi-pollutant situations. This complexity is one of the most important challenges which still need to be addressed in order to develop heavy metal bioremediation to practical and sustainable solutions (Dhangar et al., 2020). Table 2 summarizes the major challenges limiting the efficiency of biological heavy metal bioremediation, including metal toxicity to microorganisms, environmental variability (pH, temperature, salinity), low bioavailability, and competition with native microbial communities such as slow remediation rates, difficulty in biomass recovery, and reduced performance in complex, multi-metal contaminated sites.

Table 2. Key Challenges and Limitations Affecting the Efficiency of Biological Heavy Metal Bioremediation

Challenges	Limitations	Impact on Bioremediation Efficiency
Metal Toxicity to Biological Agents	Inhibition of microbial activity	High concentrations of heavy metals inhibit enzymatic functions, disrupt cellular membranes, and induce oxidative stress, reducing microbial survival and metabolic activity.
	Reduced viability of bioremediating organisms	Even metal-resistant strains may lose efficiency under extreme contamination levels or prolonged exposure.
	Metabolic burden	Energy-intensive detoxification mechanisms (e.g., efflux pumps, metal sequestration) slow microbial growth and extend remediation timelines.

continued on following page

Table 2. Continued

Challenges	Limitations	Impact on Bioremediation Efficiency
Environmental Variability	pH and temperature sensitivity	Fluctuations in pH and temperature affect microbial metabolism and alter metal speciation, impacting uptake and transformation rates.
	Salinity and oxygen availability	High salinity and low oxygen conditions limit the growth of many metal-remediating microbes, especially in coastal or subsurface environments.
	Competing ions	Presence of other ions reduces metal bioavailability through competition for binding or transport sites.
	Soil and sediment heterogeneity	Non-uniform distribution of contaminants and microbes limits consistent treatment in natural ecosystems.
Limited Metal Bioavailability	Metal complexation	Metals bound to organic matter or present as insoluble mineral complexes are inaccessible to microbial uptake.
	Low solubility forms	Precipitated or crystalline metal forms resist biological interaction, reducing remediation effectiveness.
Scalability Issues	Laboratory-to-field inconsistency	Controlled laboratory conditions do not accurately represent complex field environments, leading to unpredictable outcomes during scale-up.
	Treatment efficiency	Field-scale systems often show spatial variability in microbial activity and metal removal rates.
Risk of Secondary Pollution	Metal-laden biomass disposal	Improper handling or disposal of biomass containing accumulated metals can result in re-release of contaminants.
	Detoxification	Biotransformation reduces toxicity but does not remove metals permanently from the ecosystem.
Genetically Engineered Microorganisms (GEMs)	Ecological risks	Potential horizontal gene transfer and unintended ecological interactions raise environmental safety concerns.
	Regulatory restrictions	Strict biosafety and ethical regulations limit field application of GEMs.
	Public acceptance	Social and ethical concerns reduce stakeholder acceptance of genetically modified organisms in open environments.
Economic Constraints	High operational costs	Costs associated with microbial cultivation, nutrient supplementation, and system maintenance increase overall expenditure.
	Infrastructure requirements	Bioreactors and controlled treatment systems require significant capital investment.
	Long treatment duration	Compared to physicochemical methods, bioremediation often requires longer remediation periods.
Regulatory and Monitoring Challenges	Lack of standardized protocols	Absence of uniform guidelines for performance evaluation complicates regulatory approval.
	Long-term monitoring requirements	Continuous monitoring increases operational complexity and costs.

continued on following page

Table 2. Continued

Challenges	Limitations	Impact on Bioremediation Efficiency
Mixed-Contaminant Systems	Presence of multiple pollutants	Co-existing metals, organic compounds, and emerging contaminants interact synergistically or antagonistically.
	Altered microbial activity	Organic pollutants or toxic mixtures can suppress microbial metabolism and metal uptake.

9. FUTURE PERSPECTIVES

The future of heavy metal bioremediation lies in the development of multidisciplinary approaches that integrate biological, chemical, and physical techniques. Coupled bioremediation and adsorption or phytoremediation or electrochemical systems are likely to be more efficient and robust. Such integrated strategies can mitigate the limitations of approaches by leveraging the addition of microbial bioremediation with biochar or nanomaterial-based adsorption can enhance the elimination of metals in highly contaminated locations and stabilize leftover metals in the matrices treated (Khan et al., 2021). Advancements in synthetic biology and systems biology hold the potential to redefine the capabilities of bioremediation agents. Future research will likely focus on developing microbial chassis that balance high metal tolerance with low ecological impact and possess programmable responses to environmental signals (Patnaik and Mallick, 2021). Genome editing using CRISPR provides greater accuracy than ever before in engineering metal resistance systems, regulatory systems, and stress response systems. Furthermore, synthetic microbial consortia with complementary metabolic functions can be designed to enhance resilience and remediation performance in complex ecosystems. Such purpose-designed systems can be optimized to certain contamination profiles and customised remediation plans developed based on the needs of various industries (Bhatt et al., 2022).

Sustainability and circular economy principles are expected to be central to the design of future bioremediation technologies. Rather than viewing heavy metals solely as contaminants, next-generation approaches will focus on their recovery and reuse (González-Camejo et al., 2022). Bioleaching, biomineralization and synthesis of nanoparticles are avenues of transforming waste metals into useful products. This paradigm shift, viewing remediation as resource recovery, aligns with global sustainability goals and could improve the economic viability of bioremediation initiatives. Techno-economic analysis and life cycle assessment will find widespread use in assessing the environmental and economic trade-offs, which will guide the policy and investment decisions (Yadav et al., 2020).

Artificial intelligence, big data analytics, and remote sensing will further use advanced technology to provide the high extent of scalability and predictability of heavy metal bioremediation. AI-derived decision support systems have the potential to optimize site selection, microbial consortia design and operating parameters in real time. Climate-driven predictive modeling and land-use patterns can also aid in predicting the future contamination risks, allowing the active response to contamination. These digital technologies will especially be useful with huge and geographically remote contaminated areas (Blanco-Vieites et al., 2022). Continued collaboration among microbiologists, environmental engineers, data scientists, and policymakers will be essential for translating laboratory discoveries into effective field-ready solutions. When properly regulated, with the involvement of the community and the investment in research and development, the concept of heavy metal bioremediation can become one of the key points of sustainable environmental management in the nearest decades (Chin et al., 2020).

10. CONCLUSION

Microalgae in heavy metal bioremediation are a viable and sustainable solution to industrialization and urbanization-related wastewater contamination. Rapid growth, high surface-area-to-volume ratio, metabolic flexibility, and heavy metal removal via biosorption, bioaccumulation, and biotransformation are amazing traits of microalgae. These mechanisms efficiently detoxify dangerous contaminants, remove nutrients, generate oxygen, and enhance water quality. The integration of fungi and algal-fungal consortia improve cleanup efficiency by combining complementing biological mechanisms. Bioremediation systems are more effective and scalable to genetically altered microbes, nanobiotechnology, omics-based techniques, and enhanced bioreactor designs. Despite these benefits, metal toxicity, environmental unpredictability, economic limits, biomass collection, and secondary contamination issues prevent large-scale application. However, integrated treatment systems, artificial intelligence, and circular economy concepts may overcome these constraints and improve efficiency. The capacity to recover resources and produce biofuels and other bioproducts from algae biomass makes this strategy economically viable. In conclusion, multidisciplinary innovations and sustainable techniques can make microalgae-based bioremediation an eco-friendly and cost-effective solution for heavy metal contamination and long-term environmental and water sustainability.

REFERENCES

Abdelfattah, A., Ali, S. S., Ramadan, H., El-Aswar, E. I., Eltawab, R., Ho, S.-H., Elsamahy, T., Li, S., El-Sheekh, M. M., Schagerl, M., Kornaros, M., & Sun, J. (2023). Microalgae-based wastewater treatment: Mechanisms, challenges, recent advances, and future prospects. *Environmental Science and Ecotechnology, 13*, 100205. DOI: 10.1016/j.ese.2022.100205 PMID: 36247722

Ahmed, M., Mavukkandy, M. O., Giwa, A., Elektorowicz, M., Katsou, E., Khelifi, O., Naddeo, V., & Hasan, S. W. (2022). Recent developments in hazardous pollutants removal from wastewater and water reuse within a circular economy. *npj. NPJ Clean Water, 5*(1), 12. DOI: 10.1038/s41545-022-00154-5

Ahmed, N. E., Salem, S. S., & Hashem, A. H. (2022). Statistical optimization, partial purification, and characterization of phytase produced from *Talaromyces purpureogenus* NSA20 using potato peel waste and its application in dyes de-colorization. *Biointerface Research in Applied Chemistry, 12*(4), 4417–4431.

Ajiboye, T. O., Oyewo, O. A., & Onwudiwe, D. C. (2021). Simultaneous removal of organics and heavy metals from industrial wastewater: A review. *Chemosphere, 262*, 128379. DOI: 10.1016/j.chemosphere.2020.128379 PMID: 33182079

Arora, N., Gulati, K., Tripathi, S., & Pruthi, V. (2018). Algae as a budding tool for mitigation of arsenic from aquatic systems. In *Mechanisms of arsenic toxicity and tolerance in plants* (pp. 269–297). DOI: 10.1007/978-981-13-1292-2_12

Balzano, S., Sardo, A., Blasio, M., Chahine, T. B., Anno, F. D., Sansone, C., & Brunet, C. (2020). Microalgal metallothioneins and phytochelatins and their potential use in bioremediation. *Frontiers in Microbiology, 11*, 1–16. DOI: 10.3389/fmicb.2020.00517 PMID: 32431671

Bhatt, P., Bhandari, G., Bhatt, K., & Simsek, H. (2022). Microalgae-based removal of pollutants from wastewaters: Occurrence, toxicity and circular economy. *Chemosphere, 306*, 135576. DOI: 10.1016/j.chemosphere.2022.135576 PMID: 35803375

Bhuyan, M. S. (2023). Ecological risks associated with seaweed cultivation and identifying risk minimization approaches. *Algal Research, 69*, 102967. DOI: 10.1016/j.algal.2022.102967

Blanco-Vieites, M., Suárez-Montes, D., Delgado, F., Álvarez-Gil, M., Battez, A. H., & Rodríguez, E. (2022). Removal of heavy metals and hydrocarbons by microalgae from wastewater in the steel industry. *Algal Research, 64*, 102700. DOI: 10.1016/j.algal.2022.102700

Boretti, A., & Rosa, L. (2019). Reassessing the projections of the World Water Development Report. *npj. NPJ Clean Water, 2*(1), 15. DOI: 10.1038/s41545-019-0039-9

Brar, P. K., Örmeci, B., & Dhir, A. (2022). Algae: A cohesive tool for biodiesel production along with wastewater treatment. *Sustainable Chemistry and Pharmacy, 28*, 100730. DOI: 10.1016/j.scp.2022.100730

Briffa, J., Sinagra, E., & Blundell, R. (2020). Heavy metal pollution in the environment and their toxicological effects on humans. *Heliyon, 6*(9), e04691. DOI: 10.1016/j.heliyon.2020.e04691 PMID: 32964150

Carvalho, J. C., de, , & Irineudo, A. M. (2020). Microalgal biomass pretreatment for integrated processing into biofuels, food, and feed. *Bioresource Technology, 300*, 122719. DOI: 10.1016/j.biortech.2019.122719 PMID: 31956056

Chan, S. S., Khoo, K. S., Chew, K. W., Ling, T. C., & Show, P. L. (2022). Recent advances biodegradation and biosorption of organic compounds from wastewater: Microalgae-bacteria consortium-A review. *Bioresource Technology, 344*, 126159. DOI: 10.1016/j.biortech.2021.126159 PMID: 34673198

Chen, M., Chang, L., Zhang, J., Guo, F., Vymazal, J., He, Q., & Chen, Y. (2020). Global nitrogen input on wetland ecosystem: The driving mechanism of soil labile carbon and nitrogen on greenhouse gas emissions. *Environmental Science and Ecotechnology, 4*, 100063. DOI: 10.1016/j.ese.2020.100063 PMID: 36157707

Chin, J. Y., Chng, L. M., Leong, S. S., Yeap, S. P., Yasin, N. H. M., & Toh, P. Y. (2020). Removal of synthetic dye by *Chlorella vulgaris* microalgae as natural adsorbent. *Arabian Journal for Science and Engineering, 45*(9), 7385–7395. DOI: 10.1007/s13369-020-04557-9

Chu, W. L., & Phang, S. M. (2019). Biosorption of heavy metals and dyes from industrial effluents by microalgae. In *Microalgae biotechnology for development of biofuel and wastewater treatment* (pp. 599–634). Springer. DOI: 10.1007/978-981-13-2264-8_23

Danouche, M., El Ghachtouli, N., El Baouchi, A., & El Arroussi, H. (2020). Heavy metals phycoremediation using tolerant green microalgae: Enzymatic and non-enzymatic antioxidant systems for the management of oxidative stress. *Journal of Environmental Chemical Engineering, 8*(5), 104460. DOI: 10.1016/j.jece.2020.104460

Delanka-Pedige, H. M. K., Cheng, X., Munasinghe-Arachchige, S. P., Bandara, G. L. C. L., Zhang, Y., Xu, P., Schaub, T., & Nirmalakhandan, N. (2020). Conventional vs. algal wastewater technologies: Reclamation of microbially safe water for agricultural reuse. *Algal Research, 51*, 102022. DOI: 10.1016/j.algal.2020.102022

Demertzioglou, M., Genitsaris, S., Mazaris, A. D., Kyparissis, A., Voutsa, D., Kozari, A., Kormas, K. A., Stefanidou, N., Katsiapi, M., Michaloudi, E., & Moustaka-Gouni, M. (2022). A catastrophic change in a European protected wetland: From harmful phytoplankton blooms to fish and bird kill. *Environmental Pollution, 312*, 120038. DOI: 10.1016/j.envpol.2022.120038 PMID: 36030951

Deshaware, S., Marathe, S. J., Bedade, D., Deska, J., & Shamekh, S. (2021). Investigation on mycelial growth requirements of *Cantharellus cibarius* under laboratory conditions. *Archives of Microbiology*, *203*(4), 1539–1545. DOI: 10.1007/s00203-020-02142-0 PMID: 33399895

Dhangar, K., & Kumar, M. (2020). Tricks and tracks in removal of emerging contaminants from the wastewater through hybrid treatment systems: A review. *The Science of the Total Environment*, *738*, 140320. DOI: 10.1016/j.scitotenv.2020.140320 PMID: 32806367

Dixon, C., & Wilken, L. R. (2018). Green microalgae biomolecule separations and recovery. *Bioresources and Bioprocessing*, *5*(1), 9. DOI: 10.1186/s40643-018-0199-3

Dudhagara, D. R., Javia, B. M., & Vala, A. K. (2023). Exploiting marine fungi in the removal of hazardous pollutants and biomass valorisation. In *Marine organisms: A solution to environmental pollution? Uses in bioremediation and in biorefinery* (pp. 117–146). Springer. DOI: 10.1007/978-3-031-17226-7_7

El-Naggar, N. E.-A., Hamouda, R. A., Saddiq, A. A., & Alkinani, M. H. (2021). Simultaneous bioremediation of cationic copper ions and anionic methyl orange azo dye by brown marine alga *Fucus vesiculosus*. *Scientific Reports*, *11*(1), 3555. DOI: 10.1038/s41598-021-82827-8 PMID: 33574404

Gao, T., Li, Q., Wang, X., Liu, Y., Zhao, Z., Han, L., & Chen, X. (2022). Characteristics and diversity of microbial communities in lead-zinc tailings under heavy metal stress in north-west China. *Letters in Applied Microbiology*, *74*(2), 277–287. DOI: 10.1111/lam.13608 PMID: 34822179

Gondi, R., Kavitha, S., Kannah, R. Y., Karthikeyan, O. P., Kumar, G., Tyagi, V. K., & Banu, J. R. (2021). Algal-based system for removal of emerging pollutants from wastewater: A review. *Bioresource Technology*, *344*, 126245. DOI: 10.1016/j.biortech.2021.126245 PMID: 34743994

González-Camejo, J., Aparicio, S., Pachés, M., Borrás, L., & Seco, A. (2022). Comprehensive assessment of the microalgae-nitrifying bacteria competition in microalgae-based wastewater treatment systems: Relevant factors, evaluation methods and control strategies. *Algal Research*, *61*, 102563. DOI: 10.1016/j.algal.2021.102563

Hamida, R. S., Ali, M. A., Redhwan, A., & Bin-Meferij, M. M. (2020). Cyanobacteria–a promising platform in green nanotechnology: A review on nanoparticles fabrication and their prospective applications. *International Journal of Nanomedicine*, *15*, 6033–6066. DOI: 10.2147/IJN.S256134 PMID: 32884261

Hasanin, M. S., Hashem, A. H., Abd El-Sayed, E. S., & El-Saied, H. (2020). Green eco-friendly bio-deinking of mixed office waste paper using various enzymes from *Rhizopus microsporus* AH3: Efficiency and characteristics. *Cellulose (London, England)*, *27*(8), 4443–4453. DOI: 10.1007/s10570-020-03071-3

Hena, S., Gutierrez, L., & Croue, J. P. (2020). Removal of metronidazole from aqueous media by *C. vulgaris*. *Journal of Hazardous Materials*, *384*, 121400. DOI: 10.1016/j.jhazmat.2019.121400 PMID: 31624001

Hena, S., Gutierrez, L., & Croue, J. P. (2021). Removal of pharmaceutical and personal care products (PPCPs) from wastewater using microalgae: A review. *Journal of Hazardous Materials*, *403*, 124041. DOI: 10.1016/j.jhazmat.2020.124041 PMID: 33265054

Hosseinzadeh, A., Zhou, J. L., Li, X., Afsari, M., & Altaee, A. (2022). Techno-economic and environmental impact assessment of hydrogen production processes using bio-waste as renewable energy resource. *Renewable & Sustainable Energy Reviews*, *156*, 111991. DOI: 10.1016/j.rser.2021.111991

Husain, R., Sharma, S., Khan, S., & Mishra, A. (2022). Microbial bioremediation of heavy metals by marine bacteria. In *Development in wastewater treatment research and processes* (pp. 177–203). Elsevier. DOI: 10.1016/B978-0-323-85839-7.00014-1

Ibuot, A., Webster, R. E., Williams, L. E., & Pittman, J. K. (2020). Increased metal tolerance and bioaccumulation of zinc and cadmium in *Chlamydomonas reinhardtii* expressing a AtHMA4 C-terminal domain protein. *Biotechnology and Bioengineering*, *117*(10), 2996–3005. DOI: 10.1002/bit.27476 PMID: 32579250

Imwene, K. O., Ngumba, E., & Kairigo, P. K. (2022). Emerging technologies for enhanced removal of residual antibiotics from source-separated urine and wastewaters: A review. *Journal of Environmental Management*, *322*, 116065. DOI: 10.1016/j.jenvman.2022.116065 PMID: 36063692

Ji, H., Zhu, Y., Duan, J., Liu, W., & Zhao, D. (2019). Reductive immobilization and long-term remobilization of radioactive pertechnetate using bio-macromolecules stabilized zero valent iron nanoparticles. *Chinese Chemical Letters*, *30*(12), 2163–2168. DOI: 10.1016/j.cclet.2019.06.004

Kapahi, M., & Sachdeva, S. (2019). Bioremediation options for heavy metal pollution. *Journal of Health & Pollution*, *9*(24), 191203. DOI: 10.5696/2156-9614-9.24.191203 PMID: 31893164

Kaparapu, J., & Krishna Prasad, M. (2018). Equilibrium, kinetics and thermodynamic studies of cadmium(II) biosorption on *Nannochloropsis oculata. Applied Water Science*, *8*(6), 1–9. DOI: 10.1007/s13201-018-0810-y

Khan, N. A., Khan, A. H., Tiwari, P., Zubair, M., & Naushad, M. (2021). New insights into the integrated application of Fenton-based oxidation processes for the treatment of pharmaceutical wastewater. *Journal of Water Process Engineering*, *44*, 102440. DOI: 10.1016/j.jwpe.2021.102440

Lalucat, J., Gomila, M., Mulet, M., Zaruma, A., & García-Valdés, E. (2022). Past, present and future of the boundaries of the *Pseudomonas* genus: Proposal of *Stutzerimonas* gen. nov. *Systematic and Applied Microbiology*, *45*(1), 126289. DOI: 10.1016/j.syapm.2021.126289 PMID: 34920232

Lin, W., Chen, L., Tan, Z., Deng, Z., & Liu, H. (2022). Application of filamentous fungi in microalgae-based wastewater remediation for biomass harvesting and utilization: From mechanisms to practical application. *Algal Research*, *62*, 102614. DOI: 10.1016/j.algal.2021.102614

Luo, L., Luo, S., Wang, H., Hu, K., Lin, X., Liu, L., & Yan, B. (2021). Effect of nano-TiO2 on humic acid utilization from piggery biogas slurry by microalgae. *Bioresource Technology*, *337*, 125414. DOI: 10.1016/j.biortech.2021.125414 PMID: 34174768

Mo, Z., Tai, D., Zhang, H., & Shahab, A. (2022). A comprehensive review on the adsorption of heavy metals by zeolite imidazole framework (ZIF-8) based nanocomposite in water. *Chemical Engineering Journal*, *443*, 136320. DOI: 10.1016/j.cej.2022.136320

Mojiri, A., Zhou, J. L., Ratnaweera, H., Rezania, S., & Mansoureh, V. N. (2022). Pharmaceuticals and personal care products in aquatic environments and their removal by algae-based systems. *Chemosphere*, *288*, 132580. DOI: 10.1016/j.chemosphere.2021.132580 PMID: 34687686

Morseletto, P., Mooren, C. E., & Munaretto, S. (2022). Circular economy of water: Definition, strategies and challenges. *Circular Economy and Sustainability*, *2*(4), 1–15. DOI: 10.1007/s43615-022-00165-x

Nagarajan, D., Lee, D.-J., Varjani, S., Lam, S. S., Allakhverdiev, S. I., & Chang, J.-S. (2022). Microalgae-based wastewater treatment—Microalgae-bacteria consortia, multi-omics approaches and algal stress response. *The Science of the Total Environment, 845*, 157110. DOI: 10.1016/j.scitotenv.2022.157110 PMID: 35787906

Patnaik, R., & Mallick, N. (2021). Microalgal biodiesel production: Realizing the sustainability index. *Frontiers in Bioengineering and Biotechnology, 9*, 620777. DOI: 10.3389/fbioe.2021.620777 PMID: 34124015

Pradhan, D., Sukla, L. B., Mishra, B. B., & Devi, N. (2019). Biosorption for removal of hexavalent chromium using microalgae *Scenedesmus* sp. *Journal of Cleaner Production, 209*, 617–629. DOI: 10.1016/j.jclepro.2018.10.288

Rashidi, B., & Trindade, L. M. (2018). Detailed biochemical and morphologic characteristics of the green microalga *Neochloris oleoabundans* cell wall. *Algal Research, 35*, 152–159. DOI: 10.1016/j.algal.2018.08.033

) Saavedra, R., Muñoz, R., Taboada, M. E., Vega, M., & Bolado, S. (2018). Comparative uptake study of arsenic, boron, copper, manganese and zinc from water by

Sathya, R., Arasu, M. V., Al-Dhabi, N. A., Vijayaraghavan, P., Ilavenil, S., & Rejiniemon, T. (2023). Towards sustainable wastewater treatment by biological methods—Challenges and advantages of recent technologies. *Urban Climate, 47*, 101378. DOI: 10.1016/j.uclim.2022.101378

Sedlakova-Kadukova, J., Kopcakova, A., Gresakova, L., Godany, A., & Pristas, P. (2019). Bioaccumulation and biosorption of zinc by a novel *Streptomyces* K11 strain isolated from highly alkaline aluminium brown mud disposal site. *Ecotoxicology and Environmental Safety, 167*, 204–211. DOI: 10.1016/j.ecoenv.2018.09.123 PMID: 30340085

Solisio, C., Al Arni, S., & Converti, A. (2019). Adsorption of inorganic mercury from aqueous solutions onto dry biomass of *Chlorella vulgaris*: Kinetic and isotherm study. *Environmental Technology, 40*(5), 664–672. DOI: 10.1080/09593330.2017.1400114 PMID: 29098939

Torres, E. (2020). Biosorption: A review of the latest advances. *Processes (Basel, Switzerland), 8*(12), 1584. DOI: 10.3390/pr8121584

Tsednee, M., Castruita, M., Salomé, P. A., Sharma, A., Lewis, B. E., Schmollinger, S. R., Strenkert, D., Holbrook, K., Otegui, M. S., Khatua, K., Das, S., Datta, A., Chen, S., Ramon, C., Ralle, M., Weber, P. K., Stemmler, T. L., Pett-Ridge, J., Hoffman, B. M., & Merchant, S. S. (2019). Manganese co-localizes with calcium and phosphorus in *Chlamydomonas* acidocalcisomes and is mobilized in manganese-deficient conditions. *The Journal of Biological Chemistry*, *294*(46), 17626–17641. DOI: 10.1074/jbc.RA119.009130 PMID: 31527081

Vazirzadeh, A., Jafarifard, K., Ajdari, A., & Chisti, Y. (2022). Removal of nitrate and phosphate from simulated agricultural runoff water by *Chlorella vulgaris*. *The Science of the Total Environment*, *802*, 149988. DOI: 10.1016/j.scitotenv.2021.149988 PMID: 34525699

Wang, C., Luo, D., Zhang, X., Huang, R., Cao, Y., Liu, G., Zhang, Y., & Wang, H. (2022). Biochar-based slow-release of fertilizers for sustainable agriculture: A mini review. *Environmental Science and Ecotechnology*, *10*, 100167. DOI: 10.1016/j.ese.2022.100167 PMID: 36159737

Wang, Y., Selvamani, V., Yoo, I., Kim, T. W., & Hong, S. H. (2021). A novel strategy for the microbial removal of heavy metals: Cell-surface display of peptides. *Biotechnology and Bioprocess Engineering; BBE*, *26*(1), 1–9. DOI: 10.1007/s12257-020-0218-z

Wei, L., Xiaopan, S., & Heydari, F. (2022). Microalgae classification using improved metaheuristic algorithm. *Mathematical Problems in Engineering*, *2022*, 3783977.

Wu, H., & Vaneeckhaute, C. (2021). Nutrient recovery from wastewater: A review on the integrated physicochemical technologies of ammonia stripping, adsorption and struvite precipitation. *Chemical Engineering Journal*, *433*, 133664. DOI: 10.1016/j.cej.2021.133664

Yadav, G., Shanmugam, S., Sivaramakrishnan, R., Kumar, D., Mathimani, T., Brindhadevi, K., Pugazhendhi, A., & Rajendran, K. (2020). Mechanism and challenges behind algae as a wastewater treatment choice for bioenergy production and beyond. *Fuel*, *285*, 119093. DOI: 10.1016/j.fuel.2020.119093

Yang, Y., Fan, X., Zhang, J., Qiao, S., Wang, X., Zhang, X., Miao, L., & Hou, J. (2022). A critical review on the interaction of iron-based nanoparticles with blue-green algae and their metabolites: From mechanisms to applications. *Algal Research*, *64*, 102670. DOI: 10.1016/j.algal.2022.102670

Ye, M., & Li, Y. Y. (2023). Methanogenic treatment of dairy wastewater: A review of current obstacles and new technological perspectives. *The Science of the Total Environment*, *866*, 161447. DOI: 10.1016/j.scitotenv.2023.161447 PMID: 36621500

Yu, Z., Zhang, T., Hao, R., & Zhu, Y. (2019). Sensitivity of *Chlamydomonas reinhardtii* to cadmium stress is associated with phototaxis. *Environmental Science. Processes & Impacts*, *21*(6), 1011–1020. DOI: 10.1039/C9EM00013E PMID: 31120077

Zhan, Y., Shen, X., Chen, M., Yang, K., & Xie, H. (2022). Bioleaching of tellurium from mine tailings by indigenous *Acidithiobacillus ferrooxidans*. *Letters in Applied Microbiology*, *75*(5), 1076–1083. DOI: 10.1111/lam.13569 PMID: 34586632

Zhang, J., Zhou, F., Liu, Y., Huang, F., & Zhang, C. (2020). Effect of extracellular polymeric substances on arsenic accumulation in *Chlorella pyrenoidosa*. *The Science of the Total Environment*, *704*, 135368. DOI: 10.1016/j.scitotenv.2019.135368 PMID: 31831249

Zhang, X., & Liu, Y. (2021). Circular economy-driven ammonium recovery from municipal wastewater: State of the art, challenges and solutions forward. *Bioresource Technology*, *334*, 125231. DOI: 10.1016/j.biortech.2021.125231 PMID: 33962161

Zhang, X., & Liu, Y. (2021). Reverse osmosis concentrate: An essential link for closing loop of municipal wastewater reclamation towards urban sustainability. *Chemical Engineering Journal*, *421*, 129851. DOI: 10.1016/j.cej.2020.127773

Zhang, X., & Liu, Y. (2022). Resource recovery from municipal wastewater: A critical paradigm shift in the post era of activated sludge. *Bioresource Technology*, *363*, 127932. DOI: 10.1016/j.biortech.2022.127932 PMID: 36096327

Chapter 9
Diatom-Based Wastewater Polishing Heavy Metal and Micro-Pollutant Uptake

Sakibul Islam Ratul
http://orcid.org/0009-0000-7217-5190
Daffodil International University, Dhaka, Bangladesh

Md. Shoeab Akhter
http://orcid.org/0009-0004-4582-2937
International University of Business Agriculture and Technology, Dhaka, Bangladesh

Upoma Chowdhury
Chattogram Veterinary and Animal Sciences University, Khulshi, Bangladesh

Mohammed Ataur Rahman
International University of Business Agriculture and Technology, Dhaka, Bangladesh

ABSTRACT

The presence of heavy metals and trace organic micropollutants in the effluent water is still a significant issue, since conventional purification processes are unable to eliminate these contaminants, while advanced techniques are expensive and require substantial amounts of energy. Siliceous unicellular microalgae called diatoms have been shown to be a promising and environmentally-friendly technology to purify wastewater. Their specific structure and metabolic activity allow for effective removal of contaminants via the use of EPS, silica frustule adsorption, and intra-

DOI: 10.4018/979-8-2600-3080-6.ch009

cellular transformation. The chapter examines the mechanisms of the process and discusses the performance of the system in various types of bioreactors, including biofilm reactors, high-rate algal ponds, and algae-bacteria consortia. Innovative approaches like functionalization of the frustule and increased EPS production are also analyzed.

1. INTRODUCTION

The world of wastewater management is changing significantly in terms of the challenges it is facing. The conventional treatment regimes were to eliminate suspended solids, organic matter and pathogens. But nowadays wastewater is a complex combination of intractable pollutants not sufficiently treated by traditional infrastructure. A huge percentage of wastewater in the world are released without proper treatment, such as heavy metals, residual pharmaceuticals endocrine-disrupting compounds (EDCs), and antibiotic resistance factors into the waters (Boro et al., 2025). Some of the heavy metals like cadmium, lead, mercury, chromium, copper, zinc, nickel and arsenic are a result of industrial processes such as electroplating, mining and textile processing. These chemicals are toxic at extremely low levels, are concentrated in the biological systems and will remain in the environment. Concurrently, pharmaceuticals and personal care products have become widely used and this has brought with it a new category of micropollutants in the wastewater systems. They usually exhibit low concentrations, but can be biologically important, such as disrupting hormones, being toxic to aquatic organisms, and causing antibiotic resistance. Some of these pollutants can be dealt with using conventional tertiary treatment technologies but these are usually costly and energy consuming (Asif et al., 2026). They are also less efficient at low levels of contaminant and such levels are common in treated wastewater. Conversely, microalgal-based treatment systems are low-energy systems which are mostly run by solar power. Amongst them, diatoms have become the focus of growing attention because they have been shown to be more effective in the removal of heavy metals, as well as organic micropollutants.

Diatoms possess several unique characteristics that distinguish them from other microalgae used in wastewater treatment. Diatoms have a number of distinctive features separating them among other microalgae utilised in the treatment of wastewater. One of the most distinctive features of diatoms is their silica-based cell wall, known as the frustule. The surface of this structure is highly porous, and therefore has a high surface area which allows adsorption of metal ions to occur effectively (Hamed, 2023). Cation exchange processes can also be done on the frustule on its chemical properties, which has also helped in the removal of pollutants in the non-living biomass. The other important characteristic is that of extracellular polymeric substances (EPS). These materials create a tracery around the cell and have negatively charged functional groups which attract and bind metal ions in the surrounding water. This process helps concentrate the pollutants at the cell surface, which ultimately improves the pollutant removal efficiency. Diatoms also have very complex metabolic pathways capable of converting organic micropollutants. They have an enzymatic system that permits the breakdown of complex chemical substances and this helps in the process of detoxification of wastewater. Together, these structural, chemical, and biological characteristics create a multifunctional system capable of removing contaminants through both physicochemical and biological mechanisms. The chapter gives a general account of the diatom based wastewater treatment system with particular emphasis on the use of the systems in the elimination of heavy metals and micropollutants. It incorporates existing research of biological processes, system architecture, and implementation. The chapter starts by drawing the basic principles of wastewater treatment, and contaminant behaviour. It proceeds to discuss the biological features of the diatoms that make them effective. Later sections address the process of pollutant uptake, system configurations, and optimization strategies of operation. Additional chapters deal with analytical techniques, modelling strategies and integration of treatment systems in larger resource recovery systems. The chapter ends with a summary of the findings, discussion of implications in practice, and future research and implementation recommendations. This chapter specifically highlights the role of diatoms in the simultaneous removal of heavy metals and emerging organic micropollutants, a process that has not been the focus of previous reviews. The chapter summarizes structural characteristics of diatoms, such as frustule architecture and extracellular polymeric substances (EPS), and intracellular detoxification and biotransformation processes that control contaminant removal. Moreover, it connects the biological treatment processes with the concepts of resource recovery by emphasizing the possibility of recycling harvested biomass, recovering metals, producing biofuels, and utilizing silica-based material applications in a circular bioeconomy. The chapter also incorporates the latest advances in engineered frustule technologies, microalgae–bacteria consortia, and wastewater polishing system design, providing a cohesive view that spans the fundamental mechanisms, engineering applications, and sustainability implications. The sequential roles of diatom structural components, extracellular polymeric substances (EPS), and intracellular metabolic processes in contaminant removal are illustrated in Figure 1.

Figure 1. Mechanistic Pathway of Diatom-Based Wastewater Treatment from Contaminant Capture to Purified Effluent

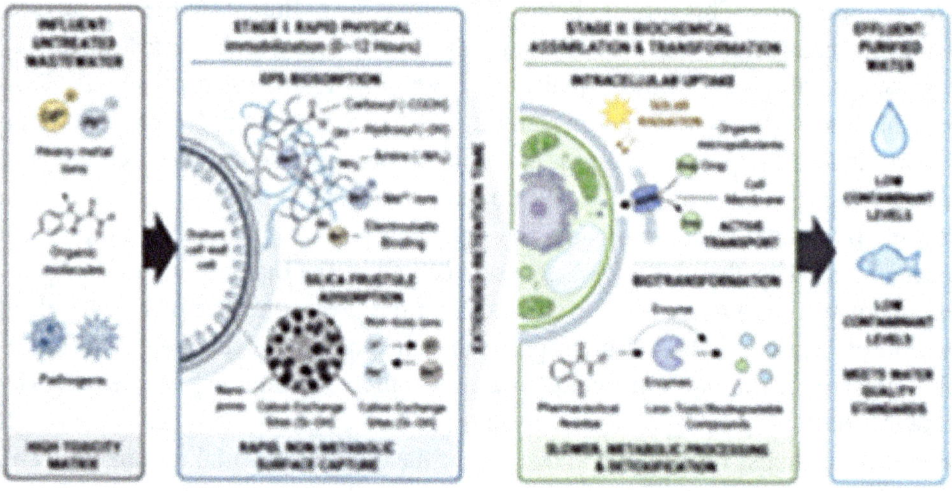

2. THEORETICAL OVERVIEW

2.1 Principles of Microalgal-Based Wastewater Treatment

Microalgal wastewater treatment is a conglomeration of a number of biological processes that work in tandem, helping to eliminate several types of contaminants. These processes are effective depending on the nature of contaminants involved, species of microalgae, and the working conditions of the system. Photoautotrophic assimilation is one of these processes, in which microalgae take in inorganic nutrients, including nitrogen and phosphorus, changing them into biomass (Maltsev et al., 2023). Through this process, the levels of nutrients in wastewater are reduced dramatically and under the appropriate conditions of light and the environment. Carbon dioxide fixation is another mechanism that provides oxygen as a by-product. This oxygen promotes aerobic bacterial activities and breakdown organic matter minimizing biochemical oxygen needs. Biosorption is a quick procedure where dissolved metals and organic substances are attached to the microalgal cells and extra-cellular structures (Bai et al., 2025). This action works based on electrostatic interactions, chemical bonding, and hydrophobic interactions, making it possible to remove contaminants efficiently within a short period. Also, microalgae can biotransform the organic micropollutants into a less damaging or more degradable state through enzymatic biotransformation. It is a process which involves a number of metabolic processes which change the chemical structure of the pollutants. All these processes must be combined in the effective design of a system, and the conditions in the reactor and the approach to biomass management must be aligned to allow all these processes to work in harmony.

2.2 Tertiary Treatment: Standards and Performance Benchmarks

The standard of water quality concerning heavy metals and micropollutants has been raised and is now very stringent because of the use of better and advanced detection technologies and the increasing awareness of the effects they have on the environment and human health. Heavy metals have a relatively low level of regulatory limits as they are toxic and remain in the environment. In the same vein, the organic micropollutants like pharmaceutical residues and synthetic hormones are controlled at very low concentrations since they may interfere with the biological systems even at low concentrations. Such high standards pose a great challenge to the waste water treatment systems, especially the less efficient conventional methods of waste water treatment that do not effectively eliminate low-concentration contaminants. Microalgal treatment systems have shown the capability of achieving significant decreases in nutrient levels as well as some contaminants, hence, it can be a viable alternative in tertiary treatment (Liu & Hong, 2021). Nevertheless, the achievement of the strictest standards of micropollutants is a primary objective and needs additional adjustment of the biological and operational processes.

2.3 Micropollutant and Heavy Metal Behaviour of Wastewater Matrices

Several physicochemical factors can determine the behaviour of heavy metals in wastewater, they are the pH, dissolved organic matter, alkalinity, and redox conditions (Liang et al., 2023). These elements dictate the chemical forms that metals are found, thus determining their availability to be removed. Metals are found at normal wastewater pH in various forms, such as free ions, inorganic complexes and organic-bound (Akhter et al., 2026). The relative abundance of each form changes with the environmental conditions, and the abundance of dissolved organic matter tends to result in greater metal complexation. Extracellular products of diatoms possess a high metal binding capacity, and can often compete with dissolved organic matter in terms of binding metals. This improves the removal of metals in complex wastewater matrices too. In the case of organic micropollutants, chemical characteristics like hydrophobicity are very important in influencing behaviour. More hydrophobic compounds are more likely to be associated with cellular components and extracellular matrices whereas more polar compounds are unlikely to accumulate and more dependent on biochemical transformation processes to be removed (Smułek & Kaczorek, 2022). Microalgal systems can also respond to exposure to contaminants, including the production of more extracellular substances and activation of protective metabolic pathways. These reactions are capable of increasing the total ability of the system to eliminate pollutants in real environmental situations.

3. DIATOMS: STRUCTURE, BIOLOGY AND BIOLOGICAL ROLE

3.1 Taxonomy and Classification

Diatoms are microalgae of the division Bacillariophyta of the kingdom Chromista, and are extremely diverse, comprising thousands of species. They can be widely categorized into two major groups in terms of their symmetry. Radially symmetrical diatoms are called centric diatoms, and are common in planktons. On the other hand, the pennate diatoms have bilateral symmetry and are commonly characterized with the surfaces or sediments. Pennate diatoms are more appropriate in treating wastewater. They are more tolerant to nutrient-rich and contaminated conditions, create stable biofilms on surfaces and can perform in different environmental conditions (Agarwal, 2025). A number of species in this category have been extensively researched owing to their effectiveness in eliminating pollutants and ability to respond to the characteristics of wastewater.

3.2 Frustule Architecture

The silica-comprising cell structure, called the frustule, is the characteristic of diatoms. This structure is obtained by a very regulated biological process that generates complex porous structures at more than one scale. The frustule harbors pores of different sizes which combined form a huge surface area to interact with dissolved substances. These pores allow the flow of nutrients and contaminants and structural strength. Silanol groups that are abundant on the surface of the frustule have the ability to react with metal ions by cation exchange reactions (Buhani et al., 2026). This will enable the frustule to be directly involved in removing pollutants, even in non-living biomass. Moreover, the silica framework of the frustule may be chemically engineered to increase the adsorption capacity and selectivity. These modifications allow the better elimination of certain contaminants, which makes the diatom-based materials interesting in a high-tech treatment.

3.3 EPS: Composition, Function and Metal Affinity

Diatoms generate extra cellular polymeric materials that encircle the cells in a gel-like structure. This matrix contains a lot of functional groups which are able to bind metal ions and other contaminants (Priya et al., 2024). These substances are constituted of polysaccharides, proteins, and charged functional groups that make them highly interact with dissolved metals. This leads to the high capacity to bind contaminants as compared to most other microalgal systems. Another effect of the extracellular matrix is the localized environment in which the contaminants become concentrated towards the cell surface. This improves efficiency of the surface adsorption and cellular uptake. Notably, encountering contaminants can provoke higher levels of production of these extracellular substances. This adaptive response enhances the overall capacity of the system to capture and retain pollutants under varying environmental conditions.

3.4 Wastewater Adaptation, Growth Kinetics and Silicon Nutrition

The diatoms development in the wastewater systems is determined by the presence of nutrients, light and dissolved silicon. Diatoms like other microalgae use nitrogen, phosphorus and carbon to grow, however it also needs silicon to make their silica cell walls. Silicon (availability) can also affect biomass generation in wastewater systems, especially when the cell density is increased. In certain systems, silicon supplementation might be required to ensure the optimal growth and performance. The growth of diatoms under low light levels and low temperature is a great advantage over most other microalgae (Dolganyuk et al., 2020). This enables diatoms to remain productive under environmental conditions that may limit the growth of other microalgal species. They are versatile to various wastewater compositions and environmental conditions, making them applicable in a broad spectrum of treatment applications. They are able to sustain constant growth rates at different nutrient concentrations and contamination conditions, which makes them applicable in the context of real-life wastewater treatment.

4. MECHANISMS OF HEAVY METAL AND MICROPOLLUTANT UPTAKE

4.1 EPS-Mediated Biosorption: Quantitative Data

The uptake of heavy metals by diatoms takes place in several compartments where all the compartments play a role in the total uptake. Sequential extraction techniques have shown that metals are localized to extracellular polymeric fractions, the cell surface and intracellular fractions. There is a considerable amount of metal absorption connected with the extracellular matrix, the first and the fastest binding site. The frustule surface adsorption adds a further fraction and the intracellular accumulation grows with time as metabolic processes become active (Zhang et al., 2025). These give out varying contributions with time of exposure. Removal is majorly governed by extracellular binding and surface interactions during the initial contact period, that takes place in a matter of hours. Increased accumulation in the cell is more pronounced over extended periods, thus increasing the overall removal efficiency. This means that short retention-time systems can realize high removal by rapid biosorption mechanisms alone, without more complex removal mechanisms. But to obtain the highest removal efficiency longer contact times are needed to enable intracellular processes to be effective. The combination of their extracellular matrix and silica-based cell wall tends to enable diatoms to have greater metal uptake capacities than other microalgae since the extracellular matrix and cell wall offer several different binding sites and contaminant removal mechanisms (Sangeetha et al., 2023). Figure 2 summarizes the principal extracellular and intracellular pathways responsible for contaminant capture, adsorption, detoxification, and transformation in diatom-based wastewater treatment systems.

Figure 2. Extracellular and Intracellular Mechanisms Governing Heavy Metal and Micropollutant Removal by Diatoms

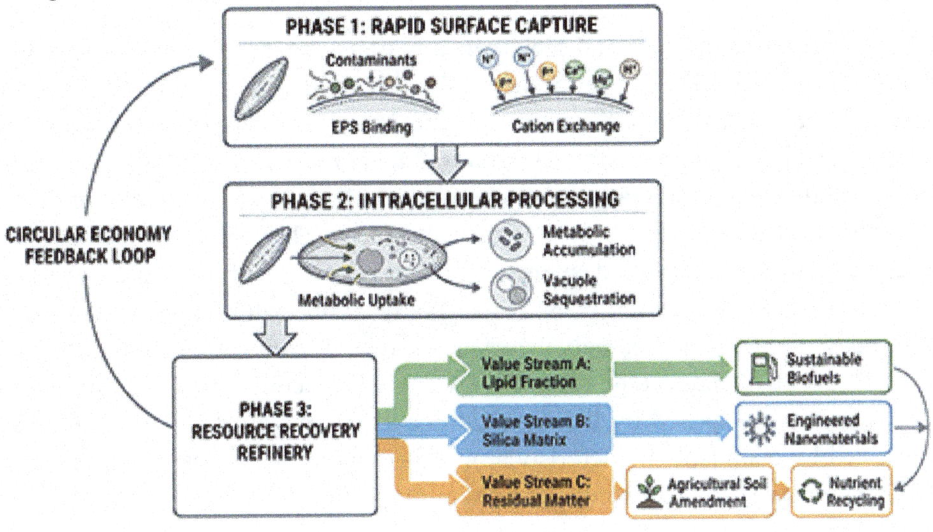

4.2 Intracellular Accumulation Phytochelatins, Metallothioneins and Vacuolar Sequestration

Once bound at the cell surface, the metal ions are taken into the cell by membrane transport systems which normally play a role in the uptake of nutrients. After entering, the cell sets up detoxification systems to avoid toxicity. One of the main aspects in this response is the production of metal-binding peptides that complex with metal ions making them less reactive (Caetano-Silva et al., 2021). The complexes are effective in keeping the metals stabilized in the cell to avoid cellular damage of structures and metabolic activities. The bound metals are then taken into intracellular storage sites, where they are stored in a stable and non-toxic form. This compartmentalization enables cells to withstand relatively high concentrations of metals within them without negative effects. These mechanisms, which occur within the cells, are important to the overall ability of diatoms to extract metals in wastewater, which supplement surface-level biosorption and lead to long-term retention of contaminants.

4.3 Micropollutant Removal: Biotransformation and Lipid Partitioning

Diatoms can accumulate and biochemically transform organic micropollutants. Hydrophobic molecules are prone to localize in lipid-rich areas of the cell and extra-cellular matrix (Capua-Shenkar et al., 2023). It is a fast dissolute concentration reduction method; especially in compounds that are more affine to organic phases. Simultaneously, the enzymatic processes in the cell would convert the micropollutants into less toxic or easier degradable substances. The changes in the chemical structure of the compounds decrease their biological activity and persistence in the environment. This twin process is especially significant due to the fact that it guarantees that it will not be only water that will be transferred to the biomass but this will also be altered on the chemical level. Consequently, there is a diminished environmental risk posed by these contaminants. Removal of micropollutants is influenced by chemical characteristics of the compounds, biomass concentration and retention time, and the longer the retention period, the greater is the effect on transformation.

4.4 Frustule Silica as an Engineered Nanomaterial Platform

In addition to the natural application of the silica structure of the frustule in biosorption, the silica structure of the frustule of the diatom can be used in further treatment as a functional material. The porous nature of the frustule and the large surface area has provided it with a good platform upon which chemical modification can be done. The adsorption properties of the material can be optimized to suit individual contaminants by introducing certain functional groups onto its surface. The altered frustules may have an increased selectivity and be more effective in binding the metals and other pollutants (Panwar & Dutta, 2019). Such engineered materials may be applied in filtration systems, or as reusable adsorbents, offering an alternative route of treatment to living biological systems. Notably, the frustules may be harvested out of the biomass of diatoms, and they can be reused as value-added products. This opens the possibilities of wastewater treatment and resource recovery, which underlies a circular and sustainable environmental management (Zhang, 2025). Table 1 shows the major diatom species and engineered diatom-based systems used in wastewater polishing, highlighting their target contaminants, removal efficiencies, and primary functional mechanisms. Figure 3 provides a detailed mechanistic illustration of extracellular structural barriers and intracellular detoxification pathways involved in heavy metal sequestration and micropollutant removal by diatoms.

Figure 3. Structural and Cellular Detoxification Mechanisms Underlying Pollutant Removal in Diatoms

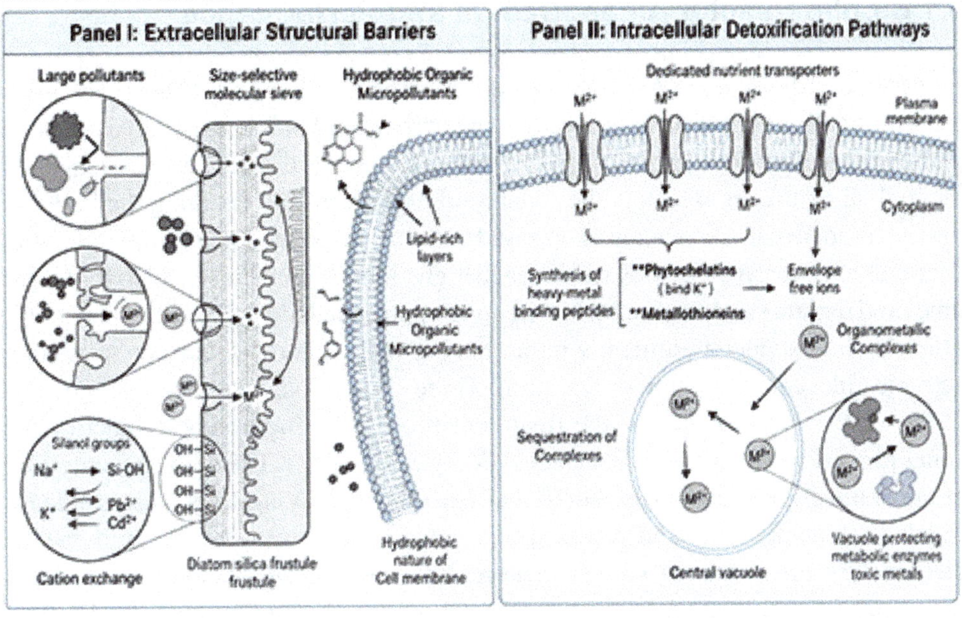

Table 1. Key Diatom Species and Systems Used in Wastewater Polishing: Target Contaminants, Removal Efficiency, and Functional Mechanisms

species	Target contaminants	Removal efficiency(%)	Functional category	Primary mechanism
Nitzschia spp. (Parikh et al., 2026a)	Cadmium (Cd), Zinc (Zn), Copper (Cu)	85–95	Heavy metal removal (High)	EPS binding and intracellular storage
Navicula spp. (Bhuyan et al., 2026)	Lead (Pb), Mercury (Hg)	75-90	Heavy metal removal (High)	Surface adsorption and intracellular retention
Cyclotella spp. (Li et al., 2021)	Copper (Cu), Chromium (Cr)	70–90	Heavy metal removal (Moderate)	Enzymatic transformation and surface interaction
Mixed diatom biofilm (Bai et al., 2025)	Pharmaceuticals, hormones	50–85	Organic pollutant removal	Biotransformation, adsorption, and photodegradation
Functionalized frustule materials (Subbaiah et al., 2021)	Mercury (Hg), Arsenic (As), Lead (Pb)	90–99	Advanced engineered system	Enhanced chemical adsorption and selective binding

5. DIATOM-BASED WASTEWATER POLISHING SYSTEMS

5.1 Biofilm Reactor Algae: Design and Performance

Algal biofilm reactor An algal biofilm reactor is a system where the diatoms grow on a solid support surface to create stable biofilm that is continually in contact with the flowing wastewater. These systems solve one of the greatest challenges of microalgal treatment, which is the separation of biomass, by enabling the biomass to stay stationary in the reactor as opposed to being suspended in the water column. Diatom biofilms have the potential to reach high biomass densities in relatively short times and remain structurally stable with long operational times (Parikh et al., 2026). This allows sustained treatment without having to use energy-intensive harvesting activities like centrifugation or filtration. Three-dimensional geometry of biofilms improves the performance of the treatment process by increasing contact of the contaminants with the reactive surfaces (Wang et al., 2023). The high concentration of the extracellular substances, surface of frustules and living cells creates a dense matrix that enables a rapid adsorption process and slower biological reactions to take place in the same system. Moreover, these reactors are capable of being supplemented with external supply of carbon, i.e., carbon dioxide, which contributes to the algal growth and the efficiency of the system. This opens possibilities of integration of wastewater treatment with carbon capture approaches.

5.2 High-Rate Algal Ponds: Pilot-Scale Performance

High-rate algal ponds are shallow continuously mixed systems used to treat large amounts of wastewater (Biliani & Manariotis, 2026). They also have rather long retention periods, which provide enough time between biomass and contaminants. These systems have proven to be able to remove nutrients consistently and moderately remove heavy metals over a large variety of environmental conditions. They can be used with varying seasons with a stable performance level provided that the appropriate parameters of their operation like retention time and mixing intensity are changed accordingly. Under cooler temperatures and reduced light intensity, diatoms tend to dominate these systems and enhance better removal efficiency because of their structure and biochemical characteristics. Their presence improves metal removal in comparison to other systems that are dominated by other microalgae. One of the most viable and scalable types of microalgal wastewater treatment is thought to be high-rate algal pond, especially in areas where land is available and sunlight is abundant (Novoveská et al., 2023).

5.3 Microalgae Bacteria Consortia: Systems of improved performance

The co-culture of microalgae and bacteria is an important improvement in the efficiency of wastewater treatment. The consortia systems are systems that pool the abilities of the two groups which lead to increased efficiencies of removal of a large variety of contaminants. Bacteria aid in removing contaminants and this is through transformation of organic compounds, release of bound metals and the production of more extracellular materials which increase sorption capacity. They also favour algal growth with the provision of vital nutrients and vitamins. Diatoms, in their turn, provide the oxygen due to the process of photosynthesis that promotes metabolic activity in bacteria. This reciprocal interaction forms a synergistic mechanism in which the two organisms complement each other. Such integrated systems have been demonstrated to have better removal efficiencies than single-species systems and are especially applicable to complex wastewater matrices where different contaminants are found (Wang et al., 2024).

5.4 Operational Optimisation: The Major Environmental Factors

Given the proper management of various operational parameters, the performance of diatom-based wastewater treatment systems is dependent. pH is a key factor in deciding the availability and affinity of contaminants. The most favorable environment is often found in a near-neutral range, with functional groups of metal binding being the most active and the solubility of the metal being very high. The intensity of light has an effect on the production of biomass and enzyme activity. A sufficient amount of light is required to maintain photosynthesis and support the metabolic activities in the transformation of contaminants. Hydraulic retention time is the time taken by the wastewater in contact with the biomass. Retention times are short to support high adsorption rates, whereas long retention times permit intracellular retention, and biochemical conversion. Temperature influences metabolic rates and productivity of the system. The diatoms are especially beneficial since they can remain active in lower temperatures than most of the other microalgae.

Silicon is necessary to ensure diatom growth and structure. Sufficient silicon concentrations promote constant biomass growth, and make sure that the adsorption processes involving the frustule remain efficient (Guo et al., 2026). Table 2 shows the comparative performance of conventional and diatom-based wastewater treatment systems in terms of heavy metal removal, micropollutant removal, and energy requirement..

Table 2. Comparative Performance of Conventional and Diatom-Based Wastewater Treatment Systems in Terms of Removal Efficiency and Energy Demand

Treatment System	Heavy Metal Removal	Micropollutant Removal	Energy Requirement (kWh/m^3)
Chemical precipitation (Lim et al., 2020)	Moderate to high at high concentrations; low at trace levels	low	0.3-0.9
Activated carbon (Dilokekunakul et al., 2020)	moderate	high	1-3
Membrane systems (Segredo-Morales et al., 2024)	high	Moderate to high	1.5-4
Diatom high-rate algal ponds (Sutherland et al., 2020)	Moderate to high	moderate	0.05-0.20
Diatombiofilmreactors (Arnaldo et al., 2024)	high	Moderate to high	0.10-0.40
Microalgae–bacteria consortia (Zhao et al., 2023)	high	high	0'08-0.30

6. CHARACTERISATION, ANALYTICAL METHODS, AND ISOTHERM MODELLING

6.1 Structural and Chemical Characterisation

The structural and chemical analytical methods need to be combined to ensure that diatom-based biosorption systems are accurately evaluated. Such techniques offer understanding of the distribution of contaminants, binding and transformation at different scales. Mapping of the spatial distribution of metals within the extracellular matrix, frustule surface, and cellular areas is performed by scanning electron microscopy coupled with energy-dispersive X-ray spectroscopy (SEM -EDS) (Riley et al., 2026). This method gives clear evidence of compartment-specific accumulation. Transmission electron microscopy (TEM) is a technique to view intracellular structures, such as metal inclusions and storage in cellular compartments. It offers high-resolution images of sub-cellular characteristics linked with the presence of contaminants.

Fourier-transform infrared spectroscopy (FTIR) is used to determine functional groups that have been used in metal binding. Alteration of characteristic absorption bands shows that carboxyl, hydroxyl, and amine groups are involved in the coordination process (Guerrero-Pérez & Patience, 2020). The oxidation state of accumulated metals is determined by X-ray photoelectron spectroscopy (XPS). This is especially vital in establishing the aspect of whether chemical transformation of the contaminants has taken place within the system. To analyze aqueous, inductively coupled plasma mass spectrometry (ICP-MS) enables the detection of various metal ions in very low concentrations with a high level of sensitivity and can therefore be used to monitor the adherence to stringent water quality requirements. In the case of organic micropollutants, liquid chromatography with tandem mass spectrometry (LC-MS/MS) can give accurate quantification and identification of parent compounds plus their transformation products. This allows distinguishing between mere accumulation and actual biochemical transformation (Smith et al., 2018).

7. CONCEPTUAL FRAMEWORK: BIOSORPTION–BIOTRANSFORMATION–RECOVERY

The concept of Diatom-based wastewater polishing could be interpreted as a multi-level treatment, which involves quick physicochemical contact, biological reactivation, and recovery of resources into one system. The integrated relationship between contaminant capture, biological transformation, biomass valorization, and circular resource recovery is conceptualized in Figure 4.

Figure 4. Integrated Biosorption–Biotransformation–Recovery Pathway for Circular Resource Recovery in Diatom-Based Wastewater Treatment

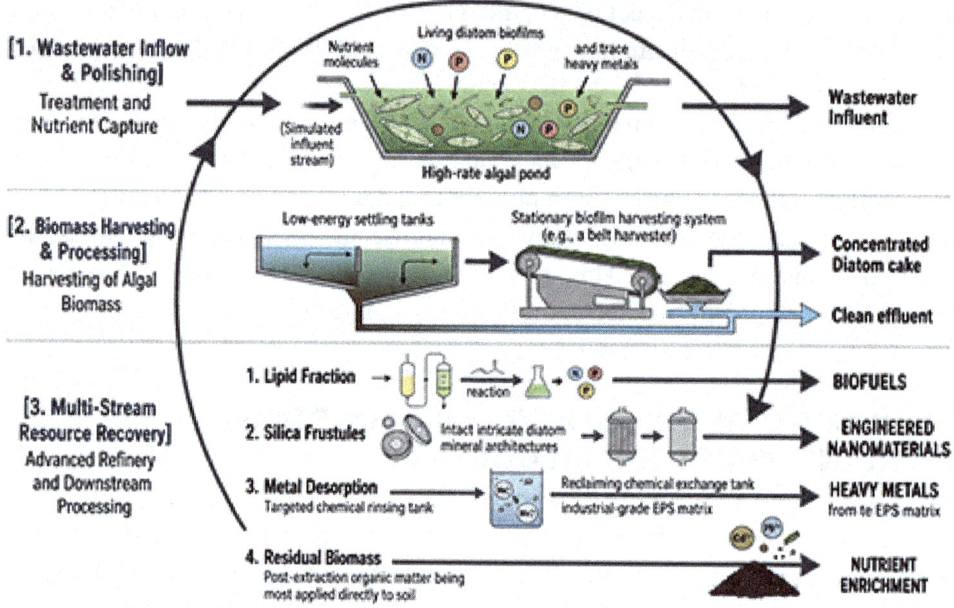

The functioning of this system is based on a series of interdependent processes, which help to achieve the overall contaminant removal and optimize the system. The initial phase is the fast surface capture that takes place within minutes to several hours. At this stage, dissolved heavy metals and the hydrophobic micropollutants are exchanged between the bulk wastewater into the extracellular polymeric matrix and on the silica-based frustule (Roychoudhury et al., 2022). This is mediated by electrostatic interactions, ion exchange and hydrophobic partitioning and does not need metabolic energy. A significant percentage of the elimination of contaminants takes place at this level because of the high binding capacity and surface area of the system. The second step involves metabolic build-up and biotransformation and is performed over more protracted durations. Initially bound metals are transported into the cell, where they are detoxified by binding with intracellular compounds and are stored in the cell in stable forms, within cellular compartments (Adhikary et al., 2024). Meanwhile, organic micro pollutants are enzymatically converted, and produce the less toxic or more biodegradable form. This step maximizes the total removal efficiency and makes sure that the contaminants are not simply transferred, but rather that they are detoxified. The third stage involves biomass harvesting and resource recovery. The biomass that has been accumulated has metals, lipids and structures of silica that can be processed and reused. Desorption methods can be used to extract metals, and biofuels can be produced out of lipids found in the biomass. The rest of the silica-based structures can either be used as advanced materials or recycled in the system. This combined system helps to shift the traditional wastewater treatment towards the circular bioeconomy framework (Marami et al., 2022). Rather than viewing wastewater management as an expensive process, deriving useful resources out of biomass opens up possibilities of financial gain. This kind of combination between the elimination of pollutants and the production of resources increases the sustainability of the systems, as well as the possibility of large-scale implementation. Moreover, the remaining biomass after processing has the potential to be reused in agricultural processes, including soil enrichment and nutrient fortification, adding additional value to the value chain. The multi-functional nature of this method makes the diatom based systems a solution to the environment and economy, able to tackle water quality issues and play a role in resource recovery and sustainable development.

8. FINDINGS

The evidence that has been accumulated shows that diatom-based wastewater treatment systems perform well and reliably in the removal of both the heavy metal and organic micropollutants in the optimized conditions (Rahman et al., 2024). These systems are a combination of quick physicochemical reactions and slower biological reactions that give rise to overall elimination of contaminants in various pathways. In the case of heavy metals, diatom systems are capable of good adsorption capacity and high removal efficiency with a wide variety of common contaminants. The metals like cadmium and lead are efficiently eliminated by a combination of extracellular binding, surface adsorption along with intracellular accumulation (Sun et al., 2025). The efficiencies of removal are usually moderate to very high with respect to the operational conditions like pH, retention time, and concentration of biomass. The uptake between 0 and 12 hours of exposure is rapid and involves extracellular and surface interactions, whereas mechanisms take place over a longer period of time to improve removal by intracellular sequestration and stabilization. The additional removal of chromium is also an example of the dual-function nature of these systems that both adsorption and intracellular transformation processes work together to minimize toxicity (Lodariya et al., 2025). Moreover, integrated systems containing diatoms and bacterial communities are always more effective because of synergistic processes leading to better availability, uptake, and transformation of contaminants. The functions of diatom structures are further utilized to create sophisticated material applications that increase the range of treatment. The silica-based frustules are highly functionalized to provide extremely high removal efficiencies of certain metals, which provides a low-energy, highly selective alternative to traditional adsorption media (Sahoo, 2025). In the case of organic micropollutants, systems based on diatoms have been shown to be highly effective in removing pollutants by physical partitioning as well as biochemical conversion. The absorption of hydrophobic compounds into lipid-rich cellular components and extra cellular matrices happens quickly decreasing the concentration of the compound in the water phase. At the same time, the cells use enzymatic activities to convert them into forms that are not so harmful or biodegradable. This conversion is significant especially in the fact that it decreases the biological activity of contaminants instead of merely transferring it to the biomass. Pharmaceutical compounds, hormones and anti-inflammatory drugs exhibit moderate to high removal efficiencies and there is evidence that a considerable decrease in toxicity occurs following treatment (Sigonya et al., 2023). There is also sensitivity of the diatom systems to exposure to micropollutants with structural changes in the community being observed. This is another benefit as such systems may serve as treatment technologies as well as environmental condition indicators. The diatom-based systems have the capacity to produce biomass that can be used to generate more value by recovering resources. Productivities can be different based on the configuration of the system but are high enough to allow meaningful co-product to be generated. Biomass can be converted to biofuels by using lipid fractions in the biomass, which can be used in generating renewable energy. Silica-based diatom biomass is a high-value product that can be used in cutting-edge technology because of its structural peculiarities (Vinayak & Gordon, 2025). Moreover, it is possible to extract metals stored in the biomass and reuse them, especially in industrial wastewater stream treatment systems. Biomass left over when it is processed can also be used in agriculture to help in soil regeneration and recycling of nutrients. All in all, these results prove that not only diatom-based wastewater treatment systems are effective in removing contaminants, they also can produce valuable by-products. This places them as comprehensive solutions that merge environmental cleanup and resource reclamation, to add to sustainability and economic viability.

Additionally, the literature contains quantitative evidence of the potential for contaminant removal in diatom-based systems. Reported efficiencies for removing heavy metals are typically 70–95% for most of the species, depending upon reactor setup and wastewater characteristics. For instance, Nitzschia spp. has removed 85-95% of cadmium, zinc, and copper, and Navicula spp. has removed 75-90% of lead and mercury. For selected heavy metals, such as arsenic and mercury, functionalized diatom frustule materials have been reported with up to 99% removal efficiency (Rahman et al., 2024). In the case of organic micropollutants, removal rates of 50-85% for pharmaceuticals and endocrine-disrupting compounds have been shown with mixed diatom biofilms. Operational conditions also have a strong impact on treatment performance, including an optimum pH range of 6.5-8.0, which is generally recommended; an optimum hydraulic retention time of several hours to a few days; and a temperature range of 20-30°C. The availability of silicon is also critical in ensuring the formation of frustules and biomass productivity. When compared with conventional tertiary treatment technologies, diatom-based systems generally have a significantly lower energy input (0.05–0.40 kWh m^{-3}) and are competitive in terms of pollutant removal, making them a potential sustainable wastewater polishing technology.

9. DISCUSSION

It is possible to explain the higher effectiveness of the systems of wastewater treatment by the diatoms by their peculiarities of the combination of structural, biochemical, and adaptive features. The diatoms, unlike other groups of microalgae, have silica-based frustule and a highly charged extracellular matrix, both of which play an important role in the removal of contaminants (Roychoudhury et al., 2022). These characteristics form a variety of binding sites and routes, which enables the effective absorption of heavy metals by both contacting the surface and the intracellular mechanisms. The frustule also gives the added adsorption surface that further increases the removal capacity as compared to the case in which the bioactivity alone is applied. Simultaneously, the extracellular matrix has a high affinity to metal ions because it contains a large number of negatively charged functional groups. These components combined, form a synergistic system, which performs better than other microalgal systems regularly. The fact that they can change with the changing environmental conditions is an important benefit of diatom-based systems. Contaminants may also trigger greater synthesis of extracellular binding materials, which are an effective response to an increase in pollutant loads (Chen et al., 2025) . This flexibility enables the system to operate under changing influent conditions, and this ability is not present in the traditional fixed-capacity treatment technologies. Diatom-based systems offer several advantages over conventional wastewater treatment technologies. They are especially good in removing contaminants that are at lower levels that the chemical precipitation and other conventional procedures are less effective. Moreover, their energy needs are considerably less, which qualifies them as more sustainable and cost-effective. In the case of organic micropollutants, diatom systems have a clear advantage in that they remove them both physically and biochemically. This two-fold mechanism lowers the concentration of contaminants, as well as their biological activity, which deals with the chemical presence of contaminants as well as their ecological effects. Although these benefits are present, there are a number of operational issues. In open systems, the environment may affect the composition of species through factors like the intensity of light and temperature and may destabilize the system. Silicon is necessary as a nutrient to facilitate growth and this is an extra operational consideration to other microalgal systems (Zhou et al., 2025). Moreover, the absence of standardized regulatory frameworks of biological tertiary treatment technologies may cause uncertainty with large scale implementation. Nevertheless, these issues are mainly concerning system design, operation, and policy and not inherent technology restrictions. The development of reactor design, nutrient management and monitoring systems will probably help solve most of these problems. New innovations are also extending the prospects of the diatom-based systems. The functionalised frustule supplies have provided novel opportunities to achieve highly selective and efficient removal of contaminants with the help of bio-derived nanostructures (Savio et al., 2022). The current developments in genetic and metabolic engineering provide the chance to boost the natural abilities of diatoms and make them more efficient and adaptable. Moreover, more specific and efficient operation of treatment systems is becoming possible due to the integration of the latest monitoring technologies and automated control systems. These innovations aid in adopting experimental applications into dependable, large scale application. Altogether, the diatom-based wastewater treatment systems are a big step on the way to sustainable water management. Their efficiency, flexibility and resource recovery potential make them a promising alternative to traditional treatment technologies, and there is great potential in their evolution and use in the future (Pandey, 2025).

9.1 Limitations and Future Challenges

Although the performance of diatom-based wastewater treatment systems is encouraging, there are still some difficulties to be addressed for their large-scale implementation. One of the major constraints is the absence of long-term field studies; most reported results are from laboratory or pilot-scale studies (Makhathini, 2026). The efficiency of the treatment process can vary under actual conditions depending on the composition of the wastewater, its temperature, light, and seasonal shifts. Therefore, stable diatom communities and stable pollutant removal performance over long periods of operation are still a big challenge. The scale-up and commercialization also need to be pursued. Harvesting biomass, maintenance of the reactors, silicon supplementation, and process monitoring can increase the complexity and costs of the operation. Moreover, the design standards and regulations of diatom-based tertiary treatment systems are not yet standardized, which is a barrier in their industrial application. Additionally, the economic viability of the contaminated biomass resource recovery (metal extraction and silica reutilization) needs to be further validated. Future work should include long-term demonstration projects, techno-economic studies, life cycle analysis, and development of automated monitoring and control systems. To overcome these challenges to better operational reliability, commercial viability, and broad implementation of diatom-based wastewater polishing technologies.

10. RECOMMENDATIONS

The diatom-based wastewater treatment system development involves concerted action in the biological growth, engineering validation, economic integration and regulatory facilitation. The use of this technology should be integrated at strategic levels in order to move it out of research and pilot projects to widespread and practical use. One of the major concerns is the creation of diatom strains that have better performance features. Building of regional and national strain libraries that specialise on wastewater-adapted species will offer a robust biological base to apply to the deployment of systems. These lines must have both inherently metal tolerant strains and genetically engineered better strains with increased metal binding, extracellular matrix production, and micropollutant transformation abilities (Akoijam & Joshi, 2024). Genetic technologies can be used today in an effort to enhance these characteristics, but with an emphasis on enhancing removal efficiency whilst ensuring stability in the real world. The validation on a pilot scale in a variety of environmental and geographic conditions is necessary to verify the system reliability and flexibility. The areas with the high levels of wastewater contamination and high solar potential should be increased to test these systems and provide the maximum benefit. The pilot projects should be subjected to standardized monitoring protocols that would guarantee the use of similar data collection measures to determine the efficiency in contaminant removal, transformation processes, biomass productivity, and resource recovery outputs (Anastopoulos et al., 2024). It will allow substantial comparison and contribute to the building of performance benchmarks that are applicable on a global level. The principles of a circular bio economy need to be integrated into the design of the system. Instead of considering biomass as a waste, treatment systems must be designed as multi-output facilities which can help extract value resources (Romero-Perdomo & González-Curbelo, 2023). Metals may also be reused and extracted, lipids may be transformed to biofuels and silica structures may be used to develop advanced materials (Chettri et al., 2025). Remaining biomass can also be used in agriculture either as a nutrient or amendment in soil. These value chains will help to design systems that will be economically viable and promote adoption. Policies and regulatory frameworks should be changed to accommodate the use of biological tertiary treatment systems. Clear guidelines and performance standards must be established which not only take care of contaminant removal, but also reduction in biological activity and toxicity. The case will be enhanced by regulatory acceptance of the resource recovery as a co-benefit of wastewater treatment. Moreover, research-industry-public utilities collaborative implementation models should be promoted. These partnerships can shorten technology development cycles, mitigate risk and help in the shift of pilot systems to full-scale infrastructure. On the whole, the effective implementation of the diatom-based wastewater treatment systems requires the combination of the biological innovation, engineering design, economic feasibility, and enabling policy frameworks into a coordinated approach.

11. CONCLUSION

The diatom-based wastewater polishing seems to be a promising and sustainable method to remove heavy metal and emerging micropollutants from wastewater. Efficient adsorption, accumulation, and transformation of contaminants are achieved due to the unique structural and biological features of diatoms, including silica frustules, extracellular polymeric substances (EPS), and intracellular detoxification mechanisms. These are all integrated processes that enable wastewater treatment to be performed effectively while avoiding the need for high-energy-consuming conventional technology. A literature review shows that diatom systems can effectively remove a broad spectrum of contaminants under optimal conditions and can be incorporated into biofilm reactors, high-rate algal ponds, and microalgae-bacteria consortia. The ability to recover pollutants and also metals, biofuels, and silica-based materials from harvested biomass also underlines the role of harvested biomass in circular bioeconomy and resource recovery measures. Despite these benefits, there are still some barriers to overcome, such as scale-up, long-term operational stability, biomass management, and commercialization. Further research should focus on pilot- and full-scale validation, standardization of performance assessment methods, techno-economic evaluation, and the development of advanced monitoring and control systems. Solving these challenges will enhance the real-world viability of diatom-based technologies and pave the way for broader implementation of these environmentally friendly solutions for advanced wastewater treatment and resource recovery. This chapter is important for the integration of biological pollutant removal mechanisms with the concepts of resource recovery and circular bioeconomy in one single framework. The chapter integrates the concepts of frustule engineering, EPS-mediated biosorption, microbial transformation of micropollutants, and biomass valorization to offer a comprehensive view that can guide the design of future wastewater polishing systems.

REFERENCES

Adhikary, S., Saha, J., Dutta, P., & Pal, A. (2024). Bacterial Homeostasis and Tolerance to Potentially Toxic Metals and Metalloids through Diverse Transporters: Metal-Specific Insights. *Geomicrobiology Journal, 41*(5), 496–518. DOI: 10.1080/01490451.2024.2340517

Agarwal, N. (2025). *Diatoms: Basic and Applied Research.* Educohack Press.

Akhter, Md. S., Ratul, S. I., Chowdhury, U., Miah, M. S., Mirza, M. J. A., & Hossain, A. (2026). AI-Powered Generative Models for Predictive and Optimized Aquaculture Water Management. In M. Lahby & R. Gaamouche (Eds.), *Generative AI and Optimization Techniques for Sustainable Water Management* (pp. 113–133). Springer Nature Switzerland. DOI: 10.1007/978-3-032-19012-3_8

Akoijam, N., & Joshi, S. R. (2024). Genome Editing and Genetically Engineered Bacteria for Bioremediation of Heavy Metals. In *Genome Editing in Bacteria (Part 2)* (pp. 184–221). Bentham Science Publishers. https://www.benthamdirect.com/content/books/9789815223798.chapter-6

Anastopoulos, I., Giannakoudakis, D. A., Frontistis, Z., Zorpas, A. A., Pashalidis, I., & Triantafyllidis, K. (2024). Biomass-derived adsorbents: Universal database development for their synthesis and remediation efficiency as a necessary step to move from laboratory- to pilot-scale applications. *Current Opinion in Green and Sustainable Chemistry, 47*, 100902. DOI: 10.1016/j.cogsc.2024.100902

Arnaldo, M. D. G., Mossion, A., Beignon, T., Vuillemin, H., Guihéneuf, F., Wielgosz-Collin, G., & Méléder, V. (2024). Diatom Biofilm. In *Diatom Photosynthesis* (pp. 449–476). John Wiley & Sons, Ltd., DOI: 10.1002/9781119842156.ch15

Asif, M. B., Baig, N., & Abdallah, M. A.-E. (2026). A Perspective on Quaternary Treatments for Micropollutant Elimination in Urban Wastewater Treatment Plants. *Earth Systems and Environment*. Advance online publication. DOI: 10.1007/s41748-026-01086-7

Bai, R. G., Chandrasekharan Nair, S., Joller-Vahter, L., & Kikas, T. (2025a). Microalgae in Mitigating Industrial Pollution: Bioremediation Strategies and Biomagnification Potential. *Biomass, 5*(4), 61. DOI: 10.3390/biomass5040061

Bai, R. G., Chandrasekharan Nair, S., Joller-Vahter, L., & Kikas, T. (2025b). Microalgae in Mitigating Industrial Pollution: Bioremediation Strategies and Biomagnification Potential. *Biomass, 5*(4), 61. DOI: 10.3390/biomass5040061

Bhuyan, P. P., Nahak, S., Sahoo, H., Sahoo, S., & Pradhan, B. (2026). Microalgal bioremediation of heavy metals and by-products recovery as lipid: The dual role of microalgae. *International Journal of Phytoremediation*, *0*(0), 1–17. DOI: 10.1080/15226514.2026.2667453 PMID: 42118003

Biliani, E., & Manariotis, I. D. (2026, May 07). S., & D. Manariotis, I. (2026). High-rate algal ponds in wastewater treatment: A critical look at recent developments. *Environmental Science. Water Research & Technology*, *12*(5), 1359–1373. Advance online publication. DOI: 10.1039/D6EW00116E

Boro, D., Chirania, M., Verma, A. K., Chettri, D., & Verma, A. K. (2025). Comprehensive approaches to managing emerging contaminants in wastewater: Identification, sources, monitoring and remediation. *Environmental Monitoring and Assessment*, *197*(4), 456. DOI: 10.1007/s10661-025-13809-w PMID: 40119196

Buhani, J., Juliasih, N. L. G. R., Wahyu Purnamasari, I., Suharso, , Noviany, , Sumadi, , Alghamdi, H. M., Elgarahy, A. M., & Elwakeel, K. Z. (2026). Selective adsorption and mechanistic insights into divalent cadmium, copper, and lead recovery using silica–magnetite modified Nitzschia sp. Biomaterial. *International Journal of Phytoremediation*, *28*(7), 1258–1277. DOI: 10.1080/15226514.2025.2607515 PMID: 41548094

Caetano-Silva, M. E., Netto, F. M., Bertoldo-Pacheco, M. T., Alegría, A., & Cilla, A. (2021). Peptide-metal complexes: Obtention and role in increasing bioavailability and decreasing the pro-oxidant effect of minerals. *Critical Reviews in Food Science and Nutrition*, *61*(9), 1470–1489. DOI: 10.1080/10408398.2020.1761770 PMID: 32370550

Capua-Shenkar, J., Varsano, N., Kruth, H., & Addadi, L. (2023). Crystals in Atherosclerosis: Crystal Cholesterol Structures, Morphologies, Formation and Dissolution. What Do We Know? In Abela, G. S., & Nidorf, S. M. (Eds.), *Cholesterol Crystals in Atherosclerosis and Other Related Diseases* (pp. 49–71). Springer International Publishing., DOI: 10.1007/978-3-031-41192-2_4

Chen, Y., Xuan, T., Dong, M., Liang, H., Cai, G., Xia, B., Tang, J., Xie, Z., & Zhao, X. (2025). Mechanistic insights into microalgal removal of ciprofloxacin–copper co-contaminants: Extracellular polymeric substances adsorption, intracellular degradation, and combined toxicity. *Journal of Hazardous Materials*, *500*, 140558. DOI: 10.1016/j.jhazmat.2025.140558 PMID: 41289666

Chettri, D., Boro, M., Malik, A. A., & Verma, A. K. (2025). Nanotechnology for a greener future: Sustainable energy and biofuel advances. *Biofuels*, *16*(6), 622–637. DOI: 10.1080/17597269.2024.2440207

Dilokekunakul, W., Teerachawanwong, P., Klomkliang, N., Supasitmongkol, S., & Chaemchuen, S. (2020). Effects of nitrogen and oxygen functional groups and pore width of activated carbon on carbon dioxide capture: Temperature dependence. *Chemical Engineering Journal, 389*, 124413. DOI: 10.1016/j.cej.2020.124413

Dolganyuk, V., Belova, D., Babich, O., Prosekov, A., Ivanova, S., Katserov, D., Patyukov, N., & Sukhikh, S. (2020). Microalgae: A Promising Source of Valuable Bioproducts. *Biomolecules, 10*(8), 1153. DOI: 10.3390/biom10081153 PMID: 32781745

Guerrero-Pérez, M. O., & Patience, G. S. (2020). Experimental methods in chemical engineering: Fourier transform infrared spectroscopy—FTIR. *Canadian Journal of Chemical Engineering, 98*(1), 25–33. DOI: 10.1002/cjce.23664

Guo, Y., Ouyang, S., Chen, J., Lin, Z., Xiao, C., Liu, D., Chen, G., & Zhang, Q. (2026). Diatomaceous bio-silicon as a sustainable alternative to mineral silicon fertilizers: Impacts on rice yield, soilfertility, and heavy metal dynamics. *Plant and Soil, 519*(1), 451–468. DOI: 10.1007/s11104-025-08121-y

Hamed, A. (2023). A brief and general overview on diatoms and their applications. *Egyptian Journal of Phycology*. DOI: 10.21608/egyjs.2023.218427.1019

Li, N., Qin, L., Jin, M., Zhang, L., Geng, W., & Xiao, X. (2021). Extracellular adsorption, intracellular accumulation and tolerance mechanisms of *Cyclotella* sp. To Cr(VI) stress. *Chemosphere, 270*, 128662. DOI: 10.1016/j.chemosphere.2020.128662 PMID: 33127109

Liang, E., Li, J., Li, B., Liu, S., Ma, R., Yang, S., Cai, H., Xue, Z., & Wang, T. (2023). Roles of dissolved organic matter (DOM) in shaping the distribution pattern of heavy metal in the Yangtze River. *Journal of Hazardous Materials, 460*, 132410. DOI: 10.1016/j.jhazmat.2023.132410 PMID: 37647662

Lim, H.-H., Oh, Y.-S., & Shin, H.-S. (2020). Determination of *N*-nitrosodimethylamine and *N*-nitrosomethylethylamine in drug substances and products of sartans, metformin and ranitidine by precipitation and solid phase extraction and gas chromatography–tandem mass spectrometry. *Journal of Pharmaceutical and Biomedical Analysis, 189*, 113460. DOI: 10.1016/j.jpba.2020.113460 PMID: 32663759

Liu, X., & Hong, Y. (2021). Microalgae-Based Wastewater Treatment and Recovery with Biomass and Value-Added Products: A Brief Review. *Current Pollution Reports, 7*(2), 227–245. DOI: 10.1007/s40726-021-00184-6

Liu, Y., Zhang, X., Du, Y., Du, X., Zhang, Y., Deng, L., Li, C., & Guo, J. (2025). Functionalization of Wood for the Removal of Heavy Metal Ions from Waster Water: A Review. *Forests*, *16*(11), 1684. DOI: 10.3390/f16111684

Lodariya, M., Bhattacharya, D., & Abhilash, K. R. (2025). Synergistic approaches to soil remediation: Engineered biochar and microbial interactions for climate-resilient remediation. *Biodegradation*, *36*(6), 110. DOI: 10.1007/s10532-025-10200-x PMID: 41139376

Makhathini, T. P. (2026). Microbial Electrochemical Technologies in Wastewater Treatment: Scale-Up Challenges, Pilot Testing, and Practical Implementation. *Water (Basel)*, *18*(8), 966. DOI: 10.3390/w18080966

Maltsev, Y., Kulikovskiy, M., & Maltseva, S. (2023). Nitrogen and phosphorus stress as a tool to induce lipid production in microalgae. *Microbial Cell Factories*, *22*(1), 239. DOI: 10.1186/s12934-023-02244-6 PMID: 37981666

Marami, H., Tsapekos, P., Khoshnevisan, B., Madsen, J. A., Andersen, J. K., Rafiee, S., & Angelidaki, I. (2022). Going beyond conventional wastewater treatment plants within circular bioeconomy concept – a sustainability assessment study. *Water Science and Technology*, *85*(6), 1878–1903. DOI: 10.2166/wst.2022.096 PMID: 35358077

Novoveská, L., Nielsen, S. L., Eroldoğan, O. T., Haznedaroglu, B. Z., Rinkevich, B., Fazi, S., Robbens, J., Vasquez, M., & Einarsson, H. (2023). Overview and Challenges of Large-Scale Cultivation of Photosynthetic Microalgae and Cyanobacteria. *Marine Drugs*, *21*(8), 445. DOI: 10.3390/md21080445 PMID: 37623726

Pandey, A. K. (2025). Sustainable water management through integrated technologies and circular resource recovery. *Environmental Science. Water Research & Technology*, *11*(8), 1822–1846. DOI: 10.1039/D5EW00103J

Panwar, V., & Dutta, T. (2019). Diatom Biogenic Silica as a Felicitous Platform for Biochemical Engineering: Expanding Frontiers. *ACS Applied Bio Materials*, *2*(6), 2295–2316. DOI: 10.1021/acsabm.9b00050 PMID: 35030724

Parikh, H., Dave, G., Adhikari, S., & Tiwari, A. (2026a). Diatoms as ecological engineers in wastewater systems: A comprehensive review. *Environmental Technology Reviews*, *15*(1), 223–248. DOI: 10.1080/21622515.2026.2669110

Parikh, H., Dave, G., Adhikari, S., & Tiwari, A. (2026b). Diatoms as ecological engineers in wastewater systems: A comprehensive review. *Environmental Technology Reviews*, *15*(1), 223–248. DOI: 10.1080/21622515.2026.2669110

Priya, A. S., Premanand, R., Ragupathi, I., Bhaviripudi, V. R., Aepuru, R., Kannan, K., & Shanmugaraj, K. (2024). Comprehensive Review of Hydrogel Synthesis, Characterization, and Emerging Applications. *Journal of Composites Science, 8*(11), 457. DOI: 10.3390/jcs8110457

Rahman, T. T., Jiang, T., Zhang, C., Rao, Y., Sims, R. C., & Hou, L. (2024a). Diatoms in wastewater treatment: Potentials, applications, and values of biomass. *Critical Reviews in Environmental Science and Technology, 54*(7), 557–580. DOI: 10.1080/10643389.2023.2259278

Rahman, T. T., Jiang, T., Zhang, C., Rao, Y., Sims, R. C., & Hou, L. (2024b). Diatoms in wastewater treatment: Potentials, applications, and values of biomass. *Critical Reviews in Environmental Science and Technology, 54*(7), 557–580. DOI: 10.1080/10643389.2023.2259278

Riley, S., Marcuccio, F., Xu, X., Brandon, N., Korchev, Y., Liu, F., Shevchuk, A., & George, C. (2026). In Situ Nanoscale Imaging of Cultivated, As-Grown Diatom Frustules Electrochemical Lithiation/Delithiation in Li Ion Battery Anodes. *ACS Applied Energy Materials*. DOI: 10.1021/acsaem.6c00689

Romero-Perdomo, F., & González-Curbelo, M. Á. (2023). Integrating Multi-Criteria Techniques in Life-Cycle Tools for the Circular Bioeconomy Transition of Agri-Food Waste Biomass: A Systematic Review. *Sustainability (Basel), 15*(6), 5026. DOI: 10.3390/su15065026

Roychoudhury, P., Bose, R., Dąbek, P., & Witkowski, A. (2022a). Photonic Nano-/Microstructured Diatom Based Biosilica in Metal Modification and Removal—A Review. *Materials (Basel), 15*(19), 6597. DOI: 10.3390/ma15196597 PMID: 36233939

Roychoudhury, P., Bose, R., Dąbek, P., & Witkowski, A. (2022b). Photonic Nano-/Microstructured Diatom Based Biosilica in Metal Modification and Removal—A Review. *Materials (Basel), 15*(19), 6597. DOI: 10.3390/ma15196597 PMID: 36233939

Sahoo, J. K. (2025). Diatomite (Diatomaceous Earth): Global Perspectives on Geology, Processing, Applications, Environmental Sustainability, and Emerging Biomedical Frontiers. *Mining, Metallurgy & Exploration, 42*(6), 3363–3378. DOI: 10.1007/s42461-025-01411-0

Sangeetha, J., Codreanu, S., & Thangadurai, D. (2023). *Microalgal Biotechnology: Bioprospecting Microalgae for Functional Metabolites towards Commercial and Sustainable Applications*. CRC Press. DOI: 10.1201/9781003332251

Savio, S., Farrotti, S., Di Giulio, A., De Santis, S., Ellwood, N. T. W., Ceschin, S., & Congestri, R. (2022). Functionalization of Frustules of the Diatom Staurosirella pinnata for Nickel (Ni) Adsorption From Contaminated Aqueous Solutions. *Frontiers in Marine Science, 9*, 889832. Advance online publication. DOI: 10.3389/fmars.2022.889832

Segredo-Morales, E., González, E., González-Martín, C., & Vera, L. (2024). Novel vertical upflow multi-column configured membrane photobioreactor with a filtration control system for outdoor microalgae-bacteria cultivation, harvesting and wastewater reclamation. *Chemical Engineering Journal, 482*, 148799. DOI: 10.1016/j.cej.2024.148799

Sigonya, S., Mokhothu, T. H., Mokhena, T. C., & Makhanya, T. R. (2023). Mitigation of Non-Steroidal Anti-Inflammatory and Antiretroviral Drugs as Environmental Pollutants by Adsorption Using Nanomaterials as Viable Solution—A Critical Review. *Applied Sciences (Basel, Switzerland), 13*(2), 772. DOI: 10.3390/app13020772

Smith, E. R., Hewitson, T. D., Hanssen, E., & Holt, S. G. (2018). Biochemical transformation of calciprotein particles in uraemia. *Bone, 110*, 355–367. DOI: 10.1016/j.bone.2018.02.023 PMID: 29499417

Smułek, W., & Kaczorek, E. (2022). Factors Influencing the Bioavailability of Organic Molecules to Bacterial Cells—A Mini-Review. *Molecules (Basel, Switzerland), 27*(19), 6579. DOI: 10.3390/molecules27196579 PMID: 36235114

Subbaiah, M. P., Kalimuthu, P., Jung, J., & Jeon, B.-H. (2021). Recent advances in effective capture of inorganic mercury from aqueous solutions through sulfurized 2D-material-based adsorbents. *Journal of Materials Chemistry. A, Materials for Energy and Sustainability, 9*(34), 18086–18101. DOI: 10.1039/D1TA01018B

Sun, S., Zhang, M., Yang, G., Wang, Y., & Chen, Z. (2025). A review on the mechanisms and applications of bacteria in responding, adsorbing and transporting heavy metals. *Archives of Microbiology, 207*(11), 286. DOI: 10.1007/s00203-025-04441-w PMID: 41051408

Sutherland, D. L., Park, J., Ralph, P. J., & Craggs, R. J. (2020). Improved microalgal productivity and nutrient removal through operating wastewater high rate algal ponds in series. *Algal Research, 47*, 101850. DOI: 10.1016/j.algal.2020.101850

Vinayak, V., & Gordon, R. (2025). *Diatom Cultivation for Biofuel, Food and High-Value Products*. John Wiley & Sons. DOI: 10.1002/9781394174980

Wang, S., Li, J., Wang, W., Zhou, C., Chi, Y., Wang, J., Li, Y., & Zhang, Q. (2023). An overview of recent advances and future prospects of three-dimensional biofilm electrode reactors (3D-BERs). *Journal of Environmental Management, 342*, 118192. DOI: 10.1016/j.jenvman.2023.118192 PMID: 37285769

Wang, Z., Hu, G., & Hong, Y. (2024). Strong Alliance of Microalgae and Bacteria: The State-of-the-Art Review and Future Prospects of Utilizing Microalgae-Bacteria Consortia for Comprehensive Treatment of Swine Wastewater. *Current Pollution Reports, 10*(4), 744–764. DOI: 10.1007/s40726-024-00325-7

Zhang, J., Jin, R., Dong, L., Yu, C., Luo, W., Hu, Y., Zhang, Y., Zhang, W., Liu, J., Zhou, W., & Shao, S. (2025). Sticky or thrifty: Divergence in extracellular glycometabolic architecture underpins contrasting fungal bioflocculation strategies. *Water Research, 285*, 124126. DOI: 10.1016/j.watres.2025.124126 PMID: 40602066

Zhang, T. (2025). Sustainable Wastewater Treatment and the Circular Economy. *Water (Basel), 17*(3), 335. DOI: 10.3390/w17030335

Zhao, D., Cheah, W. Y., Lai, S. H., Ng, E.-P., Khoo, K. S., Show, P. L., & Ling, T. C. (2023). Symbiosis of microalgae and bacteria consortium for heavy metal remediation in wastewater. *Journal of Environmental Chemical Engineering, 11*(3), 109943. DOI: 10.1016/j.jece.2023.109943

Zhou, R., Xiao, B., Zhao, L., Tang, Y., Yang, W., Liu, H., Zhang, X., Yuan, C., Wang, X., Long, X., He, Q., & Li, W. (2025). Silicon supplementation via carriers to enhance diatom competition with cyanobacterial blooms and remediate eutrophic waters. *Water Research, 284*, 123922. DOI: 10.1016/j.watres.2025.123922 PMID: 40449328

Chapter 10
Microalgae-Based Bioremediation of Aquatic Pollutants and Integrating Carbon Sequestration With Eco-Toxicological Safety and Wastewater Valorization:
Role of Microalgae Against Pollution

Muhammad Sohail
http://orcid.org/0009-0002-8066-3076
University of Education, Lahore, Pakistan

Syed Sikandar Habib
Department of Zoology, University of Sargodha, Sargodha, Pakistan

ABSTRACT

The dual challenge of escalating carbon emissions and increasing aquatic pollution demands integrated biotechnological solutions capable of addressing both environmental and resource recovery goals. Microalgae have emerged as promising candidates for sustainable carbon sequestration and wastewater treatment. This chapter explores the role of microalgae in bioremediating aquatic pollutants, including heavy metals, pesticides, and industrial effluents, while simultaneously

DOI: 10.4018/979-8-2600-3080-6.ch010

contributing to carbon fixation and biomass production. The integration of microalgal cultivation with wastewater valorization is discussed in the context of circular bioeconomy principles, highlighting opportunities for resource recovery in the form of biofertilizers, biofuels, and bioplastics. Furthermore, the chapter assesses the environmental safety and sustainability of using algal biomass derived from polluted waters. This work provides a holistic view of microalgae-driven environmental biotechnology as a key pathway toward water sustainability and carbon neutrality.

1. INTRODUCTION

Aquatic ecosystem threats are increasing continuously from multiple sources such as industrialization, intensive agriculture, and growing urbanization. From these sources, various pollutants, such as PFAS, heavy metals, dyes, and organic solvents, microplastic, nanoplastics, that are released by industrial effluents; excess nutrients, pesticides, and veterinary medications are transported by agricultural runoff; and medicines and personal care items are introduced into water bodies by municipal and hospital effluents (Okoro et al., 2023; Thakur et al., 2025; Pandey et al., 2026). Simultaneously, the ubiquitous presence of microplastics and nanoplastics contributes a persistent and weakly degradable proportion that has the ability to transport and absorb additional pollutants via food webs (Kumar et al., 2025b; Swain et al., 2025; Ling et al., 2026; Ma et al., 2026). When combined, these stresses endanger human health and food security, degrade water quality, and disturb aquatic biodiversity.

While pollution (at small level) has been reduced by conventional wastewater treatment methods, including advanced oxidation, chemical coagulation–flocculation, and activated sludge, these methods have significant limits in facing new contaminants. These systems are sometimes expensive and energy-intensive to run, produce a lot of secondary sludge that needs to be disposed of further, and are often unable to eliminate trace-level pollutants such as micro/nanoplastics, endocrine-disrupting substances, and medications (Aryanfar et al., 2025; Ren et al., 2025; Jiang et al., 2026; Saini et al., 2026). Furthermore, the traditional treatment infrastructure's contribution to the circular economy and climate mitigation objectives is limited because it is not intrinsically built for resource recovery or carbon neutrality (Gautam et al., 2025; da Silva et al., 2026). More long-lasting, multipurpose therapy approaches are therefore desperately needed.

A very diverse category of photosynthetic microorganisms, microalgae live in freshwater, marine, and harsh environments. The selection of species that are most suited for both carbon sequestration and pollution removal depends on this diversity (González Fernández et al., 2025; Guo et al., 2025; Padhi et al., 2025; Zhang et al., 2025). Lakes, rivers, ponds, and reservoirs are frequently home to freshwater microalgae like *Chlorella vulgaris, Scenedesmus obliquus*, and *Chlamydomonas reinhardtii* (Sinha et al., 2025; Sui et al., 2025). Because of their natural resistance to heavy metals and other pollutants and their quick growth in nutrient-rich effluent, these species are very beneficial (Mahlangu et al., 2024; Deepika et al., 2025; Minhas et al., 2026). *Nannochloropsis sp., Dunaliella salina*, and *Tetraselmis sp.* are examples of marine microalgae that flourish in oceans and coastal environments (Doan et al., 2026). They have a high resistance to salinity and frequently produce lipid-rich biomass that can be used as biofuel.

Microalgae-based systems have become a viable substitute, providing an integrated strategy that combines resource recovery, carbon capture, and bioremediation. Microalgae fix CO2 and produce biomass that can be converted into biofuels, biofertilizers, animal feed, or high-value bioproducts (Asad et al., 2024; Azeem et al., 2024; Mounaimi et al., 2026), while also assimilating nutrients (nitrogen, phosphorus) and some organic and inorganic contaminants through photosynthesis (Wang and Hong, 2022; El Amri et al., 2026; Zou et al., 2026). Ecotoxicological safety, however, is a crucial and understudied aspect of this strategy: little is known about the possible accumulation, transformation, and trophic transfer of pollutants (including micro- and nanoplastics and related chemicals) within algal biomass and subsequent food chains (Liao et al., 2023; Alowaisy et al., 2026; Ma et al., 2026). Closing this gap is crucial to guiding the safe application of integrated algal technologies for sustainable water management and ensuring that microalgae-based treatment does not inadvertently transfer or multiply risks.

2. MICROALGAE AS A BIOREMEDIATION TOOL

2.1 Biological Characteristics of Microalgae

Microalgae's photosynthetic efficiency, quick growth rates, and environmental adaptability make them excellent bioremediation agents (Akram et al., 2026; Anas et al., 2026; Kaur et al., 2026). Through processes like biosorption, bioaccumulation, and biodegradation, they can eliminate a variety of pollutants, such as heavy metals, organic contaminants, dyes, pesticides, medicines, and new contaminants (Tarfeen et al., 2022; Tripathi et al., 2023; Ansar et al., 2026). By absorbing ambient CO_2 and recycling nutrients, these mechanisms enable microalgae to lower pollutant concentrations to non-toxic levels, making the strategy both economically and environmentally viable (Ethiraj and Samuel, 2024; Iakovidou et al., 2024; Rajkumar et al., 2025). Strain selection, cultivation methods, and integration with nanotechnology are some of the elements that boost microalgal bioremediation's effectiveness against refractory contaminants (Maqsood et al., 2024; El-Sheekh et al., 2025; Firincă et al., 2025). Microalgae-based wastewater treatment also has the potential to produce important biomass for biofuels and other products while using little energy (Choudhary et al., 2022; de Carvalho et al., 2022). Microalgae continue to be a promising tool for sustainable environmental detoxification across a variety of industrial and municipal wastewaters, despite obstacles including improving removal processes and public acceptability of genetically modified strains.

3. MECHANISM OF POLLUTANT REMOVAL BY MICROALGAE

Through several crucial processes, microalgae eliminate pollutants from wastewater: bioaccumulation, which involves intracellular uptake; biosorption, in which pollutants attach to the cell surface; and biodegradation, in which microalgal enzymes convert pollutants into less hazardous compounds (Pandey et al., 2026; Elleuch et al., 2025; Touliabah et al., 2022). As shown in Figure 1, these mechanisms operate at the cellular level, where organic molecules and microplastics bind to functional groups on the cell wall (biosorption), contaminants are actively transported and sequestered inside the cell (bioaccumulation), and organic pollutants are enzymatically degraded (biodegradation). The diagram shows biosorption (surface binding of organic molecules and microplastics to the cell wall), bioaccumulation (active transport, vacuole sequestration, and protein binding), and biodegradation (enzymatic breakdown of pesticides and pharmaceuticals). Simultaneously, CO_2 is fixed in the chloroplast via carbon-concentrating mechanisms (CCMs) and the Calvin–Benson cycle, leading to biomass production and O_2 release.

Microalgae also successfully reduce common wastewater contaminants by assimilating nutrients like phosphorus and nitrogen (Wang and Hong, 2022; Qv et al., 2023; Yang et al., 2026). Because of the interactions between pollutant molecules and algal cell walls, biosorption frequently predominates, but biodegradation is essential for the breakdown of organic micropollutants, particularly those with electron-donating groups (Liu et al., 2021; El Messaoudi et al., 2025; Raj et al., 2025). Additionally, because microalgae-based systems are phototrophic, they can break down persistent substances like pesticides and medications more effectively than traditional activated sludge (Pandey et al., 2025; Sarker and Kaparaju, 2025). Through synergistic effects, the mixing of bacteria and microalgae in consortia further improves removal efficiency. Despite encouraging outcomes, removal performance is significantly impacted by operational parameters such as pH, light intensity, and cultivation conditions, which must be optimized for large-scale applications.

Figure 1. Cellular mechanisms of pollutant removal and carbon sequestration by microalgae.

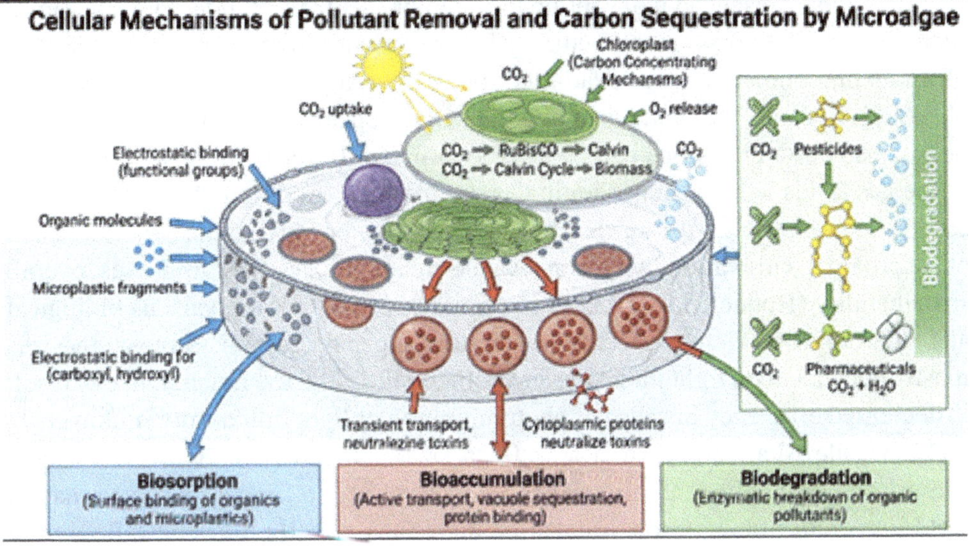

4. ROLE OF MICROALGAE IN CARBON SEQUESTRATION

Microalgae's extremely active photosynthetic mechanism and fast growth dynamics make them one of the most effective biological systems for sequestering carbon (De Silva et al., 2024; Patel, 2025; Sarmah and Panda, 2026). Microalgae use well-known carbon absorption processes, such as the Calvin–Benson cycle, where the enzyme RuBisCO catalyzes CO_2 fixation into ribulose-1,5-bisphosphate, to transform carbon dioxide (CO_2) into organic carbon through oxygenic photosynthesis (Kumar et al., 2025a; Gupta et al., 2026b; Serna-García et al., 2026). Because of their basic cellular structure, high surface-area-to-volume ratio, and direct exposure to dissolved CO2 in aquatic environments, microalgae have far better photosynthetic efficiency than terrestrial plants—often claimed to be 10–50 times larger on a per-area basis (Shekh et al., 2022; Shareefdeen et al., 2023; Dodangodage et al., 2026). Furthermore, many species have carbon-concentrating mechanisms (CCMs) that reduce photorespiration losses by increasing CO2 uptake even at low ambient concentrations (Kupriyanova et al., 2023; Kaste et al., 2024; Singh and Gaur, 2026). Because of this, microalgae can quickly transform inorganic carbon into biomass, which makes them ideal for incorporation into systems designed to capture and use carbon sustainably.

Microalgae present a viable option for removing CO2 from industrial emissions, such as flue gases from manufacturing facilities and power plants, in the context of mitigating climate change (Padhi et al., 2025; Ahuja et al., 2026; Echeverri et al., 2026). Because they can withstand high CO2 concentrations and some flue gas pollutants, waste emissions can be directly utilized, lowering atmospheric greenhouse gas levels while also generating useful biomass. Additionally, by combining carbon sequestration with nutrient removal and water purification, the inclusion of microalgal cultivation with wastewater treatment systems improves overall sustainability (Hoque et al., 2025; Song et al., 2025). However, various biological and environmental factors affect how well microalgae fix carbon. Photosynthesis is mostly dependent on light intensity, and both too little and too much light can result in decreased efficiency because of photoinhibition or light limitation (Simkin et al., 2022; Wasilewska-Dębowska et al., 2022; Tikkanen and Aro, 2026). Enzymatic activity and metabolic rates are influenced by temperature, and each species has an ideal temperature range. Growth and the ability to absorb carbon are controlled by the availability of nutrients, especially nitrogen and phosphorus, while species-specific variations determine overall performance, adaptation, and resilience under a range of environmental circumstances (Yang et al., 2025; Gan et al., 2026). Therefore, maximizing the potential of microalgae to sequester carbon in practical applications requires optimizing these parameters.

5. ROLE OF MICROALGAE IN WASTEWATER TREATMENT AND VALORIZATION

Through its incorporation into various culture technologies, such as open pond systems, photobioreactors, and hybrid configurations, microalgae-based systems have become an efficient and sustainable method for treating wastewater (Srimongkol et al., 2022; Ugwuanyi et al., 2024; Hasan and Enad, 2026). As illustrated in Figure 2, microalgae can be cultivated in both open raceway ponds and closed photobioreactors, utilizing polluted wastewater as a growth medium while simultaneously performing pollutant removal and carbon fixation. Polluted wastewater serves as a growth medium for microalgae cultivated in open raceway ponds or closed photobioreactors. Through photosynthesis, microalgae remove pollutants via biosorption and bioaccumulation while fixing CO_2 and producing biomass that is valorized into biofuels, biofertilizers, animal feed, and bioplastics under circular bioeconomy principles. Ecotoxicological safety assessment of the biomass is required before application.

These are inexpensive to build and maintain, open ponds like raceway ponds are popular for large-scale applications, especially in developing nations (Ansari et al., 2023; Gayo-Peláez et al., 2026; Gupta et al., 2026a). However, they are frequently constrained by poorer output, unstable environmental conditions, and contamination threats. By controlling elements such as light, temperature, and carbon dioxide supply, photobioreactors, on the other hand, offer a regulated environment that improves microalgal growth, photosynthetic efficiency, and pollution removal capability (Perin and Morosinotto, 2023; Shah et al., 2023; Yahaya et al., 2025; Ahuja et al., 2026). Despite being more effective, their expensive installation and maintenance costs prevent them from being widely used. In an effort to strike a balance between cost-effectiveness and operational efficiency, hybrid systems—which combine the benefits of both open and closed systems—are being investigated more and more. In these systems, microalgae are essential for eliminating organic contaminants, heavy metals, and nutrients like phosphorus and nitrogen, which improves water quality and produces useful biomass (Mahlangu et al., 2024; Gill et al., 2025; Faruque et al., 2026; Song et al., 2026).

Figure 2. Integrated microalgae-based system for bioremediation of aquatic pollutants, carbon sequestration, and wastewater valorization.

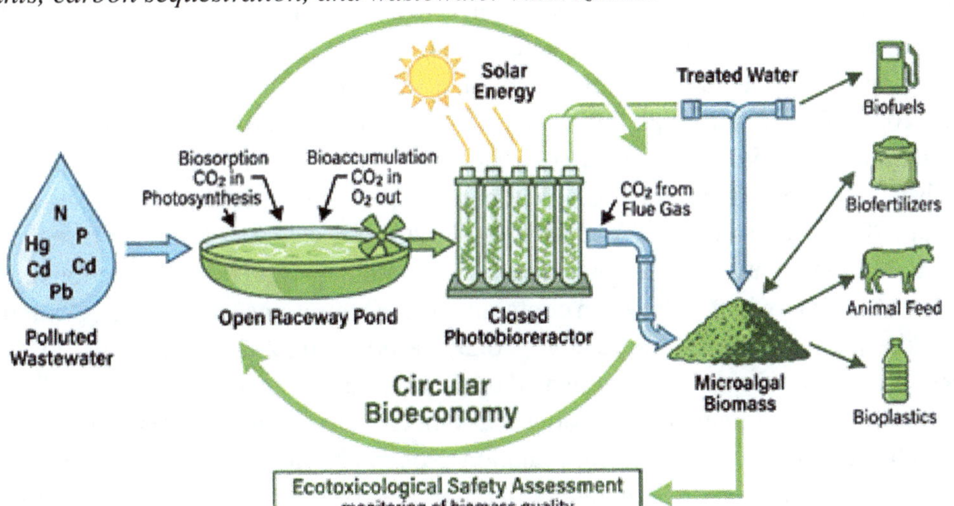

Microalgal biomass contributes to a sustainable circular bioeconomy by providing substantial prospects for resource recovery and valorization in addition to wastewater treatment (de Morais et al., 2023; Ezhumalai et al., 2024; Santos et al., 2025).

The gathered biomass can be transformed into a variety of biofuels, such as biogas through anaerobic digestion, bioethanol through fermentation of carbohydrates, and biodiesel through lipid extraction (Panahi et al., 2022; Mao and Zhang, 2023; Samal and Sukla, 2025). Apart from producing energy, microalgae are an abundant supply of valuable substances like proteins, vitamins, antioxidants, and pigments like carotenoids and chlorophylls that are used in the pharmaceutical, nutraceutical, and aquaculture sectors (Kaur et al., 2023; Osathanunkul et al., 2025; Rahman et al., 2026). Wastewater is transformed from a burden into a useful resource by this dual function of product development and pollution cleanup (Tripathy et al., 2025; Haldkar et al., 2026). By decreasing waste discharge, lowering environmental degradation, and fostering economic sustainability, this strategy is consistent with the waste-to-resource notion from the standpoint of the circular bioeconomy. Microalgae-based systems are a promising alternative for future environmental management and sustainable development since they reduce the overall cost of wastewater treatment, contribute to the creation of renewable energy, and lessen reliance on fossil fuels.

6. ECOTOXICOLOGICAL CONCERNS

An important but sometimes disregarded factor in the assessment of microalgae-based wastewater treatment systems is ecotoxicological safety. The majority of research has historically focused on the effectiveness of removing pollutants (such as nutrients, heavy metals, or organic contaminants), but this strategy is insufficient on its own to guarantee environmental safety (Hou et al., 2020; Khanam et al., 2023; Kumar and Singh, 2025). During treatment, pollutants may change into intermediate by-products that can occasionally be more hazardous than the parent substances, creating unseen ecological concerns (Nidheesh et al., 2022; Ashiwaju et al., 2023; Adam et al., 2025; Farina et al., 2026). According to recent research, complex combinations of newly developing contaminants and their interactions can cause residual toxicity even when contaminant concentrations are lowered (Goutam Mukherjee et al., 2022; Gomes et al., 2025; Rasool et al., 2025; Joda et al., 2026). Risk assessment is further complicated by the constant exposure of aquatic systems to a variety of contaminants, including heavy metals, medications, and microplastics, which can show synergistic or combination toxicity (Adeleye et al., 2024; Aib et al., 2025; Pathak et al., 2026). Therefore, in order to go beyond simple removal metrics and guarantee that treated effluents are truly safe for aquatic ecosystems, ecotoxicological evaluation must be incorporated into wastewater treatment assessment.

Both acute and chronic toxicity can result from the impacts of treated or untreated wastewater on aquatic creatures, including fish, invertebrates, and microbial ecosystems (Wu et al., 2024; Singh, 2025; Sohail, 2025; Bamaniya et al., 2026).

It is becoming more widely acknowledged that sub-lethal consequences, including altered behavior, decreased reproduction, oxidative stress, and developmental abnormalities, are sensitive markers of environmental contamination. Assessments based on biomarkers offer an effective way to identify these effects at various biological levels. The health of organisms exposed to pollutants is often assessed using hematological alterations (e.g., variations in red and white blood cell counts), biochemical markers (e.g., antioxidant enzymes, catalase, glutathione-S-transferase), and signs of oxidative stress (Habib et al., 2024; Iqbal et al., 2025; Saeed et al., 2025; Naz et al., 2026). Furthermore, histological changes in essential organs such as the kidney, liver, and gills provide unmistakable proof of tissue-level injury. Crucially, recent studies employing bioassays (such as zebrafish embryo testing) show that microalgae-based treatment can dramatically lower pollutant concentration and related toxicity, albeit effectiveness (Deepika et al., 2025) varies depending on the species and circumstances. To verify the true increase in water quality, it is therefore essential to compare toxicity before and after treatment.

Table 1. Roles of Microalgae in Environmental Sustainability

Aspect	Mechanism	Functions	Benefits	Limitations
Carbon sequestration	Photosynthetic CO_2 fixation	Conversion of CO_2 into biomass via photosynthesis	Reduces greenhouse gases, supports climate change mitigation, and produces biomass for bioenergy	Requires optimized light, CO_2 supply, and cultivation systems
Wastewater treatment	Nutrient uptake and microbial interaction	Removal of nitrogen, phosphorus, organic matter, and pathogens	Improves water quality, enables water reuse, low operational cost	Harvesting biomass is challenging, with large-scale implementation issues
Ecotoxicological Safety	Detoxification and transformation of pollutants	Reduces the toxicity of contaminants before discharge into aquatic systems	Protects aquatic organisms, improves ecosystem health, bioindicator potential	Some pollutants may bioaccumulate in algal biomass.
Bioremediation Control	Sensitivity to environmental conditions, contamination risk	Removal of heavy metals, pesticides, hydrocarbons, and nutrients (N, P)	Eco-friendly, cost-effective, and reduces pollutant load	Sensitivity to environmental conditions, contamination risk

7. CHALLENGES AND LIMITATION

Although microalgae-based systems show great promise, a number of obstacles and restrictions prevent their widespread commercialization and large-scale application. Scaling up from laboratory or pilot systems to industrial-level operations while preserving efficiency and cost-effectiveness is one of the fundamental obstacles. Large-scale farming necessitates substantial infrastructure, such as regulated systems, land, and water supplies, which can be costly, especially in poor nations. Despite being less expensive, open pond systems are very vulnerable to contamination by undesirable bacteria, predators, and rival algae species, which can cause unstable culture performance. Conversely, closed systems, like photobioreactors, have higher capital and operating expenses but provide greater control. Furthermore, because of the small cell size and diluted nature of cultures, harvesting microalgal biomass continues to be a significant bottleneck that necessitates energy-intensive procedures like centrifugation, flocculation, or filtration. The cost is further increased by post-harvest biomass processing for high-value products or biofuel. The use of genetically modified strains, the safe discharge of treated effluents, and the widespread application of microalgae technologies in natural ecosystems all lack defined norms, which makes regulatory and environmental considerations extremely important.

8. FUTURE RESEARCH DIRECTIONS

In the future, research aims to overcome these constraints through interdisciplinary integration and technological innovation. Microalgae genetic engineering has great potential to improve growth rates, pollution removal effectiveness, and stress tolerance in a variety of environmental settings. By enabling targeted pollutant removal, improving light usage, and permitting effective biomass harvesting, advances in nanotechnology can further enhance system performance. Furthermore, hybrid systems with greater efficiency and versatility can be produced by combining microalgae-based systems with cutting-edge treatment technologies, including membrane filtering, sophisticated oxidation procedures, and microbial fuel cells. The creation of thorough ecotoxicological safety frameworks is an important issue that needs more focus in order to guarantee that treated wastewater is both chemically and biologically safe for aquatic life. This involves incorporating long-term ecological monitoring and biomarker-based evaluations into routine evaluation procedures. Lastly, real-world implementation including the creation of regulatory standards, financial incentives, and public-private partnerships will require governmental backing and sustainability concerns. Microalgae-based solutions have the potential to revolutionize environmental protection and sustainable wastewater management by tackling these issues through innovation and governance.

9. CONCLUSION

In conclusion, by concurrently authorizing effective pollutant removal, carbon sequestration, and resource recovery, microalgae-based systems provide an integrated and sustainable solution for aquatic pollution. Their contribution to the advancement of circular bioeconomy systems is demonstrated by their capacity to treat wastewater while absorbing CO_2 and generating valuable biomass. But since aquatic organisms may be impacted by transformation byproducts and persistent toxicity, maintaining ecotoxicological safety is still crucial. For true environmental sustainability, it is therefore essential to combine treatment effectiveness with biomarker-based risk assessment. Microalgae have a great potential to play a significant role in the development of environmentally friendly wastewater treatment systems in the future, with further technological breakthroughs and appropriate regulatory backing. Additionally, industries and stakeholders engagement can improve the role of microalgae in pollution mitigation as well as can be more effective in contrast to other mechanisms.

REFERENCES

Adam, A. B., Mu'azu, J. B., Titilayo, O. A., Ba'aku, A. E., Gani, J., & Abubakar, M. Y. (2025). The Role of Organic Pollutants in Water Pollution A Review. *Journal of Chemical Technology*, *1*, 142–159.

Adeleye, A. T., Bahar, M. M., Megharaj, M., Fang, C., & Rahman, M. M. (2024). The unseen threat of the synergistic effects of microplastics and heavy metals in aquatic environments: A critical review. *Current Pollution Reports*, *10*(3), 478–497. DOI: 10.1007/s40726-024-00298-7

Ahuja, V., Sagar, N. A., Sharma, D., Ahlawat, G. M., Menon, V., Yang, Y.-H., Bhatia, S. K., & Dasgupta, D. (2026). A Review on Carbon Dioxide Mitigation and Sustainable Energy Production Through Algal Biorefineries. *BioEnergy Research*, *19*(1), 29. DOI: 10.1007/s12155-025-10955-4

Aib, H., Parvez, M. S., & Czédli, H. M. (2025). Pharmaceuticals and microplastics in aquatic environments: A comprehensive review of pathways and distribution, toxicological and ecological effects. *International Journal of Environmental Research and Public Health*, *22*(5), 799. DOI: 10.3390/ijerph22050799 PMID: 40427912

Akram, S., Guan, G., Hu, B., Saddique, M. A. B., Xi, Y., Luo, X., & Ren, M. (2026). Microalgae-based strategies for cadmium remediation: Insights, challenges, and future directions. *Critical Reviews in Biotechnology*, *46*(2), 318–337. DOI: 10.1080/07388551.2025.2585681 PMID: 41260883

Alowaisy, A., Allouzi, M. M. A., Chia, W. Y., Mayol, A. P., Sriariyanun, M., & Show, P. L. (2026). Interactions between phytoplankton species and micro/nano-plastics and heavy metal contamination. *Biological Reviews of the Cambridge Philosophical Society*, *101*(1), 60–73. DOI: 10.1111/brv.70071 PMID: 41058153

Anas, M., Khattak, W. A., Avci, M. A., Hakki, E. E., & Fahad, S. (2026). Role of Cyanobacteria in Bioremediation and Environmental Restoration. In *Cyanobacterial Blooms: Ecology, Evolution and Biogeochemical Impacts: Microbial Dynamics and Global Implications* (pp. 421-446). Springer. DOI: 10.1007/978-3-032-06042-6_17

Ansar, Z. A., Alappat, M. M., Subramanian, S., & Anbarasu, A. (2026). Microbial and multi-omics approaches for bioremediation of emerging contaminants: Environmental impact and future engineering solutions. *Biodegradation*, *37*(2), 52. DOI: 10.1007/s10532-026-10276-z PMID: 41832337

Ansari, F. A., Nasr, M., Rawat, I., & Bux, F. (2023). Meeting sustainable development goals (SDGs) through progression of pilot-scale algal system to commercial raceway pond (300,000 L). *Biomass Conversion and Biorefinery*, *13*(9), 7663–7676. DOI: 10.1007/s13399-021-01661-0

Aryanfar, Y., Keçebaş, A., Hassanzadeh, S., Sadabad, A. N., Rashid, F. L., Ahmad, S., Hendy, A., Irfan, H. M., Ghasemlou, S. M., & Fernandez, J. B. (2025). Integrated Evaluation of Contaminant Profiles, Detection Techniques, and Management Strategies for Tannery Sludge. *Asia-Pacific Journal of Chemical Engineering*, •••, e70191.

Asad, M., Sohail, M., Muneeb, A., Qadeer, S., Majeed, A. A. A., Rafiq, H., Rafique, A., Zubair, S. K., & Rafique, N. (2024). Strategies for enhancing animal feed quality and productivity. In *Complementary & Alternative Medicine* (pp. 364–372). Feed Additives.

Ashiwaju, B. I., Uzougbo, C. G., & Orikpete, O. F. (2023). Environmental impact of pharmaceuticals: A comprehensive review. *Matrix Science Pharma*, *7*(3), 85–94. DOI: 10.4103/mtsp.mtsp_15_23

Azeem, S., Sohail, M., Azeem, A., Haq, M. Z. U., Hassan, Z., Khan, A. A., & Shahzad, U. (2024). Guardian crops cultivating resilience against pests. In *Revolutionizing Pest Management for Sustainable Agriculture* (pp. 181–202). IGI Global Scientific Publishing. DOI: 10.4018/979-8-3693-3061-6.ch008

Bamaniya, P., Vaghela, D., Tandel, V., Chaudhari, S., & Bambhaniya, I. (2026). Linking Terrestrial and Aquatic Systems: Pollutant Transport, Bioaccumulation Cycles, and Global Food Security. In *Plastic Tides: Tackling Plastic Pollution for Global Food Security* (pp. 195-228). Springer.

Choudhary, S., Tripathi, S., & Poluri, K. M. (2022). Microalgal-based bioenergy: Strategies, prospects, and sustainability. *Energy & Fuels*, *36*(24), 14584–14612. DOI: 10.1021/acs.energyfuels.2c02922

da Silva, S. R., Borges, P. M., Tošić, N., & Andrade, J. J. O. (2026). Advances in circular valorization of construction and demolition waste (CDW) toward low-carbon and resilient construction: A Comprehensive review. *Sustainability (Basel)*, *18*(6), 2759. DOI: 10.3390/su18062759

de Carvalho, J. C., Molina-Aulestia, D. T., Martinez-Burgos, W. J., Karp, S. G., Manzoki, M. C., Medeiros, A. B. P., Rodrigues, C., Scapini, T., Vandenberghe, L. P. S., Vieira, S., Woiciechowski, A. L., Soccol, V. T., & Soccol, C. R. (2022). Agro-industrial wastewaters for algal biomass production, bio-based products, and biofuels in a circular bioeconomy. *Fermentation (Basel, Switzerland)*, *8*(12), 728. DOI: 10.3390/fermentation8120728 de Morais, E. G., da Silveira, J. T., Schüler, L. M., de Freitas, B. C. B., Costa, J. A. V., de Morais, M. G., Ferrer, I., & Barreira, L. (2023). Biomass valorization via pyrolysis in microalgae-based wastewater treatment: Challenges and opportunities for a circular bioeconomy. *Journal of Applied Phycology*, *35*(6), 2689–2708. DOI: 10.1007/s10811-023-03104-x

De Silva, A. G. S., Hashim, Z. K., Solomon, W., Zhao, J.-B., Kovács, G., Kulmány, I. M., & Molnár, Z. (2024). Unveiling the role of edaphic microalgae in soil carbon sequestration: Potential for agricultural inoculants in climate change mitigation. *Agriculture*, *14*(11), 2065. DOI: 10.3390/agriculture14112065

Deepika, C., Frascaroli, G., Gowri, A., & Silkina, A. (2025). Microalgae-based bioremediation: A sustainable approach for the removal of emergent pollutants. In *Biotechnological Interventions in the Removal of Emerging Pollutants* (pp. 101–139). Springer. DOI: 10.1007/978-981-97-9922-0_6

Doan, L., Pham, N. V., Nguyen, P. B., Le, Q. N., Phung, T. K., & Ngo, T. T. (2026). Marine Algae as Sustainable Platforms for the Green Synthesis of Metal Nanoparticles. *ACS Omega*, *11*(10), 15598–15616. DOI: 10.1021/acsomega.5c12334 PMID: 41867616

Dodangodage, C. A., Kasturiarachchi, J. C., Wijesekara, I. A., Perera, T. A., Rajapakshe, D., & Halwatura, R. (2026). Integrated Microalgal–Aquaponic Systems for Enhanced Water Treatment and Food Security: A Critical Review of Recent Advances in Process Integration and Resource Recovery. *Phycology*, *6*(1), 14. DOI: 10.3390/phycology6010014

Echeverri, D. O., Beltrán, V. V., Mesa, A. P., Ríos, L. A., & Echeverri, E. O. (2026). Financial Potential of Life Cycle Assessment of Production of Biofuels via Hydrothermal Liquefaction Using Microalgae in Cement Industry. *Energies*, *19*(5), 1153. DOI: 10.3390/en19051153

El Amri, R., Zahnoune, R., Boudouch, O., Hamdouch, M., Hattab, I., Hasib, A., Chakraborty, S., & Elkacmi, R. (2026). Microalgae as a Natural Resource: Economic Potential and Applications. In *Wastewater Treatment* (pp. 251–262). CRC Press. DOI: 10.1201/9781003624905-12

El Messaoudi, N., Georgin, J., Franco, D. S., Yılmazoğlu, M., Temur, B., Miyah, Y., Benjelloun, M., & Al-Msiedeen, A. M. (2025). Composites based on porphyran as an adsorbent for environmentally friendly water treatment. *Reactive & Functional Polymers*, •••, 106616.

El-Sheekh, M. M., El-Kassas, H. Y., & Ali, S. S. (2025). Microalgae-based bioremediation of refractory pollutants: An approach towards environmental sustainability. *Microbial Cell Factories*, *24*(1), 19. DOI: 10.1186/s12934-024-02638-0 PMID: 39810167

Ethiraj, S., & Samuel, M. S. (2024). A comprehensive review of the challenges and opportunities in microalgae-based wastewater treatment for eliminating organic, inorganic, and emerging pollutants. *Biocatalysis and Agricultural Biotechnology*, *60*, 103316. DOI: 10.1016/j.bcab.2024.103316

Ezhumalai, G., Arun, M., Manavalan, A., Rajkumar, R., & Heese, K. (2024). A holistic approach to circular bioeconomy through the sustainable utilization of microalgal biomass for biofuel and other value-added products. *Microbial Ecology*, *87*(1), 61. DOI: 10.1007/s00248-024-02376-1 PMID: 38662080

Farina, I., Scotto di Uccio, A., Moretti, A., & Colangelo, F. (2026). A review on PFAS mitigation in wastewater: Integrating removal technologies, challenges and innovative solutions. *Reviews in Environmental Science and Biotechnology*, *25*(1), 15. DOI: 10.1007/s11157-026-09766-6

Faruque, M. O., Mohammed, T., Farooq, W., Nzila, A., Hossain, M. M., & Razzak, S. A. (2026). Performance of Chlorella sorokiniana in bioremediation of organic pollutants from produced water under heavy metal stress. *Journal of Water Process Engineering*, *83*, 109736. DOI: 10.1016/j.jwpe.2026.109736

Firincă, C., Zamfir, L.-G., Constantin, M., Răut, I., Jecu, M.-L., Doni, M., Gurban, A.-M., & esan, T. E. (2025). Innovative approaches and evolving strategies in heavy metal bioremediation: Current limitations and future opportunities. *Journal of Xenobiotics*, *15*(3), 63. DOI: 10.3390/jox15030063 PMID: 40407527

Gan, W., Deng, X., Liu, Y., Liu, L., Yang, M., & Xu, Y. (2026). Potassium nutrition regulates growth, nutrient-carbon interactions and lignification during autumn in Tsoongiodendron odorum seedlings. *BMC Plant Biology*, *26*(1), 695. DOI: 10.1186/s12870-026-08497-8 PMID: 41803723

Gautam, P., Salunke, D., Lad, D., & Gautam, A. (2025). Convergent synergy of carbon capture within the circular economy paradigm: A nexus for realizing multifaceted sustainable development goals. *Carbon Research*, *4*(1), 3. DOI: 10.1007/s44246-024-00178-1

Gayo-Peláez, J. I., Oatley-Radcliffe, D. L., Silkina, A., & Barron, A. R. (2026). Cost-effective and sustainable microalgae cultivation: A low-cost artificially integrated LED photobioreactor ensuring high-quality algal biomass production from industrial CO2 flue gas in a high latitude country. *Cleaner & Circular Bioeconomy*, Article 100209.

Gill, J. M., Hussain, S. M., Ali, S., Zahoor, A. F., Alasmari, A., Munir, M., Naeem, E., Amjad, M., Yousaf, Z., & Faisal, M. (2025). Optimizing aquaculture sustainability: Microalgae-based co-culture systems for aquaculture wastewater treatment and pollution reduction. *Algal Research*, *92*, 104430. DOI: 10.1016/j.algal.2025.104430

Gomes, A. R., Matos, L. P., Silva, A. M., Guimaraes, A. T. B., Luz, T. M., Brito, R. R., Lima Rodrigues, A. S., Vaz, B. G., Oliveira, J. A., Soares, W. R., Menezes, I. P. P., & Malafaia, G. (2025). Complex interactions and ecotoxicological consequences of tannery effluents on aquatic life: Differential responses in scinax fuscovarius tadpoles and poecilia reticulata fish. *Water, Air, and Soil Pollution*, *236*(2), 105. DOI: 10.1007/s11270-024-07710-4

González Fernández, L. A., Medellín Castillo, N. A., Sánchez Polo, M., Navarro Frómeta, A. E., & Vilasó Cadre, J. E. (2025). Algal-based carbonaceous materials for environmental remediation: Advances in wastewater treatment, carbon sequestration, and biofuel applications. *Processes (Basel, Switzerland)*, *13*(2), 556. DOI: 10.3390/pr13020556

Goutam Mukherjee, A., Ramesh Wanjari, U., Eladl, M. A., El-Sherbiny, M., Elsherbini, D. M. A., Sukumar, A., Kannampuzha, S., Ravichandran, M., Renu, K., Vellingiri, B., Kandasamy, S., & Valsala Gopalakrishnan, A. (2022). Mixed contaminants: Occurrence, interactions, toxicity, detection, and remediation. *Molecules (Basel, Switzerland)*, *27*(8), 2577. DOI: 10.3390/molecules27082577 PMID: 35458775

Guo, X., Liu, Y., Xie, T., Li, Y., Liu, H., & Wang, Q. (2025). Impact of ecological restoration on carbon sink function in coastal wetlands: A review. *Water (Basel)*, *17*(4), 488. DOI: 10.3390/w17040488

Gupta, D., Gauri, R. K., Tyagi, A., Tyagi, N., Sharma, Y., Doli, K. S., Sharma, H., & Sharma, M. (2026a). Advancements in Design and Development of Lab Scale Photobiorectors for Improved Microalgal Feedstock. *Microbiology Archives, an International Journal*.

Gupta, P., Prabhakaran, R., Soni, N., Singh, S. D. K., Kamaraj, C., Selvan, S. T., Asokan, N., Pamidimarri, S. D., & Velramar, B. (2026b). Utilization of Genetically Engineered RuBisCO Enzyme for Capturing Carbon Dioxide to Address Environmental Challenges-An Overview. *Water, Air, and Soil Pollution, 237*(2), 58. DOI: 10.1007/s11270-025-08639-y

Habib, S. S., Naz, S., Fazio, F., Cravana, C., Ullah, M., Rind, K. H., Attaullah, S., Filiciotto, F., & Khayyam, K. (2024). Assessment and bioaccumulation of heavy metals in water, fish (wild and farmed) and associated human health risk. *Biological Trace Element Research, 202*(2), 725–735. DOI: 10.1007/s12011-023-03703-2 PMID: 37178449

Haldkar, A., Wali, H., Yadav, M., Dwibedi, V., & George, N. (2026). Transforming agro-waste into ecofriendly nanomaterials for wastewater treatment: Embracing the principles of green chemistry and the circular economy. In *Sustainable Agricultural Waste Management* (pp. 267–300). Elsevier. DOI: 10.1016/B978-0-443-38301-4.00017-5

Hasan, Z. S., & Enad, H. Y. (2026). Comparative Evaluation of Microalgae-Based Wastewater Treatment Using Closed Photobioreactors and Open Pond Systems for Nutrient and Heavy Metal Removal and Resource Recovery. *Water, Air, and Soil Pollution, 237*(6), 375. DOI: 10.1007/s11270-025-09070-z

Hoque, M. M., Iannelli, V., Padula, F., Radice, R. P., Saha, B. K., Martelli, G., Scopa, A., & Drosos, M. (2025). Microalgae: Green engines for achieving carbon sequestration, circular economy, and environmental sustainability—A review based on last ten years of research. *Bioengineering (Basel, Switzerland), 12*(9), 909. DOI: 10.3390/bioengineering12090909 PMID: 41007154

Hou, D., O'Connor, D., Igalavithana, A. D., Alessi, D. S., Luo, J., Tsang, D. C., Sparks, D. L., Yamauchi, Y., Rinklebe, J., & Ok, Y. S. (2020). Metal contamination and bioremediation of agricultural soils for food safety and sustainability. *Nature Reviews. Earth & Environment, 1*(7), 366–381. DOI: 10.1038/s43017-020-0061-y

Iakovidou, G., Itziou, A., Tsiotsias, A., Lakioti, E., Samaras, P., Tsanaktsidis, C., & Karayannis, V. (2024). Application of microalgae to wastewater bioremediation, with CO2 biomitigation, health product and biofuel development, and environmental biomonitoring. *Applied Sciences (Basel, Switzerland), 14*(15), 6727. DOI: 10.3390/app14156727

Iqbal, K., Sohail, M., Hussain Rind, K., & Habib, S. S. (2025). Agrochemical contamination and fish health: Eco-toxicological impacts and mitigation strategies. *Chemistry and Ecology, 41*(7), 959–993. DOI: 10.1080/02757540.2025.2508993

Jiang, J.-C., Feng, Z.-Y., Meng, L.-Y., & Jin, B. (2026). Recent advances in pretreatment technologies for the analysis of per-and polyfluoroalkyl substances. *Analytical and Bioanalytical Chemistry, 418*(7), 1–23. DOI: 10.1007/s00216-026-06317-4 PMID: 41663795

Joda, J. F., Raji, R. O., Nmadu, M. E., & Oyewole, O. A. (2026). The Interactions of Emergent Contaminants (ECs) with Soil Microbiome. *Ecotoxicology (London, England), 35*(4), 71. DOI: 10.1007/s10646-026-03058-z PMID: 41762294

Kaste, J. A., Walker, B. J., & Shachar-Hill, Y. (2024). Reaction-diffusion modeling provides insights into biophysical carbon-concentrating mechanisms in land plants. *Plant Physiology, 196*(2), 1374–1390. DOI: 10.1093/plphys/kiae324 PMID: 38857179

Kaur, M., Bhatia, S., Gupta, U., Decker, E., Tak, Y., Bali, M., Gupta, V. K., Dar, R. A., & Bala, S. (2023). Microalgal bioactive metabolites as promising implements in nutraceuticals and pharmaceuticals: Inspiring therapy for health benefits. *Phytochemistry Reviews, 22*(4), 903–933. DOI: 10.1007/s11101-022-09848-7 PMID: 36686403

Kaur, S., Kaur, S., Thakur, B., & Narwat, N. (2026). Microalgae-based bioremediation of emerging contaminants: Techniques, recent developments, and future perspectives. *Archives of Microbiology, 208*(1), 29. DOI: 10.1007/s00203-025-04567-x PMID: 41264010

Khanam, Z., Sultana, F. M., & Mushtaq, F. (2023). Environmental pollution control measures and strategies: An overview of recent developments. In *Geospatial analytics for environmental pollution modeling: Analysis, control and management* (pp. 385-414).

Kumar, A., Sharma, A., Majumder, P., Vishal, V., & Dutta, A. (2025a). Chemical and microalgal conversion of carbon dioxide into fuels and materials: A review. *Environmental Chemistry Letters, 23*(5), 1209–1229. DOI: 10.1007/s10311-025-01855-2

Kumar, P., & Singh, J. (2025). Perspective and challenges of synergistic removal of toxic contaminants from effluent using different treatment techniques. In *Microbial niche nexus sustaining environmental biological wastewater and water-energy-environment nexus* (pp. 419–451). Springer. DOI: 10.1007/978-3-031-62660-9_17

Kumar, S., Tripathi, A., Yadhav, S., Mishra, S., & Ghangrekar, M. M. (2025b). Interactions of Microplastics with Microbial Communities and the Food Web/Plants. In *Microplastics in the Environment* (pp. 283–310). Fate, Impacts, Removal, and Management. DOI: 10.1002/9781394251100.ch11

Kupriyanova, E. V., Pronina, N. A., & Los, D. A. (2023). Adapting from low to high: An update to CO2-concentrating mechanisms of cyanobacteria and microalgae. *Plants*, *12*(7), 1569. DOI: 10.3390/plants12071569 PMID: 37050194

Liao, H., Gao, D., Kong, C., Junaid, M., Li, Y., Chen, X., Zheng, Q., Chen, G., & Wang, J. (2023). Trophic transfer of nanoplastics and di (2-ethylhexyl) phthalate in a freshwater food chain (Chlorella pyrenoidosa-Daphnia magna-Micropterus salmoides) induced disturbance of lipid metabolism in fish. *Journal of Hazardous Materials*, *459*, 132294. DOI: 10.1016/j.jhazmat.2023.132294 PMID: 37591169

Ling, X., Lu, G., Jiang, R., Zhang, L., Zhang, P., Wang, M., Xue, Q., Yu, Y., & Ding, K. (2026). Enhanced Ecological Risk of Plastic-Associated Chemicals by Microplastics in the Riverine Benthic Food Web. *Environmental Science & Technology*, *60*(10), 8120–8132. DOI: 10.1021/acs.est.5c13059 PMID: 41773988

Liu, R., Li, S., Tu, Y., & Hao, X. (2021). Capabilities and mechanisms of microalgae on removing micropollutants from wastewater: A review. *Journal of Environmental Management*, *285*, 112149. DOI: 10.1016/j.jenvman.2021.112149 PMID: 33607565

Ma, W., Liang, X., Ding, C., Ye, Y., & Li, J. (2026). From Enrichment to Fate: Transport, Transformation, and Fate of Micro- and Nanoplastics in Marine Environments. *Toxics*, *14*(2), 120. DOI: 10.3390/toxics14020120 PMID: 41745794

Mahlangu, D., Mphahlele, K., De Paola, F., & Mthombeni, N. H. (2024). Microalgae-mediated biosorption for effective heavy metals removal from wastewater: A review. *Water (Basel)*, *16*(5), 718. DOI: 10.3390/w16050718

Mao, B., & Zhang, B. (2023). Combining ABE fermentation and anaerobic digestion to treat with lipid extracted algae for enhanced bioenergy production. *The Science of the Total Environment*, *875*, 162691. DOI: 10.1016/j.scitotenv.2023.162691 PMID: 36898333

Maqsood, Q., Waseem, R., Sumrin, A., Wajid, A., Tariq, M. R., Ali, S. W., & Mahnoor, M. (2024). Recent trends in bioremediation and bioaugmentation strategies for mitigation of marine based pollutants: Current perspectives and future outlook. *Discover Sustainability*, *5*(1), 524. DOI: 10.1007/s43621-024-00607-6

Minhas, A. K., Gaur, S., Sunny, S., Paladugu, C., Ravishankar, G. A., Pereira, L., & Ambati, R. R. (2026). Microalgae-Based Wastewater Treatment Processes for the Bioremediation and Valorization of Biomass: A Review. *Phycology*, *6*(1), 18. DOI: 10.3390/phycology6010018

Mounaimi, S., Al-Tawaha, A. R. M., Chippalakatti, P., Sangeetha, J., & Thangadurai, D. (2026). Microalgal Bioresources For Sustainable Agriculture. In *Microbial Fertilizer Technology for Sustainable Crop Production* (pp. 53–87). Apple Academic Press. DOI: 10.1201/9781003594826-3

Naz, S., Brandts, I., Resina, A., Tudela, P., Sugrañes, M., Expósito, N., Barrientos-Riosalido, A., Rovira, J., Balasch, J., & Teles, M. (2026). Environmentally Derived Nanoplastics Induce Oxidative and Genotoxic Responses in Gilthead Seabream (Sparus aurata). *Journal of Hazardous Materials: Plastics*, Article 100048.

Nidheesh, P. V., Couras, C., Karim, A. V., & Nadais, H. (2022). A review of integrated advanced oxidation processes and biological processes for organic pollutant removal. *Chemical Engineering Communications*, *209*(3), 390–432. DOI: 10.1080/00986445.2020.1864626

Okoro, H. K., Orosun, M. M., Oriade, F. A., Momoh-Salami, T. M., Ogunkunle, C. O., Adeniyi, A. G., Zvinowanda, C., & Ngila, J. C. (2023). Potentially toxic elements in pharmaceutical industrial effluents: A review on risk assessment, treatment, and management for human health. *Sustainability (Basel)*, *15*(8), 6974. DOI: 10.3390/su15086974

Osathanunkul, M., Thanaporn, S., Karapetsi, L., Nteve, G. M., Pratsinakis, E., Stefanidou, E., Lagiotis, G., Avramidou, E., Zorxzobokou, L., Tsintzou, G., Athanasiou, A., Mpelai, S., Constandinidis, C., Pantiora, P., Merino, M., Mullor, J. L., Dobrovic, L., Cerasino, L., Ogawa, T., & Madesis, P. (2025). Diversity of bioactive compounds in microalgae: Key classes and functional applications. *Marine Drugs*, *23*(6), 222. DOI: 10.3390/md23060222 PMID: 40559631

Padhi, D., Kashyap, S., Mohapatra, R. K., Dineshkumar, R., & Nayak, M. (2025). Microalgae-based flue gas CO2 sequestration for cleaner environment and biofuel feedstock production: A review. *Environmental Science and Pollution Research International*, *32*(23), 13539–13565. DOI: 10.1007/s11356-025-35958-8 PMID: 39888525

Panahi, H. K. S., Dehhaghi, M., Guillemin, G. J., Gupta, V. K., Lam, S. S., Aghbashlo, M., & Tabatabaei, M. (2022). Bioethanol production from food wastes rich in carbohydrates. *Current Opinion in Food Science*, *43*, 71–81. DOI: 10.1016/j.cofs.2021.11.001

Pandey, P., & Tewary, S. Angelin, & Paruchuri, S. (2026). Potential Sources of Emerging Environmental Pollutants: Fate of EPs in Environment (Soil, Water, and Air). In *Emerging Contaminants in the Environment: Occurrence, Detection and Biodegradation* (pp. 41-76). Springer.

Pandey, S., Soni, R., Devi, P., Singh, P., & Pandey, A. (2025). Innovations in microalgae-based wastewater remediation, which simultaneously recovers value-added products from industrial and non-industrial wastewater: A review. *Plant Archives*, *25*, 921–932.

Patel, A. K. (2025). Microalgae in dual role: An effective platform for biological carbon capturing and high-value carotenoid production. *Current Pollution Reports*, *11*(1), 34. DOI: 10.1007/s40726-025-00362-w

Pathak, H. K., Gandhi, K., & Singh, R. P. (2026). Potential Impact of Heavy Metals and Microplastics in River Ecosystem on Aquatic Organisms and Human Health, and Sustainable Mitigation Approaches. *Water, Air, and Soil Pollution*, *237*(4), 199. DOI: 10.1007/s11270-025-08844-9

Perin, G., & Morosinotto, T. (2023). Understanding regulation in complex environments: A route to enhance photosynthetic light-reactions in microalgae photobioreactors. *Frontiers in Photobiology*, *1*, Article 1274525.

Qv, M., Dai, D., Liu, D., Wu, Q., Tang, C., Li, S., & Zhu, L. (2023). Towards advanced nutrient removal by microalgae-bacteria symbiosis system for wastewater treatment. *Bioresource Technology*, *370*, 128574. DOI: 10.1016/j.biortech.2022.128574 PMID: 36603749

Rahman, N. A., Katayama, T., Wahid, M. E. A., Jusoh, M., Yusoff, F. M., & Takahashi, K. (2026). Microalgae as biofactories of natural antioxidants: From stress responses to industrial applications. *Journal of Applied Phycology*, *38*(2), 1–34. DOI: 10.1007/s10811-026-03804-0

Raj, A., Gauba, P., & Bhatt, E. (2025). Advanced nanoparticles for environmental remediation of emerging pollutants: A review. *Soil & Sediment Contamination*, *34*(6), 1344–1373. DOI: 10.1080/15320383.2024.2422387

Rajkumar, R., Takriff, M. S., & Veeramuthu, A. (2025). Technical insights into carbon dioxide sequestration by microalgae: A biorefinery approach towards sustainable environment. *Biomass Conversion and Biorefinery*, *15*(23), 1–16. DOI: 10.1007/s13399-022-02446-9

Rasool, B. S., Abbas, A. K., & Haddad, R. (2025). Photocatalytic innovations in environmental remediation: Mechanisms, materials, and challenges for persistent organic pollutant removal. *Environmental Monitoring and Assessment*, *197*(10), 1086. DOI: 10.1007/s10661-025-14531-3 PMID: 40911193

Ren, W., Zhang, Q., Chen, J., Xiao, X., Duan, X., Luo, X., & Wang, S. (2025). Catalytic resource recovery for transformation of the wastewater industry. *Nature Water*, *3*(11), 1–15. DOI: 10.1038/s44221-025-00530-8

Saeed, K., Abdullah, S., Yaqoob, M., Fatima, M., Majeed, Z., Ismail, M., Ghanem, S. F., Eissa, M. E., Hassanien, H., Abdul Kari, Z., & Eissa, E.-S. H. (2025). Integrating hematological profile and biomarker analysis for metals and metalloids ecotoxicological assessment on Wallago catfish and Indian carp health in Head Siphon, Mailsi, Pakistan. *Biometals : An International Journal on the Role of Metal Ions in Biology, Biochemistry, and Medicine*, *38*(6), 1897–1915. DOI: 10.1007/s10534-025-00737-y PMID: 40911255

Saini, K., Prajapati, A., Kumar, S. S., Kumar, V., & Bajar, S. (2026). A Review on the Fate of Emerging Contaminants in Landfill Leachate: Insights from Conventional Treatment Approaches. *Water, Air, and Soil Pollution*, *237*(5), 273. DOI: 10.1007/s11270-025-08937-5

Samal, D. K., & Sukla, L. B. (2025). Unveiling the dual potential of microalgae and seaweed biomass for sustainable biofuel production: A review. *RSC Advances*, *15*(41), 34160–34175. DOI: 10.1039/D5RA04845A PMID: 40979955

Santos, B., Freitas, F., Sobral, A. J., & Encarnação, T. (2025). Microalgae and circular economy: Unlocking waste to resource pathways for sustainable development. *International Journal of Sustainable Engineering*, *18*(1), 2501488. DOI: 10.1080/19397038.2025.2501488

Sarker, N. K., & Kaparaju, P. (2025). Microalgal wastewater treatment: A circular economy approach. In *Nature-Based Technologies for Wastewater Treatment and Bioenergy Production* (p. 153). DOI: 10.2166/9781789064100_0153

Sarmah, M., & Panda, M. R. (2026). Systematic review on proliferation of microalgae and carbon sequestration potential for sustainability. *World Journal of Microbiology & Biotechnology*, *42*(3), 100. DOI: 10.1007/s11274-026-04849-6 PMID: 41731188

Serna-García, R., Lanzoni, Y., García-Depraect, O., Muñoz, R., & Cantera, S. (2026). Harnessing Biogas into High-Value Chemicals: The Role of Algal–Methanotrophic Co-Cultures. *Marine Drugs*, *24*(2), 81. DOI: 10.3390/md24020081 PMID: 41745484

Shah, S. V., Lamba, B. Y., Sharma, R., Tiwari, A. K., Parveen, A., & Kumar, V. (2023). Effect of different light on microalgae growth and its integration with carbon dioxide sequestration technologies with emphasis on biofixation by microalgae. *Materials Today: Proceedings*, *73*, 92–99. DOI: 10.1016/j.matpr.2022.09.370

Shareefdeen, Z., Elkamel, A., & Babar, Z. B. (2023). Recent developments on the performance of algal bioreactors for CO2 removal: Focusing on the light intensity and photoperiods. *Biotech*, *12*(1), 10. DOI: 10.3390/biotech12010010 PMID: 36648836

Shekh, A., Sharma, A., Schenk, P. M., Kumar, G., & Mudliar, S. (2022). Microalgae cultivation: Photobioreactors, CO2 utilization, and value-added products of industrial importance. *Journal of Chemical Technology and Biotechnology*, *97*(5), 1064–1085. DOI: 10.1002/jctb.6902

Simkin, A. J., Kapoor, L., Doss, C. G. P., Hofmann, T. A., Lawson, T., & Ramamoorthy, S. (2022). The role of photosynthesis related pigments in light harvesting, photoprotection and enhancement of photosynthetic yield in planta. *Photosynthesis Research*, *152*(1), 23–42. DOI: 10.1007/s11120-021-00892-6 PMID: 35064531

Singh, A., & Gaur, R. (2026). Evaluating the Carbon Sequestration Potential of Spirulina spp.: Mechanistic Insights and Prospects for Industrial Valorization. *Waste and Biomass Valorization*, •••, 1–18. DOI: 10.1007/s12649-026-03510-5

Singh, G. (2025). Linear Alkylbenzene Sulphonate in Freshwater Ecosystems: Sources, Persistence, Bioaccumulation and Impacts on Fish Health–A Critical Review. *International Journal of Scientific Research in Modern Science and Technology*, *4*(12), 39–50. DOI: 10.59828/ijsrmst.v4i12.403

Sinha, T., Borah, D., & Rout, J. (2025). Algal communities in urban freshwater ponds: Prospects for biodiesel production. *Vegetos*, *38*(1), 202–212. DOI: 10.1007/s42535-023-00804-0

Sohail, M. (2025). Fish as Bioindicators of Pesticide-Induced Histological and Biochemical Toxicity. *International Journal of Forest Sciences*, *2*, 205–215.

Song, Q., Kong, F., Liu, B.-F., Song, X., Tan, X., Ren, N.-Q., & Ren, H.-Y. (2026). Synergistic coupling of microalgae and corn cob residue biochar for enhanced lipid production and heavy metal removal from waste leachate and corn cob hydrolysate. *Energy*, *347*, 140430. DOI: 10.1016/j.energy.2026.140430

Song, Y., Hu, Z., Yang, X., An, Y., & Lu, Y. (2025). Synergistic microalgae-duckweed systems for enhanced aquaculture wastewater treatment, biomass recovery, and CO2 sequestration: A novel approach for sustainable resource recovery. *Environmental Research*, *274*, 121271. DOI: 10.1016/j.envres.2025.121271 PMID: 40043930

Srimongkol, P., Sangtanoo, P., Songserm, P., Watsuntorn, W., & Karnchanatat, A. (2022). Microalgae-based wastewater treatment for developing economic and environmental sustainability: Current status and future prospects. *Frontiers in Bioengineering and Biotechnology, 10*, 904046. DOI: 10.3389/fbioe.2022.904046 PMID: 36159694

Sui, J., Yang, F., Huang, Z., & Li, M. (2025). Physiological responses and degradation potential of microalgae Chlorella vulgaris, Chlamydomonas sp. and Tetradesmus obliquus to organophosphorus pesticides chlorpyrifos and dimethoate. *Journal of Applied Phycology, 37*(6), 4123–4137. DOI: 10.1007/s10811-025-03676-w

Swain, P. R., Parida, P. K., Majhi, P. J., Behera, B. K., & Das, B. K. (2025). Microplastics as emerging contaminants: Challenges in inland aquatic food web. *Water (Basel), 17*(2), 201. DOI: 10.3390/w17020201

Tarfeen, N., Nisa, K. U., Hamid, B., Bashir, Z., Yatoo, A. M., Dar, M. A., Mohiddin, F. A., Amin, Z., Ahmad, R. A., & Sayyed, R. (2022). Microbial remediation: A promising tool for reclamation of contaminated sites with special emphasis on heavy metal and pesticide pollution: A review. *Processes (Basel, Switzerland), 10*(7), 1358. DOI: 10.3390/pr10071358

Thakur, S., Chandra, A., Kumar, V., & Bharti, S. (2025). Environmental pollutants: Endocrine disruptors/pesticides/reactive dyes and inorganic toxic compounds metals, radionuclides, and metalloids and their impact on the ecosystem. In *Biotechnology for environmental sustainability* (pp. 55–100). Springer. DOI: 10.1007/978-981-97-7221-6_3

Tikkanen, M., & Aro, E.-M. (2026). Interacting short-term regulatory mechanisms enable the conversion of light energy to chemical energy in photosynthesis. *Journal of Experimental Botany, 77*(4), 895–909. DOI: 10.1093/jxb/eraf451 PMID: 41124594

Tripathi, M., Singh, P., Singh, R., Bala, S., Pathak, N., Singh, S., Chauhan, R. S., & Singh, P. K. (2023). Microbial biosorbent for remediation of dyes and heavy metals pollution: A green strategy for sustainable environment. *Frontiers in Microbiology, 14*, 1168954. DOI: 10.3389/fmicb.2023.1168954 PMID: 37077243

Tripathy, S., Kar, O. P., & Pradhan, A. (2025). Challenges and innovations in industrial wastewater treatment: Safeguarding water resources and promoting sustainable practices. *Water, Air, and Soil Pollution, 236*(2), 92. DOI: 10.1007/s11270-025-07742-4

Ugwuanyi, E. D., Nwokediegwu, Z. Q. S., Dada, M. A., Majemite, M. T., & Obaigbena, A. (2024). The role of algae-based wastewater treatment systems: A comprehensive review. *World Journal of Advanced Research & Reviews, 21*(2), 937–949. DOI: 10.30574/wjarr.2024.21.2.0521

Wang, X., & Hong, Y. (2022). Microalgae biofilm and bacteria symbiosis in nutrient removal and carbon fixation from wastewater: A review. *Current Pollution Reports*, *8*(2), 128–146. DOI: 10.1007/s40726-022-00214-x

Wasilewska-Dębowska, W., Zienkiewicz, M., & Drozak, A. (2022). How light reactions of photosynthesis in C4 plants are optimized and protected under high light conditions. *International Journal of Molecular Sciences*, *23*(7), 3626. DOI: 10.3390/ijms23073626 PMID: 35408985

Wu, X., Nawaz, S., Li, Y., & Zhang, H. (2024). Environmental health hazards of untreated livestock wastewater: Potential risks and future perspectives. *Environmental Science and Pollution Research International*, *31*(17), 24745–24767. DOI: 10.1007/s11356-024-32853-6 PMID: 38499926

Yahaya, E., Yeo, W. S., Nandong, J., & Ngu, J. C. Y. (2025). CO2 fed microalgae cultivation in photobioreactor: Review on challenges and possible solutions. *Environmental Technology Reviews*, *14*(1), 540–564. DOI: 10.1080/21622515.2025.2508945

Yang, J., Zhou, Z., Qi, W., Gao, X., Wang, Y., Wang, X., Yi, X., Ma, M., & Wu, S. (2025). Phenotypic plasticity and integration synergistically enhance plant adaptability to flooding and nitrogen stresses. *Plant and Soil*, *513*(1), 1009–1030. DOI: 10.1007/s11104-025-07230-y

Yang, L., Wei, W., Sun, X., Li, X., Wen, X., Mao, B., Hao, R., Qiu, W., & Li, H. (2026). Competitive nitrogen/phosphorus scarcity triggers multi-pathway nitrogen removal and phosphorus-stress response in microalgae-bacterial symbiosis: Poly-metabolic synergy enhancement mechanisms. *Bioresource Technology*, *445*, 134138. DOI: 10.1016/j.biortech.2026.134138 PMID: 41621545

Zhang, Y., Zhang, X., Wang, M., Zhang, R., Li, Y., Zhang, G., Fang, W., & Zhang, P. (2025). Greenhouse gas emissions and carbon sequestration capacity of constructed wetlands with different hydrophytes. *Journal of Water Process Engineering*, *76*, 108292. DOI: 10.1016/j.jwpe.2025.108292

Zou, Z., Li, M., He, S., Zhou, Z., Tian, J., Cai, B., Lu, Y., Dai, Z., Zhu, X., & Han, Z. (2026). Multifunctionalized Immobilized Microalgae Technologies for Application in Advanced Environmental Remediation Systems. *Transactions of Tianjin University*, 1-28.

Chapter 11 Microalgal Biomass Valorisation Into Soil Amendments and Biofertilizers

R. Nisha
http://orcid.org/0009-0007-3231-9605
Sacred Heart College, India

V. Vaishnavi
http://orcid.org/0009-0006-8922-5231
Sacred Heart College, India

P. Mythili
http://orcid.org/0009-0001-8328-1631
Sacred Heart College, India

P. Prakathi
http://orcid.org/0009-0000-4457-9008
Sacred Heart College, India

R. Vishva
http://orcid.org/0009-0004-7004-2991
Sacred Heart College, India

S. Dayana
Sacred Heart College, India

ABSTRACT

The conversion of microalgal biomass into soil amendments and biofertilizers is a practical strategy for fostering the circular bioeconomy and sustainable agriculture. Microalgae are rich in essential nutrients such as nitrogen, phosphorous, potassium, micronutrients, amino acids, and phytohormones making them a powerful organic alternative to synthetic fertilizers. Additionally, techno-economic factors and scaling challenges are analyzed to identify feasible pathways for commercialization. Microalgal biomass offers a multifunctional, environmentally sustainable substitute to reduce dependence on chemical fertilizers, improve soil quality, and promote sus-

DOI: 10.4018/979-8-2600-3080-6.ch011

tainable resource recovery. Future climate-resilient agricultural systems' adoption of microalgae-derived soil supplements will be contingent upon continued study into strain selection, extraction methods, safety standards, and legislative support.

1. INTRODUCTION

Agriculture across the globe faces significant challenges driven by population growth, climate change, soil degradation, and the overuse of synthetic fertilizers. The intense agricultural practices have led to soil infertility, nutritional imbalance, groundwater contamination, and greenhouse gas emissions. While conventional fertilizers can enhance agricultural yields, their production is energy-intensive and contributes significantly to environmental degradation. Therefore, the development of circular, sustainable, and eco-friendly substitutes that can ensure food security and preserve environmental integrity is urgently needed. Microalgae that can be turned into valuable bioresources have emerged as a solution to these issues. These photosynthetic bacteria efficiently convert sunlight, carbon dioxide, and nutrients into biomass abundant in proteins, lipids, carbohydrates, vitamins, and bioactive substances. In contrast to conventional crops, microalgae do not require arable land, rendering them very sustainable, and may be cultivated using wastewater and industrial CO_2 emissions. The concept of valorising microalgal biomass as soil amendments and biofertilizers aligns with the ideas of circular bioeconomy by promoting nutrient recycling and resource recovery (Renuka et al., 2018). The notion of biomass valorisation entails the transformation of biological elements into more advantageous products. Microalgal biomass can be utilized in agricultural systems as bio-stimulants, compost, or biochar. Such applications enhance soil health, augment microbial activity, increase nutrient availability, and facilitate carbon sequestration. The production, composition, pathways for valuing the agronomic benefits of microalgal biomass, environmental security, techno-economic feasibility, and future of microalgal biomass in sustainable agriculture. The increasing pressures of climate change, soil erosion, water scarcity, and population growth are forcing a fundamental change in agriculture across the world's food systems. During the Green Revolution, synthetic fertilizer was widely used. This had many beneficial effects on crop productivity, but it also had unanticipated ecological repercussions, such as eutrophication, nutrient imbalance, acidification of soils, groundwater pollution, and greenhouse gas emissions (Alvarez et al., 2021).

Fertilizers, particularly those containing nitrogen, contribute to the release of nitrous oxide, a potent greenhouse gas. Moreover, phosphorus reserves are finite and unevenly distributed globally, raising concerns for future fertilizer security. These challenges need the pursuit of alternate nutrition sources that are renewable, environmentally friendly, and economically sustainable. Due to this, microalgae have emerged as promising bioresources that can simultaneously address several environmental and agricultural issues. The photosynthetic microorganisms with the fastest growth rates and the most effective nutrition absorption are microalgae. They can produce biomass with high levels of protein, carbs, lipids, vitamins, and growth-promoting metabolites from sunlight, carbon dioxide, and inorganic nutrients (Wu and Ge, 2019). Because they don't require arable ground to grow, microalgae can be grown on saline water, wastewater streams, and non-arable soil, unlike traditional crops. This unique capability renders microalgae significant contributors to nutrient recycling and climate regulation. The process of turning harvested microalgal biomass into agricultural products with additional value, such as compost additives, biochar, soil amendments, biofertilizers, and bio stimulants, is known as "valuation of microalgal biomass." Algal biomass can be integrated into the agricultural system and function as a closed-loop nutrient, as an alternative to being viewed as a byproduct of carbon capture or wastewater treatment technology (Srivastav et al., 2020). This approach aligns with the circular bioeconomy concept, necessitating the conversion of waste into resources to mitigate environmental effect and optimize economic advantage. Furthermore, microalgal biomass possesses multifunctional benefits beyond just nutrients. It enhances soil organic carbon, improves soil structure and aggregation, fosters beneficial microbial communities, and boosts plant resilience to biotic and abiotic stressors. With the rising prominence of regenerative and climate-smart agriculture, microalgal-derived soil solutions represent a promising innovation that can reduce dependence on chemical fertilizers while maintaining production. This chapter provides a comprehensive review of the use of microalgal biomass as a soil amendment and biofertilizer, focusing on its biological properties, cultivation and production methods, nutrient profiles, valorization approaches, impacts on soil health and crop productivity, environmental safety issues, economic viability, and emerging research directions (Jiang et al., 2020).

Microalgae have emerged recently as important elements in sustainable agriculture because of their versatility and suitability for a circular economy model. Microalgae not only serve as sources of nutrients but also restore soil biology, increase nutrient use efficiency, and help with carbon capture. They can be cultivated in tandem with wastewater plants, livestock effluent systems, and carbon dioxide reduction technologies. Therefore, the produced biomass can be used to create a range of agricultural materials, from biofertilizers, soil amendments, biostimulants, to biochar. Unlike traditional fertilizers, which cause excessive nutrient discharge and nutrient runoff, microalgae-based products ensure controlled nutrient delivery while improving soil physical and chemical properties. Also, the presence of phytohormones, amino acids, polysaccharides, vitamins, and antioxidants in microalgal biomass facilitates plant development and increases their tolerance to environmental stresses. As the global food production sector struggles to lower its ecological footprint while remaining productive, there is increasing recognition that microalgae present a unique solution to this challenge by virtue of their capacity to recycle nutrients, clean waste, store carbon, and enhance plant productivity. (Mutum et al., 2022).

2. MICROALGAE: BIOLOGY AND BIOMASS PRODUCTION

Microalgae exhibit evolutionary and taxonomic heterogeneity, encompassing green algae (Chlorophyta), cyanobacteria (blue-green algae), and diatoms (Bacillariophyta) within the eukaryotic microalgae category. They inhabit a wide range of ecosystems, including freshwater lakes, marine settings, desert crusts, and wastewater lagoons. This environmental adaptability signifies their remarkable physiological versatility, enabling survival under varying light conditions, temperature fluctuations, salinity levels, and nutritional environments. Microalgae have superior photosynthetic efficiency relative to terrestrial plants due to their uncomplicated cellular architecture and direct exposure to light and nutrients. Biomass proliferates within 24-48 hours under ideal conditions due to elevated growth rates (Zhang et al., 2019). Furthermore, most cyanobacteria can fix nitrogen, converting atmospheric nitrogen into bioavailable forms. This possesses a natural nitrogen enrichment characteristic that renders it very advantageous in agriculture. Another notable attribute of microalgae is their metabolic diversity. They can grow photoautotrophic (using light and CO_2), heterotrophically (utilizing organic carbon), or mixotrophically (via a combination of both methods). The approach is adaptable to tailored growing strategies based on resource availability and intended biomass utilization. Microalgae are a very diversified assemblage of photosynthetic microorganisms that exhibit ecological adaptability and inhabit freshwater, marine, and terrestrial ecosystem. These comprise prokaryotic cyanobacteria and eukaryotic algal taxa, including green algae and diatoms. Their structural simplicity, metabolic versatility, and remarkable physiological adaptability reflect their evolutionary variety. Microalgae possess a unicellular or rudimentary multicellular structure, in contrast to higher plants, and exhibit rapid and efficient cell division, along with nutrient absorption capabilities. Their photosynthetic apparatus allows them to harness solar energy and carbon dioxide to produce organic biomass with remarkable efficiency, frequently surpassing terrestrial crop yields per unit area (Mitter et al., 2021).

The biological characteristics of microalgae make it highly promising for sustainable biomass production. Due of their brief production cycles, they can significantly increase in biomass ideal climatic conditions. A significant number of species can endure extensive variations in salinity, temperature, pH, and light intensity, allowing them to thrive in many settings, including non-arable land. Certain cyanobacteria possess the ability to fix atmospheric nitrogen through specialized heterocyst, transforming inert nitrogen gas into accessible forms. The Nitrogen fixation is directly linked to soil fertility when this biomass is included into agricultural systems (Raimi et al., 2021). Microalgae exhibit metabolic plasticity, capable of growing as photoautotrophs by utilizing light and carbon dioxide, or as heterotrophs by utilizing organic carbon sources, or as mixotrophs through a combination of both methods. This adaptability facilitates the optimization of cultivation strategies according to available resources. In agriculture, photoautotrophic growth is particularly significant since it sequesters atmospheric or industrial CO_2, so contributing to climate mitigation efforts. In conjunction with wastewater treatment systems, microalgae assimilate dissolved nitrogen and phosphorus, effectively removing excess nutrients and generating valuable biomass. Microalgae cultivation systems can be broadly categorized into open and closed systems. Extensive production predominantly utilizes open pond systems, particularly raceway ponds, due to their economical construction and operational expenses (Lehmann et al., 2020). These are superficial ponds reliant on natural sunlight and mechanical circulation to maintain cell suspension.

Open systems, while economically advantageous for large-scale biomass production, are vulnerable to contamination, evaporation losses, and environmental fluctuations that can affect output. In scenarios where the value of a product is little but substantial quantities, such as biofertilizers and soil additives, are necessary, open pond farming is both feasible and scalable. Photobioreactor systems are enclosed, providing enhanced environmental regulation; hence, light exposure, nutrition supply, temperature, and gas exchange may be optimized. These technologies mitigate contamination hazards and, in most instances, enhance biomass productivity and homogeneity. Its application is, however, confined to high-value products due to elevated capital and energy needs (Keesstr et al., 2018). To attain economic viability in agricultural applications, the integration of a photobioreactor system with wastewater treatment or carbon capture systems can improve economic sustainability by offsetting production costs through diverse revenue streams. The harvesting and downstream processing of microalgal biomass has been identified as a significant technical challenge in its production. Because microalgal cells are small and typically reside in a dilute state, efficient separation procedures are required. Flotation, centrifugation, membrane filtration, flocculation, and sedimentation are widely utilized techniques. It is considered economical in flocculation, especially for the utilization of biodegradable bio-flocculants by farmers. Centrifugation is energy-intensive and highly efficient; yet it may not be economically viable to produce huge quantities of fertilizers. Post-harvest biomass can be utilized in its fresh state, sun-dried, mechanically dried, or subjected to enzymatic or chemical treatments to enhance nutrient content (Sutton et al., 2016). The employed processing technique will significantly influence the final product regarding stability, nutrient release characteristics, and agronomic efficacy. The biological features and industrial technologies of microalgae demonstrate significant potential for scalable biomass generation applicable in agriculture. The cultivation of microalgae, when integrated into circular systems such as wastewater treatment facilities or industrial CO_2 capture units, should be seen not merely as a method for biomass production but also as a comprehensive environmental strategy (Nkonya et al., 2016).

3. NUTRITIONAL AND BIOCHEMICAL COMPOSITION OF MICROBIAL BIOMASS

The biochemical makeup of microalgal biomass is varied and abundant, serving as the foundation for assessing its agronomic potential. Microalgae are characterized by significant amounts of essential macronutrients, micronutrients, proteins, and physiologically active compounds that collectively enhance soil fertility and facilitate plant growth. Their nutrient content varies by species, growing circumstances, and nutrient availability, although all exhibit elevated quantities of nitrogen, phosphorous, and potassium, the three primary macronutrients essential for plant development. Nitrogen constitutes a substantial component of microalgal biomass, typically existing as proteins and amino acids. In certain species, protein content can constitute 40-70 percent of dry weight, rendering microalgae one of the most nitrogen-rich biological materials applicable in agriculture (Qiu et al., 2021). Nitrogen in microalgal biomass is organically complexed, in contrast to manufactured nitrogen fertilizers that rapidly release nutrients and are susceptible to leaching or volatilization. This gradual release characteristic enhances nitrogen utilization efficiency and mitigates environmental pollution effects. Microalgae typically acquire phosphorus in the form of intracellular polyphosphate granules.

During soil decomposition, these phosphorus reserves are progressively mineralized into forms that can be readily utilized by plants, enhancing long-term phosphorus availability and nutrition. Given the finite nature of global phosphate rocks, the recycling of microalgal phosphorus via wastewater represents a significant advantage for the long-term sustainability of agriculture (Mabuhay et al., 2004). Potassium, despite varying concentrations, facilitates osmotic equilibrium, enzyme activity, and stress tolerance in plants. The ratios of nitrogen, phosphorus, and potassium in microalgal biomass are equilibrated, facilitating its application as a comprehensive organic fertilizer. Besides macronutrients, microalgae possess a significant concentration of critical micronutrients, including iron, zinc, copper, manganese, magnesium, and molybdenum. Plants necessitate these trace elements for enzyme reactions, chlorophyll production, photosynthesis, and metabolic regulation. Cultivating microalgae in nutrient-rich wastewater streams is an effective technique for sequestering dissolved minerals, which are accumulated in the biomass. However, this capacity for nutrient retention necessitates vigilant monitoring to prevent the introduction of excessive heavy metals into agricultural soils (Pidlisnyuk et al., 2020).

Microalgae produce several bioactive chemicals, endowing them with capabilities that extend beyond simply nutrient provision. Auxins, cytokinin, gibberellins, and abscisic acid are phytohormones identified in several microalgal species. These natural growth regulators influence cell division, root elongation, shoot development, and stress responses. These substances enable the utilization of bio-stimulants through bio-stimulating microalgal biomass and extracts to enhance plant vigor, even when applied in minimal concentrations. Furthermore, microalgal polysaccharides and extracellular polymeric compounds augment soil cell structure by enhancing its water retention capacity through promoting aggregation (Franchi et al., 2017). The release of amino acids and peptides during biomass decomposition enhances enzymatic activity and nutrient cycling efficiency, fostering the establishment of microbial communities in the soil. Additional antioxidants, such as carotenoids and phenolic substances, enhance plant resilience to oxidative stress induced by drought, salinity, or disease invasion. The multifunctional microalgal biomass comprises a synergistic combination of macronutrients, micronutrients, proteins, phytohormones, and bioactive metabolites. It serves not only as a nutritional source but also as a soil conditioner, microbial stimulant, and plant growth booster. This complex biochemical composition underpins the burgeoning research emphasis on microalgae-derived products as eco-friendly alternatives to conventional fertilizers in climate-resilient agricultural systems (Wang et al., 2016).

4. VALORISATION PATHWAYS OF MICROALGAL BIOMASS

The valorization of microalgal biomass into soil amendments and biofertilizers encompasses a range of technical pathways designed to optimize nutrient recovery and agronomic efficacy, adhering to the principles of circular bioeconomy. Microalgal biomass can be utilized by various physical, biological, chemical, or thermochemical processes, depending on the intended agricultural result, rather than being restricted to a specific application method. The valorization strategy significantly influences nutrient availability, soil dynamics, microbial interactions, environmental sustainability, and the overall feasibility of the economic approach. One of the most direct and well-studied ways is the application of fresh or dried microalgal biomass to soil (Rani et al., 2019). The biomass introduced into agricultural fields undergoes microbial degradation, leading to the slow mineralization of nutrients bound by organic molecules. The distinct advantage of this controlled release method compared to synthetic fertilizers is the provision of nutrients in soluble forms that are prone to leaching, volatilization, and runoff.

The organic matrix of microalgae retains nutrients within cellular structures, hence ensuring their availability throughout time. The incorporation of microalgal biomass enhances soil organic matter content, hence improving aggregation stability, aeration, and water retention capacity. The enhancement of soil's physical properties improves root penetration and facilitates more effective nutrient absorption. Cyanobacterial inoculation has been demonstrated to substantially reduce nitrogen fertilizer requirements in flooded rice production systems (Wang et al., 2015). Nitrogen-free microalgae enhance soil fertility by releasing organic acids, enzymes, and growth-promoting chemicals that increase nutrient solubility and facilitate plant-microbe interactions. Microalgal biofertilizers can be formulated as granules, pellets, or liquid solutions, allowing for targeted application in accordance with contemporary agricultural practices. Composting represents another potentially efficient method for the valorization of microalgal biomass.

Microalgae serve as an excellent co-substrate due to their elevated nitrogen and protein content when combined with carbon-rich agricultural residues such as straw, husks, or manure. The incorporation of microalgae reduces the carbon to nitrogen ratio in compost mixtures and enhances the rate of microbial decomposition and composting cycles shown in Figure 1. The heightened microbial activity during composting enhances nutrient stability and reduces odor emissions are shown in Figure 1. The resultant compost is enriched with humic compounds, stable organic matter, and bioavailable nutrients, rendering it a good soil conditioner. Furthermore, composting diminishes the risks associated with pathogens and organic pollutants by exposing the biomass to elevated temperatures, hence promoting sanitization (vanEekeren et al., 2020).

Figure 1. Valorisation pathway of Microalgal Biomass for Soil Amendments and Biofertilizers

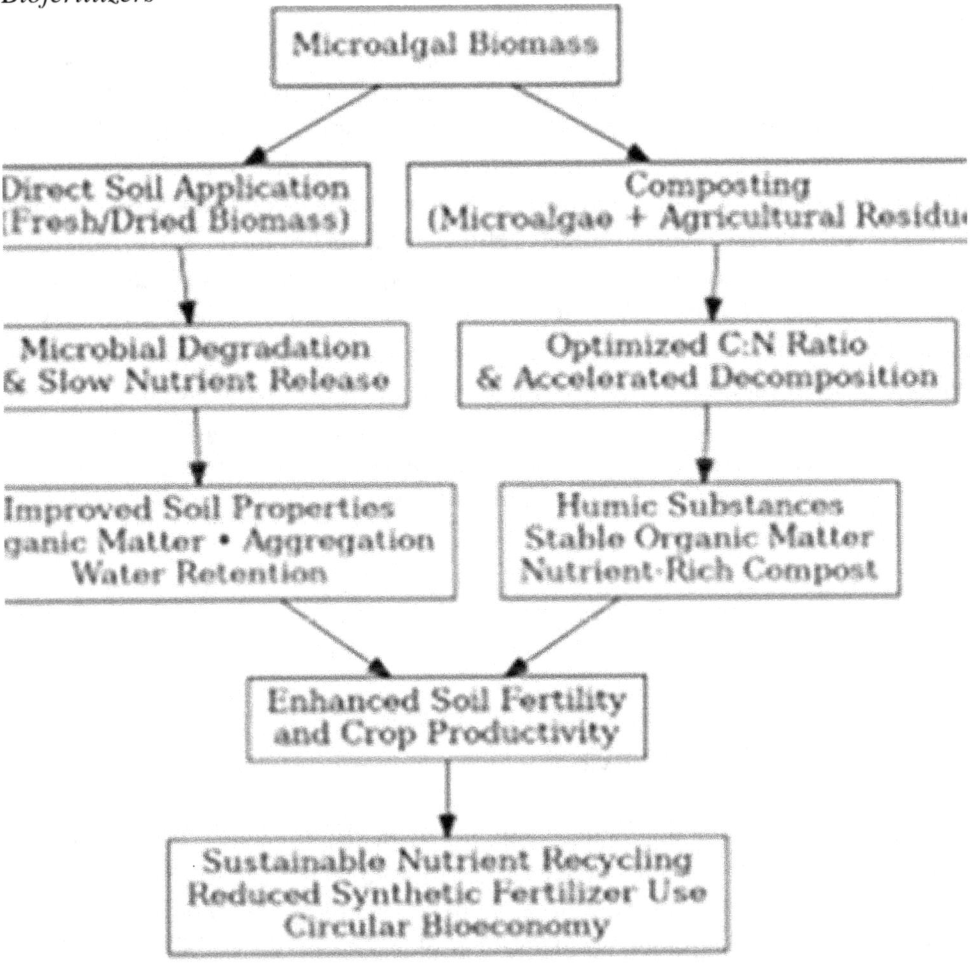

Microalgal hydrolysates were transformed into liquid bio-stimulants, representing a notably innovative method of valorization. Cell walls are degraded to get internal nutrients and bioactive substances in soluble forms by enzymatic, alkaline, or acid hydrolysis. These hydrolysates contain amino acids, peptides, phytohormones, vitamins, and minerals suitable for application by foliar sprays or soil drenches. In contrast to bulk biomass use, liquid formulations are quickly assimilated by plants and produce immediate physiological effects, including augmented chlorophyll synthesis, improved root architecture, and higher resilience to abiotic stresses such as drought and salinity. Microalgal extracts contain natural growth regulators, rendering them not just fertilizers but also plant bio-stimulants capable of improving crop output when applied in modest quantities (Prashar and Shah, 2016). A viable alternative valorization method is the thermochemical conversion of microalgal biomass into biochar, which can be utilized for long-term soil enhancement and carbon sequestration. Microalgal biomass is converted into a stable carbonaceous substance with a porous structure and large surface area via pyrolysis under restricted oxygen circumstances. Microalgal biochar improves soil aeration, augments cation exchange capacity, and enhances moisture retention. The high mineral ash content enhances nutrient availability, and the stable carbon content contributes to long-term carbon storage in soils, which mitigates climate change (Adimalla, 2020). Microalgal biochar, relative to lignocellulosic biochar, typically contains higher levels of nitrogen and phosphorus, making it particularly suitable for nutrient-enriched soil amendment initiatives. The selection of an appropriate valorization pathway is influenced by several aspects, including biomass quality, intended crop types, soil characteristics, economic constraints, and environmental regulations. The integrated biorefinery idea is being explored in several scenarios where microalgal biomass is first utilized to produce high-value products like pigments or lipids, and the residual biomass is subsequently employed as a soil amendment or fertilizer. The utilization of biomass has a ripple effect that enhances overall efficiency and economic viability of resources (Karwowska and kononiuk, 2020).

5. MICROALGAE IN WASTEWATER TREATMENT AND NUTRIENT RECOVERY

The combination of microalgae growing with the treatment of wastewaters is one of the most promising ways to ensure resource-efficiency and sustainable agriculture practices. Agricultural wastewater, as well as municipal and industrial wastewaters, contain high amounts of nitrogen, phosphorus, potassium, and micronutrients, which, when discharged into the environment, serve as pollutants due to the excessive concentration of nutrients in aquatic ecosystems. Microalgae can be used to remove such elements by assimilating them as part of their biomass. Microalgal cultivation as part of the wastewater treatment offers many benefits compared to the other types of technology. Microalgae use carbon dioxide as part of the photosynthesis and produce oxygen, which creates conditions for aerobic decomposition of organic matter by bacteria present in wastewater. Moreover, not only nutrients but also suspended substances, microbial pathogens, and some contaminants can be effectively eliminated from wastewater through microalgae-based treatment. The biomass produced with the help of this process can be used for the creation of fertilizers or biostimulants. The application of wastewater for microalgae growing can create a circular bioeconomy in agriculture. Nutrient resources contained in wastewater will be recovered and used for the purpose of producing fertilizers or biostimulants (Nur et al., 2025).

The use of microalgae for wastewater treatment has turned out to be a novel and sustainable biotechnological approach, which serves to clean the environment, recover nutrients, and produce biomass. Traditional technologies used in wastewater purification demand large energy expenditure along with additional chemicals that lead to high financial expenses and adverse impacts on the environment. On the contrary, microalgae consume solar energy as a source of energy and assimilate nutrients from wastewater; thus, the technology is eco-friendly and economically favorable. Municipal sewage water, agricultural water, livestock water, and fish farm water are rich in nitrogen and phosphorous that are vital for microalgal cultivation. Microalgae can assimilate nutrients into their biomass and decrease nutrient load in wastewater (Win et al., 2018).

5.1. Wastewater as a Growth Medium

The use of wastewater as a growth medium for microalgae exemplifies a significant convergence of environmental biotechnology and sustainable agriculture. Municipal, agricultural, and industrial effluents are generally abundant in nitrogen, phosphorous, potassium, and trace minerals nutrients vital for microalgal proliferation. Rather than regarding these nutrients as pollutants necessitating removal via energy-intensive treatment methods, microalgae present a biological alternative by immediately absorbing them into biomass. The development of microalgae in wastewater diminishes the necessity for synthetic growth media, hence substantially reducing manufacturing expenses (Garfí et al., 2017). Effluents from secondary and tertiary treatment, livestock manure, aquaculture, and agro-industrial processes have all been effectively utilized for algal biomass generation. The elevated nitrogen levels in these effluents frequently promote accelerated algal proliferation, leading to increased biomass productivity. Furthermore, the incorporation of microalgal cultivation within wastewater treatment facilities facilitates concurrent nutrient elimination and biomass production, establishing a dual-benefit system. Nevertheless, the composition of wastewater exhibits significant variability based on its origin, necessitating meticulous monitoring and, in certain instances, pre-treatment. Factors including pH, turbidity, hazardous pollutants, and organic load affect algal growth efficacy. Dilution, filtering, or pre-aeration may be required to enhance growth conditions. Notwithstanding these obstacles, wastewater-based cultivation persists as one of the most economically and environmentally viable methods for large-scale microalgal biomass production for agricultural applications (Arashiro et al., 2018).

5.2 Nutrient Removal Mechanisms

Microalgae play an important role in wastewater treatment by removing contaminants, including excess nutrients, organic matter, and various pollutants mostly by assimilation and biological uptake. Nitrogen is absorbed as ammonium, nitrate, or nitrite and is then transformed into amino acids and proteins. Ammonium is the predominant nitrogen form in most wastewater systems, and microalgae preferentially utilize it over nitrate reduction due to its lower energy requirements. Phosphorus elimination occurs through cellular absorption as orthophosphate and subsequent storage as polyphosphate granules (Saúco et al., 2021). Certain microalgal species have the capacity to accumulate nutrients more than their immediate metabolic requirements, wherein phosphorus is absorbed beyond immediate metabolic needs and subsequently retained inside the cells. This method enhances the removal of phosphorus from wastewater streams and results in biomass with elevated phosphorus content. In addition to serving as a direct agent of digestion, microalgae also contribute indirectly to nutrient elimination via symbiotic relationships with bacteria. The oxygen generated by algae in aquatic environments facilitates the decomposition of organic materials by aerobic bacteria. Bacteria release carbon dioxide, which microalgae utilize for food production. This symbiotic relationship enhances the overall efficacy of wastewater treatment. Furthermore, elevated pH levels during intense photosynthesis might result in ammonia volatilization and phosphate precipitation, thereby leading to nutritional depletion (Robles et al., 2020).

6. EFFECTS ON SOIL FERTILITY AND SOIL HEALTH

The application of microalgal biomass to soil significantly influences soil fertility, structure, biological activity, and long-term sustainability. In contrast to synthetic fertilizers that primarily supply soluble nutrients, microalgal additions serve as integrated soil conditioners that enhance the chemical and biological characteristics of soil ecosystems. The enhancement of soil organic matter is a crucial function of microalgal biomass. The degradation of biomass introduces carbon-rich elements that improve soil structure, aggregation, and water retention capacity. Enhanced soil aggregation reduces erosion and improves aeration, facilitating better root penetration and increased nutrient absorption. Microalgal biomass is crucial for improving nutrient dynamics in soils (Gupta et al., 2022). The organic nitrogen in algal proteins undergoes mineralization, a process facilitated by soil microorganisms, resulting in the production of ammonium and nitrate to fulfil plant requirements. This gradual release characteristic minimizes nutrient losses from leaching and volatilization, as observed with conventional fertilizers. Similarly, phosphorus in polyphosphate granules is released by microbial enzymes, facilitating ongoing plant growth. Microalgae contain a broad spectrum of essential macro- and micronutrients, providing a balanced nutrient source that minimizes the potential for nutrient deficiencies or imbalances from the exclusive use of specific nutrient fertilizers (Sarikhani et al., 2019).

Besides providing nutrients, microalgal biomass stimulates microbial populations in the soil. Amino acids, carbohydrates, and extracellular polymeric compounds provide substrates that promote the growth of beneficial bacteria and fungi. An increase in microbial biomass enhances enzymatic activity, accelerates nutrient cycling, and strengthens symbiotic relationships, particularly mycorrhizal associations. Increased microbial diversity improves soil resilience against diseases and environmental disturbances (Patel et al., 2024). Research has demonstrated that soil treated with microalgal biomass exhibits enhanced dehydrogenase, phosphatase, and urease activity, indicative of heightened biological activity. Additionally, biochar generated from microalgae can be utilized to improve the long-term carbon sequestration and stability of soil. Biochar possesses a porous structure that improves cation exchange capacity, moisture retention, and provides habitat niches for microorganisms. This ultimately stabilizes reservoirs of soil organic carbon and fosters climate-smart agriculture. The improvements in physical, chemical, and biological soil properties have positioned microalgal biomass as a regenerative resource capable of rehabilitating damaged soils and sustaining agricultural output over the long term (Tariq et al., 2024).

7. AGRONOMIC PERFORMANCE AND CROP PRODUCTIVITY

Microalgal biomass has been widely evaluated as a soil amendment and biofertilizer for its impact on the agronomic performance of diverse crops across varied environmental conditions. The experimental trials demonstrated an enhancement in seed germination rates, root length, shoot growth, chlorophyll content, and overall biomass accumulation. This is due to advantages that go beyond nutrient provision to include the availability of bioactive compounds such as phytohormones, amino acids, and vitamins, which directly influence plant physiological processes. The application of microalgae consistently promotes root development. Stimulated root branching and elongation augment the plant's capacity to explore soil for nutrients and water, hence increasing nutrient utilization efficiency (Dao et al., 2020). The presence of micronutrients and bio-stimulants enhances chlorophyll synthesis, resulting in improved photosynthetic performance, which translates into greater biomass production and yield. Microalgal additions have been associated with higher yields in cereal crops (rice, wheat, and maize) compared to non-fertilized controls, and in some cases, they have achieved yields comparable to typical fertilizer systems. Positive benefits of microalgal-based therapy have been noted even under stressful settings (Gonçalves et al., 2023).

The organic compounds and polysaccharides in algal biomass in saline soils improve soil structure and mitigate osmotic stress caused by salts. During drought conditions, enhanced water retention capacity facilitates the sustained growth of plants. Moreover, the antioxidant compounds present in microalgae mitigate oxidative stress in thermally stressed plants or those afflicted by diseases (Pagnussat et al., 2020). The stress-reducing advantages are particularly relevant in the context of climate change, where extreme weather conditions pose significant hazards to agricultural productivity.

Comparative evaluations indicate that while synthetic fertilizers promote rapid initial development, microalgal additions are more advantageous for sustained growth due to their gradual nitrogen release and soil conditioning effects. Integrated nutrient management systems employing reduced chemical fertilizer application alongside biofertilizers including microalgae have shown promising results, yielding substantial returns with little environmental repercussions. The adaptability of microalgal preparations, available as dry biomass, compost addition, liquid extract, or biochar, allows for their application in many cropping systems, including horticulture, cereals, legumes, and plantation crop production. The agronomic findings substantiate that microalgal biomass can enhance crop output, improve nutrient efficiency, and bolster plant resistance, thereby establishing it as a viable component of sustainable agricultural systems (Peng et al., 2020).

8. ENVIRONMENTAL AND SAFETY CONSIDERATIONS

Microalgal biomass has been widely evaluated as a soil amendment and biofertilizer for its impact on the agronomic performance of diverse crops across varied environmental conditions. The experimental trials demonstrated significant improvements in seed germination and overall plant growth, as evidenced by enhanced root and shoot development, increased chlorophyll content, and greater biomass accumulation. The beneficial effects of microalgal biomass extend beyond nutrient provision, owing to the presence of bioactive molecules such as phytohormones, amino acids, and vitamins that modulate key physiological and metabolic processes in plants. The application of microalgae consistently promotes root development. Stimulated root branching and elongation augment the plant's capacity to explore soil for nutrients and water, hence increasing nutrient utilization efficiency. The presence of micronutrients and bio-stimulants enhances chlorophyll synthesis, resulting in improved photosynthetic performance, which translates into greater biomass production and yield (Kishor et al., 2021). Microalgal additions have been associated with higher yields in cereal crops (rice, wheat, and maize) compared to non-fertilized controls, and in some cases, they have achieved yields comparable to typical fertilizer systems. Positive benefits of microalgal-based therapy have been noted even under stressful settings. The organic compounds and polysaccharides in algal biomass in saline soils improve soil structure and mitigate osmotic stress caused by salts.

During drought conditions, enhanced water retention capacity facilitates the sustained growth of plants. Moreover, the antioxidant compounds present in microalgae mitigate oxidative stress in thermally stressed plants or those afflicted by diseases (Abeyratne et al., 2023). The stress-reducing advantages are particularly relevant in the context of climate change, where extreme weather conditions pose significant hazards to agricultural productivity. Comparative evaluations indicate that while synthetic fertilizers promote rapid initial development, microalgal additions are more advantageous for sustained growth due to their gradual nitrogen release and soil conditioning effects.

Integrated nutrient management systems employing reduced chemical fertilizer application alongside biofertilizers including microalgae have shown promising results, yielding substantial returns with little environmental repercussions. The adaptability of microalgal preparations, available as dry biomass, compost addition, liquid extract, or biochar, allows for their application in many cropping systems, including horticulture, cereals, legumes, and plantation crop production. The agronomic findings substantiate that microalgal biomass can enhance crop output, improve nutrient efficiency, and bolster plant resistance, thereby establishing it as a viable component of sustainable agricultural systems (Zoppi et al., 2022).

9. FUTURE PERSPECTIVES

Valorization of microalgal biomass in agriculture will continue to grow at the nexus of policy support frameworks, circular economy integration, biotechnology innovation, and climate resilience tactics. Even though there has been a lot of effort in proving the agronomic and environmental advantages of soil amendments and biofertilizers made from microalgae, further development is needed to allow for widespread use and distribution. The long-term sustainability of this new industry will depend on ongoing studies into strain selection, genetic enhancement, crop optimization, and product standardization. Technological advancements in downstream processing and harvesting will also be crucial to future growth. Low-cost bio-flocculants, decentralized production methods, and energy-efficient dewatering technology can all lower operating expenses. The creation of stable, user-friendly formulations, like concentrated liquid extracts, encapsulated biofertilizers, and granulated biomass, will increase farmer acceptance and storage stability (Sun et al., 2026).

Furthermore, microalgal biofertilizers and precision agricultural technology could be used to maximize application rates according (Sethunathan et al., 2004) to crop need and soil nutrient mapping, improving the efficiency of nutrient utilization. Strategies for adapting to climate change further emphasize the value of microalgal biomass in agricultural systems of the future. With the rising frequency of extreme weather events, including drought, salinity, and temperature stress, soil additives that improve water retention, microbial resilience, and plant stress tolerance will gain significant importance (Rana et al., 2012). Products developed from microalgae, owing to their bio-stimulant characteristics and soil conditioning benefits, are ideally suited to promote climate-smart agriculture. Furthermore, their function in carbon sequestration and greenhouse gas reduction corresponds with international climate obligations and sustainable development objectives.

Regulatory alignment and policy incentives will be crucial to expedite adoption. Implementing quality standards, safety certificates, contamination limits, and explicit recommendations for wastewater-derived biomass will enhance trust among farmers and consumers. Public-private partnerships, research financing, and demonstration initiatives can enhance knowledge transfer and market integration. Programs for education and awareness aimed at farmers will be essential to dispel doubts and promote the shift from chemical-intensive inputs to biologically derived alternatives. Microalgal biomass integration into sustainable agricultural systems is a long-term systemic shift toward regenerative nutrient management, not just a technological advancement. The rate and scope of this shift will depend on the ongoing interdisciplinary cooperation of microbiologists, agronomists, engineers, economics, and legislators (Fedoroff et al., 2010).

10. CONCLUSION

A revolutionary approach to furthering the goals of sustainable agriculture and the circular bioeconomy is the conversion of microalgal biomass into soil amendments and biofertilizers. Microalgae present a distinctive amalgamation of rapid proliferation, elevated nutritional density, metabolic flexibility, and ecological adaptation that sets them apart from traditional organic inputs. Their capacity to assimilate nitrogen, phosphorous, and carbon dioxide from wastewater and industrial emissions facilitates concurrent waste cleanup and biomass production, establishing closed-loop nutrient cycles that mitigate environmental impacts. Microalgal biomass, when added to soil, provides important macro- and micronutrients as well as bioactive chemicals that promote plant growth, augment microbial activity, and enhance soil structure. The progressive mineralization of organic nutrients guarantees continuous nutrient availability while reducing leaching losses and greenhouse gas emissions. Furthermore, thermochemical conversion to biochar provides substantial long-term carbon sequestration potential, enhancing the climate mitigation capacity of microalgal systems. Successful large-scale deployment necessitates meticulous focus on safety monitoring, pollutant management, economic optimization, and regulatory adherence. Wastewater-derived biomass requires thorough quality evaluation to mitigate heavy metal buildup and pathogen hazards. Improvements in cultivation efficiency, harvesting technology, and integrated biorefinery models will be essential for enhancing cost-effectiveness and market competitiveness. In conclusion, the valorization of microalgal biomass offers a versatile and environmentally sustainable alternative to traditional fertilizers. Microalgae provide a comprehensive answer to critical difficulties in global food systems by integrating wastewater treatment, carbon capture, and agricultural fertilizer control. Ongoing research, technological advancements, and favorable policy frameworks position microalgae-derived soil amendments and biofertilizers as pivotal in establishing resilient, resource-efficient, and sustainable agricultural systems for the future.

REFERENCES

Abeyratne, W. M. L. K., Bayat, H., Delanka-Pedige, H. M. K., Zhang, Y., Brewer, C. E., & Nirmalakhandan, N. (2023). Multi-criteria evaluation of energy recovery from urban wastewater sludges by anaerobic digestion and hydrothermal liquefaction. *Journal of Environmental Chemical Engineering, 11*(2), 109628. DOI: 10.1016/j.jece.2023.109628

Adimalla, N. (2020). Spatial distribution, exposure, and potential health risk assessment from nitrate in drinking water from semi-arid region of South India. *Human and Ecological Risk Assessment, 26*(2), 310–334. DOI: 10.1080/10807039.2018.1508329

Alvarez, A. L., Weyers, S. L., Goemann, H. M., Peyton, B. M., & Gardner, R. D. (2021). Microalgae, soil and plants: A critical review of microalgae as renewable resources for agriculture. *Algal Research, 54*, 102200. DOI: 10.1016/j.algal.2021.102200

Arashiro, L. T., Montero, N., Ferrer, I., Acién, F. G., Gómez, C., & Garfí, M. (2018). Life cycle assessment of high-rate algal ponds for wastewater treatment and resource recovery. *The Science of the Total Environment, 622–623*, 1118–1130. DOI: 10.1016/j.scitotenv.2017.12.051 PMID: 29890581

Dao, G., Wang, S., Wang, X., Chen, Z., Wu, Y., Wu, G., Lu, Y., Liu, S., & Hu, H. (2020). Enhanced *Scenedesmus* sp. growth in response to gibberellin secretion by symbiotic bacteria. *The Science of the Total Environment, 740*, 140099. DOI: 10.1016/j.scitotenv.2020.140099 PMID: 32927572

Fedoroff, N. V., Battisti, D. S., Beachy, R. N., Cooper, P. J. M., Fischhoff, D. A., Hodges, C. N., Knauf, V. C., Lobell, D., Mazur, B. J., Molden, D., Reynolds, M. P., Ronald, P. C., Rosegrant, M. W., Sanchez, P. A., Vonshak, A., & Zhu, J. K. (2010). Radically rethinking agriculture for the 21st century. *Science, 327*(5967), 833–834. DOI: 10.1126/science.1186834 PMID: 20150494

Franchi, E., Rolli, E., Marasco, R., Agazzi, G., Borin, S., Cosmina, P., Pedron, F., Rosellini, I., Barbafieri, M., & Petruzzelli, G. (2017). Phytoremediation of a multi-contaminated soil: Mercury and arsenic phytoextraction assisted by mobilizing agent and plant growth-promoting bacteria. *Journal of Soils and Sediments, 17*(5), 1224–1236. DOI: 10.1007/s11368-015-1346-5

Garfí, M., Flores, L., & Ferrer, I. (2017). Life cycle assessment of wastewater treatment systems for small communities: Activated sludge, constructed wetlands and high-rate algal ponds. *Journal of Cleaner Production, 161*, 211–219. DOI: 10.1016/j.jclepro.2017.05.116

Gonçalves, J., Freitas, J., Fernandes, I., & Silva, P. (2023). Microalgae as Biofertilizers: A Sustainable Way to Improve Soil Fertility and Plant Growth. *Sustainability (Basel), 15*(16), 12413. DOI: 10.3390/su151612413

Gupta, R., Kumari, A., Sharma, S., Alzahrani, O. M., Noureldeen, A., & Darwish, H. (2022). Identification, characterization and optimization of phosphate solubilizing rhizobacteria from rice rhizosphere. *Saudi Journal of Biological Sciences, 29*(1), 35–42. DOI: 10.1016/j.sjbs.2021.09.075 PMID: 35002393

Jiang, Y., Chen, S., Hu, B., Zhou, Y., Liang, Z., Jia, X., & Shi, Z. (2020). A comprehensive framework for assessing the impact of potential agricultural pollution on grain security and human health in economically developed areas. *Environmental Pollution, 263*, 114653. DOI: 10.1016/j.envpol.2020.114653

Karwowska, M., & Kononiuk, A. (2020). Nitrates/nitrites in food—Risk for nitrosative stress and benefits. *Antioxidants, 9*(3), 241. DOI: 10.3390/antiox9030241 PMID: 32188080

Kishor, R., Purchase, D., Saratale, G. D., Saratale, R. G., Ferreira, L. F. R., Bilal, M., Chandra, R., & Bharagava, R. N. (2021). Ecotoxicological and health concerns of persistent coloring pollutants of textile industry wastewater and treatment approaches for environmental safety. *Journal of Environmental Chemical Engineering, 9*(2), 105012. DOI: 10.1016/j.jece.2020.105012

Lehmann, J., Bossio, D. A., Kögel-Knabner, I., & Rillig, M. C. (2020). The concept and future prospects of soil health. *Nature Reviews. Earth & Environment, 1*(10), 544–553. DOI: 10.1038/s43017-020-0080-8 PMID: 33015639

Mabuhay, J. A., Nakagoshi, N., & Isagi, Y. (2004). Influence of erosion on soil microbial biomass, abundance and community diversity. *Land Degradation & Development, 15*(2), 183–195. DOI: 10.1002/ldr.607

Mitter, E. K., Tosi, M., Obregón, D., Dunfield, K. E., & Germida, J. J. (2021). Rethinking crop nutrition in times of modern microbiology: Innovative biofertilizer technologies. *Frontiers in Sustainable Food Systems, 5*, 606815. DOI: 10.3389/fsufs.2021.606815

Mutum, L., Janda, T., Ördög, V., & Molnár, Z. (2022). Potential of Different Forms of Microalgae for Soil Improvement. *Biologia Futura, 73*(1), 1–8. DOI: 10.1007/s42977-021-00103-2 PMID: 34735698

Nkonya, E., Anderson, W., Kato, E., Koo, J., Mirzabaev, A., von Braun, J., & Meyer, S. (2016). Global cost of land degradation. In Nkonya, E., Mirzabaev, A., & von Braun, J. (Eds.), *Economics of land degradation and improvement: A global assessment for sustainable development* (pp. 117–165). Springer., DOI: 10.1007/978-3-319-19168-3_6

Nur, M. M. A., Mahreni, , Murni, S. W., Setyoningrum, T. M., Hadi, F., Widayati, T. W., Jaya, D., Sulistyawati, R. R. E., Puspitaningrum, D. A., Dewi, R. N., Hadiyanto, , & Hasanuzzaman, M. (2025). Innovative Strategies for Utilizing Microalgae as Dual-Purpose Biofertilizers and Phycoremediators in Agroecosystems. *Biotechnology Reports (Amsterdam, Netherlands)*, *45*, e00870. DOI: 10.1016/j.btre.2024.e00870 PMID: 39758973

Pagnussat, L., Maroniche, G., Curatti, L., & Creus, C. (2020). Auxin-dependent alleviation of oxidative stress and growth promotion of *Scenedesmus obliquus* C1S by *Azospirillum brasilense*. *Algal Research*, *47*, 101839. DOI: 10.1016/j.algal.2020.101839

Patel, S. K., Singh, S., Benjamin, J. C., Singh, V. R., Bisht, D., & Lal, R. K. (2024). Plant growth-promoting activities of *Serratia marcescens* and *Pseudomonas fluorescens* on *Capsicum annuum* L. *Ecological Frontiers*, *44*(4), 654–663. DOI: 10.1016/j.ecofro.2024.01.002

Peng, H., de-Bashan, L. E., Bashan, Y., & Higgins, B. T. (2020). Indole-3-acetic acid from *Azospirillum brasilense* promotes growth in green algae at the expense of energy storage products. *Algal Research*, *47*, 101845. DOI: 10.1016/j.algal.2020.101845

Pidlisnyuk, V., Mamirova, A., Pranaw, K., Shapoval, P. Y., Trögl, J., & Nurzhanova, A. (2020). Potential role of plant growth-promoting bacteria in *Miscanthus × giganteus* phytotechnology applied to trace element contaminated soils. *International Biodeterioration & Biodegradation*, *155*, 105103. DOI: 10.1016/j.ibiod.2020.105103

Prashar, P., & Shah, S. (2016). Impact of fertilizers and pesticides on soil microflora in agriculture. In Lichtfouse, E. (Ed.), *Sustainable Agriculture Reviews* (Vol. 19). Springer., DOI: 10.1007/978-3-319-26777-7_8

Qiu, L., Zhang, Q., Zhu, H., Reich, P. B., Banerjee, S., van der Heijden, M. G. A., Sadowsky, M. J., Ishii, S., Jia, X., Shao, M., Liu, B., Jiao, H., Li, H., & Wei, X. (2021). Erosion reduces soil microbial diversity, network complexity and multifunctionality. *The ISME Journal*, *15*(8), 2474–2489. DOI: 10.1038/s41396-021-00913-1 PMID: 33712698

Raimi, A., Roopnarain, A., & Adeleke, R. (2021). Biofertilizer production in Africa: Current status, factors impeding adoption and strategies for success. *Scientific African, 11*, e00694. DOI: 10.1016/j.sciaf.2021.e00694

Rana, A., Joshi, M., Prasanna, R., Shivay, Y. S., & Nain, L. (2012). Biofortification of wheat through inoculation of plant growth promoting rhizobacteria and cyanobacteria. *European Journal of Soil Biology, 50*, 118–126. DOI: 10.1016/j.ejsobi.2012.01.005

Rani, R., Kumar, V., Gupta, P., & Chandra, A. (2019). Application of plant growth promoting rhizobacteria in remediation of pesticides contaminated stressed soil. In Kumar, V. (Eds.), *New and future developments in microbial biotechnology and bioengineering* (pp. 341–353). Elsevier., DOI: 10.1016/B978-0-12-818258-1.00029-7

Renuka, N., Guldhe, A., Prasanna, R., Singh, P., & Bux, F. (2018). Microalgae as multi-functional options in modern agriculture: Current trends, prospects and challenges. *Biotechnology Advances, 36*(4), 1255–1273. DOI: 10.1016/j.biotechadv.2018.04.004 PMID: 29673972

Robles, Á., Capson-Tojo, G., Gales, A., Viruela, A., Sialve, B., Seco, A., Steyer, J.-P., & Ferrer, J. (2020). Performance of a membrane-coupled high-rate algal pond for urban wastewater treatment at demonstration scale. *Bioresource Technology, 301*, 122672. DOI: 10.1016/j.biortech.2019.122672 PMID: 31945681

Sarikhani, M. R., Khoshru, B., & Greiner, R. (2019). Isolation and identification of temperature tolerant phosphate solubilizing bacteria as a potential microbial fertilizer. *World Journal of Microbiology & Biotechnology, 35*(8), 137. DOI: 10.1007/s11274-019-2702-1 PMID: 31363938

Saúco, C., Cano, R., Marín, D., Lara, E., Rogalla, F., & Arbib, Z. (2021). Hybrid wastewater treatment system based on a combination of high-rate algae pond and vertical constructed wetland system at large scale. *Journal of Water Process Engineering, 43*, 102311. DOI: 10.1016/j.jwpe.2021.102311

Sethunathan, N., Megharaj, M., Chen, Z., Williams, B. D., Lewis, G., & Naidu, R. (2004). Algal degradation of a known endocrine disrupting insecticide, α-endosulfan, and its metabolite, endosulfan sulfate, in liquid medium and soil. *Journal of Agricultural and Food Chemistry, 52*(10), 3030–3035. DOI: 10.1021/jf035173x PMID: 15137849

Srivastav, A. L. (2020). Chemical fertilizers and pesticides: Role in groundwater contamination. In Srivastav, A. L. (Ed.), *Agrochemicals detection, treatment and remediation* (pp. 143–159). Butterworth-Heinemann., DOI: 10.1016/B978-0-08-103017-2.00006-4

Sun, Z., Liu, X., Ugya, A. Y., Liu, H., Sun, L., & Luo, G. (2026). Microalgae and Cyanobacteria as a Tool for Agricultural Sustainability: A Review of Biofertilizer and Biostimulant Potential. *Frontiers in Plant Science*, *16*, 1733394. DOI: 10.3389/fpls.2025.1733394 PMID: 41602522

Sutton, P. C., Anderson, S. J., Costanza, R., & Kubiszewski, I. (2016). The ecological economics of land degradation: Impacts on ecosystem service values. *Ecological Economics*, *129*, 182–192. DOI: 10.1016/j.ecolecon.2016.06.016

Tariq, M., Tahreem, N., Zafar, M., Raza, G., Shahid, M., Zunair, M., Iram, W., & Zahra, S. T. (2024). Occurrence of diverse plant growth promoting bacteria in soybean (*Glycine max* L. Merrill) root nodules and their prospective role in enhancing crop yield. *Biocatalysis and Agricultural Biotechnology*, *57*, 103100. DOI: 10.1016/j.bcab.2024.103072

van Eekeren, N., Jongejans, E., van Agtmaal, M., Guo, Y., van der Velden, M., Versteeg, C., & Siepel, H. (2022). Microarthropod communities and their ecosystem services restore when permanent grassland with mowing or low-intensity grazing is installed. *Agriculture, Ecosystems & Environment*, *323*, 107682. DOI: 10.1016/j.agee.2021.107682

Wang, S., Chen, H. Y. H., Tan, Y., Fan, H., & Ruan, H. (2016). Fertilizer regime impacts on abundance and diversity of soil fauna across a poplar plantation chronosequence in coastal eastern China. *Scientific Reports*, *6*(1), 20816. DOI: 10.1038/srep20816 PMID: 26857390

Wang, S., Tan, Y., Fan, H., Ruan, H., & Zheng, A. (2015). Responses of soil microarthropods to inorganic and organic fertilizers in a poplar plantation in a coastal area of eastern China. *Applied Soil Ecology*, *89*, 69–75. DOI: 10.1016/j.apsoil.2015.01.004

Win, T. T., Barone, G. D., Secundo, F., & Fu, P. (2018). Algal Biofertilizers and Plant Growth Stimulants for Sustainable Agriculture. *Industrial Biotechnology (New Rochelle, N.Y.)*, *14*(4), 203–211. DOI: 10.1089/ind.2018.0010

Wu, H., & Ge, Y. (2019). Excessive application of fertilizer, agricultural non-point source pollution, and farmers' policy choice. *Sustainability (Basel)*, *11*(4), 1165. DOI: 10.3390/su11041165

Zhang, T., Yang, Y., Ni, J., & Xie, D. (2019). Adoption behavior of cleaner production techniques to control agricultural non-point source pollution: A case study in the Three Gorges Reservoir Area. *Journal of Cleaner Production*, *223*, 897–906. DOI: 10.1016/j.jclepro.2019.03.194

Zoppi, G., Pipitone, G., Pirone, R., & Bensaid, S. (2022). Aqueous phase reforming process for the valorization of wastewater streams: Application to different industrial scenarios. *Catalysis Today*, *387*, 224–236. DOI: 10.1016/j.cattod.2021.06.002

Chapter 12

Techno-Economic Assessments and Life Cycle Analysis of Microalgal Bioenergy Systems

M. Gopika
Sacred Heart College, India

A. Revathi
http://orcid.org/0009-0005-8820-129X
Sacred Heart College, India

G. Anusuya
http://orcid.org/0009-0007-2211-4980
Sacred Heart College, India

P. Ashwitha
http://orcid.org/0009-0008-8049-9134
Sacred Heart College, India

Yuvalakshmi Logesan
http://orcid.org/0009-0005-4579-1170
Sacred Heart College, India

ABSTRACT

Microalgae-based biofuels have emerged as a highly promising renewable fuel alternative to traditional petroleum-based fuels, especially in the transportation sector

DOI: 10.4018/979-8-2600-3080-6.ch012

due to their high biomass productivity and feasibility of growth on non-arable land using wastewater resources. Despite their high promise, the large-scale commercialization of microalgal biofuels has been hampered by a number of biological, technological, and economic hurdles. The use of techno-economic analysis (TEA) and life cycle analysis (LCA) has been identified as an essential analytical tool for assessing the feasibility of microalgal biofuel systems, which have been known to estimate key parameters such as minimum fuel selling price (MFSP), global warming potential (GWP), and net energy ratio (NER). However, the ability to make inter-study comparisons has been impeded by inconsistencies in system boundaries, modeling approaches, and methodological frameworks.

1. INTRODUCTION

The pressing need to mitigate climate change, provide energy security, and achieve sustainability is the driving force behind the shift to low-carbon and renewable energy sources. In this context, the need to provide a renewable energy source to replace fossil fuels necessitates the sustainable production of bioenergy. For bioenergy to reach its full potential, the production of bioenergy must be critically evaluated based on its life cycle assessment (LCA) and techno-economic analysis (TEA), with policies to support its production in place (Osman et al., 2021). The increasing severity of the climate crisis, which is marked by weather events, increased global warming, and irreversible ecological degradation, necessitates a fundamental rethink of the world's energy systems (Chen et al., 2023). Since fossil fuels are the major cause of greenhouse emissions, the use of renewable energy sources has a positive impact on the environment as well as the economy. In a bio-based economy, microalgae have been identified as very promising feedstocks for bioenergy and bioproducts. Microalgal biomass is abundant in useful metabolites such as lipids, proteins, and carbohydrates, as well as healthful compounds including carotenoids, phycobiliproteins, and polyunsaturated fatty acids (PUFAs) (Rashwan et al., 2023). Microalgae possess better properties than traditional crops in terms of carbon fixation ability, photosynthetic efficiency, and biomass productivity. Moreover, they can be grown on saline water or wastewater in non-arable land. Due to these characteristics, microalgae are flexible feedstocks for the environmentally friendly production of energy, food, feed, and chemicals. Additionally, microalgae have a function in the bioremediation process, which involves cleaning wastewater and capturing carbon from industrial flue gas emissions, as reported by Mata et al. (2010) and Chisti (2007). However, the commercialization of microalgae-based bioproducts is constrained by the low product productivity and the high cost of manufacturing and processing microalgal biomass. During the last two decades, renewable energy has increasingly emerged as a major component of global energy policies to counter climate change, secure energy supplies, and promote economic stability. The detrimental effects of fossil fuels on the environment, including pollution, biodiversity loss, and carbon dioxide emissions, have intensified this trend (IPCC, 2021). Since 2010, improvements in wind energy, solar photovoltaic, and biofuels have decreased the costs of renewable energy by 50-80%, allowing decentralized energy systems to provide access to electricity in remote areas (IRENA, 2021). The process of transesterification of FAMEs results in biodiesel, which is a carbon-neutral substitute for conventional diesel, accounting for more than half of the total biofuel produced globally. The algae refineries are vital in the provision of energy security and sustainability. The algae have the capacity to provide an alternative to non-renewable sources of fossil fuels through various biofuels, including biodiesel, bioethanol, and biogas. The economic viability of algal biomass can be improved by its metabolic versatility due to the joint production of vital products such as bioplastics, drugs, and animal feed (Chew et al., 2017). Using wastewater, growing on non-arable land, and sequestering carbon through high photosynthetic efficiency are just a few of the environmental advantages that algae biorefineries offer that align with the idea of a circular economy (Singh & Gu, 2010).

2. TECHNO-ECONOMIC ASSESSMENT (TEA)

Recently, there has been much interest in the environmental benefits of using biomass. The main driving factor for the widespread use of chemicals and fuels derived from biomass is their comparability in terms of cost to conventional fossil resources. The important tool for cost estimation and overcoming financial barriers, TEA, is useful in the development and commercialization of biomass thermochemical conversion systems (Bridgwater, 2012). Bridgwater et al. (2002) have formulated the capital and operating costs equations after assessing the techno-economic feasibility of the three major thermo-chemical methods: fast pyrolysis, gasification, and combustion. The cost equations serve as important benchmarks for economic assessment. The production costs are greatly affected by the conversion methods and end-products. A brief account of the production costs of different biofuels, biochemicals, and energy carriers from different thermo-chemical methods is given below, with the costs in 2014 US dollars and after considering 2% inflation.

3. LIFE CYCLE ASSESSMENT (LCA)

Life Cycle Assessment (LCA) is a methodical process of assessing the environmental impacts of activities, procedures, or products, from their entire life cycle. The production, harvesting, processing, usage, and disposal stages of the entire biomass process would be part of the life cycle assessment of the products made from the biomass, as per ISO 14040, 2006. The system boundaries for the thermochemical conversion of lignocellulosic biomass would include the processes of biomass planting, harvesting, and transportation, plant operation and product upgrading, and plant decommissioning and recycling. Because of the availability of data, the first two stages are the most frequently taken into account in LCA. Among the effect categories commonly considered in LCAs are eutrophication, resource depletion, ozone depletion, acidification, and the potential for global warming. The most frequently quantified impact category is the global warming potential, expressed in CO_2 equivalents, which is followed by eutrophication and acidification (Cherubini & Strømman, 2011). System boundaries, functional units, allocation techniques, and feedstock type are significant factors that impact LCA outcomes. For instance, the usage and application of fertilizers has a significant effect on greenhouse gas emissions during the biomass production process; however, these emissions may be offset by carbon sequestration through photosynthesis (Sebastián et al., 2011). Energy-intensive preprocessing methods like drying, grinding, and pelleting have a high environmental impact. Moist biomass, such as forest residues, requires higher drying energy compared to agricultural residues (Iribarren et al., 2012).

4. BIOMASS CONVERSION PATHWAYS

Although the environmental benefits of using biomass have recently come to light, their broad use is primarily based on their cost-competitiveness with traditional fossil fuels and chemicals. The development and application of biomass thermochemical synthesis techniques depend on the production prices of chemicals and fuels, which are important TEA outcomes. Calculating these production costs is made possible by a discounted cash flow analysis of biorefineries. The products and procedures of thermochemical reactions determine the production costs. Bridgwater et al. analysed the initial and ongoing expenses of several thermochemical power generation techniques, including rapid pyrolysis, gasification, and combustion, and evaluated their techno-economic viability. This information is very useful for economic feasibility studies of various processes. Biomass thermochemical conversion processes produce power, biofuels, and biochemicals. The production costs were estimated to conduct economic studies of these products. All cost data used in this assessment are given in 2014 US dollars with a 2% inflation rate.

4.1. Liquefaction

A thermochemical process called hydrothermal liquefaction (HTL) can turn wet biomass into a liquid bio-crude. Water serves as a catalyst and reactant in HTL at high pressures and temperatures. HTL is more appropriate for high-moisture feedstocks like microalgae, as it requires less processing compared to gasification and pyrolysis (Toor et al., 2011). Fast pyrolysis generally has lower production and capital costs compared to gasification methods, although the level of technological maturity differs depending on the process. Syngas cleaning and conditioning in gasification-based liquefaction is a substantial cost factor (Trippe et al., 2011). The commercialization of gasification for biomass liquefaction is still in its infancy, as pioneer plants have shown production costs that are 60-90% higher and capital costs over twice as much as nth plants (Swanson et al., 2010). Two examples are provided to illustrate the differences between them. One example is a low-temperature fluidized bed gasifier, while the other is a high-temperature entrained flow gasifier. The latter has higher capital costs but lower production costs compared to the former, primarily due to differences in output yield. Commercialization of gasification-based liquefaction of biomass has yet to be achieved. It is important to note that for pioneer plants, production costs are 60-90% higher, and capital costs are twice as much as for nth plants. In addition to bio-oil, bio-Syncrude is another intermediate product being studied for the liquefaction of biomass. Bio-Syncrude is a mixture of char and liquid produced by fast pyrolysis of biomass, as opposed to bio-oil. Capital expenditure is 30% of total costs, and biomass feedstock is 50%. Labor and power usage are insignificant.

4.2. Combustion

Power generation from biomass combustion uses a range of boiler designs, such as circulating fluidized bed (CFB) boilers, fluidized bed combustors, and pulverized coal (PC) boilers. In comparison to pure biomass power plants, it is anticipated that cofiring biomass with coal will reduce the cost of energy (COE). The COE of supercritical boilers is generally expected to be lower than that of subcritical boilers. By lowering the cost of producing electricity from waste heat, combined heat and power (CHP) plants increase economic feasibility (McKendry, 2002).

4.3. Gasification and Pyrolysis

Gasification of biomass yields syngas (CO and H_2), which can be used for heating, electricity, or chemical/energy production. Using heat without oxygen, pyrolysis produces biochar, syngas, and bio-oil. Fast pyrolysis is meant to create the most liquid fuel, whereas slow pyrolysis is meant to produce the most charcoal. Bio-oil can be used directly for energy or transformed into transportation fuels (Bridgwater, 2012). Gasification, which can be regarded as an extension of pyrolysis, focuses on the production of gas and not liquids or solids in the presence of gasifying agents like steam, air, or oxygen. Nitrogen dilution effects are avoided by the use of oxygen as the gasifying medium, as opposed to the use of air, which produces gases with high calorific values. The design of the reactor, the rate of heating, and post-processing of the gases produced during gasification are also significant factors. Three distinct product phases are produced during pyrolysis, which is caused by various chemical reactions of the feedstock when air is not present. Pyrolysis is further classified into six groups according to the process parameters, as shown in Table 1. For maximum liquid output, fast pyrolysis with short residence durations and moderate temperatures is a common technique. Slow pyrolysis, which involves moderate heating rates and longer vapors residence times, is mainly used for charcoal production. Particle size distribution is an important factor in pyrolysis processes due to the non-conductive properties of lignocellulosic biomass.

Table 1. Biomass-to-Energy Conversion Technologies and Outputs

Technology	Process Conditions (Shortened)	Reactor Types	Liquid (wt%)	Gas (wt%)	Solid (wt%)
Fast Pyrolysis	Atm. pressure; <3 mm; 0.5–2 s; 400–550°C; no O_2	Fixed bed, tubular, bubbling & circulating fluidized bed, ablative, rotating cone, auger, cyclone	65–75	13–25	12–19
Slow Pyrolysis	Low heating; 350–750°C; long residence; no O_2	—	30–50	15–30	30–60
Intermediate Pyrolysis	<500°C; 4–10 s; atm. pressure	—	45–55	25–35	15–25
Flash Pyrolysis	<0.5 s; <0.5 mm; 400–1000°C	—	60–70	10–15	15–25
Vacuum Pyrolysis	300–500°C; <50 kPa	—	45–60	17–27	19–27
Ablative Pyrolysis	450–600°C; atm.; <3.5 mm	—	60–80	6–10	12–20

continued on following page

Table 1. Continued

Technology	Process Conditions (Shortened)	Reactor Types	Liquid (wt%)	Gas (wt%)	Solid (wt%)
Gasification	600–1200°C; catalyst ±; gasifying agent	Fixed, moving, fluidized, entrained flow	—	1–2.6 m^3/kg biomass	—
Combustion	740–1300°C; air: 0.1–0.5 kg/m^2·s	Fixed, fluidized, circulating, entrained bed	—	—	—
Hydrothermal Liquefaction	250–550°C; 5–25 MPa; 5–140°C/min	Parr high-pressure reactor	60–75	15–20	8–20
Carbonization	400–1200°C; 4–5°C rate	Steel container in furnace	20–35 (bio-char)	—	—
Co-firing	Biomass: 5–20 wt%	Boiler	Power & heat	—	—

5. ECONOMIC ANALYSIS OF BIOENERGY SYSTEMS

An important factor in the adoption of bioenergy technologies is their economic viability. The most important economic criteria are minimum fuel selling price (MFSP), net present value (NPV), internal rate of return (IRR), and payback period. The most important cost factors are determined by performing sensitivity tests on feedstock costs, investment costs, and efficiency. Feedstock prices are usually in the range of 30-50% of total production costs, and therefore, low-cost and sustainable feedstocks are required. Capital investment can vary greatly depending on the size and technology maturity. For example, first-of-a-kind biofuel plants have higher capital investment costs due to the challenges of scaling up and lack of operational experience. Carbon pricing, government policies, and renewable energy credits can improve economic feasibility (Klein-Marcuschamer et al., 2012). The three main methods of producing liquid transportation fuels from biomass include rapid pyrolysis and hydro processing, gasification and Fischer-Tropic synthesis, and biochemical processing. The factors that affect the costs of producing gasification include temperature, oxygen flow rates, and type of gasifiers. Production costs are also significantly impacted by how each alternative is developed and used. Compared to gasification or biochemical liquefaction, direct liquefaction via rapid pyrolysis has reduced production costs. Compared to gasification or biochemical liquefaction, direct liquefaction via rapid pyrolysis has reduced production costs. This is as a result of a decreased overall capital demand. Rapid pyrolysis, along with the use of commercial hydrogen, is a less expensive way to produce liquid fuels from bio-oil than pyrolysis alone. However, the process of purifying bio-oil from pyrolysis, which is used as a fuel source, is still a relatively new process. It has been found that the cost of creating liquid fuels from the use of gasification and upgrading is generally more than the cost of creating fuels from the use of pyrolysis. In a study done by Trippe et al. on the cost of manufacturing gasoline and diesel fuels using the process of gasification and Fischer-Tropsch synthesis, it has been found that the cost of the inputs used in the production of syngas is the biggest contributor to the overall cost of production. This cost is generally determined by the operational costs incurred in the process of cleaning and conditioning the syngas. Regardless of temperature, pyrolysis and biochemical procedures have lower capital costs than gasification-based liquefaction.

However, compared to biochemical conversion, which has comparatively cheap operating costs, gasification systems produce less liquid fuel. Swanson et al. used the same syngas synthesis method to compare two gasification procedures for the manufacture of transportation fuels. The first process utilizes a low-temperature fluidized bed gasifier, while the second process utilizes a high-temperature entrained flow gasifier. The high-temperature process has a higher capital cost than the low-temperature gasifier, but the production costs are lower, mainly because of the high yield of the products. Gasification-based biomass liquefaction technology is still in its infancy. The capital costs of the pioneer plants are more than twice as high as those of the nth plant, and the production costs are 60-90% higher. Another intermediate technology being researched for the production of liquid fuels from biomass, aside from bio-oil, is bio-Syncrude. It is a mixture of char and liquids produced from the rapid pyrolysis of biomass, as opposed to bio-oil. Thirty percent of the total cost is the capital cost, and fifty percent is the cost of the biomass feedstock. The cost of the power and labor is minimal.

6. LIFE CYCLE ASSESSMENT OF BIOENERGY PRODUCTION

For the successful implementation of the bioenergy projects, it is required to have a potential analysis, logistics, and analysis of biomass. This review has focused on the life cycle assessment and techno-economic analysis of the bioenergy production process. The initial part of the review has focused on the strategies, methods, and applications of life cycle assessment in the bioenergy production process. Further, the review has focused on the impact of policies in the sustainable production of bioenergy, and then it has been discussed in the context of the methodology of the comprehensive techno-economic assessment approach for the environmental implications of bioenergy (Hiloidhari et al., 2017). It is important to consider the pros and cons of the production of bioenergy by taking into consideration the environmental risks. It is necessary that a life cycle assessment method be applied in these assessments. The procedure of evaluating and quantifying the environmental costs of energy, material, and waste emissions in a product, process, or activity is called life cycle assessment. There are a number of stages in the life cycle of biofuels and bioproducts, which affect the environment and society. In these stages, primary resources such as crops, algae, or forestry residues are first produced in the form of biomass. Then, the biomass is converted into bioproducts, biofuels, or bioenergy by means of pyrolysis, enzymatic hydrolysis, or fermentation, respectively. These proceed to the use stage, where they can either be converted into other uses such as electricity generation, biodegradable polymers, or transportation fuels or be used for electricity generation. After the completion of the life cycle, biofuels and bioproducts then undergo the process of recycling, reusing, or disposing of them in a responsible manner. During the usage phase, their social and environmental impacts are measured and assessed, considering factors such as resource use, greenhouse gas emissions, and economic value. A detailed life cycle assessment of these renewable alternatives will assist in understanding their overall impacts and will also aid in making ecologically responsible and sustainable decisions. Furthermore, in any chain of bioenergy production, the greenhouse gas emissions and other environmental impacts of bioenergy production are normally negative because of the use of energy-intensive equipment in irrigation, preparation of the land, planting, harvesting, collecting crops, and transporting the feedstocks (Jeswani et al., 2020). As mentioned in the previous paragraphs, the production, use, and consumption of synthetic fertilizers and pesticides contribute to greenhouse gas emissions and affect the quality of the soil and water. It has been found that agricultural processes have the most significant impact on the environment in sugarcane-based bioenergy production.

However, sugarcane bagasse can generate bioelectricity and bioethanol, which are more sustainable alternatives. Using bagasse for cogeneration-based power generation can help achieve reduced energy-related emissions when compared to systems that rely on fossil fuels or bagasse bioethanol (Khatri & Pandit, 2022). Additionally, it has been found that combining first and second-generation technologies yields better environmental outcomes than traditional first-generation technologies when bagasse bioethanol is manufactured (Ayodele et al., 2020). In comparison to Brazilian bioethanol, Indian bioethanol typically has lower or comparable emissions pertaining to greenhouse gases, non-renewable energy, human health, and ecotoxicity, according to the cradle-to-gate life cycle evaluation carried out by Tsiropoulos et al. (2014). The main energy source for India's bioethanol industry is sugarcane molasses, a byproduct of the production of sugar. Because the primary product (sugar) and the waste product (molasses) have similar environmental consequences, decreased emissions can be ascribed to this. Still, there are methodological issues with the life cycle assessment of biofuels and bioproducts. Functional units are typically defined in terms of mass or volume in cradle-to-gate assessments, but they are typically defined in terms of driving distance or energy value in cradle-to-grave assessments. If the utilization of raw materials is prioritized, then land-area-based functional units would be pertinent. Environmental loading allocation is a big challenge, especially in the case of biomass systems, which have a high level of integration with by-products in the growth and processing stages. It is important to understand the distinction between the cradle-to-gate and cradle-to-grave life cycle evaluation. Cradle-to-gate life cycle evaluation includes the assessment from the extraction stage to the waste disposal stage, whereas the cradle-to-grave life cycle evaluation includes the assessment from the extraction stage to the factory gate.

6.1. From Feedstock to End-of-Life

The life cycle assessment of bioenergy includes the entire bioenergy chain, i.e., the production, final consumption, and disposal of bioenergy. In the production stage, environmental influences such as mechanical processing, fertilizers, pesticides, water, and land use are factors. The processing stage, which involves biomass processing, conversion, and product upgrading, involves energy and material use. Use phase emissions depend on fuel characteristics and efficiency of combustion, while end-of-life options involve waste management and recycling options (Cherubini et al., 2021). The kind of lignocellulosic biomass feedstock used in thermochemical conversion is crucial. These three substances are the major part of lignocellulosic biomass, but the percentage of cellulose, hemicellulose, and lignin depends on the biomass material. More cellulose, less lignin is desirable. In this review, we will discuss different types of lignocellulosic biomass feedstock, which are energy crops, forest residues, and agriculture residues. The way that different feedstock types are grown and harvested has an impact on the environment. Third-generation biomass, or energy crops, were developed as efficient, low-maintenance energy sources. To compete with fossil fuels, these high-yield or short-rotation crops have undergone genetic optimization to produce notable energy outputs. Their primary drawback is the application of pesticides and herbicides, which could lead to insect and weed resistance and jeopardize non-GMO crops.

Timber harvesting operations, decaying or dead wood, and waste from lumber mills are the sources of forest leftovers. While ash content may vary, forest residues resemble wood in moisture content and thermal value. Moisture content varies throughout forest environments. Agricultural residues the third feedstock category comprise remaining portions (husk or straw) from regularly harvested crops including wheat, corn, and rice. Corn stover, wheat straw, and rice husk are examples of easily accessible agricultural wastes with low moisture content and high heating properties. The demand for feedstock is raised by crop seasonality and competing uses for these wastes (soil enhancement, animal feed). Some residues are commonly employed for soil conditioning and groundwater management by protecting soil from wind, rain, and direct sunlight. Consequently, LCA must evaluate impacts of using these feedstocks for chemical and fuel production.

6.2. Allocation and System Boundaries

As biorefineries, allocation techniques are essential for distributing environmental costs among co-products. Economic allocation, energy allocation, and mass allocation are examples of allocation techniques. As it defines the processes that are included in the study, it has a great influence on the results of the LCA. While the cradle-to-grave life cycle assessment considers the phases of usage and disposal, the cradle-to-gate life cycle assessment considers the phases of natural resource extraction up to the factory gate boundary (Gnansounou et al., 2009). The process of assigning environmental implications to co-products is included in the life cycle assessment of multifunctional systems. The LCA viewpoint makes it possible to evaluate the significant environmental advantages that can be attained by using various waste management techniques. Bio electrochemical systems (BES) have the ability to remove particular pollutants from wastewater, and at the same time, they can produce energy and specialty chemicals. These include denitrification, sulfur compound removal, and Cr^{6+} reduction. At the same time that ethanol is produced as the main product of acetate reduction, hydrogen is produced in the cathode compartments of BES. In conclusion, bio electrochemical systems that can produce electricity, treat wastewater, and recover co-products have the potential to provide renewable energy, sequester CO_2, reduce greenhouse gases, and be economically viable. As bioelectrochemistry is an ever-evolving field, the technologies described above are only a part of the ever-changing technology portfolios.

7. TECHNO-ECONOMIC EVALUATION OF BIOENERGY PRODUCTION

TEA evaluates the viability of bioenergy projects by combining economic and technical performance data. This method involves process flow modeling, capital and operating cost estimation, and financial analysis. The main elements are process design and simulation, equipment sizing and costing, operating cost estimation (feedstock, utilities, and labor), financial analysis (NPV, IRR, MFSP), and sensitivity and uncertainty analysis. As stated by Brosowski et al. (2019), the heating values of biomass energy resources are approximately 3×10^6 kcal/mg. Such resources include forestry, agriculture, and municipal waste, which comprise wood, sawdust, straw, manure, paper mill waste, and municipal trash. It has been recognized as a viable alternative to fossil fuels, as it has the capability to mitigate the effects of emissions. Green technologies, which comprise biofuels and bioproducts, can address the energy needs of the world while reducing the effects of greenhouse gases and arresting the process of global warming. The calculation, comparison, and evaluation of different technical plans and strategies in a way that ensures the most optimal balance between economic viability and technology is called techno-economic analysis. This method of analysis ensures the comparison of the profit earned from the most economical and cost-effective designs and the costs of different technical plans, taking into account program costs and revenue (Beaucamp et al., 2022). The techno-economic analysis ensures the estimation of the advantages of material and immaterial elements in practical applications (Pan et al., 2023). By considering different aspects of technology, economics, and market behavior, techno-economic analysis provides a holistic evaluation that can encompass sustainability aspects (Salimbeni et al., 2023). Fossil fuels are a threat to the bioenergy sector, as the feedstock materials change from food crops to non-food crops and agricultural residues in order to not compete in the food market. Using the agro-economic and sustainability concepts, techno-economic analysis can help in the selection of the raw materials, interacting with the government, and even obtaining financial subsidies (Deep Singh et al., 2022). Techno-economic analysis is widely used in the production of bioenergy, including cash flow analysis, sensitivity analysis, technology analysis, environmental cost analysis, and market analysis. Techno-economic analysis is vital in the assessment of the economic viability of bioenergy production. Several parameters, including energy production, revenue, costs, and capital investment, are used to determine various financial parameters such as NPV, IRR, and payback period.

In bioenergy production, various parameters including costs of energy, material costs, and costs of operation are used to determine the viability of a bioenergy production project. In addition, raw material processing and chemical reactions have a significant effect on efficiency, production, reliability, and sustainability of bioenergy products. The results of techno-economic analysis are used by bioenergy production industries to determine the most profitable technology. For instance, Li et al. (2023) suggested a technology for bioenergy production through anaerobic fermentation of microalgae from sewage. With a daily productivity of 0.8 and 0.6 kg/m^3, respectively, closed and open systems for biomass dehydration treatment were distinguished with the use of techno-economic analysis (Somers et al., 2021). This implied that for bioenergy producers, open systems had lower capital costs than closed systems. According to sensitivity analysis, labor depreciation accounted for 39–42% of overall operating costs, making it the most significant factor in the production of bioenergy (Ighalo et al., 2022). It is also useful in comparing different materials and methods of processing. When compared to untreated maize straw, pre-treatment of maize straw by sodium hydroxide increased the production of methane by 73.4%. This proves the effectiveness of the pre-treatment of maize straw by sodium hydroxide (Zdeb et al., 2023). The method of alkaline treatment is best applied to materials that contain a higher percentage of lignin, as the cost of treating wastewater may be greater than the cost of increased methane production. Market research and cash flow are also useful sources of information in business development planning. Biogas, which is produced from domestic and agricultural waste, is a significant source of biofuel, which can be easily replaced by natural gas (Keerthana Devi et al., 2022). Though the methods of producing biogas are simple, the cost of acquiring and storing the materials may be relatively high. In terms of cash flow and market analysis, the construction of biogas reactors in community units was proposed to reduce production costs and provide communities with direct access to biogas (Ali et al., 2021a). The impact of techno-economic analysis on biorefinery projects can be substantial. In conclusion, techno-economic analysis is a critical tool in assessing the viability of bioenergy production. The techno-economic analysis of bioenergy production may help in decision-making regarding critical financial parameters, including net present value, internal rate of return, and return on investment, for different bioenergy production technologies by critically examining the cost and profitability of different bioenergy production technologies and their environmental impacts. The techno-economic analysis of bioenergy production is a critical tool in the production of sustainable bioenergy in the transition from fossil fuels (Hosseinzadeh et al., 2022).

8. CARBON CAPTURE, STORAGE, AND UTILIZATION (CCS/CCU) IN BIOENERGY

Bioenergy with carbon capture and storage is a viable option for mitigating carbon dioxide emissions. It has the capacity for negative emission, which may be employed in the mitigation of global warming. Aside from emission reduction technologies, negative emission technologies (NETs) are a critical component of the methods for attaining climate control targets (Rogelj et al., 2016). NETs are estimated to have the capacity for the removal of 10 billion tons of carbon dioxide in the mid-century and 20 billion tons by the end of the century (National Academies of Sciences et al., 2018). According to BECCS, ocean alkalization, subsurface mineralization, biochar, direct air capture, accelerated weathering, and other NETs may have the capacity for negative emission. The BECCS technology and biochar have attracted attention due to their viability and scalability (Heck et al., 2018). Despite the fact that fast and complete decarbonization activities bring hope to achieve the 2°C or 1.5°C temperature control objectives, the significance of negative emission technologies, specifically BECCS, has increased to achieve successful emission mitigation (Luderer et al., 2009). The abuse of BECCS technology in encouraging the consumption of fossil fuels would be harmful to the mitigation of climate change. The BECCS technology is still in its infancy stage of development, and various uncertainties are preventing its usage in the future from a social, economic, environmental, and sustainable point of view.

The main challenges are the effects of carbon capture and storage technology on local employment and safety (Pawar et al., 2015) the economic feasibility of the technology, considering the cost of biomass conversion and carbon capture and storage retrofits in coal-fired power plants (Yi et al., 2018) the effects of biomass supply chains on carbon intensity, land, and water availability (Yang et al., 2019a) and the policy to support the implementation of BECCS technology. Hence, to facilitate the large-scale and long-term implementation of BECCS, a comprehensive assessment of BECCS projects on various environmental, economic, and social issues is critically required. The concept and approach of conducting life cycle assessment on environmental issues were introduced by the International Organization for Standardization. However, the present study is affected by the scarcity of economic and social assessment indicators, data, and tools.

8.1. Recent Developments and Deployment Status

CCS/CCU technology allows the capture of CO_2 emissions from atmospheric or industrial sources, making it possible to store the gas in geological structures or use it as an input material in product manufacturing. Recently, supportive government policies have encouraged the development of projects, and the total capture capacity of CCS technology has increased to 361 Mt CO_2 per year globally. The main countries involved in this technology are the United States, United Kingdom, Canada, China, and Norway. The technology has been applied in the areas of electricity power generation, cement manufacturing, hydrogen production, and natural gas processing (IEA, 2023). Despite the introduction of the technology in the 1970s and 1980s, there were very few CCS/CCU projects that were adopted, particularly from 2010 to 2020. Economic difficulties and a lack of regulatory frameworks resulted in unanticipated expenses and delays (Energy Transitions Commission, 2022). However, the current CCS/CCU technology has improved significantly in recent years because of the significant legislative reforms that have enhanced the investment potential of CCS/CCU projects. Currently, as of July 2023, CCS/CCU is used in the global industry sector, fuel transformation sector, and power sector in 392 pipeline projects and 41 commercial capture plants. The main countries where CCS/CCU projects are being developed include the USA, the UK, Canada, China, and Norway. These countries have focused on the development of coal-to-gas plants, hydrogen production, refining, and natural gas processing. The industry has seen a rise of 102% in the total capacity from the previous year, reaching 361 Mt. Because of economies of scale and risks, CCS/CCU projects have shifted from vertical integration to networked developments. Solutions for regional CO2 sources are facilitated by the existence of capture hubs and transit networks. The sectors involved in the production of cement, power and heat, hydrogen and ammonia, biofuels, and natural gas are the ones adopting CCS technology at the fastest rate. In addition, there are now 30 DAC plants in the process of development (IEA, 2023a). Fuel transformation encompasses the processes of fuel conversion and biofuels. Power generation encompasses the production of heat and electricity, while other sectors include cement, iron and steel, natural gas, hydrogen/ammonia, and others. Collecting air directly, CO2 transit, CO2 storage, and CO2 transport and storage are methods that are exclusive to CCU/CCS. Despite these technological improvements, a significant number of CCS/CCU projects are in the early stages of development, which may imply a disconnect between the announcements of deployment (approximately 400 MtCO2/year) and the Net Zero Emissions scenario targets (approximately 1 GtCO2/year by 2030). More public engagement, government encouragement, and technological advancements are what are desperately required for the challenging 1.5°C to 2°C global warming targets.

Additionally, enormous theoretical CO_2 storage capacity—up to 1,000 Gt—offers substantial opportunity for mitigating climate change consequences. To ensure the required rates of emission reduction as set by global climate targets, CCS/CCU technology research and deployment must be accelerated (IEA, 2023a; Calvin et al., 2023). Policy issues need to be handled in a way that is unique to the country and industry in order to facilitate the adoption of CCS/CCU technology. Algeria, Brazil, Canada, Japan, Norway, Saudi Arabia, the United Arab Emirates, the United Kingdom, and the United States are the biggest producers of hydrocarbons, while the biggest importers of hydrocarbons include China, Brazil, Canada, Japan, Saudi Arabia, and the United States. These countries can serve as examples to other countries in the world (Nunes & Costa, 2021). These countries have taken into account a number of crucial factors, including enhancing international cooperation, quick scaling up with the help of flagship projects by 2030, enhancing research and development, enhancing the capture of CO2 in sectors where it is difficult to reduce, enhancing infrastructure development by 2040, and enhancing public acceptance and awareness. It is projected that by 2030, the global investment in CCS/CCU would surpass $160 billion, enhancing the contribution to the achievement of climate goals (IEA, 2023a). This has to be supplemented with effective legislative and policy instruments. Prospective assessments of CCS/CCU technology are difficult because of the rapidly changing regulatory and operational conditions.

To enable thorough research, there is a need for greater foreground technology system choices and background systems with improved temporal and geographical resolution. International cooperation, scaling up quickly through flagship projects by 2030, research and development, capture of CO2 in sectors that are difficult to mitigate, infrastructure development by 2040, and increasing public acceptance and awareness are just a few of the major issues that were taken into consideration in the national roadmaps. Global investment in CCS/CCU is expected to reach over $160 billion by 2030, further supporting its role in achieving climate goals (IEA, 2023a). Strong legislative and policy frameworks must underpin these initiatives. In the 1940s, coal or biomass gasification facilities produced physical sorbents such as Rectisol and Selexol.

The 1950s and 1960s saw the development of adsorption methods for producing hydrogen and ammonia using solid sorbents. The natural gas industry created membrane techniques to increase the effectiveness of CO2 capture in the 1980s. Although cryogenic processes (now at TRL 6) have been explored for better CO_2 capture efficiency and purity, traditional chemical absorption processes are still used in power plants and other industries (Air Liquide, 2022). Calcium looping and chemical looping combustion (CLC), which are areas of current research, are especially important for the cement industry. Calcium looping is being investigated at TRL 6-8, while CLC is being investigated at TRL 5-6 (Global CCS Institute, 2021). At various stages of TRLs, solvents for physical and chemical absorption processes are being developed. These include water-lean solvents, phase-change solvents, ionic liquids, and the chilled ammonia process. The development of membrane systems for power plants by Membrane Technology and Research at a level of TRL 4-7 is a big leap in the search for new membrane materials, as reported by DOEEPRI (2020). The great potential of high capacity at low temperatures and the ability to produce cleaner materials with fewer processing steps have resulted in tremendous progress in the field of adsorption technology over the past 10 years. TRLs for metal-organic frameworks (MOFs), steel, cement, zeolites, activated carbon, silica, alumina, ammonia, methanol, and other materials (Dziejarski et al., 2023). The potential of new adsorption technologies is demonstrated by Svante's VeloxoTherm technology, which employs fast-cycle temperature swing adsorption (TRL). The latest technology, such as the Allam-Fetvedt power generating cycle and the Calix Advanced Calciner for cement and lime production, with the potential for CO2 capture without any additional energy for CO2 separation, has the potential to reduce costs despite challenges in retrofitting existing infrastructure. The potential of direct air capture technology for achieving negative CO2 emissions has attracted a lot of interest. Due to its adaptability, DAC can be used anywhere in the world. Methods: Temperature-vacuum-swing adsorption with solid or liquid sorbents and hydroxide-carbonate adsorption have reached the pilot/demonstration stage (TRL 6-

7). According to Bouaboula et al. (2024), other methods of DAC currently undergoing investigation include electrolysis (TRL 1-4), amine-based absorption (TRL 3-4), bipolar membrane electrolysis (BPMED) (TRL 2-4), and cryogenic methods (TRL 1-3). Various infrastructure for the transportation of CO2, including roads, railways, and pipelines, as well as ships, which are both onshore and offshore, have already reached a TRL of 9 (Global CCS Institute, 2021). Although CO2 storage technologies, including CO2-EOR and saline reservoirs, are currently in commercial deployment (TRL 9), abandoned oil and gas reservoirs and EGR methods are currently undergoing investigation (TRL 6-8). For example, the implementation of unconventional storage technologies, such as CO2 carbonation in basalts, is shown through the Wallula project in the USA (McGrail et al., 2017) and the CarbFix project in Iceland (CarbFix, 2024). Through the implementation of modern technology, the fertilizer, food, beverage, and chemical industries have reached TRL 9 in the production of various products, including urea, polyurethanes, and polycarbonates based on methanol and CO2 (Hepburn et al., 2019). Energy carriers comprise, for example, methane (TRL 8), ethanol (TRL 6), formic acid (TRL 6), syngas (TRL 6), and Fischer-Tropsch products (TRL 5); raw materials are converted to dimethyl carbonate (TRL 8), dimethyl ether (TRL 3), carbamates (TRL 2), and fine chemicals (TRL 11-13).Enhancements from TRL 4 to TRL 8 are observable in concrete curing and mineral carbonation technologies for compounds such as magnesium carbonates and sodium bicarbonates (Chauvy & De Weireld, 2020).

8.2. Technology Readiness Levels (TRLs)

CCS/CCU technologies range from research stages (TRL 1-3) to commercialization (TRL 9). The list of mature technologies includes CO2 capture with amine-based absorption (TRL 9) and CO2 storage by CO2-EOR (TRL 9). The new technologies are direct air capture (DAC) (TRL 6-7), calcium looping (TRL 6-8), and electrochemical conversion (TRL 3-5).

9. ENVIRONMENTAL IMPACTS OF BIOENERGY PRODUCTION

The two most prominent environmental benefits of bioenergy technologies include fossil fuel substitution and greenhouse gas reduction. However, there are also some possible environmental concerns, including biodiversity loss, land use change, water use, and fertilizer runoff. Torrefaction, a low-intensity pyrolysis process performed under specific conditions of temperature, may lead to the formation of persistent organic pollutants, apart from improving the quality of biomass fuels. Anaerobic digestion may help in reducing methane emission from waste materials, but methane leakage should be controlled. Life cycle assessments emphasize the importance of sustainable feedstock and efficient conversion technologies in minimizing environmental impacts. For example, the use of residues instead of dedicated energy crops can help in overcoming land use impacts, and innovative pretreatment technologies can help in increasing process efficiency (Hiloidhari et al., 2017).

10. RECENT LCA STUDIES ON BIOENERGY PRODUCTS

Bio-ethanol, bio-diesel, bio-liquid fuels, biomethane, biohydrogen, bioelectricity, and advanced alcohols are some of the bioenergy products. Synthetic bioenergy has a number of very significant advantages over fossil fuels. Since most fuel is produced from renewable biomass, it is environmentally friendly. When it comes to carbon dioxide emissions from combustion, these bioenergy sources are carbon neutral, but with the right management, they may become carbon negative. Many of the biggest economies in the world have prioritized the development of synthetic bioenergy to ensure economic growth, environmental sustainability, and energy security. Published research studies on the life cycle impacts of the environment have shown very significant reductions in carbon dioxide emissions in all cases.

10.1. Bioethanol and Biodiesel

Bioethanol, mainly produced from corn (USA), sugarcane (Brazil), and lignocellulosic biomass (EU), has different levels of environmental impacts based on the raw material used. Although corn-based bioethanol usually requires higher water and fertilizer use, sugarcane bioethanol might have significant land-use change requirements. Lignocellulosic bioethanol usually has more favorable greenhouse gas emissions, even if it encounters economic and technological challenges (Soleymani Angili et al., 2021). Because there are no emissions from waste disposal, biodiesel produced from leftover cooking oil has a lower carbon footprint than biodiesel produced from vegetable oil. In esterification reactions, which are known to be important emission sources, there is a preference for alkali-catalyzed reactions over acid-catalyzed reactions (Peñarrubia Fernandez et al., 2017).

10.2. Biogas and Advanced Biofuels

Renewable energy sources, including biogas, which help to reduce methane emission from landfills, can be obtained from anaerobic digestion of waste materials. Life cycle analysis has proved the benefits of waste management and mitigation of greenhouse gases, although results vary depending on the type of waste materials used, efficiency of the anaerobic digestion process, and fuel consumption scenario (Amin et al., 2022). High-density biofuels, including hydroprocessed esters and fatty acids and biomass to liquids, can be used for existing transportation infrastructures. However, their impacts on the environment depend on the sustainability of the process and efficiency of the process, as noted by Balajii and Niju (2019).

11. CONCLUSION

Microalgae have very promising feedstock properties for the development of sustainable bioenergy production, simultaneously mitigating greenhouse gas emissions and improving energy security. However, for both economic and environmental sustainability, biorefinery concepts and technologies are essential. Life cycle and techno-economic assessments are very important tools for research and development and decision-making. Standardized reporting and methodology will improve the comparability of studies. Future studies should concentrate on developing conversion technologies, improving culture techniques, and utilizing co-products in order to fully realize the promise of microalgal bioenergy.

REFERENCES

Air Liquide. (2022). Cryocap-Carbon capture technologies. https://www.engineering-airliquide.com

Ali, S., Paul Peter, A., Chew, K. W., Munawaroh, H. S. H., & Show, P. L. (2021a). Resource recovery from industrial effluents through the cultivation of microalgae: A review. *Bioresource Technology*, *337*, 125461. Advance online publication. DOI: 10.1016/j.biortech.2021.125461 PMID: 34198241

Amin, F. R., Khalid, H., Zhang, H., & Zhang, R. (2022). Life cycle assessment of biogas production from different feedstocks. *Renewable & Sustainable Energy Reviews*, *150*, 111489.

Amin, M. A., Shukor, H., Yin, L. S., Kasim, F. H., Shoparwe, N. F., Makhtar, M. M. Z., & Yaser, A. Z. (2022). Methane biogas production in Malaysia: Challenge and future plan. *International Journal of Chemical Engineering*, *2022*(1), 2278211. Advance online publication. DOI: 10.1155/2022/2278211

Ayodele, B. V., Alsaffar, M. A., & Mustapa, S. I. (2020). An overview of integration opportunities for sustainable bioethanol production from first- and second-generation sugar-based feedstocks. *Journal of Cleaner Production*, *245*, 118857. Advance online publication. DOI: 10.1016/j.jclepro.2019.118857

Balajii, M., & Niju, S. (2019). A review on biodiesel production from microalgae: Influencing factors and future prospects. *Environmental Science and Pollution Research International*, *26*(9), 8618–8635.

Balajii, M., & Niju, S. (2019). Biochar-derived heterogeneous catalysts for biodiesel production. *Environmental Chemistry Letters*, *17*(4), 1447–1469. DOI: 10.1007/s10311-019-00885-x

Beaucamp, A., Muddasar, M., Amiinu, I. S., Moraes Leite, M., Culebras, M., Latha, K., Gutiérrez, M. C., Rodriguez-Padron, D., del Monte, F., Kennedy, T., Ryan, K. M., Luque, R., Titirici, M.-M., & Collins, M. N. (2022). Lignin for energy applications—State of the art, life cycle, technoeconomic analysis and future trends. *Green Chemistry*, *24*(21), 8193–8226. DOI: 10.1039/D2GC02724K

Bouaboula, H., Chaouki, J., Belmabkhout, Y., & Zaabout, A. (2024). Comparative review of direct air capture technologies: From technical, commercial, economic, and environmental aspects. *Chemical Engineering Journal*, *484*, 149411. Advance online publication. DOI: 10.1016/j.cej.2024.149411

Bridgwater, A. V. (2012). Review of fast pyrolysis of biomass and product upgrading. *Biomass and Bioenergy*, *38*, 68–94. DOI: 10.1016/j.biombioe.2011.01.048

Bridgwater, A. V., Toft, A. J., & Brammer, J. G. (2002). A techno-economic comparison of power production by biomass fast pyrolysis with gasification and combustion. *Renewable & Sustainable Energy Reviews*, *6*(3), 181–246. DOI: 10.1016/S1364-0321(01)00010-7

Brosowski, A., Krause, T., Mantau, U., Mahro, B., Noke, A., Richter, F., Raussen, T., Bischof, R., Hering, T., Blanke, C., Müller, P., & Thrän, D. (2019). How to measure the impact of biogenic residues, wastes and by-products: Development of a national resource monitoring based on the example of Germany. *Biomass and Bioenergy*, *127*, 105275. Advance online publication. DOI: 10.1016/j.biombioe.2019.105275

Carbfix. (2024). Carbfix—We turn CO2 into stone. https://www.carbfix.com

Chauvy, R., & De Weireld, G. (2020). CO2 utilization technologies in Europe: A short review. *Energy Technology (Weinheim)*, *8*(12), 2000627. Advance online publication. DOI: 10.1002/ente.202000627

Chen, P., Xie, Q., & Addy, M. (2023). Climate change impacts on renewable energy systems. *Nature Climate Change*, *13*(2), 112–120.

Cherubin, M. R., Carvalho, J. L., Cerri, C. E., Nogueira, L. A., Souza, G. M., & Cantarella, H. (2021). Land use and management effects on sustainable sugarcane-derived bioenergy. *Land (Basel)*, *10*(1), 72. Advance online publication. DOI: 10.3390/land10010072

Cherubini, F., & Strømman, A. H. (2011). Life cycle assessment of bioenergy systems: State of the art and future challenges. *Bioresource Technology*, *102*(2), 437–451. DOI: 10.1016/j.biortech.2010.08.010 PMID: 20832298

Chew, K. W., Yap, J. Y., Show, P. L., Suan, N. H., Juan, J. C., Ling, T. C., Lee, D.-J., & Chang, J. S. (2017). Microalgae biorefinery: High value products perspectives. *Bioresource Technology*, *229*, 53–62. DOI: 10.1016/j.biortech.2017.01.006 PMID: 28107722

Chisti, Y. (2007). Biodiesel from microalgae. *Biotechnology Advances*, *25*(3), 294–306. DOI: 10.1016/j.biotechadv.2007.02.001 PMID: 17350212

Deep Singh, A., Gajera, B., & Sarma, A. K. (2022). Appraising the availability of biomass residues in India and their bioenergy potential. *Waste Management*, *152*, 38–47. DOI: 10.1016/j.wasman.2022.08.001 PMID: 35973326

Dziejarski, B., Krzyżyńska, R., & Andersson, K. (2023). Current status of carbon capture, utilization, and storage technologies in the global economy: A survey of technical assessment. *Fuel, 342*, 127776. Advance online publication. DOI: 10.1016/j.fuel.2023.127776

Energy Transitions Commission. (2022). Carbon capture, utilisation and storage in the energy transition: Vital but limited. Energy Transitions Commission. https://www.energy-transitions.org

Global, C. C. S. Institute. (2021). Technology readiness and costs of CCS. https://www.globalccsinstitute.com/resources/publications-reports-research/technology-readiness-and-costs-of-ccs/

Global, C. C. S. Institute. (2023). The global status of CCS: 2023. Global CCS Institute. https://status23.globalccsinstitute.com/

Gnansounou, E., Dauriat, A., Villegas, J., & Panichelli, L. (2009). Life cycle assessment of biofuels: Energy and greenhouse gas balances. *Bioresource Technology, 100*(21), 4919–4930. DOI: 10.1016/j.biortech.2009.05.067 PMID: 19553106

Hepburn, C., Adlen, E., Beddington, J., Carter, E. A., Fuss, S., Mac Dowell, N., Minx, J. C., Smith, P., & Williams, C. K. (2019). The technological and economic prospects for CO2 utilization and removal. *Nature, 575*(7781), 87–97. DOI: 10.1038/s41586-019-1681-6 PMID: 31695213

Hiloidhari, M., Baruah, D. C., Singh, A., Kataki, S., Medhi, K., Kumar, S., Ramachandra, T. V., Jenkins, B. M., & Thakur, I. S. (2017). Emerging role of geographical information system (GIS), life cycle assessment (LCA) and spatial LCA (GIS-LCA) in sustainable bioenergy planning. *Bioresource Technology, 242*, 218–226. DOI: 10.1016/j.biortech.2017.03.079 PMID: 28343863

Hiloidhari, M., Das, D., & Baruah, D. C. (2017). Bioenergy potential from crop residue biomass in India. *Renewable & Sustainable Energy Reviews, 32*, 504–512. DOI: 10.1016/j.rser.2014.01.025

IEA. (2023). *Carbon Capture, Utilisation and Storage*. International Energy Agency.

Ighalo, J. O., Dulta, K., Kurniawan, S. B., Omoarukhe, F. O., Ewuzie, U., Eshiemogie, S. O., Ojo, A. U., & Abdullah, S. R. S. (2022). Progress in microalgae application for CO2 sequestration. *Cleaner Chemical Engineering, 3*, 100044. Advance online publication. DOI: 10.1016/j.clce.2022.100044

International Energy Agency. (2020). CCUS in clean energy transitions. IEA. https://www.iea.org/reports/ccus-in-clean-energy-transitions

International Energy Agency. (2023a). *CCUS policies and business models: Building a commercial market*. OECD., DOI: 10.1787/6ef05538-

International Energy Agency. (2023b). Section 45 Q credit for carbon oxide sequestration. IEA. https://www.iea.org/policies/4986-section-45q-credit-for-carbon-oxide-sequestration

IPCC. (2021). [The Physical Science Basis. Contribution of Working Group I to the Sixth Assessment Report.]. *Climatic Change*, •••, 2021.

IRENA. (2021). *Renewable Power Generation Costs in 2020*. International Renewable Energy Agency.

Iribarren, D., Peters, J. F., & Dufour, J. (2012). Life cycle assessment of transportation fuels from biomass pyrolysis. *Fuel*, *97*, 812–821. DOI: 10.1016/j.fuel.2012.02.053

Jeswani, H. K., Chilvers, A., & Azapagic, A. (2020). Environmental sustainability of biofuels: A review. Proceedings of the Royal Society A: Mathematical, Physical and Engineering Sciences, 476, Article 20200351. DOI: 10.1098/rspa.2020.0351

Keerthana Devi, M., Manikandan, S., Oviyapriya, M., Selvaraj, M., Assiri, M. A., Vickram, S., Subbaiya, R., Karmegam, N., Ravindran, B., Chang, S. W., & Awasthi, M. K. (2022). Recent advances in biogas production using agro-industrial waste: A comprehensive review outlook of techno-economic analysis. *Bioresource Technology*, *363*, 127871. Advance online publication. DOI: 10.1016/j.biortech.2022.127871 PMID: 36041677

Khatri, P., & Pandit, A. B. (2022). Systematic review of life cycle assessments applied to sugarcane bagasse utilization alternatives. *Biomass and Bioenergy*, *158*, 106365. Advance online publication. DOI: 10.1016/j.biombioe.2022.106365

Klein-Marcuschamer, D., Oleskowicz-Popiel, P., Simmons, B. A., & Blanch, H. W. (2012). The challenge of enzyme cost in the production of lignocellulosic biofuels. *Biotechnology and Bioengineering*, *109*(4), 1083–1087. DOI: 10.1002/bit.24370 PMID: 22095526

Li, S., Chang, H., Zhang, S., & Ho, S.-H. (2023). Production of sustainable biofuels from microalgae with CO2 bio-sequestration and life cycle assessment. *Environmental Research*, *227*, 115730. Advance online publication. DOI: 10.1016/j.envres.2023.115730 PMID: 36958384

Mata, T. M., Martins, A. A., & Caetano, N. S. (2010). Microalgae for biodiesel production and other applications: A review. *Renewable & Sustainable Energy Reviews*, *14*(1), 217–232. DOI: 10.1016/j.rser.2009.07.020

McGrail, B. P., Schaef, H. T., Spane, F. A., Horner, J. A., Owen, A. T., Cliff, J. B., Qafoku, O., Thompson, C. J., & Sullivan, E. C. (2017). Wallula basalt pilot demonstration project: Post-injection results and conclusions. *Energy Procedia, 114*, 5783–5790. DOI: 10.1016/j.egypro.2017.03.1716

McKendry, P. (2002). Energy production from biomass (part 1): Overview of biomass. *Bioresource Technology, 83*(1), 37–46. DOI: 10.1016/S0960-8524(01)00118-3 PMID: 12058829

. National Academies of Sciences, Engineering, and Medicine. (2023). Carbon dioxide utilization markets and infrastructure: Status and opportunities: A first report. National Academies Press. DOI: 10.17226/26703

Nunes, R. de C., & Costa, H. K. D. M. (2021). How CCS projects are developed in selected countries? *International Journal of Developmental Research, 11*, 48829–48834. DOI: 10.37118/ijdr.22314.07.2021

Osman, A. I., Qasim, U., & Jamil, F. (2021). Bioenergy and biofuels: History, status, and perspective. *Renewable & Sustainable Energy Reviews, 135*, 110183.

Pan, S., Zabed, H. M., Zhao, M., Qi, X., & Wei, Y. (2023). Techno-economic and life cycle assessments for bioenergy recovery from acid-hydrolyzed residues of sugarcane bagasse in the biobased xylose production platform. *Journal of Cleaner Production, 400*, 136718. Advance online publication. DOI: 10.1016/j.jclepro.2023.136718

Peñarrubia Fernandez, I. A., Liu, D. H., & Zhao, J. (2017). Life cycle assessment of biodiesel production from microalgae. *Journal of Environmental Management, 204*, 377–386.

Peñarrubia Fernandez, I. A., Liu, D.-H., & Zhao, J. (2017). LCA studies comparing alkaline and immobilized enzyme catalyst processes for biodiesel production under Brazilian conditions. *Resources, Conservation and Recycling, 119*, 117–127. DOI: 10.1016/j.resconrec.2016.05.009

Rashwan, A. K., Al-Addous, M., & Al-Hinti, I. (2023). Renewable energy integration in smart grids: Challenges and solutions. *Energy Reports, 9*, 124–135.

Salimbeni, A., Lombardi, G., Rizzo, A. M., & Chiaramonti, D. (2023). Techno-economic feasibility of integrating biomass slow pyrolysis in an EAF steelmaking site: A case study. *Applied Energy, 339*, 120991. Advance online publication. DOI: 10.1016/j.apenergy.2023.120991

Sebastián, F., Royo, J., & Gómez, M. (2011). Cofiring versus biomass-fired power plants: GHG emissions savings. *Biomass and Bioenergy, 35*(7), 2516–2524.

Singh, J., & Gu, S. (2010). Commercialization potential of microalgae for biofuels production. *Renewable & Sustainable Energy Reviews*, *14*(9), 2596–2610. DOI: 10.1016/j.rser.2010.06.014

Soleymani, A. T., Grzesik, K., Rödl, A., & Kaltschmitt, M. (2021). Life cycle assessment of bioethanol production: A review of feedstock, technology and methodology. *Energies*, *14*(10), 2939. Advance online publication. DOI: 10.3390/en14102939

Soleymani Angili, T., Grzesik, K., & Rödl, A. (2021). Life cycle assessment of bioethanol from microalgae and lignocellulosic biomass. *The Science of the Total Environment*, *762*, 143134.

Somers, M. D., Chen, P., Clippinger, J., Cruce, J. R., Davis, R., Lammers, P. J., & Quinn, J. C. (2021). Techno-economic and life-cycle assessment of fuel production from mixotrophic Galdieria sulphuraria microalgae on hydrolysate. *Algal Research*, *59*, 102419. Advance online publication. DOI: 10.1016/j.algal.2021.102419

Swanson, R. M., Platon, A., Satrio, J. A., & Brown, R. C. (2010). Techno economic analysis of biomass-to-liquids production based on gasification. *Fuel*, *89*, S11–S19. DOI: 10.1016/j.fuel.2010.07.027

Swanson, R. M., Satrio, J. A., & Brown, R. C. (2010). Techno-economic analysis of biofuels production via fast pyrolysis and hydroprocessing. *NREL Technical Report, NREL, TP-6*, A20–A46586.

Toor, S. S., Rosendahl, L., & Rudolf, A. (2011). Hydrothermal liquefaction of biomass: A review of subcritical water technologies. *Energy*, *36*(5), 2328–2342. DOI: 10.1016/j.energy.2011.03.013

Trippe, F., Fröhling, M., & Schultmann, F. (2011). Techno-economic analysis of fast pyrolysis for bio-oil production. *Biomass Conversion and Biorefinery*, *1*(2), 101–112.

Tsiropoulos, I., Faaij, A. P. C., Seabra, J. E. A., Lundquist, L., Schenker, U., Briois, J.-F., & Patel, M. K. (2014). Life cycle assessment of sugarcane ethanol production in India in comparison to Brazil. *The International Journal of Life Cycle Assessment*, *19*(5), 1049–1067. DOI: 10.1007/s11367-014-0714-5

Yu, S., Horing, J., Liu, Q., Dahowski, R., Davidson, C., Edmonds, J., Liu, B., Mcjeon, H., McLeod, J., Patel, P., & Clarke, L. (2019). CCUS in China's mitigation strategy: Insights from integrated assessment modeling. *International Journal of Greenhouse Gas Control*, *84*, 204–218. DOI: 10.1016/j.ijggc.2019.03.004

Zdeb, M., Bis, M., & Przywara, A. (2023). Multi-criteria analysis of the influence of lignocellulosic biomass pretreatment techniques on methane production. *Energies*, *16*(1), 468. Advance online publication. DOI: 10.3390/en16010468

Compilation of References

Abbas, S., & Halog, A. (2021). Analysis of Pakistani textile industry: Recommendations towards circular and sustainable production. In *Circular Economy and Sustainability* (pp. 77–111). Elsevier. DOI: 10.1007/978-981-16-3698-1_3

Abdalrhman, A. S., Zhang, Y., & Gamal El-Din, M. (2019). Electro oxidation by graphite anode for naphthenic acids degradation, biodegradability enhancement and toxicity reduction. *The Science of the Total Environment*, *671*, 270–279. DOI: 10.1016/j.scitotenv.2019.03.262 PMID: 30928756

Abdelfattah, A., Ali, S. S., Ramadan, H., El-Aswar, E. I., Eltawab, R., Ho, S.-H., Elsamahy, T., Li, S., El-Sheekh, M. M., Schagerl, M., Kornaros, M., & Sun, J. (2023). Microalgae-based wastewater treatment: Mechanisms, challenges, recent advances, and future prospects. *Environmental Science and Ecotechnology*, *13*, 100205. DOI: 10.1016/j.ese.2022.100205 PMID: 36247722

Abdel-Raouf, N., Al-Homaidan, A. A., & Ibraheem, I. B. (2012). Microalgae and wastewater treatment. *Saudi Journal of Biological Sciences*, *19*(3), 257–275. DOI: 10.1016/j.sjbs.2012.04.005 PMID: 24936135

Abeyratne, W. M. L. K., Bayat, H., Delanka-Pedige, H. M. K., Zhang, Y., Brewer, C. E., & Nirmalakhandan, N. (2023). Multi-criteria evaluation of energy recovery from urban wastewater sludges by anaerobic digestion and hydrothermal liquefaction. *Journal of Environmental Chemical Engineering*, *11*(2), 109628. DOI: 10.1016/j.jece.2023.109628

Abinandan, S., & Shanthakumar, S. (2015). Challenges and opportunities in application of microalgae (Chlorophyta) for wastewater treatment: A review. *Renewable & Sustainable Energy Reviews*, *52*, 123–132. DOI: 10.1016/j.rser.2015.07.086

Abinandan, S., Subashchandrabose, S. R., Venkateswarlu, K., & Megharaj, M. (2018). Microalgae–bacteria biofilms: A sustainable synergistic approach in remediation of acid mine drainage. *Applied Microbiology and Biotechnology, 102*(3), 1131–1144. DOI: 10.1007/s00253-017-8693-7 PMID: 29260261

Abraham, J., Abimbola, T., Braida, W. J., Terracciano, A., Su, T. L., Christodoulatos, C., Koutsospyros, A., RoyChowdhury, A., Smolinski, B., & Lawal, A. (2023). On-Site Pilot-Scale Microalgae Cultivation Using Industrial Wastewater for Bioenergy Production: A Case Study towards Circular Bioeconomy. *Bioengineering (Basel, Switzerland), 10*(12), 1339. DOI: 10.3390/bioengineering10121339 PMID: 38135931

Adam, A. B., Mu'azu, J. B., Titilayo, O. A., Ba'aku, A. E., Gani, J., & Abubakar, M. Y. (2025). The Role of Organic Pollutants in Water Pollution A Review. *Journal of Chemical Technology, 1*, 142–159.

Adeleye, A. T., Bahar, M. M., Megharaj, M., Fang, C., & Rahman, M. M. (2024). The unseen threat of the synergistic effects of microplastics and heavy metals in aquatic environments: A critical review. *Current Pollution Reports, 10*(3), 478–497. DOI: 10.1007/s40726-024-00298-7

Adhikary, S., Saha, J., Dutta, P., & Pal, A. (2024). Bacterial Homeostasis and Tolerance to Potentially Toxic Metals and Metalloids through Diverse Transporters: Metal-Specific Insights. *Geomicrobiology Journal, 41*(5), 496–518. DOI: 10.1080/01490451.2024.2340517

Adhithya, S., Nithya, K., Sathish, A., & Kumar, V. (2025). Uncovering the feasibility of using live Chlorella microbiomes in domestic and industrial wastewater treatment: Insights into monoculture and synergistic mixed co-cultured system. *Journal of Industrial and Engineering Chemistry, 149*, 140–156. DOI: 10.1016/j.jiec.2025.02.006

Adimalla, N. (2020). Spatial distribution, exposure, and potential health risk assessment from nitrate in drinking water from semi-arid region of South India. *Human and Ecological Risk Assessment, 26*(2), 310–334. DOI: 10.1080/10807039.2018.1508329

Agarwal, N. (2025). *Diatoms: Basic and Applied Research*. Educohack Press.

Ahmad, A., Srivastava, S., & Ali, B. M. J. (2017). A FRET based optical sensor for selective detection of heavy metal ions. *Algal Research, 26*, 354–361.

Ahmed Bakhit, S. M., & Elhadi Sulieman, A. M. (2026). Algal enzymes and their applications. In *Microbial Enzymes* (pp. 87–107). Elsevier., DOI: 10.1016/B978-0-443-40352-1.00007-6

Ahmed, M., Mavukkandy, M. O., Giwa, A., Elektorowicz, M., Katsou, E., Khelifi, O., Naddeo, V., & Hasan, S. W. (2022). Recent developments in hazardous pollutants removal from wastewater and water reuse within a circular economy. *npj. NPJ Clean Water*, *5*(1), 12. DOI: 10.1038/s41545-022-00154-5

Ahmed, N. E., Salem, S. S., & Hashem, A. H. (2022). Statistical optimization, partial purification, and characterization of phytase produced from *Talaromyces purpureogenus* NSA20 using potato peel waste and its application in dyes de-colorization. *Biointerface Research in Applied Chemistry*, *12*(4), 4417–4431.

Ahorsu, R., Medina, F., & Constantí, M. (2018). Significance and Challenges of Biomass as a Suitable Feedstock for Bioenergy and Biochemical Production: A Review. *Energies*, *11*(12), 3366. DOI: 10.3390/en11123366

Ahuja, V., Sagar, N. A., Sharma, D., Ahlawat, G. M., Menon, V., Yang, Y.-H., Bhatia, S. K., & Dasgupta, D. (2026). A Review on Carbon Dioxide Mitigation and Sustainable Energy Production Through Algal Biorefineries. *BioEnergy Research*, *19*(1), 29. DOI: 10.1007/s12155-025-10955-4

Aib, H., Parvez, M. S., & Czédli, H. M. (2025). Pharmaceuticals and microplastics in aquatic environments: A comprehensive review of pathways and distribution, toxicological and ecological effects. *International Journal of Environmental Research and Public Health*, *22*(5), 799. DOI: 10.3390/ijerph22050799 PMID: 40427912

Air Liquide. (2022). Cryocap-Carbon capture technologies. https://www.engineering-airliquide.com

Ajiboye, T. O., Oyewo, O. A., & Onwudiwe, D. C. (2021). Simultaneous removal of organics and heavy metals from industrial wastewater: A review. *Chemosphere*, *262*, 128379. DOI: 10.1016/j.chemosphere.2020.128379 PMID: 33182079

Akao, P. K., Singh, B., Kaur, P., Sor, A., Avni, A., Dhir, A., Verma, S., Kapoor, S., Phutela, U. G., Satpute, S., Sharma, S., Avisar, D., Sandha, K. S., & Mamane, H. (2021). Coupled microalgal–bacterial biofilm for enhanced wastewater treatment without energy investment. *Journal of Water Process Engineering*, *41*, 102029. DOI: 10.1016/j.jwpe.2021.102029

Akhter, Md. S., Miah, M. S., Rahman, H., Rahman, S. M. M., Karim, A. J. M. S., Ratul, S. I., & Rahman, M. A. (2025). Agricultural Applications of Chalcogenide-Based Materials: Toward Sustainable, Smart, and Nutrient-Efficient Farming. In K. Kannan & V. Tari (Eds.), Chalcogenide-Based Materials for Optoelectronics, Energy, and Sustainability (pp. 409–438). IGI Global Scientific Publishing. DOI: 10.4018/979-8-3373-3962-7.ch016

Akhter, Md. S., Ratul, S. I., Chowdhury, U., Miah, M. S., Mirza, M. J. A., & Hossain, A. (2026). AI-Powered Generative Models for Predictive and Optimized Aquaculture Water Management. In M. Lahby & R. Gaamouche (Eds.), *Generative AI and Optimization Techniques for Sustainable Water Management* (pp. 113–133). Springer Nature Switzerland. DOI: 10.1007/978-3-032-19012-3_8

Akoijam, N., & Joshi, S. R. (2024). Genome Editing and Genetically Engineered Bacteria for Bioremediation of Heavy Metals. In *Genome Editing in Bacteria (Part 2)* (pp. 184–221). Bentham Science Publishers. https://www.benthamdirect.com/content/books/9789815223798.chapter-6

Akram, S., Guan, G., Hu, B., Saddique, M. A. B., Xi, Y., Luo, X., & Ren, M. (2026). Microalgae-based strategies for cadmium remediation: Insights, challenges, and future directions. *Critical Reviews in Biotechnology*, *46*(2), 318–337. DOI: 10.1080/07388551.2025.2585681 PMID: 41260883

Aksu, Z., & Tezer, S. (2005). Biosorption of reactive dyes on the green alga *Chlorella vulgaris. Process Biochemistry (Barking, London, England)*, *40*(4), 1347–1361. DOI: 10.1016/j.procbio.2004.06.007

Alcántara, C., Posadas, E., Guieysse, B., & Muñoz, R. (2015). Microalgae-based Wastewater Treatment. In Kim, S.-K. (Ed.), *Handbook of Marine Microalgae* (pp. 439–455). Academic Press. DOI: 10.1016/B978-0-12-800776-1.00029-7

Aleruchi, O., Fubara, E., & Otto, B. (2023). Wastewater Quality-It's Impact on the Environment and Human Physiology: A Review. *International Journal of Advance Research and Innovation*, *9*(4), 43–58. DOI: 10.51976/ijari.942107

Ali Ch., B., Hussein, H. J., & Mushtak, T. J. (2022). *BOD: COD Ratio as Indicator for Wastewater and Industrial Water Pollution. 37.*

Ali, S., Paul Peter, A., Chew, K. W., Munawaroh, H. S. H., & Show, P. L. (2021a). Resource recovery from industrial effluents through the cultivation of microalgae: A review. *Bioresource Technology*, *337*, 125461. Advance online publication. DOI: 10.1016/j.biortech.2021.125461 PMID: 34198241

Ali, Z., Subeshan, B., Alam, M., Asmatulu, E., & Xu, J. (2023). Recent progress in extraction/transesterification techniques for the recovery of oil from algae biomass. *Biomass Conversion and Biorefinery*, *13*(4), 2553–2569. DOI: 10.1007/s13399-021-01326-y

Almomani, F., & Bohsale, R. R. (2021). Bio-sorption of toxic metals from industrial wastewater by algae strains Spirulina platensis and Chlorella vulgaris: Application of isotherm, kinetic models and process optimization. *The Science of the Total Environment, 755*, 142654. DOI: 10.1016/j.scitotenv.2020.142654 PMID: 33082037

Alowaisy, A., Allouzi, M. M. A., Chia, W. Y., Mayol, A. P., Sriariyanun, M., & Show, P. L. (2026). Interactions between phytoplankton species and micro/nano-plastics and heavy metal contamination. *Biological Reviews of the Cambridge Philosophical Society, 101*(1), 60–73. DOI: 10.1111/brv.70071 PMID: 41058153

Alvarez, A. L., Weyers, S. L., Goemann, H. M., Peyton, B. M., & Gardner, R. D. (2021). Microalgae, soil and plants: A critical review of microalgae as renewable resources for agriculture. *Algal Research, 54*, 102200. DOI: 10.1016/j.algal.2021.102200

Amenorfenyo, D. K., Huang, X., Zhang, Y., Zeng, Q., Zhang, N., Ren, J., & Huang, Q. (2019). Microalgae brewery wastewater treatment: Potentials, benefits, and the challenges. *International Journal of Environmental Research and Public Health, 16*(11), 1910. DOI: 10.3390/ijerph16111910 PMID: 31151156

Amin, F. R., Khalid, H., Zhang, H., & Zhang, R. (2022). Life cycle assessment of biogas production from different feedstocks. *Renewable & Sustainable Energy Reviews, 150*, 111489.

Amin, M. A., Shukor, H., Yin, L. S., Kasim, F. H., Shoparwe, N. F., Makhtar, M. M. Z., & Yaser, A. Z. (2022). Methane biogas production in Malaysia: Challenge and future plan. *International Journal of Chemical Engineering, 2022*(1), 2278211. Advance online publication. DOI: 10.1155/2022/2278211

Anand, U., Dey, S., Parial, D., Federici, S., Ducoli, S., Bolan, N. S., Dey, A., & Bontempi, E. (2023). Algae and bacteria consortia for wastewater decontamination and transformation into biodiesel, bioethanol, biohydrogen, biofertilizers and animal feed: A review. *Environmental Chemistry Letters, 21*(3), 1585–1609. DOI: 10.1007/s10311-023-01562-w PMID: 36785620

Anas, M., Khattak, W. A., Avci, M. A., Hakki, E. E., & Fahad, S. (2026). Role of Cyanobacteria in Bioremediation and Environmental Restoration. In *Cyanobacterial Blooms: Ecology, Evolution and Biogeochemical Impacts: Microbial Dynamics and Global Implications* (pp. 421-446). Springer. DOI: 10.1007/978-3-032-06042-6_17

Anastopoulos, I., Giannakoudakis, D. A., Frontistis, Z., Zorpas, A. A., Pashalidis, I., & Triantafyllidis, K. (2024). Biomass-derived adsorbents: Universal database development for their synthesis and remediation efficiency as a necessary step to move from laboratory- to pilot-scale applications. *Current Opinion in Green and Sustainable Chemistry, 47*, 100902. DOI: 10.1016/j.cogsc.2024.100902

Andrade, C., Andrade, L., Dias, M., Nascimento, C., & Mendes, M. (2018). Microalgae for bioremediation of textile wastewater: An overview. *MOJ Food Processing & Technology*, *6*(6), 432–433. DOI: 10.15406/mojfpt.2018.06.00200

Anh, H. T. H., Shahsavari, E., Bott, N. J., & Ball, A. S. (2021). Options for Improved Treatment of Saline Wastewater From Fish and Shellfish Processing. *Frontiers in Environmental Science*, *9*, 689580. DOI: 10.3389/fenvs.2021.689580

Ansa, E., Lubberding, H. J., Ampofo, J., & Gijzen, H. (2011). The role of algae in the removal of *Escherichia coli* in a tropical eutrophic lake. *Ecological Engineering*, *37*(3), 317–324. DOI: 10.1016/j.ecoleng.2010.11.023

Ansari, F. A., Nasr, M., Rawat, I., & Bux, F. (2023). Meeting sustainable development goals (SDGs) through progression of pilot-scale algal system to commercial raceway pond (300,000 L). *Biomass Conversion and Biorefinery*, *13*(9), 7663–7676. DOI: 10.1007/s13399-021-01661-0

Ansar, Z. A., Alappat, M. M., Subramanian, S., & Anbarasu, A. (2026). Microbial and multi-omics approaches for bioremediation of emerging contaminants: Environmental impact and future engineering solutions. *Biodegradation*, *37*(2), 52. DOI: 10.1007/s10532-026-10276-z PMID: 41832337

Apollo, S. (2023). A review of sludge production in South Africa municipal wastewater treatment plants, analysis of handling cost and potential minimization methods. *Physical Sciences Reviews*, *8*(11), 3847–3857. DOI: 10.1515/psr-2021-0234

Arashiro, L. T., Montero, N., Ferrer, I., Acién, F. G., Gómez, C., & Garfí, M. (2018). Life cycle assessment of high rate algal ponds for wastewater treatment and resource recovery. *The Science of the Total Environment*, *622–623*, 1118–1130. DOI: 10.1016/j.scitotenv.2017.12.051 PMID: 29890581

Arbib, Z., Ruiz, J., Álvarez-Díaz, P. A., Garrido-Pérez, C., & Perales, J. A. (2014). Capability of different microalgae species for phytoremediation processes: Wastewater tertiary treatment, CO_2 bio-fixation and low cost biofuels production. *Water Research*, *49*, 465–474. DOI: 10.1016/j.watres.2013.10.036

Armbruster, M. H., & Austin, J. B. (1938). The Adsorption of Gases on Plane Surfaces of Mica. *Journal of the American Chemical Society*, *60*(2), 467–475. DOI: 10.1021/ja01269a066

Arnaldo, M. D. G., Mossion, A., Beignon, T., Vuillemin, H., Guihéneuf, F., Wielgosz-Collin, G., & Méléder, V. (2024). Diatom Biofilm. In *Diatom Photosynthesis* (pp. 449–476). John Wiley & Sons, Ltd., DOI: 10.1002/9781119842156.ch15

Arora, N., Gulati, K., Tripathi, S., & Pruthi, V. (2018). Algae as a budding tool for mitigation of arsenic from aquatic systems. In *Mechanisms of arsenic toxicity and tolerance in plants* (pp. 269–297). DOI: 10.1007/978-981-13-1292-2_12

Arunakumara, K. K. I. U., & Zhang, X. (2008). Heavy metal bioaccumulation and toxicity with special reference to microalgae. *Journal of Ocean University of China*, *7*(1), 60–64. DOI: 10.1007/s11802-008-0060-y

Aryanfar, Y., Keçebaş, A., Hassanzadeh, S., Sadabad, A. N., Rashid, F. L., Ahmad, S., Hendy, A., Irfan, H. M., Ghasemlou, S. M., & Fernandez, J. B. (2025). Integrated Evaluation of Contaminant Profiles, Detection Techniques, and Management Strategies for Tannery Sludge. *Asia-Pacific Journal of Chemical Engineering*, •••, e70191.

Asad, M., Sohail, M., Muneeb, A., Qadeer, S., Majeed, A. A. A., Rafiq, H., Rafique, A., Zubair, S. K., & Rafique, N. (2024). Strategies for enhancing animal feed quality and productivity. In *Complementary & Alternative Medicine* (pp. 364–372). Feed Additives.

Asati, H., Rana, M. S., Prajapati, S. K., & Tripathi, K. M. (2025). Water-Soluble Carbon Nano Onions-Augmented Microalgae for Coupled Wastewater Remediation and Biofuel Production: A Circular Economy Approach. *ACS Sustainable Resource Management*, *2*(10), 1957–1965. DOI: 10.1021/acssusresmgt.5c00276

Ashiwaju, B. I., Uzougbo, C. G., & Orikpete, O. F. (2023). Environmental impact of pharmaceuticals: A comprehensive review. *Matrix Science Pharma*, *7*(3), 85–94. DOI: 10.4103/mtsp.mtsp_15_23

Asif, M. B., Baig, N., & Abdallah, M. A.-E. (2026). A Perspective on Quaternary Treatments for Micropollutant Elimination in Urban Wastewater Treatment Plants. *Earth Systems and Environment*. Advance online publication. DOI: 10.1007/s41748-026-01086-7

aus der Beek, T., Weber, F. A., Bergmann, A., Hickmann, S., Ebert, I., Hein, A., & Küster, A. (2016). Pharmaceuticals in the environment-Global occurrences and perspectives. *Environmental Toxicology and Chemistry*, *35*(4), 823–835. DOI: 10.1002/etc.3339 PMID: 26666847

Avilia, S., Pozzuoli, E., Iovinella, M., Ciniglia, C., & Papa, S. (2025). A Review on Innovative Strategies Towards Sustainable Drug Waste Management Through Algae-Based Systems. *Sci*, *7*(3), 92. DOI: 10.3390/sci7030092

Ayele, A., Suresh, A., & Benor, S. (2021). Phycoremediation of Heavy Metals, Factors Involved and Mechanisms Related to Functional Groups in the Algae Cell Surface – A Review. In Aravind, J., Kamaraj, M., Prashanthi Devi, M., & Rajakumar, S. (Eds.), *Strategies and Tools for Pollutant Mitigation* (pp. 269–289). Springer International Publishing., DOI: 10.1007/978-3-030-63575-6_13

Ayodele, B. V., Alsaffar, M. A., & Mustapa, S. I. (2020). An overview of integration opportunities for sustainable bioethanol production from first- and second-generation sugar-based feedstocks. *Journal of Cleaner Production, 245*, 118857. Advance online publication. DOI: 10.1016/j.jclepro.2019.118857

Azeem, S., Sohail, M., Azeem, A., Haq, M. Z. U., Hassan, Z., Khan, A. A., & Shahzad, U. (2024). Guardian crops cultivating resilience against pests. In *Revolutionizing Pest Management for Sustainable Agriculture* (pp. 181–202). IGI Global Scientific Publishing. DOI: 10.4018/979-8-3693-3061-6.ch008

Bai, R. G., Chandrasekharan Nair, S., Joller-Vahter, L., & Kikas, T. (2025a). Microalgae in Mitigating Industrial Pollution: Bioremediation Strategies and Biomagnification Potential. *Biomass, 5*(4), 61. DOI: 10.3390/biomass5040061

Bai, X., & Acharya, K. (2017). Algae-mediated removal of selected pharmaceutical and personal care products (PPCPs) from Lake Mead water. *The Science of the Total Environment, 581–582*, 734–740. DOI: 10.1016/j.scitotenv.2016.12.192 PMID: 28089530

Balajii, M., & Niju, S. (2019). A review on biodiesel production from microalgae: Influencing factors and future prospects. *Environmental Science and Pollution Research International, 26*(9), 8618–8635.

Balajii, M., & Niju, S. (2019). Biochar-derived heterogeneous catalysts for biodiesel production. *Environmental Chemistry Letters, 17*(4), 1447–1469. DOI: 10.1007/s10311-019-00885-x

Balzano, S., Sardo, A., Blasio, M., Chahine, T. B., Anno, F. D., Sansone, C., & Brunet, C. (2020). Microalgal metallothioneins and phytochelatins and their potential use in bioremediation. *Frontiers in Microbiology, 11*, 1–16. DOI: 10.3389/fmicb.2020.00517 PMID: 32431671

Bamaniya, P., Vaghela, D., Tandel, V., Chaudhari, S., & Bambhaniya, I. (2026). Linking Terrestrial and Aquatic Systems: Pollutant Transport, Bioaccumulation Cycles, and Global Food Security. In *Plastic Tides: Tackling Plastic Pollution for Global Food Security* (pp. 195-228). Springer.

Banu, J. R., Kannah, R. Y., Kavitha, S., Ashikvivek, A., Bhosale, R. R., & Kumar, G. (2020). Cost-effective biomethanation via surfactant-coupled ultrasonic liquefaction of mixed microalgal biomass harvested from an open raceway pond. *Bioresource Technology, 304*, 123021. DOI: 10.1016/j.biortech.2020.123021 PMID: 32086031

Bashir, K., Khan, S., Ali, R., Yasmin, H., Gaafar, A.-R. Z., Khilgee, F. E. A., Butt, S., & Ullah, A. (2023). Bioremediation of Metal-Polluted Industrial Wastewater with Algal-Bacterial Consortia: A Sustainable Strategy. *Sustainability (Basel), 15*(19), 14056. DOI: 10.3390/su151914056

Batista, A. P., Ambrosano, L., Graça, S., Sousa, C., Marques, P. A. S. S., Ribeiro, B., Botrel, E. P., Castro Neto, P., & Gouveia, L. (2015). Combining urban wastewater treatment with biohydrogen production – An integrated microalgae-based approach. *Bioresource Technology, 184*, 230–235. DOI: 10.1016/j.biortech.2014.10.064 PMID: 25453433

Beaucamp, A., Muddasar, M., Amiinu, I. S., Moraes Leite, M., Culebras, M., Latha, K., Gutiérrez, M. C., Rodriguez-Padron, D., del Monte, F., Kennedy, T., Ryan, K. M., Luque, R., Titirici, M.-M., & Collins, M. N. (2022). Lignin for energy applications—State of the art, life cycle, technoeconomic analysis and future trends. *Green Chemistry, 24*(21), 8193–8226. DOI: 10.1039/D2GC02724K

Behera, M., Nayak, J., Banerjee, S., Chakrabortty, S., & Tripathy, S. K. (2021). A review on the treatment of textile industry waste effluents towards the development of efficient mitigation strategy: An integrated system design approach. *Journal of Environmental Chemical Engineering, 9*(4), 105277. DOI: 10.1016/j.jece.2021.105277

Bezerra, S. S., Fontana, L., Arantes, C. C., & de Jesus, T. A. (2025). Phosphorus removal from wastewater by microalgal cultivation in photobioreactors: A systematic literature review and multivariate analysis. *Environmental Monitoring and Assessment, 197*(11), 1182. DOI: 10.1007/s10661-025-14524-2 PMID: 41055785

Bhattacharya, P., Lin, S., Turner, J. P., & Ke, P. C. (2010). Physical adsorption of charged plastic nanoparticles affects algal photosynthesis. *The Journal of Physical Chemistry. C, Nanomaterials and Interfaces, 114*(39), 16556–16561. DOI: 10.1021/jp1054759

Bhatt, P., Bhandari, G., Bhatt, K., & Simsek, H. (2022). Microalgae-based removal of pollutants from wastewaters: Occurrence, toxicity and circular economy. *Chemosphere, 306*, 135576. DOI: 10.1016/j.chemosphere.2022.135576 PMID: 35803375

Bhatt, P., Pandey, S. C., Joshi, S., Chaudhary, P., Pathak, V. M., Huang, Y., Wu, X., Zhou, Z., & Chen, S. (2022). Nanobioremediation: A sustainable approach for the removal of toxic pollutants from the environment. *Journal of Hazardous Materials, 427*, 128033. DOI: 10.1016/j.jhazmat.2021.128033 PMID: 34999406

Bhuyan, M. S. (2023). Ecological risks associated with seaweed cultivation and identifying risk minimization approaches. *Algal Research, 69*, 102967. DOI: 10.1016/j.algal.2022.102967

Bhuyan, P. P., Nahak, S., Sahoo, H., Sahoo, S., & Pradhan, B. (2026). Microalgal bioremediation of heavy metals and by-products recovery as lipid: The dual role of microalgae. *International Journal of Phytoremediation, 0*(0), 1–17. DOI: 10.1080/15226514.2026.2667453 PMID: 42118003

Bilad, M. R., Discart, V., Vandamme, D., Foubert, I., Muylaert, K., & Vankelecom, I. F. (2014). Coupled cultivation and pre-harvesting of microalgae in a membrane photobioreactor (MPBR). *Bioresource Technology, 155*, 410–417. DOI: 10.1016/j.biortech.2013.05.026 PMID: 24559585

Biliani, E., & Manariotis, I. D. (2026, May 07). S., & D. Manariotis, I. (2026). High-rate algal ponds in wastewater treatment: A critical look at recent developments. *Environmental Science. Water Research & Technology, 12*(5), 1359–1373. Advance online publication. DOI: 10.1039/D6EW00116E

Blanco-Vieites, M., Suárez-Montes, D., Delgado, F., Álvarez-Gil, M., Battez, A. H., & Rodríguez, E. (2022). Removal of heavy metals and hydrocarbons by microalgae from wastewater in the steel industry. *Algal Research, 64*, 102700. DOI: 10.1016/j.algal.2022.102700

Bohutskyi, P., & Bouwer, E. (2013). Biogas Production from Algae and Cyanobacteria Through Anaerobic Digestion: A Review, Analysis, and Research Needs. In Lee, J. W. (Ed.), *Advanced Biofuels and Bioproducts* (pp. 873–975). Springer., DOI: 10.1007/978-1-4614-3348-4_36

Bora, J., Saha, I., Vaibhav, V., Singh, M., & Malik, S. (2023). Combined Applications of Physico-Chemical Treatments in Treatment of Industrial Wastewater. In Shah, M. P. (Ed.), *Advanced and Innovative Approaches of Environmental Biotechnology in Industrial Wastewater Treatment* (pp. 399–418). Springer Nature Singapore., DOI: 10.1007/978-981-99-2598-8_18

Boretti, A., & Rosa, L. (2019). Reassessing the projections of the World Water Development Report. *npj. NPJ Clean Water, 2*(1), 15. DOI: 10.1038/s41545-019-0039-9

Boro, D., Chirania, M., Verma, A. K., Chettri, D., & Verma, A. K. (2025). Comprehensive approaches to managing emerging contaminants in wastewater: Identification, sources, monitoring and remediation. *Environmental Monitoring and Assessment*, *197*(4), 456. DOI: 10.1007/s10661-025-13809-w PMID: 40119196

Borowitzka, M. A. (1998). Limits to growth. In *Wastewater treatment with algae* (pp. 203–226). DOI: 10.1007/978-3-662-10863-5_12

Bouaboula, H., Chaouki, J., Belmabkhout, Y., & Zaabout, A. (2024). Comparative review of direct air capture technologies: From technical, commercial, economic, and environmental aspects. *Chemical Engineering Journal*, *484*, 149411. Advance online publication. DOI: 10.1016/j.cej.2024.149411

Boutin, C., & Eme, C. (2016). Domestic Wastewater Characterization by Emission Source. *13eme Congres Spécialisé IWA on Small Water and Wastewater Systems.*

Brar, P. K., Örmeci, B., & Dhir, A. (2022). Algae: A cohesive tool for biodiesel production along with wastewater treatment. *Sustainable Chemistry and Pharmacy*, *28*, 100730. DOI: 10.1016/j.scp.2022.100730

Bridgwater, A. V. (2012). Review of fast pyrolysis of biomass and product upgrading. *Biomass and Bioenergy*, *38*, 68–94. DOI: 10.1016/j.biombioe.2011.01.048

Bridgwater, A. V., Toft, A. J., & Brammer, J. G. (2002). A techno-economic comparison of power production by biomass fast pyrolysis with gasification and combustion. *Renewable & Sustainable Energy Reviews*, *6*(3), 181–246. DOI: 10.1016/S1364-0321(01)00010-7

Briffa, J., Sinagra, E., & Blundell, R. (2020). Heavy metal pollution in the environment and their toxicological effects on humans. *Heliyon*, *6*(9), e04691. DOI: 10.1016/j.heliyon.2020.e04691 PMID: 32964150

Brosowski, A., Krause, T., Mantau, U., Mahro, B., Noke, A., Richter, F., Raussen, T., Bischof, R., Hering, T., Blanke, C., Müller, P., & Thrän, D. (2019). How to measure the impact of biogenic residues, wastes and by-products: Development of a national resource monitoring based on the example of Germany. *Biomass and Bioenergy*, *127*, 105275. Advance online publication. DOI: 10.1016/j.biombioe.2019.105275

Buhani, J., Juliasih, N. L. G. R., Wahyu Purnamasari, I., Suharso, , Noviany, , Sumadi, , Alghamdi, H. M., Elgarahy, A. M., & Elwakeel, K. Z. (2026). Selective adsorption and mechanistic insights into divalent cadmium, copper, and lead recovery using silica–magnetite modified Nitzschia sp. Biomaterial. *International Journal of Phytoremediation*, *28*(7), 1258–1277. DOI: 10.1080/15226514.2025.2607515 PMID: 41548094

Caetano-Silva, M. E., Netto, F. M., Bertoldo-Pacheco, M. T., Alegría, A., & Cilla, A. (2021). Peptide-metal complexes: Obtention and role in increasing bioavailability and decreasing the pro-oxidant effect of minerals. *Critical Reviews in Food Science and Nutrition, 61*(9), 1470–1489. DOI: 10.1080/10408398.2020.1761770 PMID: 32370550

Calijuri, M. L., Do Couto, E. D. A., Assemany, P. P., Ribeiro, V. J., Lorentz, J. F., Castro, J. D. S., Assis, L. R. D., Oliveira, A. P. D. S., Pereira, A. S. A. D. P., Marangon, B. B., Magalhães, I. B., Silva, T. A., Ferreira, J., Braga, M. Q., Gama, R. C. N. D., & Henriques, B. S. (2025). Microalgae-Based Wastewater Treatment and Biomass Valorization: Insights, Challenges, and Opportunities from 15 Years of Research. *ACS Omega, 10*(42), 49273–49299. DOI: 10.1021/acsomega.5c04364 PMID: 41179135

Cao, W., Wang, X., Sun, S., Hu, C., & Zhao, Y. (2017). Simultaneously upgrading biogas and purifying biogas slurry using cocultivation of *Chlorella vulgaris* and three different fungi under various mixed light wavelength and photoperiods. *Bioresource Technology, 241*, 701–709. DOI: 10.1016/j.biortech.2017.05.194

Cao, X., Li, H., Zhou, Y., & Song, C. (2020). The shift of mutualistic relationships among algae, free living and attached bacteria through different nutrient addition mode: A mesocosm study. *Journal of Freshwater Ecology, 35*(1), 535–548. DOI: 10.1080/02705060.2020.1858984

Capua-Shenkar, J., Varsano, N., Kruth, H., & Addadi, L. (2023). Crystals in Atherosclerosis: Crystal Cholesterol Structures, Morphologies, Formation and Dissolution. What Do We Know? In Abela, G. S., & Nidorf, S. M. (Eds.), *Cholesterol Crystals in Atherosclerosis and Other Related Diseases* (pp. 49–71). Springer International Publishing., DOI: 10.1007/978-3-031-41192-2_4

Carbfix. (2024). Carbfix—We turn CO2 into stone. https://www.carbfix.com

Carvalho, J. C., de, , & Irineudo, A. M. (2020). Microalgal biomass pretreatment for integrated processing into biofuels, food, and feed. *Bioresource Technology, 300*, 122719. DOI: 10.1016/j.biortech.2019.122719 PMID: 31956056

Chai, W. S., Tan, W. G., Munawaroh, H. S. H., Gupta, V. K., Ho, S.-H., & Show, P. L. (2021). Multifaceted roles of microalgae in the application of wastewater biotreatment: A review. *Environmental Pollution, 269*, 116236. DOI: 10.1016/j.envpol.2020.116236 PMID: 33333449

Chandra, R., Pradhan, S., Patel, A., & Ghosh, U. K. (2021). An approach for dairy wastewater remediation using mixture of microalgae and biodiesel production for sustainable transportation. *Journal of Environmental Management, 297*, 113210. DOI: 10.1016/j.jenvman.2021.113210

Chan, S. S., Khoo, K. S., Chew, K. W., Ling, T. C., & Show, P. L. (2022). Recent advances biodegradation and biosorption of organic compounds from wastewater: Microalgae-bacteria consortium-A review. *Bioresource Technology, 344*, 126159. DOI: 10.1016/j.biortech.2021.126159 PMID: 34673198

Chaudhary, R., Dikshit, A. K., & Tong, Y. W. (2018). Carbon-dioxide biofixation and phycoremediation of municipal wastewater using Chlorella vulgaris and Scenedesmus obliquus. *Environmental Science and Pollution Research International, 25*(21), 20399–20406. DOI: 10.1007/s11356-017-9575-3 PMID: 28656576

Chaudhary, R., Tong, Y. W., & Dikshit, A. K. (2018). CO2-assisted removal of nutrients from municipal wastewater by microalgae Chlorella vulgaris and Scenedesmus obliquus. *International Journal of Environmental Science and Technology, 15*(10), 2183–2192. DOI: 10.1007/s13762-017-1571-0

Chaudhuri, R., & Balasubramanian, P. (2025). Enhancing nutrient removal from hydroponic effluent with simultaneous production of lipid-rich biomass through mixotrophic cultivation of microalgae. *Journal of Applied Phycology, 37*(2), 957–971. DOI: 10.1007/s10811-024-03425-5

Chauvy, R., & De Weireld, G. (2020). CO2 utilization technologies in Europe: A short review. *Energy Technology (Weinheim), 8*(12), 2000627. Advance online publication. DOI: 10.1002/ente.202000627

Cheah, W. Y., Show, P. L., Chang, J.-S., Ling, T. C., & Juan, J. C. (2015). Biosequestration of atmospheric CO2 and flue gas-containing CO2 by microalgae. *Bioresource Technology, 184*, 190–201. DOI: 10.1016/j.biortech.2014.11.026 PMID: 25497054

Chen, C. Y., Kuo, E. W., Nagarajan, D., Dong, C. D., Lee, D. J., Varjani, S., Lam, S. S., & Chang, J. S. (2021). Semi-batch cultivation of *Chlorella sorokiniana* AK-1 with dual carriers for the effective treatment of full strength piggery wastewater treatment. *Bioresource Technology, 326*, 124773. DOI: 10.1016/j.biortech.2021.124773

Cheng, J., Li, K., & Yang, Z. (2023). Digital twins for photobioreactor optimization: From lab to pilot scale. *Algal Research, 69*, 102943. DOI: 10.1016/j.algal.2022.102943

Chen, M., Chang, L., Zhang, J., Guo, F., Vymazal, J., He, Q., & Chen, Y. (2020). Global nitrogen input on wetland ecosystem: The driving mechanism of soil labile carbon and nitrogen on greenhouse gas emissions. *Environmental Science and Ecotechnology, 4*, 100063. DOI: 10.1016/j.ese.2020.100063 PMID: 36157707

Chen, P., Xie, Q., & Addy, M. (2023). Climate change impacts on renewable energy systems. *Nature Climate Change, 13*(2), 112–120.

Chen, Q., Zhang, L., Han, Y., Fang, J., & Wang, H. (2020). Degradation and metabolic pathways of sulfamethazine and enrofloxacin in *Chlorella vulgaris* and *Scenedesmus obliquus* treatment systems. *Environmental Science and Pollution Research International, 27*(22), 28198–28208. DOI: 10.1007/s11356-020-09008-4

Chen, R., Liu, Y., & Liao, W. (2016). Using an environmentally friendly process combining electrocoagulation and algal cultivation to treat high-strength wastewater. *Algal Research, 16*, 330–337. DOI: 10.1016/j.algal.2016.03.032

Chen, Y., Xuan, T., Dong, M., Liang, H., Cai, G., Xia, B., Tang, J., Xie, Z., & Zhao, X. (2025). Mechanistic insights into microalgal removal of ciprofloxacin–copper co-contaminants: Extracellular polymeric substances adsorption, intracellular degradation, and combined toxicity. *Journal of Hazardous Materials, 500*, 140558. DOI: 10.1016/j.jhazmat.2025.140558 PMID: 41289666

Chen, Z., Shao, S., He, Y., Luo, Q., Zheng, M., Zheng, M., Chen, B., & Wang, M. (2020). Nutrient removal from piggery wastewater coupled to lipid production by a newly isolated self-flocculating microalga *Desmodesmus* sp. PW1. *Bioresource Technology, 302*, 122806. DOI: 10.1016/j.biortech.2020.122806 PMID: 31982846

Chen, Z., Xie, Y., Qiu, S., Li, M., & Ge, S. (2024). Enriched functional exoproteins and increased exopolysaccharides with altered molecular conformation mutually promoted indigenous microalgal-bacterial consortium biofilm growth under high light intensity. *Chemical Engineering Journal, 480*, 148056. DOI: 10.1016/j.cej.2023.148056

Cherubini, F., & Strømman, A. H. (2011). Life cycle assessment of bioenergy systems: State of the art and future challenges. *Bioresource Technology, 102*(2), 437–451. DOI: 10.1016/j.biortech.2010.08.010 PMID: 20832298

Cherubin, M. R., Carvalho, J. L., Cerri, C. E., Nogueira, L. A., Souza, G. M., & Cantarella, H. (2021). Land use and management effects on sustainable sugarcane-derived bioenergy. *Land (Basel), 10*(1), 72. Advance online publication. DOI: 10.3390/land10010072

Chettri, D., Boro, M., Malik, A. A., & Verma, A. K. (2025). Nanotechnology for a greener future: Sustainable energy and biofuel advances. *Biofuels*, *16*(6), 622–637. DOI: 10.1080/17597269.2024.2440207

Chew, K. W., Yap, J. Y., Show, P. L., Suan, N. H., Juan, J. C., Ling, T. C., Lee, D.-J., & Chang, J. S. (2017). Microalgae biorefinery: High value products perspectives. *Bioresource Technology*, *229*, 53–62. DOI: 10.1016/j.biortech.2017.01.006 PMID: 28107722

Chia, W. Y., Ying Tang, D. Y., Khoo, K. S., Kay Lup, A. N., & Chew, K. W. (2020). Nature's fight against plastic pollution: Algae for plastic biodegradation and bioplastics production. *Environmental Science and Ecotechnology*, *4*, 100065. DOI: 10.1016/j.ese.2020.100065 PMID: 36157709

Chin, J. Y., Chng, L. M., Leong, S. S., Yeap, S. P., Yasin, N. H. M., & Toh, P. Y. (2020). Removal of synthetic dye by *Chlorella vulgaris* microalgae as natural adsorbent. *Arabian Journal for Science and Engineering*, *45*(9), 7385–7395. DOI: 10.1007/s13369-020-04557-9

Chisti, Y. (2007). Biodiesel from microalgae. *Biotechnology Advances*, *25*(3), 294–306. DOI: 10.1016/j.biotechadv.2007.02.001 PMID: 17350212

Choi, K. R., Jang, W. D., Yang, D., Cho, J. S., Park, D., & Lee, S. Y. (2019). Systems Metabolic Engineering Strategies: Integrating Systems and Synthetic Biology with Metabolic Engineering. *Trends in Biotechnology*, *37*(8), 817–837. DOI: 10.1016/j.tibtech.2019.01.003 PMID: 30737009

Choi, Y.-Y., Baek, S.-R., Kim, J.-I., Choi, J.-W., Hur, J., Lee, T.-U., Park, C.-J., & Lee, B. (2017). Characteristics and Biodegradability of Wastewater Organic Matter in Municipal Wastewater Treatment Plants Collecting Domestic Wastewater and Industrial Discharge. *Water (Basel)*, *9*(6), 409. DOI: 10.3390/w9060409

Chojnacka, K. (2010). Biosorption and bioaccumulation – the prospects for practical applications. *Environment International*, *36*(3), 299–307. DOI: 10.1016/j.envint.2009.12.001 PMID: 20051290

Chong, A., Wong, Y., & Tam, N. (2000). Performance of different microalgal species in removing nickel and zinc from industrial wastewater. *Chemosphere*, *41*(2), 251–257. DOI: 10.1016/S0045-6535(99)00418-X PMID: 10819208

Choudhary, S., Tripathi, S., & Poluri, K. M. (2022). Microalgal-based bioenergy: Strategies, prospects, and sustainability. *Energy & Fuels*, *36*(24), 14584–14612. DOI: 10.1021/acs.energyfuels.2c02922

Chugh, M., Kumar, L., Bhardwaj, D., & Bharadvaja, N. (2022). Bioaccumulation and detoxification of heavy metals. In *Development in Wastewater Treatment Research and Processes* (pp. 243–264). Elsevier., DOI: 10.1016/B978-0-323-85657-7.00013-4

Chu, W. L., & Phang, S. M. (2019). Biosorption of heavy metals and dyes from industrial effluents by microalgae. In Alam, M., & Wang, Z. (Eds.), *Microalgae Biotechnology for Development of Biofuel and Wastewater Treatment* (pp. 599–634). Springer., DOI: 10.1007/978-981-13-2264-8_23

Craggs, R. J., Heubeck, S., Lundquist, T. J., & Benemann, J. R. (2011). Algal biofuels from wastewater treatment high rate algal ponds. *Water Science and Technology, 63*(4), 660–665. DOI: 10.2166/wst.2011.100 PMID: 21330711

Craine, J. M., & Dybzinski, R. (2013). Mechanisms of plant competition for nutrients, water and light. *Functional Ecology, 27*(4), 833–840. DOI: 10.1111/1365-2435.12081

Crini, G., & Badot, P. M. (2008). Application of chitosan, a natural aminopolysaccharide, for dye removal from aqueous solutions by adsorption processes using batch studies: A review of recent literature. *Progress in Polymer Science, 33*(4), 399–447. DOI: 10.1016/j.progpolymsci.2007.11.001

Crini, G., & Lichtfouse, E. (2019). Advantages and disadvantages of techniques used for wastewater treatment. *Environmental Chemistry Letters, 17*(1), 145–155. DOI: 10.1007/s10311-018-0785-9

Crippa, I., Dolci, G., Grosso, M., & Rigamonti, L. (2025). Life Cycle Assessment of Microalgal Biomass Valorization from a Wastewater Treatment Process. *Waste and Biomass Valorization, 16*(1), 525–541. DOI: 10.1007/s12649-024-02695-x

Da Fontoura, J. T., Rolim, G. S., Farenzena, M., & Gutterres, M. (2017). Influence of light intensity and tannery wastewater concentration on biomass production and nutrient removal by microalgae Scenedesmus sp. *Process Safety and Environmental Protection, 111*, 355–362. DOI: 10.1016/j.psep.2017.07.024

da Silva, S. R., Borges, P. M., Tošić, N., & Andrade, J. J. O. (2026). Advances in circular valorization of construction and demolition waste (CDW) toward low-carbon and resilient construction: A Comprehensive review. *Sustainability (Basel), 18*(6), 2759. DOI: 10.3390/su18062759

Danouche, M., El Ghachtouli, N., El Baouchi, A., & El Arroussi, H. (2020). Heavy metals phycoremediation using tolerant green microalgae: Enzymatic and non-enzymatic antioxidant systems for the management of oxidative stress. *Journal of Environmental Chemical Engineering, 8*(5), 104460. DOI: 10.1016/j.jece.2020.104460

Dao, G., Wang, S., Wang, X., Chen, Z., Wu, Y., Wu, G., Lu, Y., Liu, S., & Hu, H. (2020). Enhanced *Scenedesmus* sp. growth in response to gibberellin secretion by symbiotic bacteria. *The Science of the Total Environment, 740*, 140099. DOI: 10.1016/j.scitotenv.2020.140099 PMID: 32927572

Das, P. P., Samanta, N. S., Dhara, S., & Purkait, M. K. (2023). Biofuel production from algal biomass. In *Green Approach to Alternative Fuel for a Sustainable Future* (pp. 167–179). Elsevier., DOI: 10.1016/B978-0-12-824318-3.00009-6

Daughton, C. G., & Ternes, T. A. (1999). Pharmaceuticals and personal care products in the environment: Agents of subtle change? *Environmental Health Perspectives, 107*(SUPPL. 6), 907–938. DOI: 10.1289/ehp.99107s6907 PMID: 10592150

Daverey, A., Pandey, D., Verma, P., Verma, S., Shah, V., Dutta, K., & Arunachalam, K. (2019). Recent advances in energy efficient biological treatment of municipal wastewater. *Bioresource Technology Reports, 7*, 100252. DOI: 10.1016/j.biteb.2019.100252

Davis, T. A., Volesky, B., & Mucci, A. (2003). A review of the biochemistry of heavy metal biosorption by brown algae. *Water Research, 37*(18), 4311–4330. DOI: 10.1016/S0043-1354(03)00293-8 PMID: 14511701

de Carvalho, J. C., Molina-Aulestia, D. T., Martinez-Burgos, W. J., Karp, S. G., Manzoki, M. C., Medeiros, A. B. P., Rodrigues, C., Scapini, T., Vandenberghe, L. P. S., Vieira, S., Woiciechowski, A. L., Soccol, V. T., & Soccol, C. R. (2022). Agro-industrial wastewaters for algal biomass production, bio-based products, and biofuels in a circular bioeconomy. *Fermentation (Basel, Switzerland), 8*(12), 728. DOI: 10.3390/fermentation8120728

De La Mora-Orozco, C., González-Acuña, I. J., Saucedo-Terán, R. A., Flores-López, H. E., Rubio-Arias, H. O., & Ochoa-Rivero, J. M. (2018). Removing Organic Matter and Nutrients from Pig Farm Wastewater with a Constructed Wetland System. *International Journal of Environmental Research and Public Health, 15*(5), 1031. DOI: 10.3390/ijerph15051031 PMID: 29883370

de Morais, E. G., da Silveira, J. T., Schüler, L. M., de Freitas, B. C. B., Costa, J. A. V., de Morais, M. G., Ferrer, I., & Barreira, L. (2023). Biomass valorization via pyrolysis in microalgae-based wastewater treatment: Challenges and opportunities for a circular bioeconomy. *Journal of Applied Phycology, 35*(6), 2689–2708. DOI: 10.1007/s10811-023-03104-x

De Philippis, R., Colica, G., & Micheletti, E. (2011). Exopolysaccharide-producing cyanobacteria in heavy metal removal from water: Molecular basis and practical applicability of the biosorption process. *Applied Microbiology and Biotechnology, 92*(4), 697–708. DOI: 10.1007/s00253-011-3601-z PMID: 21983706

De Silva, A. G. S., Hashim, Z. K., Solomon, W., Zhao, J.-B., Kovács, G., Kulmány, I. M., & Molnár, Z. (2024). Unveiling the role of edaphic microalgae in soil carbon sequestration: Potential for agricultural inoculants in climate change mitigation. *Agriculture, 14*(11), 2065. DOI: 10.3390/agriculture14112065

De-Bashan, L. E., & Bashan, Y. (2010). Immobilized microalgae for removing pollutants: Review of practical aspects. *Bioresource Technology, 101*(6), 1611–1627. DOI: 10.1016/j.biortech.2009.09.043 PMID: 19931451

Deblonde, T., Cossu-Leguille, C., & Hartemann, P. (2011). Emerging pollutants in wastewater: A review of the literature. *International Journal of Hygiene and Environmental Health, 214*(6), 442–448. DOI: 10.1016/j.ijheh.2011.08.002 PMID: 21885335

Deep Singh, A., Gajera, B., & Sarma, A. K. (2022). Appraising the availability of biomass residues in India and their bioenergy potential. *Waste Management, 152*, 38–47. DOI: 10.1016/j.wasman.2022.08.001 PMID: 35973326

Deepika, C., Frascaroli, G., Gowri, A., & Silkina, A. (2025). Microalgae-based bioremediation: A sustainable approach for the removal of emergent pollutants. In *Biotechnological Interventions in the Removal of Emerging Pollutants* (pp. 101–139). Springer. DOI: 10.1007/978-981-97-9922-0_6

Delanka-Pedige, H. M. K., Cheng, X., Munasinghe-Arachchige, S. P., Bandara, G. L. C. L., Zhang, Y., Xu, P., Schaub, T., & Nirmalakhandan, N. (2020). Conventional vs. algal wastewater technologies: Reclamation of microbially safe water for agricultural reuse. *Algal Research, 51*, 102022. DOI: 10.1016/j.algal.2020.102022

Delrue, F., Álvarez-Díaz, P. D., Fon-Sing, S., Fleury, G., & Sassi, J.-F. (2016). The environmental biorefinery: Using microalgae to remediate wastewater, a win-win paradigm. *Energies, 9*(3), 132. DOI: 10.3390/en9030132

Demertzioglou, M., Genitsaris, S., Mazaris, A. D., Kyparissis, A., Voutsa, D., Kozari, A., Kormas, K. A., Stefanidou, N., Katsiapi, M., Michaloudi, E., & Moustaka-Gouni, M. (2022). A catastrophic change in a European protected wetland: From harmful phytoplankton blooms to fish and bird kill. *Environmental Pollution, 312*, 120038. DOI: 10.1016/j.envpol.2022.120038 PMID: 36030951

Deshaware, S., Marathe, S. J., Bedade, D., Deska, J., & Shamekh, S. (2021). Investigation on mycelial growth requirements of *Cantharellus cibarius* under laboratory conditions. *Archives of Microbiology*, *203*(4), 1539–1545. DOI: 10.1007/s00203-020-02142-0 PMID: 33399895

Dhangar, K., & Kumar, M. (2020). Tricks and tracks in removal of emerging contaminants from the wastewater through hybrid treatment systems: A review. *The Science of the Total Environment*, *738*, 140320. DOI: 10.1016/j.scitotenv.2020.140320 PMID: 32806367

Dias, R. R., Deprá, M. C., De Menezes, C. R., Zepka, L. Q., & Jacob-Lopes, E. (2025). Microalgae Cultivation in Wastewater: How Realistic Is This Approach for Value-Added Product Production? *Processes (Basel, Switzerland)*, *13*(7), 2052. DOI: 10.3390/pr13072052

Diep, P., Mahadevan, R., & Yakunin, A. F. (2018). Heavy Metal Removal by Bioaccumulation Using Genetically Engineered Microorganisms. *Frontiers in Bioengineering and Biotechnology*, *6*, 157. DOI: 10.3389/fbioe.2018.00157 PMID: 30420950

Dilokekunakul, W., Teerachawanwong, P., Klomkliang, N., Supasitmongkol, S., & Chaemchuen, S. (2020). Effects of nitrogen and oxygen functional groups and pore width of activated carbon on carbon dioxide capture: Temperature dependence. *Chemical Engineering Journal*, *389*, 124413. DOI: 10.1016/j.cej.2020.124413

Ding, H., Luo, C., Li, Y., Li, Q., & Dong, Y. (2024). Impact of *Bacillus subtilis* and *Pseudomonas fluorescens* beneficial bacterial agents on soil-borne diseases, growth, and economics of continuous cropping of flue-cured tobacco. *Crop Protection (Guildford, Surrey)*, *177*, 106556. DOI: 10.1016/j.cropro.2023.106556

Dixon, C., & Wilken, L. R. (2018). Green microalgae biomolecule separations and recovery. *Bioresources and Bioprocessing*, *5*(1), 9. DOI: 10.1186/s40643-018-0199-3

Djaoudi, K., Van Wambeke, F., Barani, A., Bhairy, N., Chevaillier, S., Desboeufs, K., Nunige, S., Labiadh, M., Henry des Tureaux, T., Lefèvre, D., Nouara, A., Panagiotopoulos, C., Tedetti, M., & Pulido-Villena, E. (2020). Potential bioavailability of organic matter from atmospheric particles to marine heterotrophic bacteria. *Biogeosciences*, *17*(24), 6271–6285. DOI: 10.5194/bg-17-6271-2020

do Carmo, T. S., Moreira, F. S., Cabral, B. V., Dantas, R. C. C., de Resende, M. M., Cardoso, V. L., & Ribeiro, E. J. (2019). Phosphorus Recovery from Phosphate Rocks Using Phosphate-Solubilizing Bacteria. *Geomicrobiology Journal*, *36*(3), 195–203. DOI: 10.1080/01490451.2018.1534901

Doan, L., Pham, N. V., Nguyen, P. B., Le, Q. N., Phung, T. K., & Ngo, T. T. (2026). Marine Algae as Sustainable Platforms for the Green Synthesis of Metal Nanoparticles. *ACS Omega*, *11*(10), 15598–15616. DOI: 10.1021/acsomega.5c12334 PMID: 41867616

Dodangodage, C. A., Gamage, G. N., Wijesekara, I. A., Kasturiarachchi, J. C., Perera, T. A., Rajapakshe, D., & Halwatura, R. U. (2026). Valorization of Canteen Wastewater Through Optimized Spirulina Platensis Cultivation for Enhanced Carotenoid Production and Nutrient Removal. *Phycology*, *6*(1), 15. DOI: 10.3390/phycology6010015

Dodangodage, C. A., Kasturiarachchi, J. C., Wijesekara, I. A., Perera, T. A., Rajapakshe, D., & Halwatura, R. (2026). Integrated Microalgal–Aquaponic Systems for Enhanced Water Treatment and Food Security: A Critical Review of Recent Advances in Process Integration and Resource Recovery. *Phycology*, *6*(1), 14. DOI: 10.3390/phycology6010014

Dolganyuk, V., Belova, D., Babich, O., Prosekov, A., Ivanova, S., Katserov, D., Patyukov, N., & Sukhikh, S. (2020). Microalgae: A Promising Source of Valuable Bioproducts. *Biomolecules*, *10*(8), 1153. DOI: 10.3390/biom10081153 PMID: 32781745

Dongyu, C., Lingchao, K., Yi, W., Yuanqing, Z., & Chuanlun, Z. (2022). In situ identification of environmental microorganisms with Raman. *Environmental Science and Ecotechnology*, *11*, 100187. DOI: 10.1016/j.ese.2022.100187 PMID: 36158754

Dubinin, M. M. (1947). The equation of the characteristic curve of activated charcoal. *Doklady Akademii Nauk SSSR*, *55*, 327–329.

Dudhagara, D. R., Javia, B. M., & Vala, A. K. (2023). Exploiting marine fungi in the removal of hazardous pollutants and biomass valorisation. In *Marine organisms: A solution to environmental pollution? Uses in bioremediation and in biorefinery* (pp. 117–146). Springer. DOI: 10.1007/978-3-031-17226-7_7

Dutta, N., Kundu, P., Lee, J. T. E., & Bhattacharya, S. (2023). Implementation and optimization of algal biomass in value-added products recovery: A step towards algae-based green economy. *Hydrobiology*, *2*(2), 326–346. DOI: 10.3390/hydrobiology2020021

Dziejarski, B., Krzyżyńska, R., & Andersson, K. (2023). Current status of carbon capture, utilization, and storage technologies in the global economy: A survey of technical assessment. *Fuel*, *342*, 127776. Advance online publication. DOI: 10.1016/j.fuel.2023.127776

Echeverri, D. O., Beltrán, V. V., Mesa, A. P., Ríos, L. A., & Echeverri, E. O. (2026). Financial Potential of Life Cycle Assessment of Production of Biofuels via Hydrothermal Liquefaction Using Microalgae in Cement Industry. *Energies*, *19*(5), 1153. DOI: 10.3390/en19051153

Edelstein, M., & Ben-Hur, M. (2018). Heavy metals and metalloids: Sources, risks and strategies to reduce their accumulation in horticultural crops. *Scientia Horticulturae*, *234*, 431–444. DOI: 10.1016/j.scienta.2017.12.039

Eheneden, I., Wang, R., & Zhao, J. (2023). Antibiotic removal by microalgae-bacteria consortium: Metabolic pathways and microbial responses. *The Science of the Total Environment*, *891*, 164489. DOI: 10.1016/j.scitotenv.2023.164489 PMID: 37279806

El Amri, R., Zahnoune, R., Boudouch, O., Hamdouch, M., Hattab, I., Hasib, A., Chakraborty, S., & Elkacmi, R. (2026). Microalgae as a Natural Resource: Economic Potential and Applications. In *Wastewater Treatment* (pp. 251–262). CRC Press. DOI: 10.1201/9781003624905-12

El Messaoudi, N., Georgin, J., Franco, D. S., Yılmazoğlu, M., Temur, B., Miyah, Y., Benjelloun, M., & Al-Msiedeen, A. M. (2025). Composites based on porphyran as an adsorbent for environmentally friendly water treatment. *Reactive & Functional Polymers*, •••, 106616.

El-Naggar, N. E.-A., Hamouda, R. A., Saddiq, A. A., & Alkinani, M. H. (2021). Simultaneous bioremediation of cationic copper ions and anionic methyl orange azo dye by brown marine alga *Fucus vesiculosus*. *Scientific Reports*, *11*(1), 3555. DOI: 10.1038/s41598-021-82827-8 PMID: 33574404

El-Sheekh, M. M., El-Kassas, H. Y., & Ali, S. S. (2025). Microalgae-based bioremediation of refractory pollutants: An approach towards environmental sustainability. *Microbial Cell Factories*, *24*(1), 19. DOI: 10.1186/s12934-024-02638-0 PMID: 39810167

Emparan, Q., Harun, R., & Danquah, M. (2019). Role of phycoremediation for nutrient removal from wastewaters: A review. *Applied Ecology and Environmental Research*, *17*(1), 889–915. DOI: 10.15666/aeer/1701_889915

Energy Transitions Commission. (2022). Carbon capture, utilisation and storage in the energy transition: Vital but limited. Energy Transitions Commission. https://www.energy-transitions.org

Escapa, C., Coimbra, R. N., Paniagua, S., García, A. I., & Otero, M. (2016). Comparative assessment of diclofenac removal from water by different microalgae strains. *Algal Research*, *18*, 127–134. DOI: 10.1016/j.algal.2016.06.008

Estrada-Arriaga, E. B., & Mijaylova, P. N. (2011). Influence of operational parameters (sludge retention time and hydraulic residence time) on the removal of estrogens by membrane bioreactor. *Environmental Science and Pollution Research International, 18*(7), 1121–1128. DOI: 10.1007/s11356-011-0461-0 PMID: 21331588

EswaryDevi, T., Parthiban, R., Arun, J., & Gopinath, K. P.EswaryDevi. (2023). Processing of marine microalgae biomass via hydrothermal liquefaction for bio-oil production: Study on algae cultivation, harvesting, and process parameters. *Biomass Conversion and Biorefinery, 13*(16), 15281–15294. DOI: 10.1007/s13399-022-03446-5

Ethiraj, S., & Samuel, M. S. (2024). A comprehensive review of the challenges and opportunities in microalgae-based wastewater treatment for eliminating organic, inorganic, and emerging pollutants. *Biocatalysis and Agricultural Biotechnology, 60*, 103316. DOI: 10.1016/j.bcab.2024.103316

Ezhumalai, G., Arun, M., Manavalan, A., Rajkumar, R., & Heese, K. (2024). A holistic approach to circular bioeconomy through the sustainable utilization of microalgal biomass for biofuel and other value-added products. *Microbial Ecology, 87*(1), 61. DOI: 10.1007/s00248-024-02376-1 PMID: 38662080

Fan, H., Wang, Y., Wang, Y., & Coyte, P. C. (2022). The impact of environmental pollution on the physical health of middle aged and older adults in China. *Environmental Science and Pollution Research International, 29*(3), 4219–4231. DOI: 10.1007/s11356-021-15832-z PMID: 34403062

Farazaki, M., & Gikas, P. (2019). Nitrification-denitrification of municipal wastewater without recirculation, using encapsulated microorganisms. *Journal of Environmental Management, 242*, 258–265. DOI: 10.1016/j.jenvman.2019.04.054 PMID: 31048231

Farhadian, M., Vachelard, C., Duchez, D., & Larroche, C. (2008). In situ bioremediation of monoaromatic pollutants in groundwater: A review. *Bioresource Technology, 99*(14), 5296–5308. DOI: 10.1016/j.biortech.2007.10.025 PMID: 18054222

Farina, I., Scotto di Uccio, A., Moretti, A., & Colangelo, F. (2026). A review on PFAS mitigation in wastewater: Integrating removal technologies, challenges and innovative solutions. *Reviews in Environmental Science and Biotechnology, 25*(1), 15. DOI: 10.1007/s11157-026-09766-6

Farooq, W., Suh, W. I., Park, M. S., & Yang, J.-W. (2015). Water use and its recycling in microalgae cultivation for biofuel application. *Bioresource Technology, 184*, 73–81. DOI: 10.1016/j.biortech.2014.10.140 PMID: 25465788

Faruque, M. O., Mohammed, T., Farooq, W., Nzila, A., Hossain, M. M., & Razzak, S. A. (2026). Performance of Chlorella sorokiniana in bioremediation of organic pollutants from produced water under heavy metal stress. *Journal of Water Process Engineering*, *83*, 109736. DOI: 10.1016/j.jwpe.2026.109736

Fedoroff, N. V., Battisti, D. S., Beachy, R. N., Cooper, P. J. M., Fischhoff, D. A., Hodges, C. N., Knauf, V. C., Lobell, D., Mazur, B. J., Molden, D., Reynolds, M. P., Ronald, P. C., Rosegrant, M. W., Sanchez, P. A., Vonshak, A., & Zhu, J. K. (2010). Radically rethinking agriculture for the 21st century. *Science*, *327*(5967), 833–834. DOI: 10.1126/science.1186834 PMID: 20150494

Fernández-Reiriz, M. J., Perez-Camacho, A., Ferreiro, M. J., Blanco, J., Planas, M., Campos, M. J., & Labarta, U. (1989). Biomass production and variation in the biochemical profile (total protein, carbohydrates, RNA, lipids and fatty acids) of seven species of marine microalgae. *Aquaculture (Amsterdam, Netherlands)*, *83*(1), 17–37. DOI: 10.1016/0044-8486(89)90057-4

Firincă, C., Zamfir, L.-G., Constantin, M., Răut, I., Jecu, M.-L., Doni, M., Gurban, A.-M., & esan, T. E. (2025). Innovative approaches and evolving strategies in heavy metal bioremediation: Current limitations and future opportunities. *Journal of Xenobiotics*, *15*(3), 63. DOI: 10.3390/jox15030063 PMID: 40407527

Fomina, M., & Gadd, G. M. (2014). Biosorption: Current perspectives on concept, definition and application. *Bioresource Technology*, *160*, 3–14. DOI: 10.1016/j.biortech.2013.12.102 PMID: 24468322

Foo, K. Y., & Hameed, B. H. (2010). Insights into the modeling of adsorption isotherm systems. *Chemical Engineering Journal*, *156*(1), 2–10. DOI: 10.1016/j.cej.2009.09.013

Formosa-Dague, C., Speziale, P., Foster, T. J., Geoghegan, J. A., & Dufrêne, Y. F. (2016). Zinc-dependent mechanical properties of *Staphylococcus aureus* biofilm-forming surface protein SasG. *Proceedings of the National Academy of Sciences of the United States of America*, *113*(2), 410–415. DOI: 10.1073/pnas.1519265113

Franchi, E., Rolli, E., Marasco, R., Agazzi, G., Borin, S., Cosmina, P., Pedron, F., Rosellini, I., Barbafieri, M., & Petruzzelli, G. (2017). Phytoremediation of a multi-contaminated soil: Mercury and arsenic phytoextraction assisted by mobilizing agent and plant growth-promoting bacteria. *Journal of Soils and Sediments*, *17*(5), 1224–1236. DOI: 10.1007/s11368-015-1346-5

Freundlich, H. (1906/1907), “Über die Adsorption in Lösungen,” Zeitschrift für Physikalische Chemie.

Fu, F., & Wang, Q. (2011). Removal of heavy metal ions from wastewaters: A review. *Journal of Environmental Management, 92*(3), 407–418. DOI: 10.1016/j.jenvman.2010.11.011 PMID: 21138785

Gadd, G. M. (2009). Biosorption: Critical review of scientific rationale, environmental importance and significance for pollution treatment. *Journal of Chemical Technology and Biotechnology, 84*(1), 13–28. DOI: 10.1002/jctb.1999

Gahlout, M., Prajapati, H., Tandel, N., & Patel, Y. (2021). Biosorption: An Eco-Friendly Technology for Pollutant Removal. In Panpatte, D. G., & Jhala, Y. K. (Eds.), *Microbial Rejuvenation of Polluted Environment* (Vol. 26, pp. 207–227). Springer Singapore., DOI: 10.1007/978-981-15-7455-9_9

Gan, W., Deng, X., Liu, Y., Liu, L., Yang, M., & Xu, Y. (2026). Potassium nutrition regulates growth, nutrient-carbon interactions and lignification during autumn in Tsoongiodendron odorum seedlings. *BMC Plant Biology, 26*(1), 695. DOI: 10.1186/s12870-026-08497-8 PMID: 41803723

Gan, Y., Ji, X., & Yang, R. (2024). Metagenomic profiling of antibiotic resistance genes/bacteria removal in urban water: Algal-bacterial consortium treatment system. *Bioresource Technology, 404*, 130905. DOI: 10.1016/j.biortech.2024.130905 PMID: 38801952

Gao, T., Li, Q., Wang, X., Liu, Y., Zhao, Z., Han, L., & Chen, X. (2022). Characteristics and diversity of microbial communities in lead-zinc tailings under heavy metal stress in north-west China. *Letters in Applied Microbiology, 74*(2), 277–287. DOI: 10.1111/lam.13608 PMID: 34822179

Garcia Alba, L., Torri, C., Samorì, C., Van Der Spek, J., Fabbri, D., Kersten, S. R. A., & Brilman, D. W. F. (2012). Hydrothermal Treatment (HTT) of Microalgae: Evaluation of the Process As Conversion Method in an Algae Biorefinery Concept. *Energy & Fuels, 26*(1), 642–657. DOI: 10.1021/ef201415s

Garfí, M., Flores, L., & Ferrer, I. (2017). Life cycle assessment of wastewater treatment systems for small communities: Activated sludge, constructed wetlands and high-rate algal ponds. *Journal of Cleaner Production, 161*, 211–219. DOI: 10.1016/j.jclepro.2017.05.116

Garrido-Baserba, M., Sedlak, D. L., Molinos-Senante, M., Barnosell, I., Schraa, O., Rosso, D., Verdaguer, M., & Poch, M. (2024). Using water and wastewater decentralization to enhance the resilience and sustainability of cities. *Nature Water, 2*(10), 953–974. DOI: 10.1038/s44221-024-00303-9

Gautam, P., Salunke, D., Lad, D., & Gautam, A. (2025). Convergent synergy of carbon capture within the circular economy paradigm: A nexus for realizing multifaceted sustainable development goals. *Carbon Research*, *4*(1), 3. DOI: 10.1007/s44246-024-00178-1

Gayo-Peláez, J. I., Oatley-Radcliffe, D. L., Silkina, A., & Barron, A. R. (2026). Cost-effective and sustainable microalgae cultivation: A low-cost artificially integrated LED photobioreactor ensuring high-quality algal biomass production from industrial CO2 flue gas in a high latitude country. *Cleaner & Circular Bioeconomy*, Article 100209.

Ge, L., Deng, H., Wu, F., & Deng, N. (2009). Microalgae-promoted photodegradation of two endocrine disrupters in aqueous solutions. *Journal of Chemical Technology and Biotechnology*, *84*(3), 331–336. DOI: 10.1002/jctb.2043

Geng, Y., Shaukat, A., Azhar, W., Raza, Q.-U.-A., Tahir, A., Abideen, M. Z. U., Zia, M. A. B., Bashir, M. A., & Rehim, A. (2025). Microalgal biorefineries: A systematic review of technological trade-offs and innovation pathways. *Biotechnology for Biofuels and Bioproducts*, *18*(1), 93. DOI: 10.1186/s13068-025-02694-7 PMID: 40817213

Gentili, F. G., & Fick, J. (2017). Algal cultivation in urban wastewater: An efficient way to reduce pharmaceutical pollutants. *Journal of Applied Phycology*, *29*(1), 255–262. DOI: 10.1007/s10811-016-0950-0

Ghimire, U., Sarpong, G., & Gude, V. G. (2021). Transitioning Wastewater Treatment Plants toward Circular Economy and Energy Sustainability. *ACS Omega*, *6*(18), 11794–11803. DOI: 10.1021/acsomega.0c05827 PMID: 34056333

Gill, J. M., Hussain, S. M., Ali, S., Zahoor, A. F., Alasmari, A., Munir, M., Naeem, E., Amjad, M., Yousaf, Z., & Faisal, M. (2025). Optimizing aquaculture sustainability: Microalgae-based co-culture systems for aquaculture wastewater treatment and pollution reduction. *Algal Research*, *92*, 104430. DOI: 10.1016/j.algal.2025.104430

Global, C. C. S. Institute. (2021). Technology readiness and costs of CCS. https://www.globalccsinstitute.com/resources/publications-reports-research/technology-readiness-and-costs-of-ccs/

Global, C. C. S. Institute. (2023). The global status of CCS: 2023. Global CCS Institute. https://status23.globalccsinstitute.com/

Gnansounou, E., Dauriat, A., Villegas, J., & Panichelli, L. (2009). Life cycle assessment of biofuels: Energy and greenhouse gas balances. *Bioresource Technology*, *100*(21), 4919–4930. DOI: 10.1016/j.biortech.2009.05.067 PMID: 19553106

Gomes, A. R., Matos, L. P., Silva, A. M., Guimaraes, A. T. B., Luz, T. M., Brito, R. R., Lima Rodrigues, A. S., Vaz, B. G., Oliveira, J. A., Soares, W. R., Menezes, I. P. P., & Malafaia, G. (2025). Complex interactions and ecotoxicological consequences of tannery effluents on aquatic life: Differential responses in scinax fuscovarius tadpoles and poecilia reticulata fish. *Water, Air, and Soil Pollution, 236*(2), 105. DOI: 10.1007/s11270-024-07710-4

Gonçalves, A. L., Pires, J. C. M., & Simões, M. (2017). A review on the use of microalgal consortia for wastewater treatment. *Algal Research, 24*, 403–415. DOI: 10.1016/j.algal.2016.11.008

Gonçalves, J., Freitas, J., Fernandes, I., & Silva, P. (2023). Microalgae as Biofertilizers: A Sustainable Way to Improve Soil Fertility and Plant Growth. *Sustainability (Basel), 15*(16), 12413. DOI: 10.3390/su151612413

Gondi, R., Kavitha, S., Kannah, R. Y., Karthikeyan, O. P., Kumar, G., & Tyagi, V. K.. (2022). Algal-based system for removal of emerging pollutants from wastewater: A review. *Bioresource Technology, 344*, 126245. DOI: 10.1016/j.biortech.2021.126245 PMID: 34743994

González Fernández, L. A., Medellín Castillo, N. A., Sánchez Polo, M., Navarro Frómeta, A. E., & Vilasó Cadre, J. E. (2025). Algal-based carbonaceous materials for environmental remediation: Advances in wastewater treatment, carbon sequestration, and biofuel applications. *Processes (Basel, Switzerland), 13*(2), 556. DOI: 10.3390/pr13020556

González-Camejo, J., Aparicio, S., Pachés, M., Borrás, L., & Seco, A. (2022). Comprehensive assessment of the microalgae-nitrifying bacteria competition in microalgae-based wastewater treatment systems: Relevant factors, evaluation methods and control strategies. *Algal Research, 61*, 102563. DOI: 10.1016/j.algal.2021.102563

González-Camejo, J., Viruela, A., Ruano, M. V., Barat, R., Seco, A., & Ferrer, J. (2019). Effect of light intensity, light duration and photoperiods in the performance of an outdoor photobioreactor for urban wastewater treatment. *Algal Research, 40*, 101511. DOI: 10.1016/j.algal.2019.101511

Gouda, S. A., & Taha, A. (2023). Biosorption of heavy metals as a new alternative method for wastewater treatment. *RE:view*.

Goutam Mukherjee, A., Ramesh Wanjari, U., Eladl, M. A., El-Sherbiny, M., Elsherbini, D. M. A., Sukumar, A., Kannampuzha, S., Ravichandran, M., Renu, K., Vellingiri, B., Kandasamy, S., & Valsala Gopalakrishnan, A. (2022). Mixed contaminants: Occurrence, interactions, toxicity, detection, and remediation. *Molecules (Basel, Switzerland), 27*(8), 2577. DOI: 10.3390/molecules27082577 PMID: 35458775

Grobbelaar, J. U., Soeder, C. J., Groeneweg, J., Stengel, E., & Hartig, P. (1988). Rates of biogenic oxygen production in mass cultures of microalgae, absorption of atmospheric oxygen and oxygen availability for wastewater treatment. *Water Research, 22*(11), 1459–1464. DOI: 10.1016/0043-1354(88)90105-4

Guedes, A. C., Amaro, H. M., Sousa-Pinto, I., & Malcata, F. X. (2019). Algal spent biomass—A pool of applications. In *Biofuels from Algae* (pp. 397–433). Elsevier., DOI: 10.1016/B978-0-444-64192-2.00016-0

Guerrero-Pérez, M. O., & Patience, G. S. (2020). Experimental methods in chemical engineering: Fourier transform infrared spectroscopy—FTIR. *Canadian Journal of Chemical Engineering, 98*(1), 25–33. DOI: 10.1002/cjce.23664

Gultom, S., Zamalloa, C., & Hu, B. (2014). Microalgae harvest through fungal pelletization—Co-culture of *Chlorella vulgaris* and *Aspergillus niger. Energies, 7*(7), 4417–4429. DOI: 10.3390/en7074417

Guo, X., Liu, Y., Xie, T., Li, Y., Liu, H., & Wang, Q. (2025). Impact of ecological restoration on carbon sink function in coastal wetlands: A review. *Water (Basel), 17*(4), 488. DOI: 10.3390/w17040488

Guo, Y., Ouyang, S., Chen, J., Lin, Z., Xiao, C., Liu, D., Chen, G., & Zhang, Q. (2026). Diatomaceous bio-silicon as a sustainable alternative to mineral silicon fertilizers: Impacts on rice yield, soilfertility, and heavy metal dynamics. *Plant and Soil, 519*(1), 451–468. DOI: 10.1007/s11104-025-08121-y

Gupta, D., Gauri, R. K., Tyagi, A., Tyagi, N., Sharma, Y., Doli, K. S., Sharma, H., & Sharma, M. (2026a). Advancements in Design and Development of Lab Scale Photobiorectors for Improved Microalgal Feedstock. *Microbiology Archives, an International Journal.*

Gupta, R., Mishra, N., Mukherjee, T., Mishra, S., & Bhattacharya, S. (2025). Microalgal biorefinery and circular bioeconomy model for wastewater treatment. In *Microbial Biotechnology: Integrated Microbial Engineering for B3 – Bioenergy, Bioremediation, and Bioproducts* (pp. 415–441). Elsevier. DOI: 10.1016/B978-0-443-23796-6.00016-X

Gupta, P., Prabhakaran, R., Soni, N., Singh, S. D. K., Kamaraj, C., Selvan, S. T., Asokan, N., Pamidimarri, S. D., & Velramar, B. (2026b). Utilization of Genetically Engineered RuBisCO Enzyme for Capturing Carbon Dioxide to Address Environmental Challenges-An Overview. *Water, Air, and Soil Pollution, 237*(2), 58. DOI: 10.1007/s11270-025-08639-y

Gupta, R., Kumari, A., Sharma, S., Alzahrani, O. M., Noureldeen, A., & Darwish, H. (2022). Identification, characterization and optimization of phosphate solubilizing rhizobacteria from rice rhizosphere. *Saudi Journal of Biological Sciences, 29*(1), 35–42. DOI: 10.1016/j.sjbs.2021.09.075 PMID: 35002393

Gupta, V. K., & Rastogi, A. (2009). Biosorption of hexavalent chromium by raw and acid-treated green alga Oedogonium hatei from aqueous solutions. *Journal of Hazardous Materials, 163*(1), 396–402. DOI: 10.1016/j.jhazmat.2008.06.104 PMID: 18691812

Habib, S. S., Naz, S., Fazio, F., Cravana, C., Ullah, M., Rind, K. H., Attaullah, S., Filiciotto, F., & Khayyam, K. (2024). Assessment and bioaccumulation of heavy metals in water, fish (wild and farmed) and associated human health risk. *Biological Trace Element Research, 202*(2), 725–735. DOI: 10.1007/s12011-023-03703-2 PMID: 37178449

Haldkar, A., Wali, H., Yadav, M., Dwibedi, V., & George, N. (2026). Transforming agro-waste into ecofriendly nanomaterials for wastewater treatment: Embracing the principles of green chemistry and the circular economy. In *Sustainable Agricultural Waste Management* (pp. 267–300). Elsevier. DOI: 10.1016/B978-0-443-38301-4.00017-5

Hamed, A. (2023). A brief and general overview on diatoms and their applications. *Egyptian Journal of Phycology*. DOI: 10.21608/egyjs.2023.218427.1019

Hamida, R. S., Ali, M. A., Redhwan, A., & Bin-Meferij, M. M. (2020). Cyanobacteria–a promising platform in green nanotechnology: A review on nanoparticles fabrication and their prospective applications. *International Journal of Nanomedicine, 15*, 6033–6066. DOI: 10.2147/IJN.S256134 PMID: 32884261

Hammond, C. R., Hernández, M. S. G., & Loge, F. J. (2025). Microalgal-bacterial aggregates for wastewater treatment: Origins, challenges, and future directions. *Water Environment Research : A Research Publication of the Water Environment Federation, 97*(2), e70018. DOI: 10.1002/wer.70018 PMID: 39900561

Hamouda, A. S., Mahmoud, H. M., El-Belely, E. F., Labena, A., & Husien, S. (2019). Absorption of hexavalent chromium by green micro algae *Chlorella sorokiniana*: Live planktonic cells. *Water Practice & Technology, 14*(3), 515–529. DOI: 10.2166/wpt.2019.034

Han, D., Hu, Z., Li, D., & Tang, R. (2022). Nitrogen Removal of Water and Sediment in Grass Carp Aquaculture Ponds by Mixed Nitrifying and Denitrifying Bacteria and Its Effects on Bacterial Community. *Water (Basel), 14*(12), 1855. DOI: 10.3390/w14121855

Han, J., Zhang, L., Wang, S., Yang, G., Zhao, L., & Pan, K. (2016). Co-culturing bacteria and microalgae in organic carbon containing medium. *Journal of Biological Research (Thessaloniki)*, *23*(1), 8. DOI: 10.1186/s40709-016-0047-6 PMID: 27119075

Hasan, H. A., Muhamad, M. H., Ji, B., Nazairi, N. A., Jiat, K. W., Sim, S. I. S. W. A., & Poh, A. F. M. S. (2023). Revolutionizing wastewater treatment with microalgae: Unveiling resource recovery, mechanisms, challenges, and future possibilities. *Ecological Engineering*, *197*, 107117. DOI: 10.1016/j.ecoleng.2023.107117

Hasanin, M. S., Hashem, A. H., Abd El-Sayed, E. S., & El-Saied, H. (2020). Green eco-friendly bio-deinking of mixed office waste paper using various enzymes from *Rhizopus microsporus* AH3: Efficiency and characteristics. *Cellulose (London, England)*, *27*(8), 4443–4453. DOI: 10.1007/s10570-020-03071-3

Hasan, Z. S., & Enad, H. Y. (2026). Comparative Evaluation of Microalgae-Based Wastewater Treatment Using Closed Photobioreactors and Open Pond Systems for Nutrient and Heavy Metal Removal and Resource Recovery. *Water, Air, and Soil Pollution*, *237*(6), 375. DOI: 10.1007/s11270-025-09070-z

Hayat, S., Hayat, Q., Alyemeni, M. N., Wani, A. S., Pichtel, J., & Ahmad, A. (2012). Role of proline under changing environments: A review. *Plant Signaling & Behavior*, *7*(11), 1456–1466. DOI: 10.4161/psb.21949 PMID: 22951402

Hedayatkhah, A., Cretoiu, M. S., Emtiazi, G., Stal, L. J., & Bolhuis, H. (2018). Bioremediation of chromium contaminated water by diatoms with concomitant lipid accumulation for biofuel production. *Journal of Environmental Management*, *227*, 313–320. DOI: 10.1016/j.jenvman.2018.09.011

He, J., & Chen, J. P. (2014). A comprehensive review on biosorption of heavy metals by algal biomass: Materials, performances, chemistry, and modeling simulation tools. *Bioresource Technology*, *160*, 67–78. DOI: 10.1016/j.biortech.2014.01.068 PMID: 24630371

Hejna, M., Kapuścińska, D., & Aksmann, A. (2022). Pharmaceuticals in the Aquatic Environment: A Review on Eco-Toxicology and the Remediation Potential of Algae. *International Journal of Environmental Research and Public Health*, *19*(13), 7717. DOI: 10.3390/ijerph19137717 PMID: 35805373

Hena, S., Gutierrez, L., & Croue, J. P. (2020). Removal of metronidazole from aqueous media by *C. vulgaris*. *Journal of Hazardous Materials*, *384*, 121400. DOI: 10.1016/j.jhazmat.2019.121400 PMID: 31624001

Hena, S., Gutierrez, L., & Croue, J. P. (2021). Removal of pharmaceutical and personal care products (PPCPs) from wastewater using microalgae: A review. *Journal of Hazardous Materials, 403*, 124041. DOI: 10.1016/j.jhazmat.2020.124041 PMID: 33265054

Henze, M. (2008). *Biological wastewater treatment: Principles, modelling and design.*

Hepburn, C., Adlen, E., Beddington, J., Carter, E. A., Fuss, S., Mac Dowell, N., Minx, J. C., Smith, P., & Williams, C. K. (2019). The technological and economic prospects for CO2 utilization and removal. *Nature, 575*(7781), 87–97. DOI: 10.1038/s41586-019-1681-6 PMID: 31695213

Hernandez, J.-P., de-Bashan, L. E., Rodriguez, D. J., Rodriguez, Y., & Bashan, Y. (2009). Growth promotion of the freshwater microalga *Chlorella vulgaris* by the nitrogen-fixing, plant growth-promoting bacterium *Bacillus pumilus* from arid zone soils. *European Journal of Soil Biology* [PGPB]. *European Journal of Soil Biology, 45*(1), 88–93. DOI: 10.1016/j.ejsobi.2008.08.004

Hiloidhari, M., Baruah, D. C., Singh, A., Kataki, S., Medhi, K., Kumar, S., Ramachandra, T. V., Jenkins, B. M., & Thakur, I. S. (2017). Emerging role of geographical information system (GIS), life cycle assessment (LCA) and spatial LCA (GIS-LCA) in sustainable bioenergy planning. *Bioresource Technology, 242*, 218–226. DOI: 10.1016/j.biortech.2017.03.079 PMID: 28343863

Hiloidhari, M., Das, D., & Baruah, D. C. (2017). Bioenergy potential from crop residue biomass in India. *Renewable & Sustainable Energy Reviews, 32*, 504–512. DOI: 10.1016/j.rser.2014.01.025

Hlongwane, G. N., Sekoai, P. T., Meyyappan, M., & Moothi, K. (2019). Simultaneous removal of pollutants from water using nanoparticles: A shift from single pollutant control to multiple pollutant control. *The Science of the Total Environment, 656*, 808–833. DOI: 10.1016/j.scitotenv.2018.11.257 PMID: 30530150

Ho, A., Di Lonardo, D. P., & Bodelier, P. L. E. (2017). Revisiting life strategy concepts in environmental microbial ecology. *FEMS Microbiology Ecology, 93*(3), fix006. Advance online publication. DOI: 10.1093/femsec/fix006 PMID: 28115400

Hom-Diaz, A., Llorca, M., Rodríguez-Mozaz, S., Vicent, T., Barceló, D., & Blánquez, P. (2015). Microalgae cultivation on wastewater digestate: B-estradiol and 17α-ethynylestradiol degradation and transformation products identification. *Journal of Environmental Management, 155*, 106–113. DOI: 10.1016/j.jenvman.2015.03.003 PMID: 25785785

Hoque, M. M., Iannelli, V., Padula, F., Radice, R. P., Saha, B. K., Martelli, G., Scopa, A., & Drosos, M. (2025). Microalgae: Green engines for achieving carbon sequestration, circular economy, and environmental sustainability—A review based on last ten years of research. *Bioengineering (Basel, Switzerland)*, *12*(9), 909. DOI: 10.3390/bioengineering12090909 PMID: 41007154

Hossain, Md. S., Ahamed Anik, Md. S., Rafin, F.-A.-, Alam, S., Abir, A. A., Mourshed, M., & Ahmed, Md. R. (2026). Microalgae for sustainable energy: Techno-economic and life cycle insights into biodiesel and biohydrogen production. *Energy Conversion and Management: X*, *30*, 101840. DOI: 10.1016/j.ecmx.2026.101840

Hosseinzadeh, A., Zhou, J. L., Li, X., Afsari, M., & Altaee, A. (2022). Techno-economic and environmental impact assessment of hydrogen production processes using bio-waste as renewable energy resource. *Renewable & Sustainable Energy Reviews*, *156*, 111991. DOI: 10.1016/j.rser.2021.111991

Hou, D., O'Connor, D., Igalavithana, A. D., Alessi, D. S., Luo, J., Tsang, D. C., Sparks, D. L., Yamauchi, Y., Rinklebe, J., & Ok, Y. S. (2020). Metal contamination and bioremediation of agricultural soils for food safety and sustainability. *Nature Reviews. Earth & Environment*, *1*(7), 366–381. DOI: 10.1038/s43017-020-0061-y

Ho, Y. S., & McKay, G. (1999). Pseudo-second order model for sorption processes. *Process Biochemistry (Barking, London, England)*, *34*(5), 451–465. DOI: 10.1016/S0032-9592(98)00112-5

Huang, L., Jin, Y., Zhou, D., Liu, L., Huang, S., Zhao, Y., & Chen, Y. (2022). A Review of the Role of Extracellular Polymeric Substances (EPS) in Wastewater Treatment Systems. *International Journal of Environmental Research and Public Health*, *19*(19), 12191. DOI: 10.3390/ijerph191912191 PMID: 36231490

Huang, Y., Yu, J., Huang, X., Sun, J., Tang, W., Wei, M., Yang, G., & Fu, Y. (2025). Algal-bacterial symbiotic system for treatment of heavy metal containing wastewater: Performance, mechanisms and applications. npj. *NPJ Clean Water*, *8*(1), 86. DOI: 10.1038/s41545-025-00475-1

Hunter, R. J. (1981). *Zeta potential in colloid science: Principles and applications*. Academic Press.

Husain, R., Sharma, S., Khan, S., & Mishra, A. (2022). Microbial bioremediation of heavy metals by marine bacteria. In *Development in wastewater treatment research and processes* (pp. 177–203). Elsevier. DOI: 10.1016/B978-0-323-85839-7.00014-1

Hussain, A., Kumari, R., Sachan, S. G., & Sachan, A. (2021). Biological wastewater treatment technology: Advancement and drawbacks. In *Microbial Ecology of Wastewater Treatment Plants* (pp. 175–192). Elsevier., DOI: 10.1016/B978-0-12-822503-5.00002-3

Huy, M., Kumar, G., Kim, H., & Kim, S. (2018). Photoautotrophic cultivation of mixed microalgae consortia using various organic waste streams towards remediation and resource recovery. *Bioresource Technology*, *247*, 576–581. DOI: 10.1016/j.biortech.2017.09.108 PMID: 28982087

Iakovidou, G., Itziou, A., Tsiotsias, A., Lakioti, E., Samaras, P., Tsanaktsidis, C., & Karayannis, V. (2024). Application of microalgae to wastewater bioremediation, with CO2 biomitigation, health product and biofuel development, and environmental biomonitoring. *Applied Sciences (Basel, Switzerland)*, *14*(15), 6727. DOI: 10.3390/app14156727

Ibrahim, M. A., MacAdam, J., Autin, O., & Jefferson, B. (2014). Evaluating the impact of LED bulb development on the economic viability of ultraviolet technology for disinfection. *Environmental Technology*, *35*(4), 400–406. DOI: 10.1080/09593330.2013.829858 PMID: 24600880

Ibuot, A., Webster, R. E., Williams, L. E., & Pittman, J. K. (2020). Increased metal tolerance and bioaccumulation of zinc and cadmium in *Chlamydomonas reinhardtii* expressing a AtHMA4 C-terminal domain protein. *Biotechnology and Bioengineering*, *117*(10), 2996–3005. DOI: 10.1002/bit.27476 PMID: 32579250

IEA. (2023). *Carbon Capture, Utilisation and Storage*. International Energy Agency.

Ighalo, J. O., Dulta, K., Kurniawan, S. B., Omoarukhe, F. O., Ewuzie, U., Eshiemogie, S. O., Ojo, A. U., & Abdullah, S. R. S. (2022). Progress in microalgae application for CO2 sequestration. *Cleaner Chemical Engineering*, *3*, 100044. Advance online publication. DOI: 10.1016/j.clce.2022.100044

Ijaz, S., & Ahmed, M. (n.d.). Harnessing Microalgae for Wastewater Treatment: Optimizing Cultivation, Pollutant Removal, and Photobioreactor Technologies for a Sustainable Future. In *Green Technologies for Wastewater Treatment and Bioenergy Production* (pp. 64–109).

Iman Shayan, S., Agblevor, F. A., Bertin, L., & Sims, R. C. (2016). Hydraulic retention time effects on wastewater nutrient removal and bioproduct production via rotating algal biofilm reactor. *Bioresource Technology*, *211*, 527–533. DOI: 10.1016/j.biortech.2016.03.104 PMID: 27038261

Imwene, K. O., Ngumba, E., & Kairigo, P. K. (2022). Emerging technologies for enhanced removal of residual antibiotics from source-separated urine and wastewaters: A review. *Journal of Environmental Management, 322*, 116065. DOI: 10.1016/j.jenvman.2022.116065 PMID: 36063692

International Energy Agency. (2020). CCUS in clean energy transitions. IEA. https://www.iea.org/reports/ccus-in-clean-energy-transitions

International Energy Agency. (2023a). *CCUS policies and business models: Building a commercial market.* OECD., DOI: 10.1787/6ef05538-

International Energy Agency. (2023b). Section 45 Q credit for carbon oxide sequestration. IEA. https://www.iea.org/policies/4986-section-45q-credit-for-carbon-oxide-sequestration

IPCC. (2021). [The Physical Science Basis. Contribution of Working Group I to the Sixth Assessment Report.]. *Climatic Change*, •••, 2021.

Iqbal, K., Sohail, M., Hussain Rind, K., & Habib, S. S. (2025). Agrochemical contamination and fish health: Eco-toxicological impacts and mitigation strategies. *Chemistry and Ecology, 41*(7), 959–993. DOI: 10.1080/02757540.2025.2508993

IRENA. (2021). *Renewable Power Generation Costs in 2020.* International Renewable Energy Agency.

Iribarren, D., Peters, J. F., & Dufour, J. (2012). Life cycle assessment of transportation fuels from biomass pyrolysis. *Fuel, 97*, 812–821. DOI: 10.1016/j.fuel.2012.02.053

Issaka, E., Danso-Boateng, E., Fazal, A., Melville, L., & Adams, M. (2025). Algae biomass for energy generation: A comprehensive review of recent advances, mechanisms and bottlenecks. *Sustainable & Green Materials, 1*(2–4), 32–94. DOI: 10.1080/29965292.2025.2515823

James, S. N., & Vijayanandan, A. (2023). Recent advances in simultaneous nitrification and denitrification for nitrogen and micropollutant removal: A review. *Biodegradation, 34*(2), 103–123. DOI: 10.1007/s10532-023-10015-8 PMID: 36899211

Jayabal, R. (2025). Microalgae-bacteria systems in wastewater treatment and biofuel production: Challenges, advances, and future perspectives. *Algal Research, 91*, 104228. DOI: 10.1016/j.algal.2025.104228

Jeswani, H. K., Chilvers, A., & Azapagic, A. (2020). Environmental sustainability of biofuels: A review. Proceedings of the Royal Society A: Mathematical, Physical and Engineering Sciences, 476, Article 20200351. DOI: 10.1098/rspa.2020.0351

Jia, H., & Yuan, Q. (2018). Ammonium removal using algae bacteria consortia: The effect of ammonium concentration, algae biomass, and light. *Biodegradation, 29*(2), 105–115. DOI: 10.1007/s10532-017-9816-7 PMID: 29236193

Jiang, J.-C., Feng, Z.-Y., Meng, L.-Y., & Jin, B. (2026). Recent advances in pretreatment technologies for the analysis of per-and polyfluoroalkyl substances. *Analytical and Bioanalytical Chemistry, 418*(7), 1–23. DOI: 10.1007/s00216-026-06317-4 PMID: 41663795

Jiang, R., Qin, L., Feng, S., Huang, D., Wang, Z., & Zhu, S. (2021). The joint effect of ammonium and pH on the growth of *Chlorella vulgaris* and ammonium removal in artificial liquid digestate. *Bioresource Technology, 325*, 124690. DOI: 10.1016/j.biortech.2021.124690 PMID: 33465643

Jiang, Y., Chen, S., Hu, B., Zhou, Y., Liang, Z., Jia, X., & Shi, Z. (2020). A comprehensive framework for assessing the impact of potential agricultural pollution on grain security and human health in economically developed areas. *Environmental Pollution, 263*, 114653. DOI: 10.1016/j.envpol.2020.114653

Ji, H., Zhu, Y., Duan, J., Liu, W., & Zhao, D. (2019). Reductive immobilization and long-term remobilization of radioactive pertechnetate using bio-macromolecules stabilized zero valent iron nanoparticles. *Chinese Chemical Letters, 30*(12), 2163–2168. DOI: 10.1016/j.cclet.2019.06.004

Ji, L., Song, W., Wei, D., Jiang, D., Cai, L., Wang, Y., Guo, J., & Zhang, H. (2019). Modified mussel shell powder for microalgae immobilization to remove N and P from eutrophic wastewater. *Bioresource Technology, 284*, 36–42. DOI: 10.1016/j.biortech.2019.03.112

Ji, M.-K., Abou-Shanab, R. A., Kim, S.-H., Salama, E.-S., Lee, S.-H., Kabra, A. N., Lee, Y.-S., Hong, S., & Jeon, B.-H. (2013). Cultivation of microalgae species in tertiary municipal wastewater supplemented with CO2 for nutrient removal and biomass production. *Ecological Engineering, 58*, 142–148. DOI: 10.1016/j.ecoleng.2013.06.020

Jin, D., Zhang, X., Zhou, L., Zhang, X., & Wu, P. (2024). Emerging applications and mechanisms of algal-bacterial symbiosis on sustainable wastewater treatment: A comprehensive review. *Journal of Water Process Engineering, 65*, 105912. DOI: 10.1016/j.jwpe.2024.105912

Joda, J. F., Raji, R. O., Nmadu, M. E., & Oyewole, O. A. (2026). The Interactions of Emergent Contaminants (ECs) with Soil Microbiome. *Ecotoxicology (London, England), 35*(4), 71. DOI: 10.1007/s10646-026-03058-z PMID: 41762294

Johnson, C. A., & Amarasekare, P. (2013). Competition for benefits can promote the persistence of mutualistic interactions. *Journal of Theoretical Biology*, *328*, 54–64. DOI: 10.1016/j.jtbi.2013.03.016 PMID: 23542049

Kaloudas, D., Pavlova, N., & Penchovsky, R. (2021). Phytoremediation of wastewater by microalgae: A review. *Environmental Chemistry Letters*, *19*(4), 2905–2920. DOI: 10.1007/s10311-021-01203-0

Kanamarlapudi, S. L. R. K., & Muddada, S. (2017). Characterization of Exopolysaccharide Produced by *Streptococcus thermophilus* CC30. *BioMed Research International*, *2017*, 1–11. DOI: 10.1155/2017/4201809 PMID: 28815181

Kannah, R. Y., Kavitha, S., Karthikeyan, O. P., Rene, E. R., Kumar, G., & Banu, J. R. (2021). A review on anaerobic digestion of energy and cost-effective microalgae pretreatment for biogas production. *Bioresource Technology*, *332*, 125055. DOI: 10.1016/j.biortech.2021.125055 PMID: 33813179

Kapahi, M., & Sachdeva, S. (2019). Bioremediation options for heavy metal pollution. *Journal of Health & Pollution*, *9*(24), 191203. DOI: 10.5696/2156-9614-9.24.191203 PMID: 31893164

Kaparapu, J., & Krishna Prasad, M. (2018). Equilibrium, kinetics and thermodynamic studies of cadmium(II) biosorption on *Nannochloropsis oculata*. *Applied Water Science*, *8*(6), 1–9. DOI: 10.1007/s13201-018-0810-y

Kargi, F., & Eker, S. (2005). Removal of 2,4-dichlorophenol and toxicity from synthetic wastewater in a rotating perforated tube biofilm reactor. *Process Biochemistry (Barking, London, England)*, *40*(6), 2105–2111. DOI: 10.1016/j.procbio.2004.07.013

Karimi-Maleh, H., Shafieizadeh, M., Taher, M. A., Opoku, F., Kiarii, E. M., Govender, P. P., Ranjbari, S., Rezapour, M., & Orooji, Y. (2020). The role of magnetite/graphene oxide nano-composite as a high-efficiency adsorbent for removal of phenazopyridine residues from water samples, an experimental/theoretical investigation. *Journal of Molecular Liquids*, *298*, 112040. DOI: 10.1016/j.molliq.2019.112040

Karwowska, M., & Kononiuk, A. (2020). Nitrates/nitrites in food—Risk for nitrosative stress and benefits. *Antioxidants*, *9*(3), 241. DOI: 10.3390/antiox9030241 PMID: 32188080

Kasavan, S., Yusoff, S., Guan, N. C., Zaman, N. Q. K., & Fakri, M. F. R. (2021). Global trends of textile waste research from 2005 to 2020 using bibliometric analysis. *Environmental Science and Pollution Research International*, *28*(33), 44780–44794. DOI: 10.1007/s11356-021-15303-5 PMID: 34235692

Kaste, J. A., Walker, B. J., & Shachar-Hill, Y. (2024). Reaction-diffusion modeling provides insights into biophysical carbon-concentrating mechanisms in land plants. *Plant Physiology*, *196*(2), 1374–1390. DOI: 10.1093/plphys/kiae324 PMID: 38857179

Katam, K., Shimizu, T., Soda, S., & Bhattacharyya, D. (2020). Performance evaluation of two trickling filters removing LAS and caffeine from wastewater: Light reactor (algal-bacterial consortium) vs dark reactor (bacterial consortium). *The Science of the Total Environment*, *707*, 135987. DOI: 10.1016/j.scitotenv.2019.135987 PMID: 31869619

Kato, S., & Kansha, Y. (2024). Comprehensive review of industrial wastewater treatment techniques. *Environmental Science and Pollution Research International*, *31*(39), 51064–51097. DOI: 10.1007/s11356-024-34584-0 PMID: 39107648

Kaur, M., Bhatia, S., Gupta, U., Decker, E., Tak, Y., Bali, M., Gupta, V. K., Dar, R. A., & Bala, S. (2023). Microalgal bioactive metabolites as promising implements in nutraceuticals and pharmaceuticals: Inspiring therapy for health benefits. *Phytochemistry Reviews*, *22*(4), 903–933. DOI: 10.1007/s11101-022-09848-7 PMID: 36686403

Kaur, S., Kaur, S., Thakur, B., & Narwat, N. (2026). Microalgae-based bioremediation of emerging contaminants: Techniques, recent developments, and future perspectives. *Archives of Microbiology*, *208*(1), 29. DOI: 10.1007/s00203-025-04567-x PMID: 41264010

Keerthana Devi, M., Manikandan, S., Oviyapriya, M., Selvaraj, M., Assiri, M. A., Vickram, S., Subbaiya, R., Karmegam, N., Ravindran, B., Chang, S. W., & Awasthi, M. K. (2022). Recent advances in biogas production using agro-industrial waste: A comprehensive review outlook of techno-economic analysis. *Bioresource Technology*, *363*, 127871. Advance online publication. DOI: 10.1016/j.biortech.2022.127871 PMID: 36041677

Kehrein, P., Van Loosdrecht, M., Osseweijer, P., Garfí, M., Dewulf, J., & Posada, J. (2020). A critical review of resource recovery from municipal wastewater treatment plants – market supply potentials, technologies and bottlenecks. *Environmental Science. Water Research & Technology*, *6*(4), 877–910. DOI: 10.1039/C9EW00905A

Khalid, S., Chaudhary, K., Aziz, H., Amin, S., Sipra, H. M., Ansar, S., Rasheed, H., Naeem, M., & Onyeaka, H. (2025). Trends in extracting protein from microalgae *Spirulina platensis*, using innovative extraction techniques: Mechanisms, potentials, and limitations. *Critical Reviews in Food Science and Nutrition*, *65*(22), 4293–4309. DOI: 10.1080/10408398.2024.2386448 PMID: 39096052

Khalili, A., Najafpour, G. D., Amini, G., & Samkhaniyani, F. (2015). Influence of nutrients and LED light intensities on biomass production of microalgae Chlorella vulgaris. *Biotechnology and Bioprocess Engineering; BBE, 20*(2), 284–290. DOI: 10.1007/s12257-013-0845-8

Khanam, Z., Sultana, F. M., & Mushtaq, F. (2023). Environmental pollution control measures and strategies: An overview of recent developments. In *Geospatial analytics for environmental pollution modeling: Analysis, control and management* (pp. 385-414).

Khan, N. A., Khan, A. H., Tiwari, P., Zubair, M., & Naushad, M. (2021). New insights into the integrated application of Fenton-based oxidation processes for the treatment of pharmaceutical wastewater. *Journal of Water Process Engineering, 44*, 102440. DOI: 10.1016/j.jwpe.2021.102440

Khan, N. A., Nancharaiah, Y. V., & Venkata Mohan, S. (2021). Microalgae-based treatment of emerging contaminants in wastewater. *Bioresource Technology, 337*, 125429. DOI: 10.1016/j.biortech.2021.125429

Khatri, P., & Pandit, A. B. (2022). Systematic review of life cycle assessments applied to sugarcane bagasse utilization alternatives. *Biomass and Bioenergy, 158*, 106365. Advance online publication. DOI: 10.1016/j.biombioe.2022.106365

Kim, K.. (2014). Simultaneous removal of organics and nutrients from municipal wastewater using a photosynthetic bacterial and microalgal consortium. *Journal of Microbiology and Biotechnology, 24*(11), 1548–1554.

Kishor, R., Purchase, D., Saratale, G. D., Saratale, R. G., Ferreira, L. F. R., Bilal, M., Chandra, R., & Bharagava, R. N. (2021). Ecotoxicological and health concerns of persistent coloring pollutants of textile industry wastewater and treatment approaches for environmental safety. *Journal of Environmental Chemical Engineering, 9*(2), 105012. DOI: 10.1016/j.jece.2020.105012

Klein-Marcuschamer, D., Oleskowicz-Popiel, P., Simmons, B. A., & Blanch, H. W. (2012). The challenge of enzyme cost in the production of lignocellulosic biofuels. *Biotechnology and Bioengineering, 109*(4), 1083–1087. DOI: 10.1002/bit.24370 PMID: 22095526

Klinthong, W., Yang, Y.-H., Huang, C.-H., & Tan, C.-S. (2015). A review: Microalgae and their applications in CO2 capture and renewable energy. *Aerosol and Air Quality Research, 15*(2), 712–742. DOI: 10.4209/aaqr.2014.11.0299

Koelmans, A. A., Bakir, A., Burton, G. A., & Janssen, C. R. (2016). Microplastic as a Vector for Chemicals in the Aquatic Environment: Critical Review and Model-Supported Reinterpretation of Empirical Studies. *Environmental Science & Technology, 50*(7), 3315–3326. DOI: 10.1021/acs.est.5b06069 PMID: 26946978

Kong, W., Kong, J., Feng, S., Yang, T., Xu, L., Shen, B., Bi, Y., & Lyu, H. (2024). Cultivation of microalgae-bacteria consortium by waste gas-waste water to achieve CO_2 fixation, wastewater purification and bioproducts production. *Biotechnology for Biofuels and Bioproducts, 17*(1), 26. Advance online publication. DOI: 10.1186/s13068-023-02409-w PMID: 38360745

Koupaie, E. H., Eskicioglu, C., & Azizi, S. (2021). Role of microalgae in removal of emerging contaminants from wastewater: Mechanisms and applications. *Chemosphere, 263*, 128305. DOI: 10.1016/j.chemosphere.2020.128305

Koyande, A. K., Chew, K. W., Rambabu, K., Tao, Y., Chu, D.-T., & Show, P.-L. (2019). Microalgae: A potential alternative to health supplementation for humans. *Food Science and Human Wellness, 8*(1), 16–24. DOI: 10.1016/j.fshw.2019.03.001

Kratochvil, D., & Volesky, B. (1998). Advances in the biosorption of heavy metals. *Trends in Biotechnology, 16*(7), 291–300. DOI: 10.1016/S0167-7799(98)01218-9

Kruglova, A., Kråkström, M., Riska, M., Mikola, A., Rantanen, P., Vahala, R., & Kronberg, L. (2016). Comparative study of emerging micropollutants removal by aerobic activated sludge of large laboratory-scale membrane bioreactors and sequencing batch reactors under low-temperature conditions. *Bioresource Technology, 214*, 81–88. DOI: 10.1016/j.biortech.2016.04.037 PMID: 27128192

Krzeminski, P., Tomei, M. C., Karaolia, P., Langenhoff, A., Almeida, C. M. R., Felis, E., Gritten, F., Andersen, H. R., Fernandes, T., Manaia, C. M., Rizzo, L., & Fatta-Kassinos, D. (2019). Performance of secondary wastewater treatment methods for the removal of contaminants of emerging concern implicated in crop uptake and antibiotic resistance spread: A review. *The Science of the Total Environment, 648*, 1052–1081. DOI: 10.1016/j.scitotenv.2018.08.130

Kube, M., Jefferson, B., Fan, L., & Roddick, F. (2018). The impact of wastewater characteristics, algal species selection and immobilisation on simultaneous nitrogen and phosphorus removal. *Algal Research, 31*, 478–488. DOI: 10.1016/j.algal.2018.01.009

Kumar, A., Ergas, S., Yuan, X., Sahu, A., Zhang, Q., Dewulf, J., Malcata, F. X., & van Langenhove, H. (2010). Enhanced CO2 fixation and biofuel production via microalgae: Recent developments and future directions. *Trends in Biotechnology, 28*(7), 371–380. DOI: 10.1016/j.tibtech.2010.04.004 PMID: 20541270

Kumar, A., Sharma, A., Majumder, P., Vishal, V., & Dutta, A. (2025a). Chemical and microalgal conversion of carbon dioxide into fuels and materials: A review. *Environmental Chemistry Letters*, *23*(5), 1209–1229. DOI: 10.1007/s10311-025-01855-2

Kumar, G., Huy, M., & Kim, S. (2018). Evaluation of gradual adaptation of mixed microalgae consortia cultivation using textile wastewater via fed batch operation. *Biotechnology Reports (Amsterdam, Netherlands)*, *20*, e00289. DOI: 10.1016/j.btre.2018.e00289 PMID: 30416980

Kumar, M., Singh, A. K., & Sikandar, M. (2020). Biosorption of Hg (II) from aqueous solution using algal biomass: Kinetics and isotherm studies. *Heliyon*, *6*(1), e03321. DOI: 10.1016/j.heliyon.2020.e03321 PMID: 32042987

Kumar, P., & Singh, J. (2025). Perspective and challenges of synergistic removal of toxic contaminants from effluent using different treatment techniques. In *Microbial niche nexus sustaining environmental biological wastewater and water-energy-environment nexus* (pp. 419–451). Springer. DOI: 10.1007/978-3-031-62660-9_17

Kumar, S., Tripathi, A., Yadhav, S., Mishra, S., & Ghangrekar, M. M. (2025b). Interactions of Microplastics with Microbial Communities and the Food Web/Plants. In *Microplastics in the Environment* (pp. 283 310). Fate, Impacts, Removal, and Management. DOI: 10.1002/9781394251100.ch11

Kumar, V., & Mishra, S. K. (2025). Conclusion and Roadmap for Implementation. In Kumar Arya, S., Khatri, M., Singh, G., & Sahu, S. (Eds.), *Algal Bioengineering and Microbial Synergy to Green Remediation* (pp. 371–393). Springer Nature Singapore., DOI: 10.1007/978-981-96-8054-2_15

Kundu, P., Dutta, N., & Bhattacharya, S. (2024). Application of microalgae in wastewater treatment with special reference to emerging contaminants: A step towards sustainability. *Frontiers in Analytical Science*, *4*, 1513153. DOI: 10.3389/frans.2024.1513153

Kuppan, P., Sudharsanam, A., Venkateswarlu, K., & Megharaj, M. (2023). Solar technology–closed loop synergy facilitates low-carbon circular bioeconomy in microalgal wastewater treatment. *NPJ Clean Water*, *6*(1), 43. DOI: 10.1038/s41545-023-00256-8

Kupriyanova, E. V., Pronina, N. A., & Los, D. A. (2023). Adapting from low to high: An update to CO2-concentrating mechanisms of cyanobacteria and microalgae. *Plants*, *12*(7), 1569. DOI: 10.3390/plants12071569 PMID: 37050194

La Farre, M., Pérez, S., Kantiani, L., & Barceló, D. (2008). Fate and toxicity of emerging pollutants, their metabolites and transformation products in the aquatic environment. *Trends in Analytical Chemistry, 27*(11), 991–1007. DOI: 10.1016/j.trac.2008.09.010

Lakshmana Prabu, S., Umamaheswari, A., & Puratchikody, A. (2021). Bioremediation approaches for degradation of textile dye effluents. In Marimuthu, P. D., Sundaram, R., Jeyaseelan, A., & Kaliannan, T. (Eds.), *Bioremediation and Green Technologies* (pp. 113–133). Springer., DOI: 10.1007/978-3-030-64122-1_9

Lalucat, J., Gomila, M., Mulet, M., Zaruma, A., & García-Valdés, E. (2022). Past, present and future of the boundaries of the *Pseudomonas* genus: Proposal of *Stutzerimonas* gen. nov. *Systematic and Applied Microbiology, 45*(1), 126289. DOI: 10.1016/j.syapm.2021.126289 PMID: 34920232

Langmuir, I. (1918). The adsorption of gases on plane surfaces of glass, mica and platinum. *Journal of the American Chemical Society, 40*(9), 1361–1403.

Lapworth, D. J., Baran, N., Stuart, M. E., & Ward, R. S. (2012). Emerging organic contaminants in groundwater: A review of sources, fate and occurrence. *Environmental Pollution, 163*, 287–303. DOI: 10.1016/j.envpol.2011.12.034 PMID: 22306910

Lee, C. S., Oh, H.-S., Oh, H.-M., Kim, H.-S., & Ahn, C.-Y. (2016). Two-phase photoperiodic cultivation of algal–bacterial consortia for high biomass production and efficient nutrient removal from municipal wastewater. *Bioresource Technology, 200*, 867–875. DOI: 10.1016/j.biortech.2015.11.007 PMID: 26595666

Lee, J., Hyeon, D. Y., & Hwang, D. (2020). Single cell multiomics: Technologies and data analysis methods. *Experimental & Molecular Medicine, 52*(9), 1428–1442. DOI: 10.1038/s12276-020-0420-2 PMID: 32929225

Lehmann, J., Bossio, D. A., Kögel-Knabner, I., & Rillig, M. C. (2020). The concept and future prospects of soil health. *Nature Reviews. Earth & Environment, 1*(10), 544–553. DOI: 10.1038/s43017-020-0080-8 PMID: 33015639

Lei, A.-P., Hu, Z.-L., Wong, Y.-S., & Tam, N. F.-Y. (2007). Removal of fluoranthene and pyrene by different microalgal species. *Bioresource Technology, 98*(2), 273–280. DOI: 10.1016/j.biortech.2006.01.012 PMID: 16517155

Liang, E., Li, J., Li, B., Liu, S., Ma, R., Yang, S., Cai, H., Xue, Z., & Wang, T. (2023). Roles of dissolved organic matter (DOM) in shaping the distribution pattern of heavy metal in the Yangtze River. *Journal of Hazardous Materials, 460*, 132410. DOI: 10.1016/j.jhazmat.2023.132410 PMID: 37647662

Liao, H., Gao, D., Kong, C., Junaid, M., Li, Y., Chen, X., Zheng, Q., Chen, G., & Wang, J. (2023). Trophic transfer of nanoplastics and di (2-ethylhexyl) phthalate in a freshwater food chain (Chlorella pyrenoidosa-Daphnia magna-Micropterus salmoides) induced disturbance of lipid metabolism in fish. *Journal of Hazardous Materials*, *459*, 132294. DOI: 10.1016/j.jhazmat.2023.132294 PMID: 37591169

Li, J., Ou, R., Liao, H., Ma, J., Sun, L., Jin, Q., He, D., & Wang, Q. (2022). Natural lighting enhancing the algae proliferation and nitrogen removal in membrane-aerated bacterial-algal biofilm reactor. *The Science of the Total Environment*, *851*(Pt 1), 158063. DOI: 10.1016/j.scitotenv.2022.158063 PMID: 35981577

Li, K.-Y., Zhou, J.-L., Vadiveloo, A., Li, J.-N., Zhou, D., Wang, J.-M., Zhang, C., Xie, Y.-L., Chen, D.-Z., & Gao, F. (2025). Signal molecule-mediated algae-bacteria interactions and membrane-enabled separation synergize wastewater bioremediation. *Chemical Engineering Journal*, *523*, 168368. DOI: 10.1016/j.cej.2025.168368

Lim, H.-H., Oh, Y.-S., & Shin, H.-S. (2020). Determination of *N*-nitrosodimethylamine and *N*-nitrosomethylethylamine in drug substances and products of sartans, metformin and ranitidine by precipitation and solid phase extraction and gas chromatography–tandem mass spectrometry. *Journal of Pharmaceutical and Biomedical Analysis*, *189*, 113460. DOI: 10.1016/j.jpba.2020.113460 PMID: 32663759

Lim, J., Kim, T., & Hwang, S. (2003). Treatment of fish-processing wastewater by co-culture of *Candida rugopelliculosa* and *Brachionus plicatilis*. *Water Research*, *37*(9), 2228–2232. DOI: 10.1016/S0043-1354(02)00586-9

Li, N., Qin, L., Jin, M., Zhang, L., Geng, W., & Xiao, X. (2021). Extracellular adsorption, intracellular accumulation and tolerance mechanisms of *Cyclotella* sp. To Cr(VI) stress. *Chemosphere*, *270*, 128662. DOI: 10.1016/j.chemosphere.2020.128662 PMID: 33127109

Ling, X., Lu, G., Jiang, R., Zhang, L., Zhang, P., Wang, M., Xue, Q., Yu, Y., & Ding, K. (2026). Enhanced Ecological Risk of Plastic-Associated Chemicals by Microplastics in the Riverine Benthic Food Web. *Environmental Science & Technology*, *60*(10), 8120–8132. DOI: 10.1021/acs.est.5c13059 PMID: 41773988

Lin, W., Chen, L., Tan, Z., Deng, Z., & Liu, H. (2022). Application of filamentous fungi in microalgae-based wastewater remediation for biomass harvesting and utilization: From mechanisms to practical application. *Algal Research*, *62*, 102614. DOI: 10.1016/j.algal.2021.102614

Li, P., Luo, Y., & Yuan, X. (2022). Life cycle and techno-economic assessment of source-separated wastewater-integrated microalgae biofuel production plant: A nutrient organization approach. *Bioresource Technology, 344*, 126230. DOI: 10.1016/j.biortech.2021.126230 PMID: 34732373

Li, Q., Xu, Y., Liang, C., Peng, L., & Zhou, Y. (2023a). Nitrogen removal by algal-bacterial consortium during mainstream wastewater treatment: Transformation mechanisms and potential N2O mitigation. *Water Research, 235*, 119890. DOI: 10.1016/j.watres.2023.119890 PMID: 36958220

Li, S., Chang, H., Zhang, S., & Ho, S.-H. (2023). Production of sustainable biofuels from microalgae with CO2 bio-sequestration and life cycle assessment. *Environmental Research, 227*, 115730. Advance online publication. DOI: 10.1016/j.envres.2023.115730 PMID: 36958384

Li, S., Li, X., & Ho, S.-H. (2021). Microalgae as a solution of third world energy crisis for biofuels production from wastewater toward carbon neutrality: An updated review. *Chemosphere, 291*, 132863. DOI: 10.1016/j.chemosphere.2021.132863 PMID: 34774903

Liu, H., Al-Dhabi, N. A., Jiang, H., Liu, B., Qing, T., Feng, B., Ma, T., Tang, W., & Zhang, P. (2024). Toward nitrogen recovery: Co-cultivation of microalgae and bacteria enhances the production of high-value nitrogen-rich cyanophycin. *Water Research, 256*, 121624. DOI: 10.1016/j.watres.2024.121624 PMID: 38669903

Liu, H., Lu, Q., Wang, Q., Liu, W., Wei, Q., Ren, H., Ming, C., Min, M., Chen, P., & Ruan, R. (2017). Isolation of a bacterial strain, Acinetobacter sp. from centrate wastewater and study of its cooperation with algae in nutrients removal. *Bioresource Technology, 235*, 59–69. DOI: 10.1016/j.biortech.2017.03.111 PMID: 28364634

Liu, J., Booma Devi, P., Chinnathambi, A., & Ali Alharbi, S. (2024). Mitigating fossil fuel deficiency and environmental impacts: Performance analysis of *Scenedesmus obliquus* microalgae biodiesel in a diesel engine. *Fuel, 364*, 131033. DOI: 10.1016/j.fuel.2024.131033

Liu, R., Li, S., Tu, Y., & Hao, X. (2021). Capabilities and mechanisms of microalgae on removing micropollutants from wastewater: A review. *Journal of Environmental Management, 285*, 112149. DOI: 10.1016/j.jenvman.2021.112149 PMID: 33607565

Liu, S., Kong, Z., Guo, H., Zhang, Y., Han, X., Gao, Y., Daigger, G. T., Zhang, G., Li, R., Liu, Y., Zhang, P., & Song, G. (2024). Performance, mechanism regulation and resource recycling of bacteria-algae symbiosis system for wastewater treatment: A review. *Environmental Pollution, 362*, 125019. DOI: 10.1016/j.envpol.2024.125019 PMID: 39326826

Liu, X., & Hong, Y. (2021). Microalgae-Based Wastewater Treatment and Recovery with Biomass and Value-Added Products: A Brief Review. *Current Pollution Reports*, *7*(2), 227–245. DOI: 10.1007/s40726-021-00184-6

Liu, Y., Zhang, X., Du, Y., Du, X., Zhang, Y., Deng, L., Li, C., & Guo, J. (2025). Functionalization of Wood for the Removal of Heavy Metal Ions from Waster Water: A Review. *Forests*, *16*(11), 1684. DOI: 10.3390/f16111684

Li, X., Wang, X., Zhang, Y., Ding, N., & Zhou, Q. (2020). Algal bacterial consortia for enhanced bioelectricity generation and wastewater treatment in microbial fuel cells. *The Science of the Total Environment*, *727*, 138691. DOI: 10.1016/j.scitotenv.2020.138691

Li, Y., Xu, Y., Fu, Z., Li, W., Zheng, L., & Li, M. (2021). Assessment of energy use and environmental impacts of wastewater treatment plants in the entire life cycle: A system meta analysis. *Environmental Research*, *198*, 110458. DOI: 10.1016/j.envres.2020.110458 PMID: 33188763

Lock, G. A., & Hoyer, W. A. (1974). Carbon-Oxygen (C/O) Log: Use and Interpretation. *Journal of Petroleum Technology*, *26*(09), 1044–1054. DOI: 10.2118/4639-PA

Lodariya, M., Bhattacharya, D., & Abhilash, K. R. (2025). Synergistic approaches to soil remediation: Engineered biochar and microbial interactions for climate-resilient remediation. *Biodegradation*, *36*(6), 110. DOI: 10.1007/s10532-025-10200-x PMID: 41139376

López-Sánchez, A., Silva-Gálvez, A. L., Aguilar-Juárez, Ó., Senés-Guerrero, C., Orozco-Nunnelly, D. A., Carrillo-Nieves, D., & Gradilla-Hernández, M. S. (2022). Microalgae-based livestock wastewater treatment (MbWT) as a circular bioeconomy approach: Enhancement of biomass productivity, pollutant removal and high-value compound production. *Journal of Environmental Management*, *308*, 114612. DOI: 10.1016/j.jenvman.2022.114612 PMID: 35149401

Lu, H., Zhang, G., Wan, T., & Lu, Y. (2011). Influences of light and oxygen conditions on photosynthetic bacteria macromolecule degradation: Different metabolic pathways. *Bioresource Technology*, *102*(20), 9503–9508. DOI: 10.1016/j.biortech.2011.07.114 PMID: 21871794

Lu, L., Guest, J. S., Peters, C. A., Zhu, X., Rau, G. H., & Ren, Z. J. (2018). Wastewater treatment for carbon capture and utilization. *Nature Sustainability*, *1*(12), 750–758. DOI: 10.1038/s41893-018-0187-9

Luo, H., Cheng, Y., He, D., Liu, Y., Wang, L., & Lin, H. (2022). Recent advances in microalgae-based technologies for emerging contaminants removal: A review. *The Science of the Total Environment, 836*, 155643. DOI: 10.1016/j.scitotenv.2022.155643

Luo, L., Luo, S., Wang, H., Hu, K., Lin, X., Liu, L., & Yan, B. (2021). Effect of nano-TiO2 on humic acid utilization from piggery biogas slurry by microalgae. *Bioresource Technology, 337*, 125414. DOI: 10.1016/j.biortech.2021.125414 PMID: 34174768

Luo, Y., Guo, W., Ngo, H. H., Nghiem, L. D., Hai, F. I., Zhang, J., Liang, S., & Wang, X. C. (2014). A review on the occurrence of micropollutants in the aquatic environment and their fate and removal during wastewater treatment. *The Science of the Total Environment, 473–474*, 619–641. DOI: 10.1016/j.scitotenv.2013.12.065 PMID: 24394371

Luo, Y., Le-Clech, P., & Henderson, R. K. (2020). Characterisation of microalgae-based monocultures and mixed cultures for biomass production and wastewater treatment. *Algal Research, 49*, 101963. DOI: 10.1016/j.algal.2020.101963

Mabuhay, J. A., Nakagoshi, N., & Isagi, Y. (2004). Influence of erosion on soil microbial biomass, abundance and community diversity. *Land Degradation & Development, 15*(2), 183–195. DOI: 10.1002/ldr.607

Madhav, S., Ahamad, A., Singh, P., & Mishra, P. K. (2018). A review of textile industry: Wet processing, environmental impacts, and effluent treatment methods. *Environmental Quality Management, 27*(3), 31–41. DOI: 10.1002/tqem.21538

Mahapatra, D. M., & Murthy, G. S. (2021). Long term evaluation of a pilot scale multimodal algal bioprocess for treatment of municipal wastewater. *Journal of Cleaner Production, 311*, 127690. DOI: 10.1016/j.jclepro.2021.127690

Mahlangu, D., Mphahlele, K., De Paola, F., & Mthombeni, N. H. (2024). Microalgae-mediated biosorption for effective heavy metals removal from wastewater: A review. *Water (Basel), 16*(5), 718. DOI: 10.3390/w16050718

Mahmood, T., Hussain, N., Shahbaz, A., Mulla, S. I., Iqbal, H. M. N., & Bilal, M. (2023). Sustainable production of biofuels from the algae-derived biomass. *Bioprocess and Biosystems Engineering, 46*(8), 1077–1097. DOI: 10.1007/s00449-022-02796-8 PMID: 36331626

Makhathini, T. P. (2026). Microbial Electrochemical Technologies in Wastewater Treatment: Scale-Up Challenges, Pilot Testing, and Practical Implementation. *Water (Basel), 18*(8), 966. DOI: 10.3390/w18080966

Makut, B. B., Das, D., & Goswami, G. (2019). Production of microbial biomass feedstock via co-cultivation of microalgae-bacteria consortium coupled with effective wastewater treatment: A sustainable approach. *Algal Research*, *37*, 228–239. DOI: 10.1016/j.algal.2018.11.020

Makuwa, S., Tlou, M., Fosso-Kankeu, E., & Green, E. (2020). Evaluation of Fecal Coliform Prevalence and Physicochemical Indicators in the Effluent from a Wastewater Treatment Plant in the North-West Province, South Africa. *International Journal of Environmental Research and Public Health*, *17*(17), 6381. DOI: 10.3390/ijerph17176381 PMID: 32887228

Maltsev, Y., Kulikovskiy, M., & Maltseva, S. (2023). Nitrogen and phosphorus stress as a tool to induce lipid production in microalgae. *Microbial Cell Factories*, *22*(1), 239. DOI: 10.1186/s12934-023-02244-6 PMID: 37981666

Mao, B., & Zhang, B. (2023). Combining ABE fermentation and anaerobic digestion to treat with lipid extracted algae for enhanced bioenergy production. *The Science of the Total Environment*, *875*, 162691. DOI: 10.1016/j.scitotenv.2023.162691 PMID: 36898333

Maqsood, Q., Waseem, R., Sumrin, A., Wajid, A., Tariq, M. R., Ali, S. W., & Mahnoor, M. (2024). Recent trends in bioremediation and bioaugmentation strategies for mitigation of marine based pollutants: Current perspectives and future outlook. *Discover Sustainability*, *5*(1), 524. DOI: 10.1007/s43621-024-00607-6

Marami, H., Tsapekos, P., Khoshnevisan, B., Madsen, J. A., Andersen, J. K., Rafiee, S., & Angelidaki, I. (2022). Going beyond conventional wastewater treatment plants within circular bioeconomy concept – a sustainability assessment study. *Water Science and Technology*, *85*(6), 1878–1903. DOI: 10.2166/wst.2022.096 PMID: 35358077

Marín, D., Méndez, L., Suero, I., Díaz, I., Blanco, S., Fdz-Polanco, M., & Muñoz, R. (2022). Anaerobic digestion of food waste coupled with biogas upgrading in an outdoors algal-bacterial photobioreactor at pilot scale. *Fuel*, *324*, 124554. DOI: 10.1016/j.fuel.2022.124554

Markai, S., Andrès, Y., Montavon, G., & Grambow, B. (2003). Study of the interaction between europium (III) and Bacillus subtilis: Fixation sites, biosorption modeling and reversibility. *Journal of Colloid and Interface Science*, *262*(2), 351–361. DOI: 10.1016/S0021-9797(03)00096-1 PMID: 16256615

Markou, G., Vandamme, D., & Muylaert, K. (2014). Microalgal and cyanobacterial cultivation: The supply of nutrients. *Water Research*, *65*, 186–202. DOI: 10.1016/j.watres.2014.07.025 PMID: 25113948

Martinez-Burgos, W. J., Bittencourt Sydney, E., Bianchi Pedroni Medeiros, A., Magalhães, A. I., De Carvalho, J. C., Karp, S. G., Porto De Souza Vandenberghe, L., Junior Letti, L. A., Thomaz Soccol, V., De Melo Pereira, G. V., Rodrigues, C., Lorenci Woiciechowski, A., & Soccol, C. R. (2021). Agro-industrial wastewater in a circular economy: Characteristics, impacts and applications for bioenergy and biochemicals. *Bioresource Technology, 341*, 125795. DOI: 10.1016/j.biortech.2021.125795 PMID: 34523570

Masojídek, J., Torzillo, G., & Koblížek, M. (2013). Photosynthesis in Microalgae. In *Handbook of Microalgal Culture* (pp. 21–36). John Wiley & Sons, Ltd., DOI: 10.1002/9781118567166.ch2

Massoud, M. A., Tarhini, A., & Nasr, J. A. (2009). Decentralized approaches to wastewater treatment and management: Applicability in developing countries. *Journal of Environmental Management, 90*(1), 652–659. DOI: 10.1016/j.jenvman.2008.07.001 PMID: 18701206

Matamoros, V., Gutiérrez, R., Ferrer, I., García, J., & Bayona, J. M. (2015). Capability of microalgae-based wastewater treatment systems to remove emerging organic contaminants: A pilot-scale study. *Journal of Hazardous Materials, 288*, 34–42. DOI: 10.1016/j.jhazmat.2015.02.002

Matamoros, V., Rodríguez, Y., & Albaigés, J. (2016). A comparative assessment of intensive and extensive wastewater treatment technologies for removing emerging contaminants in small communities. *Water Research, 88*, 777–785. DOI: 10.1016/j.watres.2015.10.058

Matamoros, V., Uggetti, E., García, J., & Bayona, J. M. (2016). Assessment of the mechanisms involved in the removal of emerging contaminants by microalgae from wastewater: A laboratory scale study. *Journal of Hazardous Materials, 301*, 197–205. DOI: 10.1016/j.jhazmat.2015.08.050

Mata, T. M., Martins, A. A., & Caetano, N. S. (2010). Microalgae for biodiesel production and other applications: A review. *Renewable & Sustainable Energy Reviews, 14*(1), 217–232. DOI: 10.1016/j.rser.2009.07.020

Ma, W., Liang, X., Ding, C., Ye, Y., & Li, J. (2026). From Enrichment to Fate: Transport, Transformation, and Fate of Micro-and Nanoplastics in Marine Environments. *Toxics, 14*(2), 120. DOI: 10.3390/toxics14020120 PMID: 41745794

Ma, Y., Rajkumar, M., Zhang, C., & Freitas, H. (2016). Inoculation of Brassica oxyrrhina with plant growth promoting bacteria for the improvement of heavy metal phytoremediation under drought conditions. *Journal of Hazardous Materials, 320*, 36–44. DOI: 10.1016/j.jhazmat.2016.08.009 PMID: 27508309

McGrail, B. P., Schaef, H. T., Spane, F. A., Horner, J. A., Owen, A. T., Cliff, J. B., Qafoku, O., Thompson, C. J., & Sullivan, E. C. (2017). Wallula basalt pilot demonstration project: Post-injection results and conclusions. *Energy Procedia, 114*, 5783–5790. DOI: 10.1016/j.egypro.2017.03.1716

McKendry, P. (2002). Energy production from biomass (part 1): Overview of biomass. *Bioresource Technology, 83*(1), 37–46. DOI: 10.1016/S0960-8524(01)00118-3 PMID: 12058829

Mehta, S. K., & Gaur, J. P. (2005). Use of Algae for Removing Heavy Metal Ions From Wastewater: Progress and Prospects. *Critical Reviews in Biotechnology, 25*(3), 113–152. DOI: 10.1080/07388550500248571 PMID: 16294830

Meichen, W. (2020). *Experimental study on removal of iron and manganese in groundwater by microalgae* [Master's thesis, Anhui University of Technology].

Meng, F., Xi, L., Liu, D., Huang, W., Lei, Z., Zhang, Z., & Huang, W. (2019). Effects of light intensity on oxygen distribution, lipid production and biological community of algal-bacterial granules in photo-sequencing batch reactors. *Bioresource Technology, 272*, 473–481. DOI: 10.1016/j.biortech.2018.10.059 PMID: 30390540

Mennaa, F. Z., Arbib, Z., & Perales, J. A. (2015). Urban wastewater treatment by seven species of microalgae and an algal bloom: Biomass production, N and P removal kinetics and harvestability. *Water Research, 83*, 42–51. DOI: 10.1016/j.watres.2015.06.007

Mestrot, A., Xie, W.-Y., Xue, X., & Zhu, Y.-G. (2013). Arsenic volatilization in model anaerobic biogas digesters. *Applied Geochemistry, 33*, 294–297. DOI: 10.1016/j.apgeochem.2013.02.023

Michalak, I., Chojnacka, K., & Witek-Krowiak, A. (2013). State of the art for the biosorption process - A review. *Applied Biochemistry and Biotechnology, 170*(6), 1389–1416. DOI: 10.1007/s12010-013-0269-0 PMID: 23666641

Michelon, W., Zuchi, I. D. P., Reis, J. G., Matthiensen, A., Viancelli, A., da Cruz, A. C. C., Silva, I. T., Fongaro, G., & Soares, H. M. (2022). Virucidal activity of microalgae extracts harvested during phycoremediation of swine wastewater. *Environmental Science and Pollution Research International, 29*(19), 13427–13436. DOI: 10.1007/s11356-021-17912-6

Minhas, A. K., Gaur, S., Sunny, S., Paladugu, C., Ravishankar, G. A., Pereira, L., & Ambati, R. R. (2026). Microalgae-Based Wastewater Treatment Processes for the Bioremediation and Valorization of Biomass: A Review. *Phycology, 6*(1), 18. DOI: 10.3390/phycology6010018

Mitter, E. K., Tosi, M., Obregón, D., Dunfield, K. E., & Germida, J. J. (2021). Rethinking crop nutrition in times of modern microbiology: Innovative biofertilizer technologies. *Frontiers in Sustainable Food Systems*, *5*, 606815. DOI: 10.3389/fsufs.2021.606815

Mohamed, Z. A. (2008). Polysaccharides as a protective response against microcystin-induced oxidative stress in *Chlorella vulgaris* and *Scenedesmus quadricauda* and their possible significance in the aquatic ecosystem. *Ecotoxicology (London, England)*, *17*(6), 504–516. DOI: 10.1007/s10646-008-0204-2 PMID: 18389369

Mohanty, K. (with Mohanty, S. S.). (2025). *A Biorefinery Approach to Algal Biomass Conversion for Biofuels and Bioproducts* (1st ed). Elsevier.

Mohit, A., Mishel, V. K., & Remya, N. (2024). Life cycle assessment and technoeconomic analysis of biofuels produced from polyculture microalgae cultivated in greywater. *Journal of Environmental Management*, *356*, 120711. DOI: 10.1016/j.jenvman.2024.120711 PMID: 38537468

Mojiri, A., Zhou, J. L., Ratnaweera, H., Rezania, S., & Mansoureh, V. N. (2022). Pharmaceuticals and personal care products in aquatic environments and their removal by algae-based systems. *Chemosphere*, *288*, 132580. DOI: 10.1016/j.chemosphere.2021.132580 PMID: 34687686

Molazadeh, M., Ahmadzadeh, H., Pourianfar, H. R., Lyon, S., & Rampelotto, P. H. (2019). The use of microalgae for coupling wastewater treatment with CO2 biofixation. *Frontiers in Bioengineering and Biotechnology*, *7*, 42. DOI: 10.3389/fbioe.2019.00042 PMID: 30941348

Mondal, S., & Palit, D. (2022). Challenges in natural resource management for ecological sustainability. In *Natural Resources Conservation and Advances for Sustainability* (pp. 29–59). Elsevier., DOI: 10.1016/B978-0-12-822976-7.00004-1

Monteiro, C. M., Castro, P. M. L., & Malcata, F. X. (2012). Metal uptake by microalgae: Underlying mechanisms and practical applications. *Biotechnology Progress*, *28*(2), 299–311. DOI: 10.1002/btpr.1504 PMID: 22228490

Moreno Osorio, J. H., Pinto, G., Pollio, A., Frunzo, L., Lens, P. N. L., & Esposito, G. (2019). Startup of a nutrient removal system using *Scenedesmus vacuolatus* and *Chlorella vulgaris* biofilms. *Bioresources and Bioprocessing*, *6*(1), 1–16. DOI: 10.1186/s40643-019-0259-3

Moreno, R., & Rojo, F. (2013). The contribution of proteomics to the unveiling of the survival strategies used by Pseudomonas putida in changing and hostile environments. *Proteomics, 13*(18–19), 2822–2830. DOI: 10.1002/pmic.201200503 PMID: 23625785

Morseletto, P., Mooren, C. E., & Munaretto, S. (2022). Circular economy of water: Definition, strategies and challenges. *Circular Economy and Sustainability, 2*(4), 1–15. DOI: 10.1007/s43615-022-00165-x

Mounaimi, S., Al-Tawaha, A. R. M., Chippalakatti, P., Sangeetha, J., & Thangadurai, D. (2026). Microalgal Bioresources For Sustainable Agriculture. In *Microbial Fertilizer Technology for Sustainable Crop Production* (pp. 53–87). Apple Academic Press. DOI: 10.1201/9781003594826-3

Moutik, B., Summerscales, J., Graham-Jones, J., & Pemberton, R. (2023). Life Cycle Assessment Research Trends and Implications: A Bibliometric Analysis. *Sustainability (Basel), 15*(18), 13408. DOI: 10.3390/su151813408

Mo, Z., Tai, D., Zhang, H., & Shahab, A. (2022). A comprehensive review on the adsorption of heavy metals by zeolite imidazole framework (ZIF-8) based nanocomposite in water. *Chemical Engineering Journal, 443*, 136320. DOI: 10.1016/j.cej.2022.136320

Mujtaba, G., Rizwan, M., Kim, G., & Lee, K. (2018). Removal of nutrients and COD through co-culturing activated sludge and immobilized *Chlorella vulgaris. Chemical Engineering Journal, 343*, 155–162. DOI: 10.1016/j.cej.2018.03.007

Mujtaba, G., Rizwan, M., & Lee, K. (2015). Simultaneous removal of inorganic nutrients and organic carbon by symbiotic co-culture of Chlorella vulgaris and Pseudomonas putida. *Biotechnology and Bioprocess Engineering; BBE, 20*(6), 1114–1122. DOI: 10.1007/s12257-015-0421-5

Mujtaba, G., Rizwan, M., & Lee, K. (2017). Removal of nutrients and COD from wastewater using symbiotic co-culture of bacterium *Pseudomonas putida* and immobilized microalga *Chlorella vulgaris. Journal of Industrial and Engineering Chemistry, 49*, 145–151. DOI: 10.1016/j.jiec.2017.01.021

Mukherjee, S., & Bairwa, D. H. (2025). Environmental Issues Arising from Urbanization: A Study on the Ecological Consequences of Rapid Urban Growth. *The Journal of Humanistic Education and Development, 7*(3), 12–17. DOI: 10.22161/jhed.7.3.3

Mulla, S. I., Bagewadi, Z. K., Faniband, Z., Bilal, M., Chae, J. C., Bankole, P. O., Saratale, G. D., Bhargava, R. N., & Gurumurthy, D. M. (2021). Various strategies applied for the removal of emerging micropollutant sulfamethazine: A systematic review. *Environmental Science and Pollution Research International, 28*(28), 56–72. DOI: 10.1007/s11356-021-14259-w

Munir, N., Sarwar, Z., Abideen, Z., Saleem, F., Hasanuzzaman, M., Siddiqui, Z. S., & El-Keblawy, A. (2025). Algae-based bioremediation of soil, water, and air: A solution to polluted environment. *Environmental Science and Pollution Research International, 32*(36), 21338–21357. DOI: 10.1007/s11356-025-36880-9 PMID: 40952577

Mutum, L., Janda, T., Ördög, V., & Molnár, Z. (2022). Potential of Different Forms of Microalgae for Soil Improvement. *Biologia Futura, 73*(1), 1–8. DOI: 10.1007/s42977-021-00103-2 PMID: 34735698

Nagarajan, D., Lee, D.-J., Chen, C.-Y., & Chang, J.-S. (2020). Resource recovery from wastewaters using microalgae-based approaches: A circular bioeconomy perspective. *Bioresource Technology, 302*, 122817. DOI: 10.1016/j.biortech.2020.122817 PMID: 32007309

Nagarajan, D., Lee, D.-J., Varjani, S., Lam, S. S., Allakhverdiev, S. I., & Chang, J.-S. (2022). Microalgae-based wastewater treatment—Microalgae-bacteria consortia, multi-omics approaches and algal stress response. *The Science of the Total Environment, 845*, 157110. DOI: 10.1016/j.scitotenv.2022.157110 PMID: 35787906

Nanda, M., Jaiswal, K. K., Kumar, V. M., Vlaskin, S., Gautam, P., Bahuguna, V., & Chauhan, P. K. (2021). Micro-pollutant Pb(II) mitigation and lipid induction in oleaginous microalgae *Chlorella sorokiniana* UUIND6. *Environmental Technology & Innovation, 23*, 101613. DOI: 10.1016/j.eti.2021.101613

Narayanan, I., Rajamanickam, R., Singh, R. K., Singh, P., Prabhu, A. A., Pandey, A., Pradhan, N., Nancharaiah, Y. V., Kumar, V., Selvasembian, R., & Mal, J. (2025). Microalgae-based remediation of pharmaceutical contaminants: Emerging strategies and technological synergies. *Discover Applied Sciences, 7*(11), 1288. DOI: 10.1007/s42452-025-07226-4

Naz, S., Brandts, I., Resina, A., Tudela, P., Sugrañes, M., Expósito, N., Barrientos-Riosalido, A., Rovira, J., Balasch, J., & Teles, M. (2026). Environmentally Derived Nanoplastics Induce Oxidative and Genotoxic Responses in Gilthead Seabream (Sparus aurata). *Journal of Hazardous Materials: Plastics*, Article 100048.

Nguyen, H. T., Yoon, Y., Ngo, H. H., & Jang, A. (2021). The application of microalgae in removing organic micropollutants in wastewater. *Critical Reviews in Environmental Science and Technology*, *51*(12), 1187–1220. DOI: 10.1080/10643389.2020.1753633

Nguyen, T.-T.-D., Nguyen, T.-T., An Binh, Q., Bui, X.-T., Ngo, H. H., Vo, H. N. P., Andrew Lin, K.-Y., Vo, T.-D.-H., Guo, W., Lin, C., & Breider, F. (2020). Co-culture of microalgae-activated sludge for wastewater treatment and biomass production: Exploring their role under different inoculation ratios. *Bioresource Technology*, *314*, 123754. DOI: 10.1016/j.biortech.2020.123754 PMID: 32650264

Nidheesh, P. V., Couras, C., Karim, A. V., & Nadais, H. (2022). A review of integrated advanced oxidation processes and biological processes for organic pollutant removal. *Chemical Engineering Communications*, *209*(3), 390–432. DOI: 10.1080/00986445.2020.1864626

Nkonya, E., Anderson, W., Kato, E., Koo, J., Mirzabaev, A., von Braun, J., & Meyer, S. (2016). Global cost of land degradation. In Nkonya, E., Mirzabaev, A., & von Braun, J. (Eds.), *Economics of land degradation and improvement: A global assessment for sustainable development* (pp. 117–165). Springer., DOI: 10.1007/978-3-319-19168-3_6

Nobre, M. L. F., Tavares, D., Fraga, C., Oliveira, B., Dias, M., Mesquita, S., Oliveira, C. M., & Pires, J. C. M. (2024). Techno-economic analysis of a circular microalgal approach for enhanced wastewater treatment and resource recovery in Northern Portugal. *Journal of Cleaner Production*, *434*, 140389. DOI: 10.1016/j.jclepro.2023.140389

Norvill, Z. N., Shilton, A., & Guieysse, B. (2016). Emerging contaminant degradation and removal in algal wastewater treatment ponds: Identifying the research gaps. *Journal of Hazardous Materials*, *313*, 291–309. DOI: 10.1016/j.jhazmat.2016.03.085 PMID: 27135171

Novoveská, L., Nielsen, S. L., Eroldoğan, O. T., Haznedaroglu, B. Z., Rinkevich, B., Fazi, S., Robbens, J., Vasquez, M., & Einarsson, H. (2023). Overview and Challenges of Large-Scale Cultivation of Photosynthetic Microalgae and Cyanobacteria. *Marine Drugs*, *21*(8), 445. DOI: 10.3390/md21080445 PMID: 37623726

Nunes, R. de C., & Costa, H. K. D. M. (2021). How CCS projects are developed in selected countries? *International Journal of Developmental Research*, *11*, 48829–48834. DOI: 10.37118/ijdr.22314.07.2021

Nur, M. M. A., Mahreni, , Murni, S. W., Setyoningrum, T. M., Hadi, F., Widayati, T. W., Jaya, D., Sulistyawati, R. R. E., Puspitaningrum, D. A., Dewi, R. N., Hadiyanto, , & Hasanuzzaman, M. (2025). Innovative Strategies for Utilizing Microalgae as Dual-Purpose Biofertilizers and Phycoremediators in Agroecosystems. *Biotechnology Reports (Amsterdam, Netherlands)*, *45*, e00870. DOI: 10.1016/j.btre.2024.e00870 PMID: 39758973

Oh, K. H., & Tuovinen, O. H. (1994). Biodegradation of the phenoxy herbicides MCPP and 2,4-D in fixed-film column reactors. *International Biodeterioration & Biodegradation*, *33*(1), 93–99. DOI: 10.1016/0964-8305(94)90057-4

Okoro, H. K., Orosun, M. M., Oriade, F. A., Momoh-Salami, T. M., Ogunkunle, C. O., Adeniyi, A. G., Zvinowanda, C., & Ngila, J. C. (2023). Potentially toxic elements in pharmaceutical industrial effluents: A review on risk assessment, treatment, and management for human health. *Sustainability (Basel)*, *15*(8), 6974. DOI: 10.3390/su15086974

Olawale, S. A. (2021). Biosorption of heavy metals from aqueous solutions: An insight and review. *Archives of Industrial Engineering*, *3*, 1–31.

Olguín, E. J. (2012). Dual purpose microalgae-bacteria-based systems that treat wastewater and produce biodiesel and chemical products within a biorefinery. *Biotechnology Advances*, *30*(5), 1031–1046. DOI: 10.1016/j.biotechadv.2012.05.001

Olmos-Espejel, J. J., de Llasera, M. P. G., & Velasco-Cruz, M. (2012). Extraction and analysis of polycyclic aromatic hydrocarbons and benzo[a]pyrene metabolites in microalgae cultures by off-line/on-line methodology based on matrix solid-phase dispersion, solid-phase extraction, and high-performance liquid chromatography. *Journal of Chromatography. A*, *1262*, 138–147. DOI: 10.1016/j.chroma.2012.09.015 PMID: 23031438

Omar, H. H. (2002). Bioremoval of zinc ions by *Scenedesmus obliquus* and *Scenedesmus quadricauda* and its effect on growth and metabolism. *International Biodeterioration & Biodegradation*, *50*(2), 95–100. DOI: 10.1016/S0964-8305(02)00048-3

Orandi, S., Lewis, D. M., & Moheimani, N. R. (2012). Biofilm establishment and heavy metal removal capacity of an indigenous mining algal-microbial consortium in a photo-rotating biological contactor. *Journal of Industrial Microbiology & Biotechnology*, *39*(9), 1321–1331. DOI: 10.1007/s10295-012-1142-9 PMID: 22644382

Oruganti, R. K., Katam, K., Show, P. L., Gadhamshetty, V., Upadhyayula, V. K. K., & Bhattacharyya, D. (2022). A comprehensive review on the use of algal-bacterial systems for wastewater treatment with emphasis on nutrient and micropollutant removal. *Bioengineered, 13*(4), 10412–10453. DOI: 10.1080/21655979.2022.2056823 PMID: 35441582

Osathanunkul, M., Thanaporn, S., Karapetsi, L., Nteve, G. M., Pratsinakis, E., Stefanidou, E., Lagiotis, G., Avramidou, E., Zorxzobokou, L., Tsintzou, G., Athanasiou, A., Mpelai, S., Constandinidis, C., Pantiora, P., Merino, M., Mullor, J. L., Dobrovic, L., Cerasino, L., Ogawa, T., & Madesis, P. (2025). Diversity of bioactive compounds in microalgae: Key classes and functional applications. *Marine Drugs, 23*(6), 222. DOI: 10.3390/md23060222 PMID: 40559631

Osman, A. I., Qasim, U., & Jamil, F. (2021). Bioenergy and biofuels: History, status, and perspective. *Renewable & Sustainable Energy Reviews, 135*, 110183.

Ozkan, A., & Berberoglu, H. (2013). Physico-chemical surface properties of microalgae. *Colloids and Surfaces. B, Biointerfaces, 112*, 287–293. DOI: 10.1016/j.colsurfb.2013.08.001 PMID: 24001448

Padhi, D., Kashyap, S., Mohapatra, R. K., Dineshkumar, R., & Nayak, M. (2025). Microalgae-based flue gas CO2 sequestration for cleaner environment and biofuel feedstock production: A review. *Environmental Science and Pollution Research International, 32*(23), 13539–13565. DOI: 10.1007/s11356-025-35958-8 PMID: 39888525

Pagnussat, L., Maroniche, G., Curatti, L., & Creus, C. (2020). Auxin-dependent alleviation of oxidative stress and growth promotion of *Scenedesmus obliquus* C1S by *Azospirillum brasilense. Algal Research, 47*, 101839. DOI: 10.1016/j.algal.2020.101839

Panahi, H. K. S., Dehhaghi, M., Guillemin, G. J., Gupta, V. K., Lam, S. S., Aghbashlo, M., & Tabatabaei, M. (2022). Bioethanol production from food wastes rich in carbohydrates. *Current Opinion in Food Science, 43*, 71–81. DOI: 10.1016/j.cofs.2021.11.001

Pandey, P., & Tewary, S. Angelin, & Paruchuri, S. (2026). Potential Sources of Emerging Environmental Pollutants: Fate of EPs in Environment (Soil, Water, and Air). In *Emerging Contaminants in the Environment: Occurrence, Detection and Biodegradation* (pp. 41-76). Springer.

Pandey, A. K. (2025). Sustainable water management through integrated technologies and circular resource recovery. *Environmental Science. Water Research & Technology, 11*(8), 1822–1846. DOI: 10.1039/D5EW00103J

Pandey, A., & Sinha, A. (2022). IoT enabled smart algal bioreactors for wastewater treatment and bioenergy production. *Journal of Cleaner Production, 365*, 132675. DOI: 10.1016/j.jclepro.2022.132675

Pandey, S., Soni, R., Devi, P., Singh, P., & Pandey, A. (2025). Innovations in microalgae-based wastewater remediation, which simultaneously recovers value-added products from industrial and non-industrial wastewater: A review. *Plant Archives, 25*, 921–932.

Pan, S., Zabed, H. M., Zhao, M., Qi, X., & Wei, Y. (2023). Techno-economic and life cycle assessments for bioenergy recovery from acid-hydrolyzed residues of sugarcane bagasse in the biobased xylose production platform. *Journal of Cleaner Production, 400*, 136718. Advance online publication. DOI: 10.1016/j.jclepro.2023.136718

Panwar, V., & Dutta, T. (2019). Diatom Biogenic Silica as a Felicitous Platform for Biochemical Engineering: Expanding Frontiers. *ACS Applied Bio Materials, 2*(6), 2295–2316. DOI: 10.1021/acsabm.9b00050 PMID: 35030724

Parikh, H., Dave, G., Adhikari, S., & Tiwari, A. (2026a). Diatoms as ecological engineers in wastewater systems: A comprehensive review. *Environmental Technology Reviews, 15*(1), 223–248. DOI: 10.1080/21622515.2026.2669110

Park, J. B. K., Craggs, R. J., & Shilton, A. N. (2011). Wastewater treatment high rate algal ponds for biofuel production. *Bioresource Technology, 102*(1), 35–42. DOI: 10.1016/j.biortech.2010.06.158 PMID: 20674341

Parmar, K. S., & Patel, K. M. (2025). Biosorption and bioremediation of heavy metal ions from wastewater using algae: A comprehensive review. *World Journal of Microbiology & Biotechnology, 41*(7), 262. DOI: 10.1007/s11274-025-04424-5 PMID: 40640643

Passos, F., Gutiérrez, R., Uggetti, E., Garfí, M., García, J., & Ferrer, I. (2017). Towards energy neutral microalgae-based wastewater treatment plants. *Algal Research, 28*, 235–243. DOI: 10.1016/j.algal.2017.11.006

Patel, A. K. (2025). Microalgae in dual role: An effective platform for biological carbon capturing and high-value carotenoid production. *Current Pollution Reports, 11*(1), 34. DOI: 10.1007/s40726-025-00362-w

Patel, S. K., Singh, S., Benjamin, J. C., Singh, V. R., Bisht, D., & Lal, R. K. (2024). Plant growth-promoting activities of *Serratia marcescens* and *Pseudomonas fluorescens* on *Capsicum annuum* L. *Ecological Frontiers, 44*(4), 654–663. DOI: 10.1016/j.ecofro.2024.01.002

Pathak, H. K., Gandhi, K., & Singh, R. P. (2026). Potential Impact of Heavy Metals and Microplastics in River Ecosystem on Aquatic Organisms and Human Health, and Sustainable Mitigation Approaches. *Water, Air, and Soil Pollution*, *237*(4), 199. DOI: 10.1007/s11270-025-08844-9

Patil, A., Chakraborty, S., Yadav, Y., Sharma, B., Singh, S., & Arya, M. (2025). Bioremediation strategies and mechanisms of bacteria for resistance against heavy metals: A review. *Bioremediation Journal*, *29*(4), 448–480. DOI: 10.1080/10889868.2024.2375204

Patnaik, R., & Mallick, N. (2021). Microalgal biodiesel production: Realizing the sustainability index. *Frontiers in Bioengineering and Biotechnology*, *9*, 620777. DOI: 10.3389/fbioe.2021.620777 PMID: 34124015

Pekkoh, J., Wichaphian, A., Kamngoen, A., Sriket, N., Zin, M. T., Lomakool, S., Maneechote, W., Chromkaew, Y., Pathom-aree, W., Cheirsilp, B., & Srinuanpan, S. (2024). Heterotrophic upcycling of hydroponic wastewater supplemented with glucose and indole-3-acetic acid into high-quality Chlorella biomass for zero-waste multiproduct microalgal biorefinery. *Environmental Technology & Innovation*, *36*, 103813. DOI: 10.1016/j.eti.2024.103813

Peñarrubia Fernandez, I. A., Liu, D. H., & Zhao, J. (2017). Life cycle assessment of biodiesel production from microalgae. *Journal of Environmental Management*, *204*, 377–386.

Peñarrubia Fernandez, I. A., Liu, D.-H., & Zhao, J. (2017). LCA studies comparing alkaline and immobilized enzyme catalyst processes for biodiesel production under Brazilian conditions. *Resources, Conservation and Recycling*, *119*, 117–127. DOI: 10.1016/j.resconrec.2016.05.009

Peng, H., de-Bashan, L. E., Bashan, Y., & Higgins, B. T. (2020). Indole-3-acetic acid from *Azospirillum brasilense* promotes growth in green algae at the expense of energy storage products. *Algal Research*, *47*, 101845. DOI: 10.1016/j.algal.2020.101845

Peng, Z., Liu, G., & Huang, K. (2020). Cold Adaptation Mechanisms of a Snow Alga Chlamydomonas nivalis During Temperature Fluctuations. *Frontiers in Microbiology*, *11*, 611080. DOI: 10.3389/fmicb.2020.611080 PMID: 33584575

Pérez Mesa, A., Céspedes Grattz, P. A., Vidal Vargas, J. J., Ríos, L. A., & Ocampo Echeverri, D. (2025). Techno-Economic Assessment of Microalgae-Based Biofertilizer Production from Municipal Wastewater Using Scenedesmus sp. *Water (Basel)*, *17*(20), 2941. DOI: 10.3390/w17202941

Pérez-García, O., & Bashan, Y. (2015). Microalgal heterotrophic and mixotrophic culturing for bio-refining: From metabolic routes to techno-economics. In *Algal Biorefineries: Volume 2 Production, Refining, Design* (pp. 61–131). DOI: 10.1007/978-3-319-20200-6_3

Pérez-Rama, M. (2002). Cadmium removal by living cells of the marine microalga Tetraselmis suecica. *Bioresource Technology, 84*(3), 265–270. DOI: 10.1016/S0960-8524(02)00045-7 PMID: 12118704

Perin, G., & Morosinotto, T. (2023). Understanding regulation in complex environments: A route to enhance photosynthetic light-reactions in microalgae photobioreactors. *Frontiers in Photobiology, 1*, Article 1274525.

Peters, R. W., Ku, Y., Boardman, G. D., Walker, T. J., Hoffman, C. P., Chang, T.-K., & Parks, J. T. (1984). Wastewater Treatment-Physical and Chemical Methods. *Journal - Water Pollution Control Federation, 56*(6), 553–568.

Pholchan, M. K., Baptista, J. de C., Davenport, R. J., & Curtis, T. P. (2010). Systematic study of the effect of operating variables on reactor performance and microbial diversity in laboratory-scale activated sludge reactors. *Water Research, 44*(5), 1341–1352. DOI: 10.1016/j.watres.2009.11.005 PMID: 19945137

Pidlisnyuk, V., Mamirova, A., Pranaw, K., Shapoval, P. Y., Trögl, J., & Nurzhanova, A. (2020). Potential role of plant growth-promoting bacteria in *Miscanthus* × *giganteus* phytotechnology applied to trace element contaminated soils. *International Biodeterioration & Biodegradation, 155*, 105103. DOI: 10.1016/j.ibiod.2020.105103

Pittman, J. K., Dean, A. P., & Osundeko, O. (2011). The potential of sustainable algal biofuel production using wastewater resources. *Bioresource Technology, 102*(1), 17–25. DOI: 10.1016/j.biortech.2010.06.035 PMID: 20594826

Posadas, E., Bochon, S., Coca, M., García-González, M., García-Encina, P., & Muñoz, R. (2014). Microalgae-based agro-industrial wastewater treatment: A preliminary screening of biodegradability. *Journal of Applied Phycology, 26*(6), 2335–2345. DOI: 10.1007/s10811-014-0263-0

Pradhan, D., Sukla, L. B., Mishra, B. B., & Devi, N. (2019). Biosorption for removal of hexavalent chromium using microalgae *Scenedesmus* sp. *Journal of Cleaner Production, 209*, 617–629. DOI: 10.1016/j.jclepro.2018.10.288

Prashar, P., & Shah, S. (2016). Impact of fertilizers and pesticides on soil microflora in agriculture. In Lichtfouse, E. (Ed.), *Sustainable Agriculture Reviews* (Vol. 19). Springer., DOI: 10.1007/978-3-319-26777-7_8

Priya, A. S., Premanand, R., Ragupathi, I., Bhaviripudi, V. R., Aepuru, R., Kannan, K., & Shanmugaraj, K. (2024). Comprehensive Review of Hydrogel Synthesis, Characterization, and Emerging Applications. *Journal of Composites Science*, *8*(11), 457. DOI: 10.3390/jcs8110457

Priyadarshini, S., Jagatee, S., Tudu, J., Kumbhar, M. M., & Das, A. P. (2026). Microbe-mediated wastewater treatment: Harnessing nature's engineers for sustainable remedies. In *Decarbonization of Wastewater Pollutants as a Sustainable Solution* (pp. 103–136). Elsevier., DOI: 10.1016/B978-0-443-33062-9.00008-4

Puhakka, J. A., Melin, E. S., Järvinen, K. T., Koro, P. M., Rintala, J. A., Hartikainen, P., Shieh, W. K., & Ferguson, J. F. (1995). Fluidized-bed biofilms for chlorophenol mineralization. *Water Science and Technology*, *31*(1), 227–235. DOI: 10.2166/wst.1995.0051

Qian, W., Yang, Y., Chou, S., Ge, S., Li, P., Wang, X., Zhuang, L.-L., & Zhang, J. (2024). Effect of N/P ratio on attached microalgae growth and the differentiated metabolism along the depth of biofilm. *Environmental Research*, *240*, 117428. DOI: 10.1016/j.envres.2023.117428 PMID: 37875171

Qiao, S., Hou, C., Wang, X., & Zhou, J. (2020). Minimizing greenhouse gas emission from wastewater treatment process by integrating activated sludge and microalgae processes. *The Science of the Total Environment*, *732*, 139032. DOI: 10.1016/j.scitotenv.2020.139032

Qi, L., Liu, X., Gao, Y., Wang, Z., Zhang, N., & Su, X. (2024). Interactions between bacteria and microalgae in microalgal-bacterial symbiotic wastewater treatment systems: Mechanisms and influencing factors. *Aquatic Microbial Ecology*, *90*, 41–60. DOI: 10.3354/ame02007

Qin, C., Liu, H., Liu, L., Smith, S., Sedlak, D. L., & Gu, A. Z. (2015). Bioavailability and characterization of dissolved organic nitrogen and dissolved organic phosphorus in wastewater effluents. *The Science of the Total Environment*, *511*, 47–53. DOI: 10.1016/j.scitotenv.2014.11.005 PMID: 25527968

Qiu, L., Zhang, Q., Zhu, H., Reich, P. B., Banerjee, S., van der Heijden, M. G. A., Sadowsky, M. J., Ishii, S., Jia, X., Shao, M., Liu, B., Jiao, H., Li, H., & Wei, X. (2021). Erosion reduces soil microbial diversity, network complexity and multifunctionality. *The ISME Journal*, *15*(8), 2474–2489. DOI: 10.1038/s41396-021-00913-1 PMID: 33712698

Quiroz, D., Greene, J. M., Limb, B. J., & Quinn, J. C. (2023). Global Life Cycle and Techno-Economic Assessment of Algal-Based Biofuels. *Environmental Science & Technology*, *57*(31), 11541–11551. DOI: 10.1021/acs.est.3c02892 PMID: 37499260

Qu, W., Zhang, C., Chen, X., & Ho, S. H. (2021). New concept in swine wastewater treatment: Development of a self-sustaining synergetic microalgae-bacteria symbiosis (ABS) system to achieve environmental sustainability. *Journal of Hazardous Materials*, *418*, 126264. DOI: 10.1016/j.jhazmat.2021.126264

Qv, M., Dai, D., Liu, D., Wu, Q., Tang, C., Li, S., & Zhu, L. (2023). Towards advanced nutrient removal by microalgae-bacteria symbiosis system for wastewater treatment. *Bioresource Technology*, *370*, 128574. DOI: 10.1016/j.biortech.2022.128574 PMID: 36603749

Rahman, N. A., Katayama, T., Wahid, M. E. A., Jusoh, M., Yusoff, F. M., & Takahashi, K. (2026). Microalgae as biofactories of natural antioxidants: From stress responses to industrial applications. *Journal of Applied Phycology*, *38*(2), 1–34. DOI: 10.1007/s10811-026-03804-0

Rahman, T. T., Jiang, T., Zhang, C., Rao, Y., Sims, R. C., & Hou, L. (2024a). Diatoms in wastewater treatment: Potentials, applications, and values of biomass. *Critical Reviews in Environmental Science and Technology*, *54*(7), 557–580. DOI: 10.1080/10643389.2023.2259278

Raimi, A., Roopnarain, A., & Adeleke, R. (2021). Biofertilizer production in Africa: Current status, factors impeding adoption and strategies for success. *Scientific African*, *11*, e00694. DOI: 10.1016/j.sciaf.2021.e00694

Raj, A., Gauba, P., & Bhatt, E. (2025). Advanced nanoparticles for environmental remediation of emerging pollutants: A review. *Soil & Sediment Contamination*, *34*(6), 1344–1373. DOI: 10.1080/15320383.2024.2422387

Rajkumar, R., Takriff, M. S., & Veeramuthu, A. (2025). Technical insights into carbon dioxide sequestration by microalgae: A biorefinery approach towards sustainable environment. *Biomass Conversion and Biorefinery*, *15*(23), 1–16. DOI: 10.1007/s13399-022-02446-9

Rajta, A., Bhatia, R., Setia, H., & Pathania, P. (2020). Role of heterotrophic aerobic denitrifying bacteria in nitrate removal from wastewater. *Journal of Applied Microbiology*, *128*(5), 1261–1278. DOI: 10.1111/jam.14476 PMID: 31587489

Ramírez Mérida, L. G., & Rodríguez Padrón, R. A. (2023). Application of microalgae in wastewater: Opportunity for sustainable development. *Frontiers in Environmental Science*, *11*, 1238640. DOI: 10.3389/fenvs.2023.1238640

Ramzan, M. N., Shen, D., Wei, Y., Emmanuel, A., Nicholaus, R., Yang, W., & Zheng, Z. (2025). Nitrogen and phosphorus-related functional genes enhance nutrient removal in the integrated aquaculture wastewater bioremediation system in the presence of photosynthetic bacteria. *Aquaculture International*, *33*(2), 131. DOI: 10.1007/s10499-024-01809-5

Rana, A., Joshi, M., Prasanna, R., Shivay, Y. S., & Nain, L. (2012). Biofortification of wheat through inoculation of plant growth promoting rhizobacteria and cyanobacteria. *European Journal of Soil Biology*, *50*, 118–126. DOI: 10.1016/j.ejsobi.2012.01.005

Rani, R., Kumar, V., Gupta, P., & Chandra, A. (2019). Application of plant growth promoting rhizobacteria in remediation of pesticides contaminated stressed soil. In Kumar, V. (Eds.), *New and future developments in microbial biotechnology and bioengineering* (pp. 341–353). Elsevier., DOI: 10.1016/B978-0-12-818258-1.00029-7

Ranjit, P., Jhansi, V., & Reddy, K. V. (2021). Conventional Wastewater Treatment Processes. In Maddela, N. R., García Cruzatty, L. C., & Chakraborty, S. (Eds.), *Advances in the Domain of Environmental Biotechnology* (pp. 455–479). Springer Singapore., DOI: 10.1007/978-981-15-8999-7_17

Rashidi, B., & Trindade, L. M. (2018). Detailed biochemical and morphologic characteristics of the green microalga *Neochloris oleoabundans* cell wall. *Algal Research*, *35*, 152–159. DOI: 10.1016/j.algal.2018.08.033

Rashid, N., Bharathiraja, B., Ali, M., & Yun, J. (2023). Microalgae in wastewater treatment: Current status and future prospects. *Journal of Environmental Management*, *326*, 116673. DOI: 10.1016/j.jenvman.2022.116673

Rashwan, A. K., Al-Addous, M., & Al-Hinti, I. (2023). Renewable energy integration in smart grids: Challenges and solutions. *Energy Reports*, *9*, 124–135.

Rasool, B. S., Abbas, A. K., & Haddad, R. (2025). Photocatalytic innovations in environmental remediation: Mechanisms, materials, and challenges for persistent organic pollutant removal. *Environmental Monitoring and Assessment*, *197*(10), 1086. DOI: 10.1007/s10661-025-14531-3 PMID: 40911193

Rasoul-Amini, S., Montazeri-Najafabady, N., Shaker, S., Safari, A., Kazemi, A., Mousavi, P., Mobasher, M. A., & Ghasemi, Y. (2014). Removal of nitrogen and phosphorus from wastewater using microalgae free cells in a bath culture system. *Biocatalysis and Agricultural Biotechnology*, *3*(2), 126–131. DOI: 10.1016/j.bcab.2013.09.003

Rathod, H. (2014). *Algae-based wastewater treatment* [Master's seminar report, University of Roorkee, India].

Rawat, I., Kumar, R. R., Mutanda, T., & Bux, F. (2016). Dual role of microalgae: Phycoremediation of domestic wastewater and biomass production for sustainable biofuels production. *Applied Energy, 88*(10), 3411–3424. DOI: 10.1016/j.apenergy.2010.11.025

Ren, H. Y., Liu, B. F., Kong, F., Zhao, L., & Ren, N. (2015). Hydrogen and lipid production from starch wastewater by co-culture of anaerobic sludge and oleaginous microalgae with simultaneous COD, nitrogen and phosphorus removal. *Water Research, 85*, 404–412. DOI: 10.1016/j.watres.2015.08.057

Renuka, N., Guldhe, A., Prasanna, R., Singh, P., & Bux, F. (2018). Microalgae as multi-functional options in modern agriculture: Current trends, prospects and challenges. *Biotechnology Advances, 36*(4), 1255–1273. DOI: 10.1016/j.biotechadv.2018.04.004 PMID: 29673972

Ren, W., Zhang, Q., Chen, J., Xiao, X., Duan, X., Luo, X., & Wang, S. (2025). Catalytic resource recovery for transformation of the wastewater industry. *Nature Water, 3*(11), 1–15. DOI: 10.1038/s44221-025-00530-8

Ren, Z., Li, H., Sun, P., Fu, R., Bai, Z., Zhang, G., Sun, L., & Wei, Y. (2024). Development and challenges of emerging biological technologies for algal-bacterial symbiosis systems: A review. *Bioresource Technology, 413*, 131459. DOI: 10.1016/j.biortech.2024.131459 PMID: 39255948

Reymann, T., Kerner, M., & Kümmerer, K. (2020). Assessment of the biotic and abiotic elimination processes of five micropollutants during cultivation of the green microalgae *Acutodesmus obliquus. Bioresource Technology Reports, 11*, 100512. DOI: 10.1016/j.biteb.2020.100512

Reza, M., Iskakova, Z., Afroze, S., Kuterbekov, K., Kabyshev, A., Bekmyrza, K., Kubenova, M., Bakar, M., Azad, A., Roy, H., & Islam, M. (2023). Influence of Catalyst on the Yield and Quality of Bio-Oil for the Catalytic Pyrolysis of Biomass: A Comprehensive Review. *Energies, 16*(14), 5547. DOI: 10.3390/en16145547

Rezania, S., Taib, S. M., Din, M. F. M., Dahalan, F. A., & Kamyab, H. (2016). Comprehensive review on phytotechnology: Heavy metals removal by diverse aquatic plant species from wastewater. *Journal of Hazardous Materials, 318*, 587–599. DOI: 10.1016/j.jhazmat.2016.07.053 PMID: 27474848

Ri, C., Tao, Y., Tu, J., Li, X., She, S., Hou, L., Fu, Y., & Chen, L. (2023). Improvement of the Cd removal efficiency of a filamentous cyanobacterium Leptolyngbya sp. XZ1 through co-culture with Bacillus sp. S1. *Journal of Applied Phycology, 35*(6), 2935–2944. DOI: 10.1007/s10811-023-03078-w

Richa, A., Touil, S., Fizir, M., & Martinez, V. (2020). Recent advances and perspectives in the treatment of hydroponic wastewater: A review. *Reviews in Environmental Science and Biotechnology*, *19*(4), 945–966. DOI: 10.1007/s11157-020-09555-9

Richardson, S. D., & Kimura, S. Y. (2020). Water analysis: Emerging contaminants and current issues. *Analytical Chemistry*, *92*(1), 473–505. DOI: 10.1021/acs.analchem.9b05269 PMID: 31825597

Riley, S., Marcuccio, F., Xu, X., Brandon, N., Korchev, Y., Liu, F., Shevchuk, A., & George, C. (2026). In Situ Nanoscale Imaging of Cultivated, As-Grown Diatom Frustules Electrochemical Lithiation/Delithiation in Li Ion Battery Anodes. *ACS Applied Energy Materials*. DOI: 10.1021/acsaem.6c00689

Rizzo, L., Malato, S., Antakyali, D., Beretsou, V. G., Dolic, M. B., Gernjak, W., Heath, E., Ivancev-Tumbas, I., Karaolia, P., Ribeiro, A. R. L., Mascolo, G., McArdell, C. S., Schaar, H., Silva, A. M. T., & Fatta-Kassinos, D. (2018). Consolidated vs new advanced treatment methods for the removal of contaminants of emerging concern from urban wastewater. *The Science of the Total Environment*, *655*, 986–1008. DOI: 10.1016/j.scitotenv.2018.11.265

Robles, Á., Capson-Tojo, G., Gales, A., Viruela, A., Sialve, B., Seco, A., Steyer, J.-P., & Ferrer, J. (2020). Performance of a membrane-coupled high-rate algal pond for urban wastewater treatment at demonstration scale. *Bioresource Technology*, *301*, 122672. DOI: 10.1016/j.biortech.2019.122672 PMID: 31945681

Rochman, C. M., Hoh, E., Kurobe, T., & Teh, S. J. (2013). Ingested plastic transfers hazardous chemicals to fish and induces hepatic stress. *Scientific Reports*, *3*(1), 3263. DOI: 10.1038/srep03263 PMID: 24263561

Rodriguez-Narvaez, O. M., Peralta-Hernandez, J. M., Goonetilleke, A., & Bandala, E. R. (2017). Treatment technologies for emerging contaminants in water: A review. *Chemical Engineering Journal*, *323*, 361–380. DOI: 10.1016/j.cej.2017.04.106

Romera, E., González, F., Ballester, A., Blázquez, M. L., & Muñoz, J. A. (2006). Biosorption with algae: A statistical review. *Critical Reviews in Biotechnology*, *26*(4), 223–235. DOI: 10.1080/07388550600972153 PMID: 17095433

Romera, E., González, F., Ballester, A., Blázquez, M. L., & Muñoz, J. A. (2007). Comparative study of biosorption of heavy metals using different types of algae. *Bioresource Technology*, *98*(17), 3344–3353. DOI: 10.1016/j.biortech.2006.09.026 PMID: 17624771

Romero-Perdomo, F., & González-Curbelo, M. Á. (2023). Integrating Multi-Criteria Techniques in Life-Cycle Tools for the Circular Bioeconomy Transition of Agri-Food Waste Biomass: A Systematic Review. *Sustainability (Basel), 15*(6), 5026. DOI: 10.3390/su15065026

Rosli, S. S., Amalina Kadir, W. N. A., Wong, C. Y., Han, F. Y., Lim, J. W., Lam, M. K., Yusup, S., Kiatkittipong, W., Kiatkittipong, K., & Usman, A. (2020). Insight review of attached microalgae growth focusing on support material packed in photobioreactor for sustainable biodiesel production and wastewater bioremediation. *Renewable & Sustainable Energy Reviews, 134*, 110358. DOI: 10.1016/j.rser.2020.110306

Roy, A., Ganguly, N., Banerjee, S., & Pandit, S. (2024). From Linear Economy to Circular Bio-economy: A Paradigm Shift in Wastewater Management. In Pandit, S., Kumar, L., Mathuriya, A. S., Jadhav, D. A., & Ray, S. (Eds.), *Advancements in Bio-systems and Technologies for Wastewater Treatment* (Vol. 118, pp. 135–154). Springer International Publishing., DOI: 10.1007/978-3-031-58331-5_8

Roychoudhury, P., Bose, R., Dąbek, P., & Witkowski, A. (2022a). Photonic Nano-/Microstructured Diatom Based Biosilica in Metal Modification and Removal—A Review. *Materials (Basel), 15*(19), 6597. DOI: 10.3390/ma15196597 PMID: 36233939

Ruales, E., Bellver, M., Álvarez-González, A., Carvalho Fontes Sampaio, I., Garfi, M., & Ferrer, I. (2025). Carotenoids and biogas recovery from microalgae treating wastewater. *Bioresource Technology, 428*, 132427. DOI: 10.1016/j.biortech.2025.132427 PMID: 40120990

Saeed, K., Abdullah, S., Yaqoob, M., Fatima, M., Majeed, Z., Ismail, M., Ghanem, S. F., Eissa, M. E., Hassanien, H., Abdul Kari, Z., & Eissa, E.-S. H. (2025). Integrating hematological profile and biomarker analysis for metals and metalloids ecotoxicological assessment on Wallago catfish and Indian carp health in Head Siphon, Mailsi, Pakistan. *Biometals : An International Journal on the Role of Metal Ions in Biology, Biochemistry, and Medicine, 38*(6), 1897–1915. DOI: 10.1007/s10534-025-00737-y PMID: 40911255

Safitri, R., Priadie, B., Miranti, M., & Astuti, A. W. (2015). Ability of bacterial consortium Bacillus coagulans Bacilus licheniformis Bacillus pumilus Bacillus subtilis Nitrosomonas sp and Pseudomonas putida in bioremediation of waste water in cisirung waste water treatment plant. *AgroLife Scientific Journal, 4*(1), 146–152. https://www.agrolifejournal.usamv.ro/index.php/agrolife/article/view/402. DOI: 10.17930/AGL2015122

Safonova, E. T., Dmitrieva, I. A., & Kvitko, K. V. (1999). The interaction of algae with alkanotrophic bacteria in black oil decomposition. *Resources, Conservation and Recycling*, *27*(1-2), 193–201. DOI: 10.1016/S0921-3449(99)00014-2

Saha, I., Datta, S., & Biswas, D. (2023). Mechanistic approach to comparative biosorption of dyes by the extracellular polymer of *Ochrobactrum pseudintermedium* C1 utilizing waste mineral lubricating oil. *Bioremediation Journal*, *28*(3), 266–285. DOI: 10.1080/10889868.2023.2269220

Sahoo, J. K. (2025). Diatomite (Diatomaceous Earth): Global Perspectives on Geology, Processing, Applications, Environmental Sustainability, and Emerging Biomedical Frontiers. *Mining, Metallurgy & Exploration*, *42*(6), 3363–3378. DOI: 10.1007/s42461-025-01411-0

Saini, K., Prajapati, A., Kumar, S. S., Kumar, V., & Bajar, S. (2026). A Review on the Fate of Emerging Contaminants in Landfill Leachate: Insights from Conventional Treatment Approaches. *Water, Air, and Soil Pollution*, *237*(5), 273. DOI: 10.1007/s11270-025-08937-5

Salimbeni, A., Lombardi, G., Rizzo, A. M., & Chiaramonti, D. (2023). Techno-economic feasibility of integrating biomass slow pyrolysis in an EAF steelmaking site: A case study. *Applied Energy*, *339*, 120991. Advance online publication. DOI: 10.1016/j.apenergy.2023.120991

Salimi, M., Esrafili, A., Gholami, M., Jafari, A. J., Kalantary, R. R., Farzadkia, M., Kermani, M., & Sobhi, H. R. (2017). Contaminants of emerging concern: A review of new approach in AOP technologies. *Environmental Monitoring and Assessment*, *189*(8), 414. DOI: 10.1007/s10661-017-6097-x

Samal, D. K., & Sukla, L. B. (2025). Unveiling the dual potential of microalgae and seaweed biomass for sustainable biofuel production: A review. *RSC Advances*, *15*(41), 34160–34175. DOI: 10.1039/D5RA04845A PMID: 40979955

Sangeetha, J., Codreanu, S., & Thangadurai, D. (2023). *Microalgal Biotechnology: Bioprospecting Microalgae for Functional Metabolites towards Commercial and Sustainable Applications*. CRC Press. DOI: 10.1201/9781003332251

Santos, B., Freitas, F., Sobral, A. J., & Encarnação, T. (2025). Microalgae and circular economy: Unlocking waste to resource pathways for sustainable development. *International Journal of Sustainable Engineering*, *18*(1), 2501488. DOI: 10.1080/19397038.2025.2501488

Saravanan, A., Kumar, P. S., Varjani, S., Jeevanantham, S., Yaashikaa, P. R., Thamarai, P., Abirami, B., & George, C. S. (2021). A review on algal-bacterial symbiotic system for effective treatment of wastewater. *Chemosphere, 271*, 129540. DOI: 10.1016/j.chemosphere.2021.129540 PMID: 33434824

Sarikhani, M. R., Khoshru, B., & Greiner, R. (2019). Isolation and identification of temperature tolerant phosphate solubilizing bacteria as a potential microbial fertilizer. *World Journal of Microbiology & Biotechnology, 35*(8), 137. DOI: 10.1007/s11274-019-2702-1 PMID: 31363938

Sarker, N. K., & Kaparaju, P. (2025). Microalgal wastewater treatment: A circular economy approach. In *Nature-Based Technologies for Wastewater Treatment and Bioenergy Production* (p. 153). DOI: 10.2166/9781789064100_0153

Sarmah, M., & Panda, M. R. (2026). Systematic review on proliferation of microalgae and carbon sequestration potential for sustainability. *World Journal of Microbiology & Biotechnology, 42*(3), 100. DOI: 10.1007/s11274-026-04849-6 PMID: 41731188

Sathya, R., Arasu, M. V., Al-Dhabi, N. A., Vijayaraghavan, P., Ilavenil, S., & Rejiniemon, T. (2023). Towards sustainable wastewater treatment by biological methods—Challenges and advantages of recent technologies. *Urban Climate, 47*, 101378. DOI: 10.1016/j.uclim.2022.101378

Saúco, C., Cano, R., Marín, D., Lara, E., Rogalla, F., & Arbib, Z. (2021). Hybrid wastewater treatment system based on a combination of high-rate algae pond and vertical constructed wetland system at large scale. *Journal of Water Process Engineering, 43*, 102311. DOI: 10.1016/j.jwpe.2021.102311

Savio, S., Farrotti, S., Di Giulio, A., De Santis, S., Ellwood, N. T. W., Ceschin, S., & Congestri, R. (2022). Functionalization of Frustules of the Diatom Staurosirella pinnata for Nickel (Ni) Adsorption From Contaminated Aqueous Solutions. *Frontiers in Marine Science, 9*, 889832. Advance online publication. DOI: 10.3389/fmars.2022.889832

Schiewer, S., & Wong, M. H. (2000). Ionic strength effects in biosorption of metals by marine algae. *Chemosphere, 41*(1–2), 271–282. DOI: 10.1016/S0045-6535(99)00421-X PMID: 10819211

Schlembach, I., Grünberger, A., Rosenbaum, M. A., & Regestein, L. (2021). Measurement techniques to resolve and control population dynamics of mixed culture processes. *Trends in Biotechnology, 39*(10), 1093–1109. DOI: 10.1016/j.tibtech.2021.01.006 PMID: 33573846

Scognamiglio, V., Giardi, M. T., Zappi, D., Touloupakis, E., & Antonacci, A. (2021). Photoautotrophs–Bacteria Co-Cultures: Advances, Challenges and Applications. *Materials (Basel), 14*(11), 3027. DOI: 10.3390/ma14113027 PMID: 34199583

Sebastián, F., Royo, J., & Gómez, M. (2011). Cofiring versus biomass-fired power plants: GHG emissions savings. *Biomass and Bioenergy, 35*(7), 2516–2524.

Sedlakova-Kadukova, J., Kopcakova, A., Gresakova, L., Godany, A., & Pristas, P. (2019). Bioaccumulation and biosorption of zinc by a novel *Streptomyces* K11 strain isolated from highly alkaline aluminium brown mud disposal site. *Ecotoxicology and Environmental Safety, 167*, 204–211. DOI: 10.1016/j.ecoenv.2018.09.123 PMID: 30340085

Segredo-Morales, E., González, E., González-Martín, C., & Vera, L. (2024). Novel vertical upflow multi-column configured membrane photobioreactor with a filtration control system for outdoor microalgae-bacteria cultivation, harvesting and wastewater reclamation. *Chemical Engineering Journal, 482*, 148799. DOI: 10.1016/j.cej.2024.148799

Selvaraj, D., Dhayabaran, N. K., & Mahizhnan, A. (2022). An insight on pollutant removal mechanisms in phycoremediation of textile wastewater. *Environmental Science and Pollution Research International, 30*(60), 124714–124734. DOI: 10.1007/s11356-022-21307-6 PMID: 35708812

Semple, K. T., Cain, R. B., & Schmidt, S. (1999). Biodegradation of aromatic compounds by microalgae. *FEMS Microbiology Letters, 170*(2), 291–300. DOI: 10.1111/j.1574-6968.1999.tb13386.x

Serna-García, R., Lanzoni, Y., García-Depraect, O., Muñoz, R., & Cantera, S. (2026). Harnessing Biogas into High-Value Chemicals: The Role of Algal–Methanotrophic Co-Cultures. *Marine Drugs, 24*(2), 81. DOI: 10.3390/md24020081 PMID: 41745484

Sethunathan, N., Megharaj, M., Chen, Z., Williams, B. D., Lewis, G., & Naidu, R. (2004). Algal degradation of a known endocrine disrupting insecticide, α-endosulfan, and its metabolite, endosulfan sulfate, in liquid medium and soil. *Journal of Agricultural and Food Chemistry, 52*(10), 3030–3035. DOI: 10.1021/jf035173x PMID: 15137849

Shah, S. V., Lamba, B. Y., Sharma, R., Tiwari, A. K., Parveen, A., & Kumar, V. (2023). Effect of different light on microalgae growth and its integration with carbon dioxide sequestration technologies with emphasis on biofixation by microalgae. *Materials Today: Proceedings, 73*, 92–99. DOI: 10.1016/j.matpr.2022.09.370

Shareefdeen, Z., Elkamel, A., & Babar, Z. B. (2023). Recent developments on the performance of algal bioreactors for CO2 removal: Focusing on the light intensity and photoperiods. *Biotech*, *12*(1), 10. DOI: 10.3390/biotech12010010 PMID: 36648836

Sharma, V., & Gupta, R. (2022). Biomass Digestion for Production of Electrical Energy. In (pp. 151-178). DOI: 10.1007/978-981-16-9135-5_7

Sharma, A., & Bhatti, M. S. (2022). Simultaneous Removal of Organic Matter and Nutrients from High Strength Organic Wastewater Using Sequencing Batch Reactor (SBR). *Processes (Basel, Switzerland)*, *10*(10), 1903. https://www.mdpi.com/2227-9717/10/10/1903. DOI: 10.3390/pr10101903

Shekh, A., Sharma, A., Schenk, P. M., Kumar, G., & Mudliar, S. (2022). Microalgae cultivation: Photobioreactors, CO2 utilization, and value-added products of industrial importance. *Journal of Chemical Technology and Biotechnology*, *97*(5), 1064–1085. DOI: 10.1002/jctb.6902

Shekharaiah, P. C., Sanyal, D., Dasgupta, S., Sapre, A., & Banik, A. (2020). Heavy metal mitigation with special reference to bioremediation by mixotrophic algae bacterial protocooperation. In *Cellular and Molecular Phytotoxicity of Heavy Metals* (pp. 305–334). Springer., DOI: 10.1007/978-3-030-45975-8_15

Sheng, P. X., Ting, Y. P., Chen, J. P., & Hong, L. (2004). Sorption of lead, copper, cadmium, zinc, and nickel by marine algal biomass: Characterization of biosorptive capacity and investigation of mechanisms. *Journal of Colloid and Interface Science*, *275*(1), 131–141. DOI: 10.1016/j.jcis.2004.01.036 PMID: 15158390

Shen, Q.-H., Zhi, T.-T., Cheng, L.-H., Xu, X.-H., & Chen, H.-L. (2013). Hexavalent chromium detoxification by nonliving *Chlorella vulgaris* cultivated under tuned conditions. *Chemical Engineering Journal*, *228*, 993–1002. DOI: 10.1016/j.cej.2013.05.074

Shen, Y. (2014a). Carbon dioxide bio-fixation and wastewater treatment via algae photochemical synthesis for biofuels production. *RSC Advances*, *4*(91), 49672–49722. DOI: 10.1039/C4RA06441K

Shen, Y., Gao, J., & Li, L. (2017). Municipal wastewater treatment via co-immobilized microalgal-bacterial symbiosis: Microorganism growth and nutrients removal. *Bioresource Technology*, *243*, 905–913. DOI: 10.1016/j.biortech.2017.07.041 PMID: 28738545

Shi, J., Podola, B., & Melkonian, M. (2014). Application of a prototype-scale twin-layer photobioreactor for effective N and P removal from different process stages of municipal wastewater by immobilized microalgae. *Bioresource Technology*, *154*, 260–266. DOI: 10.1016/j.biortech.2013.11.100

Shitu, A., Tadda, M. A., Zhao, J., Danhassan, U. A., Ye, Z., Liu, D., Chen, W., & Zhu, S. (2024). Review of recent advances in utilising aquaculture wastewater for algae cultivation and microalgae-based bioproduct recovery. *Environmental Geochemistry and Health*, *46*(12), 485. DOI: 10.1007/s10653-024-02286-8 PMID: 39508916

Shi, Z., Du, H., Wang, C., & Xu, H. (2022). Quantifying the bioaccumulation of Pb to Chlorella vulgaris in the presence of dissolved organic matters with different molecular weights. *Environmental Science and Pollution Research International*, *29*(47), 70921–70932. DOI: 10.1007/s11356-022-19699-6 PMID: 35593980

Shriwastav, A., Thomas, J., & Bose, P. (2017). A comprehensive mechanistic model for simulating algal growth dynamics in photobioreactors. *Bioresource Technology*, *233*, 7–14. DOI: 10.1016/j.biortech.2017.02.080

Sial, A., Zhang, B., Zhang, A., Liu, K., Imtiaz, S. A., & Yashir, N. (2021). Microalgal–Bacterial Synergistic Interactions and Their Potential Influence in Wastewater Treatment: A Review. *BioEnergy Research*, *14*(3), 723–738. DOI. 10.1007/s12155-020-10213-9

Siew Moi, P. (2004). Handbook of Microalgal Culture. Biotechnology and Applied Phycology. *Journal of Applied Phycology*, *16*(2), 159–160. DOI: 10.1023/B:-JAPH.0000044943.45511.5e

Sigonya, S., Mokhothu, T. H., Mokhena, T. C., & Makhanya, T. R. (2023). Mitigation of Non-Steroidal Anti-Inflammatory and Antiretroviral Drugs as Environmental Pollutants by Adsorption Using Nanomaterials as Viable Solution—A Critical Review. *Applied Sciences (Basel, Switzerland)*, *13*(2), 772. DOI: 10.3390/app13020772

Silva, J. A. (2023). Wastewater Treatment and Reuse for Sustainable Water Resources Management: A Systematic Literature Review. *Sustainability (Basel)*, *15*(14), 10940. https://www.mdpi.com/2071-1050/15/14/10940. DOI: 10.3390/su151410940

Simkin, A. J., Kapoor, L., Doss, C. G. P., Hofmann, T. A., Lawson, T., & Ramamoorthy, S. (2022). The role of photosynthesis related pigments in light harvesting, photoprotection and enhancement of photosynthetic yield in planta. *Photosynthesis Research*, *152*(1), 23–42. DOI: 10.1007/s11120-021-00892-6 PMID: 35064531

Singh, A., & Gaur, R. (2026). Evaluating the Carbon Sequestration Potential of Spirulina spp.: Mechanistic Insights and Prospects for Industrial Valorization. *Waste and Biomass Valorization*, •••, 1–18. DOI: 10.1007/s12649-026-03510-5

Singh, G. (2025). Linear Alkylbenzene Sulphonate in Freshwater Ecosystems: Sources, Persistence, Bioaccumulation and Impacts on Fish Health–A Critical Review. *International Journal of Scientific Research in Modern Science and Technology*, *4*(12), 39–50. DOI: 10.59828/ijsrmst.v4i12.403

Singh, J., & Gu, S. (2010). Commercialization potential of microalgae for biofuels production. *Renewable & Sustainable Energy Reviews*, *14*(9), 2596–2610. DOI: 10.1016/j.rser.2010.06.014

Singh, N. K., Raghubanshi, A. S., Upadhyay, A. K., & Rai, U. N. (2016). Arsenic and other heavy metal accumulation in plants and algae growing naturally in contaminated area of West Bengal, India. *Ecotoxicology and Environmental Safety*, *130*, 224–233. DOI: 10.1016/j.ecoenv.2016.04.024

Singh, R., Singh, A., Balomajumder, C., & Vidyarthi, A. K. (2025). Assessment of industrial effluent discharges contributing to Ganga water pollution through a multivariate statistical framework: Investigating the context of Indian industries. *Environmental Science and Pollution Research International*, *32*(59), 31625–31643. DOI: 10.1007/s11356-024-35823-0 PMID: 39760835

Singh, S. P., & Singh, P. (2014). Effect of CO2 concentration on algal growth: A review. *Renewable & Sustainable Energy Reviews*, *38*, 172–179. DOI: 10.1016/j.rser.2014.05.043

Singh, U. B., & Ahluwalia, A. S. (2013). Microalgae: A promising tool for carbon sequestration. *Mitigation and Adaptation Strategies for Global Change*, *18*(1), 73–95. DOI: 10.1007/s11027-012-9393-3

Sinha, T., Borah, D., & Rout, J. (2025). Algal communities in urban freshwater ponds: Prospects for biodiesel production. *Vegetos*, *38*(1), 202–212. DOI: 10.1007/s42535-023-00804-0

Sivakumar, S., Nurgat, M. Z., Srinivasan, L. V., Bora, A., Wagh, M. S., Kungwani, N. A., & Narayan, M. (2025). Bioengineered Nano-sorbents: Microbial Allies in Pollutant Sequestration. In Mohanta, Y. K., Mishra, B., & Pudake, R. N. (Eds.), *Nano-microbiology for Sustainable Development* (pp. 331–372). Springer Nature Switzerland., DOI: 10.1007/978-3-031-78845-1_13

Smith, E. R., Hewitson, T. D., Hanssen, E., & Holt, S. G. (2018). Biochemical transformation of calciprotein particles in uraemia. *Bone*, *110*, 355–367. DOI: 10.1016/j.bone.2018.02.023 PMID: 29499417

Smułek, W., & Kaczorek, E. (2022). Factors Influencing the Bioavailability of Organic Molecules to Bacterial Cells—A Mini-Review. *Molecules (Basel, Switzerland)*, *27*(19), 6579. DOI: 10.3390/molecules27196579 PMID: 36235114

Sohail, M. (2025). Fish as Bioindicators of Pesticide-Induced Histological and Biochemical Toxicity. *International Journal of Forest Sciences*, *2*, 205–215.

Solé, A., & Matamoros, V. (2016). Removal of endocrine disrupting compounds from wastewater by microalgae co-immobilized in alginate beads. *Chemosphere*, *164*, 516–523. DOI: 10.1016/j.chemosphere.2016.08.047

Soleymani Angili, T., Grzesik, K., & Rödl, A. (2021). Life cycle assessment of bioethanol from microalgae and lignocellulosic biomass. *The Science of the Total Environment*, *762*, 143134.

Soleymani, A. T., Grzesik, K., Rödl, A., & Kaltschmitt, M. (2021). Life cycle assessment of bioethanol production: A review of feedstock, technology and methodology. *Energies*, *14*(10), 2939. Advance online publication. DOI: 10.3390/en14102939

Solisio, C., Al Arni, S., & Converti, A. (2019). Adsorption of inorganic mercury from aqueous solutions onto dry biomass of *Chlorella vulgaris*: Kinetic and isotherm study. *Environmental Technology*, *40*(5), 664–672. DOI: 10.1080/09593330.2017.1400114 PMID: 29098939

Somers, M. D., Chen, P., Clippinger, J., Cruce, J. R., Davis, R., Lammers, P. J., & Quinn, J. C. (2021). Techno-economic and life-cycle assessment of fuel production from mixotrophic Galdieria sulphuraria microalgae on hydrolysate. *Algal Research*, *59*, 102419. Advance online publication. DOI: 10.1016/j.algal.2021.102419

Song, Q., Kong, F., Liu, B.-F., Song, X., Tan, X., Ren, N.-Q., & Ren, H.-Y. (2026). Synergistic coupling of microalgae and corn cob residue biochar for enhanced lipid production and heavy metal removal from waste leachate and corn cob hydrolysate. *Energy*, *347*, 140430. DOI: 10.1016/j.energy.2026.140430

Song, Y., Hu, Z., Yang, X., An, Y., & Lu, Y. (2025). Synergistic microalgae-duckweed systems for enhanced aquaculture wastewater treatment, biomass recovery, and CO2 sequestration: A novel approach for sustainable resource recovery. *Environmental Research*, *274*, 121271. DOI: 10.1016/j.envres.2025.121271 PMID: 40043930

Sonune, M., Mungal, M., & Kamble, S. (2015). *Study of physico-chemical characteristics of domestic wastewater in Vishnupuri*. Nanded.

Soo, A., & Shon, H. K. (2024). A nutrient circular economy framework for wastewater treatment plants. *Desalination, 592*, 118090. DOI: 10.1016/j.desal.2024.118090

Soto, M. F., Diaz, C. A., Zapata, A. M., & Higuita, J. C. (2021). BOD and COD removal in vinasses from sugarcane alcoholic distillation by *Chlorella vulgaris*: Environmental evaluation. *Biochemical Engineering Journal, 176*, 108209. DOI: 10.1016/j.bej.2021.108191

Sriket, N., Wichaphian, A., Kamngoen, A., Pekkoh, J., Pathom-aree, W., Chromkaew, Y., Cheirsilp, B., Maneechote, W., Leong, Y. K., & Srinuanpan, S. (2025). Converting hydroponic effluent into high PUFA-rich biomass by mixotrophically cultivated microalga Chlorella sp. Through light regulation for sustainable production of bioingredient and bioenergy. *Journal of Environmental Chemical Engineering, 13*(3), 117187. DOI: 10.1016/j.jece.2025.117187

Srimongkol, P., Sangtanoo, P., Songserm, P., Watsuntorn, W., & Karnchanatat, A. (2022). Microalgae-based wastewater treatment for developing economic and environmental sustainability: Current status and future prospects. *Frontiers in Bioengineering and Biotechnology, 10*, 904046. DOI: 10.3389/fbioe.2022.904046 PMID: 36159694

Srivastav, A. L. (2020). Chemical fertilizers and pesticides: Role in groundwater contamination. In Srivastav, A. L. (Ed.), *Agrochemicals detection, treatment and remediation* (pp. 143–159). Butterworth-Heinemann., DOI: 10.1016/B978-0-08-103017-2.00006-4

Subbaiah, M. P., Kalimuthu, P., Jung, J., & Jeon, B.-H. (2021). Recent advances in effective capture of inorganic mercury from aqueous solutions through sulfurized 2D-material-based adsorbents. *Journal of Materials Chemistry. A, Materials for Energy and Sustainability, 9*(34), 18086–18101. DOI: 10.1039/D1TA01018B

Sui, J., Yang, F., Huang, Z., & Li, M. (2025). Physiological responses and degradation potential of microalgae Chlorella vulgaris, Chlamydomonas sp. and Tetradesmus obliquus to organophosphorus pesticides chlorpyrifos and dimethoate. *Journal of Applied Phycology, 37*(6), 4123–4137. DOI: 10.1007/s10811-025-03676-w

Sun, S., Zhang, M., Yang, G., Wang, Y., & Chen, Z. (2025). A review on the mechanisms and applications of bacteria in responding, adsorbing and transporting heavy metals. *Archives of Microbiology, 207*(11), 286. DOI: 10.1007/s00203-025-04441-w PMID: 41051408

Sun, Z., Liu, X., Ugya, A. Y., Liu, H., Sun, L., & Luo, G. (2026). Microalgae and Cyanobacteria as a Tool for Agricultural Sustainability: A Review of Biofertilizer and Biostimulant Potential. *Frontiers in Plant Science*, *16*, 1733394. DOI: 10.3389/fpls.2025.1733394 PMID: 41602522

Sutherland, D. L., Howard-Williams, C., Turnbull, M. H., Broady, P. A., & Craggs, R. J. (2015). Enhancing microalgal photosynthesis and productivity in wastewater treatment high-rate algal ponds for biofuel production. *Bioresource Technology*, *184*, 222–229. DOI: 10.1016/j.biortech.2014.10.074 PMID: 25453429

Sutherland, D. L., Park, J., Ralph, P. J., & Craggs, R. J. (2020). Improved microalgal productivity and nutrient removal through operating wastewater high rate algal ponds in series. *Algal Research*, *47*, 101850. DOI: 10.1016/j.algal.2020.101850

Sutherland, D. L., & Ralph, P. J. (2019). Microalgal bioremediation of emerging contaminants: Opportunities and challenges. *Water Research*, *164*, 114921. DOI: 10.1016/j.watres.2019.114921 PMID: 31382151

Sutton, P. C., Anderson, S. J., Costanza, R., & Kubiszewski, I. (2016). The ecological economics of land degradation: Impacts on ecosystem service values. *Ecological Economics*, *129*, 182–192. DOI: 10.1016/j.ecolecon.2016.06.016

Su, Y., & Jacobsen, C. (2021). Treatment of clean in place (CIP) wastewater using microalgae: Nutrient upcycling and value-added byproducts production. *The Science of the Total Environment*, *785*, 147277. DOI: 10.1016/j.scitotenv.2021.147337

Swain, P. R., Parida, P. K., Majhi, P. J., Behera, B. K., & Das, B. K. (2025). Microplastics as emerging contaminants: Challenges in inland aquatic food web. *Water (Basel)*, *17*(2), 201. DOI: 10.3390/w17020201

Swaminaathan, P., Shaji, A., Saravanan, A., & Yaashikaa, P. R. (2024). Innovative Approaches in Extremophile-Mediated Remediation of Toxic Pollutants: A Comprehensive Review. *Water Conservation Science and Engineering*, *9*(2), 39. DOI: 10.1007/s41101-024-00274-8

Swanson, R. M., Platon, A., Satrio, J. A., & Brown, R. C. (2010). Techno-economic analysis of biomass-to-liquids production based on gasification. *Fuel*, *89*, S11–S19. DOI: 10.1016/j.fuel.2010.07.027

Swanson, R. M., Satrio, J. A., & Brown, R. C. (2010). Techno-economic analysis of biofuels production via fast pyrolysis and hydroprocessing. *NREL Technical Report, NREL*, *TP-6*, A20–A46586.

Tangahu, B., Berlianto, M., & Kartika, A. (2020). Deconcentration of Chromium Contained in Wastewater Using a Bacteria and Microalgae Consortia with a High Rate Algal Reactor System. *Journal of Ecological Engineering*, *21*(8), 272–284. DOI: 10.12911/22998993/126878

Tang, L., Gao, M., Liang, S., Wang, S., & Wang, X. (2024). Enhanced biological phosphorus removal sustained by aeration-free filamentous microalgal-bacterial granular sludge. *Water Research*, *253*, 121315. DOI: 10.1016/j.watres.2024.121315 PMID: 38382289

Tang, W., Xu, X., Ye, B.-C., Cao, P., & Ali, A. (2019). Decolorization and degradation analysis of Disperse Red 3B by a consortium of the fungus Aspergillus sp. XJ-2 and the microalgae Chlorella sorokiniana XJK. *RSC Advances*, *9*(25), 14558–14566. DOI: 10.1039/C9RA01169B PMID: 35519313

Tarfeen, N., Nisa, K. U., Hamid, B., Bashir, Z., Yatoo, A. M., Dar, M. A., Mohiddin, F. A., Amin, Z., Ahmad, R. A., & Sayyed, R. (2022). Microbial remediation: A promising tool for reclamation of contaminated sites with special emphasis on heavy metal and pesticide pollution: A review. *Processes (Basel, Switzerland)*, *10*(7), 1358. DOI: 10.3390/pr10071358

Tariq, M., Tahreem, N., Zafar, M., Raza, G., Shahid, M., Zunair, M., Iram, W., & Zahra, S. T. (2024). Occurrence of diverse plant growth promoting bacteria in soybean (*Glycine max* L. Merrill) root nodules and their prospective role in enhancing crop yield. *Biocatalysis and Agricultural Biotechnology*, *57*, 103100. DOI: 10.1016/j.bcab.2024.103072

Tegegne, H. A., & Savidge, T. C. (2025). Gut microbiome metagenomics in clinical practice: Bridging the gap between research and precision medicine. *Gut Microbes*, *17*(1), 2569739. DOI: 10.1080/19490976.2025.2569739 PMID: 41137523

Temkin, M. I., & Pyzhev, V. (1940), . Kinetics of ammonia synthesis on promoted iron catalyst. *Acta Phys. Chim. USSR*, *12*, 327–356.

Thakur, S., Chandra, A., Kumar, V., & Bharti, S. (2025). Environmental pollutants: Endocrine disruptors/pesticides/reactive dyes and inorganic toxic compounds metals, radionuclides, and metalloids and their impact on the ecosystem. In *Biotechnology for environmental sustainability* (pp. 55–100). Springer. DOI: 10.1007/978-981-97-7221-6_3

Tikkanen, M., & Aro, E.-M. (2026). Interacting short-term regulatory mechanisms enable the conversion of light energy to chemical energy in photosynthesis. *Journal of Experimental Botany*, *77*(4), 895–909. DOI: 10.1093/jxb/eraf451 PMID: 41124594

Tiwari, B., Sellamuthu, B., Ouarda, Y., Drogui, P., Tyagi, R. D., & Buelna, G. (2017). Review on fate and mechanism of removal of pharmaceutical pollutants from wastewater using biological approach. *Bioresource Technology, 224*, 1–12. DOI: 10.1016/j.biortech.2016.11.042 PMID: 27889353

Toor, S. S., Rosendahl, L., & Rudolf, A. (2011). Hydrothermal liquefaction of biomass: A review of subcritical water technologies. *Energy, 36*(5), 2328–2342. DOI: 10.1016/j.energy.2011.03.013

Torres, E. (2020). Biosorption: A Review of the Latest Advances. *Processes (Basel, Switzerland), 8*(12), 1584. DOI: 10.3390/pr8121584

Torres, M. A., Barros, M. P., Campos, S. C., Pinto, E., Rajamani, S., Sayre, R. T., & Colepicolo, P. (2008). Biochemical biomarkers in algae and marine pollution: A review. *Ecotoxicology and Environmental Safety, 71*(1), 1–15. DOI: 10.1016/j.ecoenv.2008.05.009 PMID: 18599121

Touliabah, H. E.-S., El-Sheekh, M. M., Ismail, M. M., & El-Kassas, H. (2022). A Review of Microalgae- and Cyanobacteria-Based Biodegradation of Organic Pollutants. *Molecules (Basel, Switzerland), 27*(3), 1141. DOI: 10.3390/molecules27031141 PMID: 35164405

Tournier, V., Topham, C. M., Gilles, A., David, B., Folgoas, C., Moya-Leclair, E., & Pelloux-Prayer, H. (2020). An engineered PET depolymerizing enzyme breaks down PET plastics. *Nature Chemical Biology, 16*(7), 776–782. DOI: 10.1038/s41589-020-0550-4 PMID: 32367018

Tripathi, M., Singh, P., Singh, R., Bala, S., Pathak, N., Singh, S., Chauhan, R. S., & Singh, P. K. (2023). Microbial biosorbent for remediation of dyes and heavy metals pollution: A green strategy for sustainable environment. *Frontiers in Microbiology, 14*, 1168954. DOI: 10.3389/fmicb.2023.1168954 PMID: 37077243

Tripathy, S., Kar, O. P., & Pradhan, A. (2025). Challenges and innovations in industrial wastewater treatment: Safeguarding water resources and promoting sustainable practices. *Water, Air, and Soil Pollution, 236*(2), 92. DOI: 10.1007/s11270-025-07742-4

Trippe, F., Fröhling, M., & Schultmann, F. (2011). Techno-economic analysis of fast pyrolysis for bio-oil production. *Biomass Conversion and Biorefinery, 1*(2), 101–112.

Trivedi, J., Aila, M., Bangwal, D. P., Kaul, S., & Garg, M. O. (2015). Algae based biorefinery - How to make sense? *Renewable & Sustainable Energy Reviews, 47*, 295–307. DOI: 10.1016/j.rser.2015.03.052

Tsednee, M., Castruita, M., Salomé, P. A., Sharma, A., Lewis, B. E., Schmollinger, S. R., Strenkert, D., Holbrook, K., Otegui, M. S., Khatua, K., Das, S., Datta, A., Chen, S., Ramon, C., Ralle, M., Weber, P. K., Stemmler, T. L., Pett-Ridge, J., Hoffman, B. M., & Merchant, S. S. (2019). Manganese co-localizes with calcium and phosphorus in *Chlamydomonas* acidocalcisomes and is mobilized in manganese-deficient conditions. *The Journal of Biological Chemistry*, *294*(46), 17626–17641. DOI: 10.1074/jbc.RA119.009130 PMID: 31527081

Tsiropoulos, I., Faaij, A. P. C., Seabra, J. E. A., Lundquist, L., Schenker, U., Briois, J.-F., & Patel, M. K. (2014). Life cycle assessment of sugarcane ethanol production in India in comparison to Brazil. *The International Journal of Life Cycle Assessment*, *19*(5), 1049–1067. DOI: 10.1007/s11367-014-0714-5

Tuvikene, R., Truus, K., Robal, M., Volobujeva, O., Mellikov, E., Pehk, T., Kollist, A., Kailas, T., & Vaher, M. (2010). The extraction, structure, and gelling properties of hybrid galactan from the red alga Furcellaria lumbricalis (Baltic Sea, Estonia). *Journal of Applied Phycology*, *22*(1), 51–63. DOI: 10.1007/s10811-009-9425-x

Ugwuanyi, E., Nwokediegwu, Z., Dada, M., Majemite, M., & Obaigbena, A. (2024). The role of algae-based wastewater treatment systems: A comprehensive review. *World Journal of Advanced Research and Reviews*, *21*(2), 937–949. DOI: 10.30574/wjarr.2024.21.2.0521

Uludag-Demirer, S., Bustamante, M., Liu, Y., & Liao, W. (2022). Techno-economic analysis and life cycle assessment of algal cultivation on liquid anaerobic digestion effluent for algal biomass production and wastewater treatment. In *Integrated Wastewater Management and Valorization Using Algal Cultures* (pp. 149–164). Elsevier., DOI: 10.1016/B978-0-323-85859-5.00009-9

Umamaheswari, J., & Shanthakumar, S. (2016). Efficacy of microalgae for industrial wastewater treatment: A review on operating conditions, treatment efficiency, and biomass productivity. *Reviews in Environmental Science and Biotechnology*, *15*(2), 265–284. DOI: 10.1007/s11157-016-9397-7

Ummalyma, S. B., Mathew, A. K., Pandey, A., & Sukumaran, R. K. (2016). Harvesting of microalgal biomass: Efficient method for flocculation through pH modulation. *Bioresource Technology*, *213*, 216–221. DOI: 10.1016/j.biortech.2016.03.114 PMID: 27036330

Upadhyay, A. K., Singh, N. K., Singh, R., & Rai, U. N. (2016). Amelioration of arsenic toxicity in rice: Comparative effect of inoculation of *Chlorella vulgaris* and *Nannochloropsis* sp. on growth, biochemical changes and arsenic uptake. *Ecotoxicology and Environmental Safety*, *124*, 68–73. DOI: 10.1016/j.ecoenv.2015.10.002

Upadhyay, A. K., Singh, R., & Singh, D. P. (2019). Phycotechnological Approaches Toward Wastewater Management. In Bharagava, R. N., & Chowdhary, P. (Eds.), *Emerging and Eco-Friendly Approaches for Waste Management* (pp. 423–435). Springer Singapore., DOI: 10.1007/978-981-10-8669-4_18

Urbina-Suarez, N. A., Machuca-Martínez, F., & Barajas-Solano, A. F. (2021). Advanced oxidation processes and biotechnological alternatives for the treatment of tannery wastewater. *Molecules (Basel, Switzerland)*, *26*(11), 3222. DOI: 10.3390/molecules26113222

van Eekeren, N., Jongejans, E., van Agtmaal, M., Guo, Y., van der Velden, M., Versteeg, C., & Siepel, H. (2022). Microarthropod communities and their ecosystem services restore when permanent grassland with mowing or low-intensity grazing is installed. *Agriculture, Ecosystems & Environment*, *323*, 107682. DOI: 10.1016/j.agee.2021.107682

Van Ho, A., Ward, D. M., & Kaplan, J. (2002). Transition Metal Transport in Yeast. *Annual Review of Microbiology*, *56*(1), 237–261. DOI: 10.1146/annurev.micro.56.012302.160847 PMID: 12142483

Vardhan, K. H., Kumar, P. S., & Panda, R. C. (2019). A review on heavy metal pollution, toxicity, and remedial measures: Current trends and future perspectives. *Journal of Molecular Liquids*, *290*, 111197. DOI: 10.1016/j.molliq.2019.111197

Vassalle, L., García-Galán, M. J., Aquino, S. F., Afonso, R., Ferrer, I., Passos, F., & R Mota, C. (2020). Can high rate algal ponds be used as post-treatment of UASB reactors to remove micropollutants? *Chemosphere*, *248*, 126003. DOI: 10.1016/j.chemosphere.2020.125969

Vazirzadeh, A., Jafarifard, K., Ajdari, A., & Chisti, Y. (2022). Removal of nitrate and phosphate from simulated agricultural runoff water by *Chlorella vulgaris*. *The Science of the Total Environment*, *802*, 149988. DOI: 10.1016/j.scitotenv.2021.149988 PMID: 34525699

Velásquez-Orta, S. B., Yáñez-Noguez, I., Ramírez, I. M., & Ledesma, M. T. O. (2024). Pilot-scale microalgae cultivation and wastewater treatment using high-rate ponds: A meta-analysis. *Environmental Science and Pollution Research International*, *31*(34), 46994–47021. DOI: 10.1007/s11356-024-34000-7 PMID: 38985422

Villar-Navarro, E., García-Galán, M. J., Pérez, S., & Rodríguez-Mozaz, S. (2020). Microalgae-based processes for the removal of pharmaceuticals from wastewater: A review. *Environmental Pollution*, *265*, 115114. DOI: 10.1016/j.envpol.2020.115114

Vinayak, V., & Gordon, R. (2025). *Diatom Cultivation for Biofuel, Food and High-Value Products*. John Wiley & Sons. DOI: 10.1002/9781394174980

Viswanaathan, S., Perumal, P. K., & Sundaram, S. (2022). Integrated Approach for Carbon Sequestration and Wastewater Treatment Using Algal–Bacterial Consortia: Opportunities and Challenges. *Sustainability (Basel)*, *14*(3), 1075. DOI: 10.3390/su14031075

Vo, H. N., Ngo, H. H., Guo, W., Nguyen, T., Liu, Y., Liu, Y., Nguyen, D. D., & Chang, S. W. (2019). A critical review on designs and applications of microalgae-based photobioreactors for pollutants treatment. *The Science of the Total Environment*, *651*(Pt 1), 1549–1568. DOI: 10.1016/j.scitotenv.2018.09.282 PMID: 30360283

Volesky, B. (2007). Biosorption and me. *Water Research*, *41*(18), 4017–4029. DOI: 10.1016/j.watres.2007.05.062 PMID: 17632204

Volesky, B., & Holan, Z. R. (1995). Biosorption of Heavy Metals. *Biotechnology Progress*, *11*(3), 235–250. DOI: 10.1021/bp00033a001 PMID: 7619394

Wahid, M. H., Eroglu, E., Chen, X., Smith, S. M., & Raston, C. L. (2013). Functional multi-layer graphene-algae hybrid material formed using vortex fluidics. *Green Chemistry*, *15*(3), 650–655. DOI: 10.1039/c2gc36892g

Walls, L. E., Velasquez-Orta, S. B., Romero-Frasca, E., Leary, P., Yáñez Noguez, I., & Orta Ledesma, M. T. (2019). Non-sterile heterotrophic cultivation of native wastewater yeast and microalgae for integrated municipal wastewater treatment and bioethanol production. *Biochemical Engineering Journal*, *151*, 107319. DOI: 10.1016/j.bej.2019.107319

Wanda, E. M., Nyoni, H., Mamba, B. B., & Msagati, T. A. (2017). Occurrence of emerging micropollutants in water systems in Gauteng, Mpumalanga, and North West provinces, South Africa. *International Journal of Environmental Research and Public Health*, *14*(1), 79. DOI: 10.3390/ijerph14010079 PMID: 28098799

Wang, C., Jiang, C., Gao, T., Peng, X., Ma, S., Sun, Q., Xia, B., Xie, X., Bai, Z., Xu, S., & Zhuang, X. (2022). Improvement of fish production and water quality in a recirculating aquaculture pond enhanced with bacteria-microalgae association. *Aquaculture (Amsterdam, Netherlands)*, *547*, 737440. DOI: 10.1016/j.aquaculture.2021.737420

Wang, C., Luo, D., Zhang, X., Huang, R., Cao, Y., Liu, G., Zhang, Y., & Wang, H. (2022). Biochar-based slow-release of fertilizers for sustainable agriculture: A mini review. *Environmental Science and Ecotechnology*, *10*, 100167. DOI: 10.1016/j.ese.2022.100167 PMID: 36159737

Wang, C.-Y., Fu, C.-C., & Liu, Y.-C. (2007). Effects of using light-emitting diodes on the cultivation of *Spirulina platensis. Biochemical Engineering Journal, 37*(1), 21–25. DOI: 10.1016/j.bej.2007.03.004

Wang, H., Deng, L., Qi, Z., & Wang, W. (2022). Constructed microalgal-bacterial symbiotic (MBS) system: Classification, performance, partnerships and perspectives. *The Science of the Total Environment, 803*, 150082. DOI: 10.1016/j.scitotenv.2021.150082 PMID: 34525774

Wang, H., Qi, M., Bo, Y., Zhou, C., Yan, X., Wang, G., & Cheng, P. (2021). Treatment of fishery wastewater by co-culture of *Thalassiosira pseudonana* with *Isochrysis galbana* and evaluation of their active components. *Algal Research, 60*, 102541. DOI: 10.1016/j.algal.2021.102498

Wang, J., & Chen, C. (2009). Biosorbents for heavy metals removal and their future. *Biotechnology Advances, 27*(2), 195–226. DOI: 10.1016/j.biotechadv.2008.11.002 PMID: 19103274

Wang, J., Chen, R., Fan, L., Cui, L., Zhang, Y., Cheng, J., Wu, X., Zeng, W., Tian, Q., & Shen, L. (2021). Construction of fungi-microalgae symbiotic system and adsorption study of heavy metal ions. *Separation and Purification Technology, 268*, 118689. DOI: 10.1016/j.seppur.2021.118689

Wang, J., Chen, X., Sun, H., Li, X., Lu, Q., Qin, P., Yang, Y., Lai, D., Luo, L., Peng, X., Yang, Y., & Wu, Z. (2025). Accelerated removal and mechanism of tetracycline from water using immobilized bacteria combined with microalgae. *Journal of Environmental Chemical Engineering, 13*(2), 115424. DOI: 10.1016/j.jece.2025.115424

Wang, J., Deng, Y., Chen, W., Song, H., Zan, F., Wu, X., & Ren, Y. (2023). Revealing the role of algae in algae enhanced bacteria consortia for municipal wastewater treatment: Performance, characteristics, and microbial pathways. *Journal of Water Process Engineering, 53*, 103640. DOI: 10.1016/j.jwpe.2023.103640

Wang, J., Liu, W., & Yang, J. (2020). Microalgae-yeast co-culture for wastewater treatment: Challenges and perspectives. *Bioresource Technology, 315*, 123789. DOI: 10.1016/j.biortech.2020.123789

Wang, L., Min, M., Li, Y., Chen, P., Chen, Y., Liu, Y., Wang, Y., & Ruan, R. (2010). Cultivation of green algae Chlorella sp. in different wastewaters from municipal wastewater treatment plant. *Applied Biochemistry and Biotechnology, 162*(4), 1174–1186. DOI: 10.1007/s12010-009-8866-7 PMID: 19937154

Wang, S., Chen, H. Y. H., Tan, Y., Fan, H., & Ruan, H. (2016). Fertilizer regime impacts on abundance and diversity of soil fauna across a poplar plantation chronosequence in coastal eastern China. *Scientific Reports*, *6*(1), 20816. DOI: 10.1038/srep20816 PMID: 26857390

Wang, S., Ji, B., Zhang, M., Ma, Y., Gu, J., & Liu, Y. (2020). Defensive responses of microalgal bacterial granules to tetracycline in municipal wastewater treatment. *Bioresource Technology*, *312*, 123605. DOI: 10.1016/j.biortech.2020.123605 PMID: 32504951

Wang, S., Li, J., Wang, W., Zhou, C., Chi, Y., Wang, J., Li, Y., & Zhang, Q. (2023). An overview of recent advances and future prospects of three-dimensional biofilm electrode reactors (3D-BERs). *Journal of Environmental Management*, *342*, 118192. DOI: 10.1016/j.jenvman.2023.118192 PMID: 37285769

Wang, S., Tan, Y., Fan, H., Ruan, H., & Zheng, A. (2015). Responses of soil microarthropods to inorganic and organic fertilizers in a poplar plantation in a coastal area of eastern China. *Applied Soil Ecology*, *89*, 69–75. DOI: 10.1016/j.apsoil.2015.01.004

Wang, T., Liu, W., Xiong, L., Xu, N., & Ni, J. (2013). Influence of pH, ionic strength and humic acid on competitive adsorption of Pb(II), Cd(II) and Cr(III) onto titanate nanotubes. *Chemical Engineering Journal*, *215–216*, 366–374. DOI: 10.1016/j.cej.2012.11.029

Wang, X., & Hong, Y. (2022). Microalgae biofilm and bacteria symbiosis in nutrient removal and carbon fixation from wastewater: A review. *Current Pollution Reports*, *8*(2), 128–146. DOI: 10.1007/s40726-022-00214-x

Wang, Y., Ho, S.-H., Cheng, C.-L., Guo, W.-Q., Nagarajan, D., Ren, N.-Q., Lee, D.-J., & Chang, J.-S. (2016). Perspectives on the feasibility of using microalgae for industrial wastewater treatment. *Bioresource Technology*, *222*, 485–497. DOI: 10.1016/j.biortech.2016.09.106 PMID: 27765375

Wang, Y., Selvamani, V., Yoo, I., Kim, T. W., & Hong, S. H. (2021). A novel strategy for the microbial removal of heavy metals: Cell-surface display of peptides. *Biotechnology and Bioprocess Engineering; BBE*, *26*(1), 1–9. DOI: 10.1007/s12257-020-0218-z

Wang, Z., Hu, G., & Hong, Y. (2024). Strong Alliance of Microalgae and Bacteria: The State-of-the-Art Review and Future Prospects of Utilizing Microalgae-Bacteria Consortia for Comprehensive Treatment of Swine Wastewater. *Current Pollution Reports*, *10*(4), 744–764. DOI: 10.1007/s40726-024-00325-7

Wasilewska-Dębowska, W., Zienkiewicz, M., & Drozak, A. (2022). How light reactions of photosynthesis in C4 plants are optimized and protected under high light conditions. *International Journal of Molecular Sciences*, *23*(7), 3626. DOI: 10.3390/ijms23073626 PMID: 35408985

Wei, L., Xiaopan, S., & Heydari, F. (2022). Microalgae classification using improved metaheuristic algorithm. *Mathematical Problems in Engineering*, *2022*, 3783977.

Weimer, A., Kohlstedt, M., Volke, D. C., Nikel, P. I., & Wittmann, C. (2020). Industrial biotechnology of Pseudomonas putida: Advances and prospects. *Applied Microbiology and Biotechnology*, *104*(18), 7745–7766. DOI: 10.1007/s00253-020-10811-9 PMID: 32789744

Wen, L., Yang, F., Li, X., Liu, S., Lin, Y., Hu, E., Gao, L., & Li, M. (2023). Composition of dissolved organic matter (DOM) in wastewater treatment plants influent affects the efficiency of carbon and nitrogen removal. *The Science of the Total Environment*, *857*, 159541. DOI: 10.1016/j.scitotenv.2022.159541 PMID: 36265625

Wen, X., Sun, X., & Liu, X. (2025). Synergizing mechanistic and AI models for deeper insights into algal-bacterial systems in sustainable wastewater treatment. *Journal of Water Process Engineering*, *76*, 108169. DOI: 10.1016/j.jwpe.2025.108169

Whitton, R., Ometto, F., Pidou, M., Jarvis, P., Villa, R., & Jefferson, B. (2015). Microalgae for municipal wastewater nutrient remediation: Mechanisms, reactors, and outlook for tertiary treatment. *Environmental Technology Reviews*, *4*(1), 133–148. DOI: 10.1080/21622515.2015.1105308

Winkler, M.-K. H., & Van Loosdrecht, M. C. M. (2022). Intensifying existing urban wastewater. *Science*, *375*(6579), 377–378. DOI: 10.1126/science.abm3900 PMID: 35084964

Win, T. T., Barone, G. D., Secundo, F., & Fu, P. (2018). Algal Biofertilizers and Plant Growth Stimulants for Sustainable Agriculture. *Industrial Biotechnology (New Rochelle, N.Y.)*, *14*(4), 203–211. DOI: 10.1089/ind.2018.0010

Wollmann, F., Dietze, S., Ackermann, J. U., Bley, T., Walther, T., Steingroewer, J., & Krujatz, F. (2019). Microalgae wastewater treatment: Biological and technological approaches. *Engineering in Life Sciences*, *19*(12), 860–871. DOI: 10.1002/elsc.201900071 PMID: 32624978

Wu, B., Deng, J., Niu, H., Liang, J., Arslan, M., Gamal El-Din, M., Wang, Q., Guo, S., & Chen, C. (2021). Establishing and Optimizing a Bacterial Consortia for Effective Biodegradation of Petroleum Contaminants: Advancing Classical Microbiology via Experimental and Mathematical Approach. *Water (Basel), 13*(22), 3311. https://www.mdpi.com/2073-4441/13/22/3311. DOI: 10.3390/w13223311

Wu, H., & Ge, Y. (2019). Excessive application of fertilizer, agricultural non-point source pollution, and farmers' policy choice. *Sustainability (Basel), 11*(4), 1165. DOI: 10.3390/su11041165

Wu, H., & Vaneeckhaute, C. (2021). Nutrient recovery from wastewater: A review on the integrated physicochemical technologies of ammonia stripping, adsorption and struvite precipitation. *Chemical Engineering Journal, 433*, 133664. DOI: 10.1016/j.cej.2021.133664

Wu, X., Nawaz, S., Li, Y., & Zhang, H. (2024). Environmental health hazards of untreated livestock wastewater: Potential risks and future perspectives. *Environmental Science and Pollution Research International, 31*(17), 24745–24767. DOI: 10.1007/s11356-024-32853-6 PMID: 38499926

Wu, Z., Shi, W., Yuan, W., Chen, Z., Xie, Y., Lv, Z., Xu, J., Amadu, A. A., Qiu, S., & Ge, S. (2025). Development and operation of indigenous microalgal-bacterial consortium system treating eutrophic lake water: Consortium identification and system demonstration. *Bioresource Technology, 132462*, 132496. Advance online publication. DOI: 10.1016/j.biortech.2025.132496 PMID: 40204026

Xiao, M., Ma, H., Sun, M., Yin, X., Feng, Q., Song, H., & Gai, H. (2019). Characterization of cometabolic degradation of p-cresol with phenol as growth substrate by *Chlorella vulgaris. Bioresource Technology, 281*, 296–302. DOI: 10.1016/j.biortech.2019.02.079 PMID: 30826515

Xiao, Z., Meng, H., Li, S., Ning, W., Song, Y., Han, J., Chang, J.-S., Wang, Y., & Ho, S.-H. (2024). Insights into the removal of antibiotics from livestock and aquaculture wastewater by algae-bacteria symbiosis systems. *Environmental Research, 257*, 119326. DOI: 10.1016/j.envres.2024.119326 PMID: 38849002

Xie, P., Chen, C., Zhang, C., Su, G., Ren, N., & Ho, S.-H. (2020). Revealing the role of adsorption in ciprofloxacin and sulfadiazine elimination routes in microalgae. *Water Research, 172*, 115475. DOI: 10.1016/j.watres.2020.115475 PMID: 31972413

Xiong, J. Q., Govindwar, S., Kurade, M. B., Paeng, K. J., Roh, H. S., Khan, M. A., & Jeon, B. H. (2019). Toxicity of sulfamethazine and sulfamethoxazole and their removal by a green microalga, *Scenedesmus obliquus. Chemosphere, 218*, 551–558. DOI: 10.1016/j.chemosphere.2018.11.146

Xiong, J.-Q., Kurade, M. B., Abou-Shanab, R. A., Ji, M.-K., Choi, J., Kim, J. O., & Jeon, B.-H. (2016). Biodegradation of carbamazepine using freshwater microalgae *Chlamydomonas mexicana* and *Scenedesmus obliquus* and the determination of its metabolic fate. *Bioresource Technology, 205*, 183–190. DOI: 10.1016/j.biortech.2016.01.038 PMID: 26826958

Xiong, J.-Q., Kurade, M. B., & Jeon, B.-H. (2018). Can microalgae remove pharmaceutical contaminants from water? *Trends in Biotechnology, 36*(1), 30–44. DOI: 10.1016/j.tibtech.2017.09.003 PMID: 28993012

Xiong, W., Wang, S., Jin, Y., Wu, Z., Liu, D., & Su, H. (2023). Insights into nitrogen and phosphorus metabolic mechanisms of algal-bacterial aerobic granular sludge via metagenomics: Performance, microbial community and functional genes. *Bioresource Technology, 369*, 128442. DOI: 10.1016/j.biortech.2022.128442 PMID: 36470490

Yadavalli, R., & Heggers, G. R. V. N. (2013). Two stage treatment of dairy effluent using immobilized Chlorella pyrenoidosa. *Journal of Environmental Health Science & Engineering, 11*(1), 36. DOI: 10.1186/2052-336X-11-36 PMID: 24355316

Yadav, G., Shanmugam, S., Sivaramakrishnan, R., Kumar, D., Mathimani, T., Brindhadevi, K., Pugazhendhi, A., & Rajendran, K. (2020). Mechanism and challenges behind algae as a wastewater treatment choice for bioenergy production and beyond. *Fuel, 285*, 119093. DOI: 10.1016/j.fuel.2020.119093

Yagub, M. T., Sen, T. K., Afroze, S., & Ang, H. M. (2014). Dye and its removal from aqueous solution by adsorption: A review. *Advances in Colloid and Interface Science, 209*, 172–184. DOI: 10.1016/j.cis.2014.04.002 PMID: 24780401

Yahaya, E., Yeo, W. S., Nandong, J., & Ngu, J. C. Y. (2025). CO2 fed microalgae cultivation in photobioreactor: Review on challenges and possible solutions. *Environmental Technology Reviews, 14*(1), 540–564. DOI: 10.1080/21622515.2025.2508945

Yan, C., Qu, Z., Wang, J., Cao, L., & Han, Q. (2021). Microalgal bioremediation of heavy metal pollution in water: Recent advances, challenges, and prospects. *Chemosphere, 286*, 131870. DOI: 10.1016/j.chemosphere.2021.131870

Yang, I.-S., Salama, E.-S., Kim, J.-O., Govindwar, S. P., Kurade, M. B., Lee, M., Roh, H.-S., & Jeon, B.-H. (2016). Cultivation and harvesting of microalgae in photobioreactor for biodiesel production and simultaneous nutrient removal. *Energy Conversion and Management, 117*, 54–62. DOI: 10.1016/j.enconman.2016.03.017

Yang, J., Gou, Y., Fang, F., Guo, J., Ma, H., Wei, X., & Shahmoradi, B. (2018). Impacts of sludge retention time on the performance of an algal-bacterial bioreactor. *Chemical Engineering Journal, 343*, 37–43. DOI: 10.1016/j.cej.2018.02.118

Yang, J., Zhou, Z., Qi, W., Gao, X., Wang, Y., Wang, X., Yi, X., Ma, M., & Wu, S. (2025). Phenotypic plasticity and integration synergistically enhance plant adaptability to flooding and nitrogen stresses. *Plant and Soil*, *513*(1), 1009–1030. DOI: 10.1007/s11104-025-07230-y

Yang, L., Wei, W., Sun, X., Li, X., Wen, X., Mao, B., Hao, R., Qiu, W., & Li, H. (2026). Competitive nitrogen/phosphorus scarcity triggers multi-pathway nitrogen removal and phosphorus-stress response in microalgae-bacterial symbiosis: Poly-metabolic synergy enhancement mechanisms. *Bioresource Technology*, *445*, 134138. DOI: 10.1016/j.biortech.2026.134138 PMID: 41621545

Yang, Y., Fan, X., Zhang, J., Qiao, S., Wang, X., Zhang, X., Miao, L., & Hou, J. (2022). A critical review on the interaction of iron-based nanoparticles with blue-green algae and their metabolites: From mechanisms to applications. *Algal Research*, *64*, 102670. DOI: 10.1016/j.algal.2022.102670

Yang, Y., Ge, S., Pan, Y., Qian, W., Wang, S., Zhang, J., & Zhuang, L.-L. (2023). Screening of microalgae species and evaluation of algal lipid stimulation strategies for biodiesel production. *The Science of the Total Environment*, *857*, 159281. DOI: 10.1016/j.scitotenv.2022.159281 PMID: 36216060

Ye, M., & Li, Y. Y. (2023). Methanogenic treatment of dairy wastewater: A review of current obstacles and new technological perspectives. *The Science of the Total Environment*, *866*, 161447. DOI: 10.1016/j.scitotenv.2023.161447 PMID: 36621500

Yu, Q., Chen, J., Ye, M., Wei, Y., Zhang, C., & Ge, Y. (2024). N-acyl homoserine lactones (AHLs) enhanced removal of cadmium and other pollutants by algae-bacteria consortia. *Journal of Environmental Management*, *366*, 121792. DOI: 10.1016/j.jenvman.2024.121792 PMID: 39002459

Yu, S., Horing, J., Liu, Q., Dahowski, R., Davidson, C., Edmonds, J., Liu, B., Mcjeon, H., McLeod, J., Patel, P., & Clarke, L. (2019). CCUS in China's mitigation strategy: Insights from integrated assessment modeling. *International Journal of Greenhouse Gas Control*, *84*, 204–218. DOI: 10.1016/j.ijggc.2019.03.004

Yu, T., Wang, S., Yang, H., Sun, Y., Chen, Z., Xu, G., & Zhang, C. (2024). Carbon Capture and Resource Utilization by Algal–Bacterial Consortium in Wastewater Treatment: A Mini-Review. *Water (Basel)*, *16*(15), 2197. DOI: 10.3390/w16152197

Yu, Z., Zhang, T., Hao, R., & Zhu, Y. (2019). Sensitivity of *Chlamydomonas reinhardtii* to cadmium stress is associated with phototaxis. *Environmental Science. Processes & Impacts*, *21*(6), 1011–1020. DOI: 10.1039/C9EM00013E PMID: 31120077

Zabochnicka, M., Krzywonos, M., Romanowska-Duda, Z., Szufa, S., Darkalt, A., & Mubashar, M. (2022). Algal Biomass Utilization toward Circular Economy. *Life (Basel, Switzerland)*, *12*(10), 1480. DOI: 10.3390/life12101480 PMID: 36294915

Zdeb, M., Bis, M., & Przywara, A. (2023). Multi-criteria analysis of the influence of lignocellulosic biomass pretreatment techniques on methane production. *Energies*, *16*(1), 468. Advance online publication. DOI: 10.3390/en16010468

Zhang, B., Wei, L., Yuan, G., Zhang, Z., Shi, W., Cui, F., Lens, P. N. L., & Tay, J. H. (2020). Microalgal-bacterial consortia: From interspecies interactions to biotechnological applications. *Renewable & Sustainable Energy Reviews*, *118*, 109563. DOI: 10.1016/j.rser.2019.109563

Zhang, C., Wang, X., Ma, Z., Luan, Z., Wang, Y., Wang, Z., & Wang, L. (2020a). Removal of phenolic substances from wastewater by algae: A review. *Environmental Chemistry Letters*, *18*(2), 377–392. DOI: 10.1007/s10311-019-00953-2

Zhang, H., Wu, T., Sun, L., Han, Z., Wang, Q., & Li, D. (2024). The construction of a microalgal-bacterial biofilm reactor for enhanced swine wastewater treatment. *Algal Research*, *79*, 103494. DOI: 10.1016/j.algal.2024.103494

Zhang, J., Jin, R., Dong, L., Yu, C., Luo, W., Hu, Y., Zhang, Y., Zhang, W., Liu, J., Zhou, W., & Shao, S. (2025). Sticky or thrifty: Divergence in extracellular glycometabolic architecture underpins contrasting fungal bioflocculation strategies. *Water Research*, *285*, 124126. DOI: 10.1016/j.watres.2025.124126 PMID: 40602066

Zhang, J., Zhou, F., Liu, Y., Huang, F., & Zhang, C. (2020). Effect of extracellular polymeric substances on arsenic accumulation in *Chlorella pyrenoidosa. The Science of the Total Environment*, *704*, 135368. DOI: 10.1016/j.scitotenv.2019.135368 PMID: 31831249

Zhang, L., Ali, A., Su, J., Huang, T., & Wang, Z. (2024a). Ammonium nitrogen and phosphorus removal by bacterial-algal symbiotic dynamic sponge bioremediation system in micropolluted water: Operational mechanism and transformation pathways. *The Science of the Total Environment*, *947*, 174636. DOI: 10.1016/j.scitotenv.2024.174636 PMID: 38992368

Zhang, S., Sun, Z., Zheng, T., He, C., & Lin, D. (2024). Nanoplastics increase algal absorption and toxicity of Cd through alterations in cell wall structure and composition. *Water Research*, *254*, 121394. DOI: 10.1016/j.watres.2024.121394 PMID: 38442610

Zhang, S., Xiao, J., Wang, G., & Chen, G. (2020b). Enzymatic hydrolysis of lignin by ligninolytic enzymes and analysis of the hydrolyzed lignin products. *Bioresource Technology*, *304*, 122975. DOI: 10.1016/j.biortech.2020.122975 PMID: 32086036

Zhang, S.-Y., Sun, G.-X., Yin, X.-X., Rensing, C., & Zhu, Y.-G. (2013). Biomethylation and volatilization of arsenic by the marine microalgae *Ostreococcus tauri*. *Chemosphere*, *93*(1), 47–53. DOI: 10.1016/j.chemosphere.2013.04.063 PMID: 23726009

Zhang, T. (2025). Sustainable Wastewater Treatment and the Circular Economy. *Water (Basel)*, *17*(3), 335. DOI: 10.3390/w17030335

Zhang, T., Yang, Y., Ni, J., & Xie, D. (2019). Adoption behavior of cleaner production techniques to control agricultural non-point source pollution: A case study in the Three Gorges Reservoir Area. *Journal of Cleaner Production*, *223*, 897–906. DOI: 10.1016/j.jclepro.2019.03.194

Zhang, X., & Liu, Y. (2021). Circular economy-driven ammonium recovery from municipal wastewater: State of the art, challenges and solutions forward. *Bioresource Technology*, *334*, 125231. DOI: 10.1016/j.biortech.2021.125231 PMID: 33962161

Zhang, X., & Liu, Y. (2021). Reverse osmosis concentrate: An essential link for closing loop of municipal wastewater reclamation towards urban sustainability. *Chemical Engineering Journal*, *421*, 129851. DOI: 10.1016/j.cej.2020.127773

Zhang, X., & Liu, Y. (2022). Resource recovery from municipal wastewater: A critical paradigm shift in the post era of activated sludge. *Bioresource Technology*, *363*, 127932. DOI: 10.1016/j.biortech.2022.127932 PMID: 36096327

Zhang, Y., Caldwell, G. S., Zealand, A. M., & Sallis, P. J. (2019). Anaerobic co-digestion of microalgae Chlorella vulgaris and potato processing waste: Effect of mixing ratio, waste type and substrate to inoculum ratio. *Biochemical Engineering Journal*, *143*, 91–100. DOI: 10.1016/j.bej.2018.12.021

Zhang, Y., Habteselassie, M. Y., Resurreccion, E. P., Mantripragada, V., Peng, S., Bauer, S., & Colosi, L. M. (2014). Evaluating the removal of steroid estrogens by a model alga as a possible sustainability benefit of hypothetical integrated algae cultivation and wastewater treatment systems. *ACS Sustainable Chemistry & Engineering*, *2*(11), 2544–2553. DOI: 10.1021/sc5004538

Zhang, Y., Li, A., Dai, T., & Wang, Y. (2021). Fate of antibiotic resistance genes in an algal bacterial system treating wastewater: Insights from metagenomic analysis. *Environmental Science & Technology*, *55*(14), 9910–9919. DOI: 10.1021/acs.est.1c01471

Zhang, Y., Roostaei, J., Asadi, A., Doosti, Z., Ewing, M., Guan, J., Uddin, T., & van Wagnen, J. (2018). *Stratified multilayer algal-biofilm reclamation technology. USA Science and Engineering Festival*. Wayne State University.

Zhang, Y., Zhang, X., Wang, M., Zhang, R., Li, Y., Zhang, G., Fang, W., & Zhang, P. (2025). Greenhouse gas emissions and carbon sequestration capacity of constructed wetlands with different hydrophytes. *Journal of Water Process Engineering, 76*, 108292. DOI: 10.1016/j.jwpe.2025.108292

Zhang, Y., Zheng, L., Wang, S., Zhao, Y., Xu, X., Han, B., & Hu, T. (2021). Quorum Sensing Bacteria in the Phycosphere of HAB Microalgae and Their Ecological Functions Related to Cross-Kingdom Interactions. *International Journal of Environmental Research and Public Health, 19*(1), 163. Advance online publication. DOI: 10.3390/ijerph19010163 PMID: 35010421

Zhan, Y., Shen, X., Chen, M., Yang, K., & Xie, H. (2022). Bioleaching of tellurium from mine tailings by indigenous *Acidithiobacillus ferrooxidans. Letters in Applied Microbiology, 75*(5), 1076–1083. DOI: 10.1111/lam.13569 PMID: 34586632

Zhao, D., Cheah, W. Y., Lai, S. H., Ng, E.-P., Khoo, K. S., Show, P. L., & Ling, T. C. (2023). Symbiosis of microalgae and bacteria consortium for heavy metal remediation in wastewater. *Journal of Environmental Chemical Engineering, 11*(3), 109943. DOI: 10.1016/j.jece.2023.109943

Zhao, W., Tian, K., Zhang, L., Tang, Y., Chen, R., Zheng, X., & Zhao, M. (2025). Harnessing an Algae–Bacteria Symbiosis System: Innovative Strategies for Enhancing Complex Wastewater Matrices Treatment. *Sustainability (Basel), 17*(15), 7104. https://www.mdpi.com/2071-1050/17/15/7104. DOI: 10.3390/su17157104

Zhibei, C., Wenlong, Y., Mingzhu, D., & Yingjin, Y. (2022). Construction of microbial consortia for microbial degradation of complex compounds. *Frontiers in Bioengineering and Biotechnology, 10*, 1051233. DOI: 10.3389/fbioe.2022.1051233 PMID: 36561050

Zhizhan, L. (2020). *Application of microalgae yeast co-culture system in shrimp culture wastewater treatment*. Guangzhou University.

Zhou, J., Lyu, Y., Richlen, M. L., Anderson, D. M., & Cai, Z. (2020). Quorum sensing mediated microbial interactions in algal bacterial consortia: A review. *Bioresource Technology, 312*, 123589. DOI: 10.1016/j.biortech.2020.123589

Zhou, R., Xiao, B., Zhao, L., Tang, Y., Yang, W., Liu, H., Zhang, X., Yuan, C., Wang, X., Long, X., He, Q., & Li, W. (2025). Silicon supplementation via carriers to enhance diatom competition with cyanobacterial blooms and remediate eutrophic waters. *Water Research*, *284*, 123922. DOI: 10.1016/j.watres.2025.123922 PMID: 40449328

Zhou, Y., Zhou, Y., Yao, L., Luo, X., Kong, Q., Leung, K. S.-Y., & Yang, X. (2025). Insights into DOM-Enhanced Periodate Degradation of Emerging Contaminants: The Organic Fenton-like Reactions. *Environmental Science & Technology*, *59*(40), 21682–21690. DOI: 10.1021/acs.est.5c09780 PMID: 41036757

Zoppi, G., Pipitone, G., Pirone, R., & Bensaid, S. (2022). Aqueous phase reforming process for the valorization of wastewater streams: Application to different industrial scenarios. *Catalysis Today*, *387*, 224–236. DOI: 10.1016/j.cattod.2021.06.002

Zou, Z., Li, M., He, S., Zhou, Z., Tian, J., Cai, B., Lu, Y., Dai, Z., Zhu, X., & Han, Z. (2026). Multifunctionalized Immobilized Microalgae Technologies for Application in Advanced Environmental Remediation Systems. *Transactions of Tianjin University*, 1-28.

About the Contributors

Imran Ahmad is a dedicated researcher, academic, and scientific consultant deeply committed to environmental sustainability and algae biotechnology. With an MPhil and PhD from Universiti Teknologi Malaysia, his expertise lies in biological wastewater treatment, microalgal bioremediation, and resource recovery. His work bridges research and application, focusing on innovative solutions that address environmental challenges while contributing to the principles of a circular bioeconomy. He has been actively involved in designing and optimizing photobioreactors, anaerobic reactors, and microalgae-based treatment systems that offer sustainable alternatives for wastewater management and carbon sequestration. He works as a Scientific Consultant and Bioprocess Engineer with ProAlgaeTech, a startup focused on microalgal carbon sequestration and wastewater bioremediation. His contributions extend beyond academic research, as he engages in industrial applications that translate scientific advancements into real-world impact. He has been integral to international projects, including the Indo-European Horizon 2020 PAVITR project and Liquid Trees, where his insights have helped shape strategies for efficient wastewater treatment and bioenergy production. He also obtained a utility innovation for a modified grease trap design for restaurant wastewater, further demonstrating his ability to develop practical engineering solutions. His research has yielded numerous high-impact publications, reflecting his dedication to advancing knowledge in environmental engineering. His work has been recognized at both national and international levels. He has received the My Membrane Award from the Malaysia Membrane Society, UTM, and has been acknowledged by Japan's Algae Industry Incubation Consortium. He was also nominated as the Secretary of the International Conference on Environmental Science and Green Technology in 2022. Alongside his research contributions, he has spent more than eight years in academia, mentoring students, designing laboratory infrastructure, and fostering interdisciplinary collaborations. His experience in teaching and research has allowed him to contribute to shaping the future of environmental engineering education.

Beyond his professional achievements, he remains committed to developing sustainable technologies that bring about meaningful change. His work is driven by the conviction that scientific research should yield practical, lasting solutions that benefit both the environment and society. Whether through academia, consultancy, or industry partnerships, he continues exploring ways to integrate innovation with sustainability, ensuring that his contributions have a lasting impact on green technology.

Faizal Bux, Director of the Institute for Water and Wastewater Technology (IWWT,) DUT has ~28 years of experience in the field water and wastewater treatment technology. His focus research areas include wastewater technology, algal biotechnology and bioenergy production. Prof Bux has also assumed the position of NRF SARChI (South African Research Chair Initiative) Chair from 2018. As a NRF B rated researcher and Academy of Science of South Africa (ASSAf) member, he has supervised 64 Masters and Doctoral students, hosted 23 PDF's and received numerous institutional awards. The Water Research Commission has been awarded Prof Bux the 'Knowledge Tree Award for Transformation and Redress' for his impact in growing PDI involvement in 2017. More recently, Prof Bux received the 2018 National Research Foundation (NRF) award, "Champion of Research Capacity Development and Transformation at South African Higher Education Institutions". He has acted/current editor for CLEAN Soil Air Water (John Wiley & sons, Germany), Environmental Science and Health Part A (Taylor Francis, USA), Biofuels Research Journal and Water Science and Technology and as a reviewer for >50 ISI Journals. He has contributed to 148 journal articles, 18 book chapters and >150 conference presentations and led 11 WRC projects. His citations are in excess of 5000 with an H Index of 33. Prof Bux currently holds the positions of the National coordinator and Chair of International Thematic Group (ITG): Water Resource and Pollution Research of the BRICS Networks University. In 2018 Prof Bux was nominated as Chair of the ITG BRICS-NU for 2 years with the option of a second term. He served 2 terms as a Member of IWA Management Committee specialist group Microbial Ecology and Water Engineering (MEWE) and on the Executive committee of IWA-SA as Treasurer. He was been elected as a member of the Academy of Science of South Africa in 2016. Prof Bux was admitted as a Fellow of the Water Institute of Southern Africa, Royal Society of Chemistry and more recently International Water Association. He has served on the scientific program committee/Advisory board/Chair/Invited speaker of various national and International conferences and holds memberships to many professional bodies. He has an extensive network of university collaborators globally and served as scientific advisor to NGOs both locally & internationally. He has been involved in bilateral projects with India, Brazil, Egypt, China, Algeria with continuing collaboration.

He currently has bilateral projects with Japan, Oman, Poland (subject to approval) and global projects involving Oklahoma University (USA) and Aalborg University (Denmark). He has played a key role in assisting the University to increase its research capacity and contribute to the Human Resource Capital Development of the country.

Nilamjyoti Kalita is an Assistant Professor & Head of the Department of Botany at Murkong Selek College, Assam, India. He earned his Ph.D. from Gauhati University, focusing on heavy metal remediation using cyanobacterial strains from oil-contaminated sites of Upper Assam. A university topper in his M.Sc. in Botany, he has strong expertise in molecular biology, algal taxonomy, and stress physiology. Dr. Kalita has published extensively in reputable Q1 international journals such as Journal of Hazardous Materials Advances, Food Biosciences, Journal of Polymers and the Environment, Heliyon, Plant Physiology and Biochemistry, Microbiological Research, etc., and has contributed to several Springer Nature-edited volumes. His research emphasizes the application of cyanobacteria and microalgae in bioremediation and sustainable technologies. He is a recipient of the Dr. Deba Kumar Bordoloi Memorial Young Scientist Award (2026) from the Botanical Society of Assam.

***

Md. Shoeab Akhter is a passionate researcher and practicum student with a B.Sc. in the Agriculture Department of the College of Agricultural Sciences at IUBAT—International University of Business Agriculture and Technology, Dhaka 1230, Bangladesh. He is doing an internship at the Bangladesh Atomic Energy Commission. His research title is 'Green Synthesis of Nanoparticles for Agricultural Use.' Loved by nature and with a passion for the environment, his pursuits include areas of interest like agricultural sciences, food security and food science, environmental science, waste management, green energy, AI, IoT-based farming, and agritourism, among others. Along with his objective of creating an impact through writing and research, he combines new, innovative, suitable technologies and nature conservation with real-world issues. He has contributed book chapters (Emerald, CABI, Bentham Science, Springer, Cambridge Scholars) and has participated in various major projects on research. With the hope of bringing about a change in the world, he aspires to contribute to global knowledge and create a lasting impact on sustainability in society.

Eshraq Aljamal holds a PhD in Chemistry and is an academic researcher from Jordan with expertise in analytical chemistry, adsorption-based water treatment, green chemistry, and environmental sustainability. Her research focuses on the analysis and removal of emerging pollutants, biosorption and adsorption mechanisms, microalgal adsorption systems, and sustainable analytical approaches for environmental applications. She has teaching and research experience in chemistry, with academic interests spanning green analytical methods, adsorbent characterization, wastewater treatment, and sustainability-oriented environmental research.

Md. Ratul Bhuyan is currently pursuing a Bachelor of Science degree in Agricultural Science at Daffodil International University, Bangladesh. He completed his Higher Secondary Certificate and Secondary School Certificate examinations in the Science stream under the Dhaka Board with excellent academic performance. His research interests are focused on soil science, sustainable agriculture, crop stress management, biofertilizers, climate-resilient farming systems, and eco-friendly agricultural technologies. He has been actively engaged in leadership and organizational activities and previously served as a Cadet Corporal in the Bangladesh National Cadet Corps (BNCC), participating in structured training and field-based programs. He possesses strong communication, teamwork, and technical skills, with working proficiency in Microsoft Office applications. His academic and extracurricular involvement reflects a dedicated commitment to agricultural research and professional development.

Alexander Machado Cardoso, Department of Biology, Rio de Janeiro State University (UERJ), Rio de Janeiro, RJ, Brazil.

Upoma Chowdhury is a dedicated and passionate student pursuing her B.Sc. in Food Science and Technology at Chattogram Veterinary and Animal Sciences University (CVASU), Bangladesh. Inspired by a profound interest in food, nutrition, and health, she is committed to contributing to advancements in the food industry and ensuring food security for future generations. Her academic pursuits revolve around innovative research in food processing, food safety, and biotechnology, with a special focus on sustainable solutions to combat malnutrition. Along with her passion for scientific research, she actively engages in academic projects that explore the intersection of food science with public health and environmental sustainability. Driven by a vision to make meaningful contributions to the global food system, Upoma aspires to combine scientific innovation with practical solutions for addressing food-related challenges. With a strong commitment to knowledge sharing, she aims to contribute to the academic community through research publications and active participation in seminars and workshops. With the hope of creating a positive impact on both local and global scales, Upoma continues to pursue her academic journey with dedication and curiosity, aspiring to play a significant role in shaping a healthier, more sustainable future for society.

Atin Kumar is an Assistant Professor and Soil Scientist at Uttaranchal University, Dehradun, with expertise in soil fertility, soil remediation, water quality, and heavy metal contamination. He is the author and editor of 3 academic books, a patent holder, and an active reviewer for several recognized international journals. With over 50 publications and multiple book chapters, he has been honored with the Young Scientist Awards. His research integrates innovative approaches such as nanotechnology to improve soil-water-environment interactions, with a focus on sustainable agriculture and environmental resilience.

Meshabul Ferdous Mithun is a passionate researcher and undergraduate student in the Department of Agricultural Science at Daffodil International University, Bangladesh. He is deeply interested in plant health, post-harvest management, and eco-friendly approaches to crop protection that promote sustainable food systems. A nature lover at heart, he is committed to integrating biodiversity conservation with agricultural innovation, ensuring that farming practices benefit both people and the environment. Alongside his academic coursework, he is developing experience in laboratory research, fieldwork, and scholarly writing, which help her connect theoretical knowledge with practical agricultural challenges. In the future, he aspires to contribute to sustainable farming technologies that empower rural communities and address global food security challenges.

Ashish Kumar Nayak is an Assistant Professor at PVP Siddhartha Institute of Technology, Vijayawada, Andhra Pradesh. He earned his Ph.D. from IIT Kharagpur, M.Tech. from IIT Guwahati, and B.Tech. from NIT Rourkela. With 36 research articles and 7 book chapters published by reputed publishers, his research focuses on water and wastewater treatment, adsorption, sustainable environmental technologies, nanomaterial-based pollutant remediation, process optimization using soft computing techniques, solid waste management, waste recycling, waste minimization, and composting technology. He serves as an editor for nearly 20 international journals published by Springer, Taylor & Francis, Wiley, Elsevier, Bentham Science, Nature, BMC, and MDPI and actively reviews for various SCI/SCIE/Scopus-indexed journals.

Mythili P. is an accomplished in B. Sc Microbiology at Sacred Heart College (Autonomous), Tirupatur, with a strong academic foundation.

Prakathi P. is an accomplished in M. Sc Microbiology at Sacred Heart College (Autonomous), Tirupatur, with a strong academic foundation including a M.Sc.

Microbiology graduate from Sacred Heart College (Autonomous), Tirupattur, India. Her research interests include pharmaceutical microbiology, nanotechnology, and environmental microbiology. She has worked on the green synthesis of nanoparticles for wound-healing applications and has participated in laboratory training and research activities. Her innovative research spans diverse areas such as feather degradation and Probiotic chocolate demonstrating her multidisciplinary expertise. With an impressive publication record including 5 book chapters with Springer and wiley. A dynamic presenter, Prakathi has won multiple awards including consolation prize in the Sacred Heart Young Innovator Award and Best Oral Presentation at 2024 for her work on Feather degradation. Her 5 completed online certifications from prestigious institutions like AGMOOCS and Future eLearning Center further complement her research skills. Fluent in both Tamil and English, Prakathi actively disseminates knowledge through various digital platforms including Google Scholar, Blogger and YouTube, while continuing to explore novel microbiological solutions to pressing environmental and health challenges.

Saranraj P. holds a Ph.D. in Microbiology from Annamalai University, Chidambaram, India, and is NET (ASRB–ICAR) qualified in Microbiology. He is currently serving as Assistant Professor and Head, Department of Microbiology, at Sacred Heart College (Autonomous), Tirupattur, Tamil Nadu, India. In addition, he is the Proprietor and Managing Director of JPS Scientific Publications, contributing actively to academic publishing and research dissemination. Dr. Saranraj has over 15 years of research experience and 10 years of teaching experience in the fields of Environmental Microbiology, Agricultural Microbiology, Industrial Microbiology, Bioremediation, Microbial Biotechnology, and Sustainable Ecosystem Management. His research work has focused on Microbial degradation of pollutants, Effluent treatment, Soil health improvement, Plant–microbe interactions, and Applied environmental sustainability. He has an extensive publication record, with more than 240 international research articles, over 90 books (international and national), and 70+ book chapters published with reputed academic publishers.

Mohammed Ataur Rahman is the Chairman, Department of Environmental Science, Director of IUBAT Institute of SDG Studies (IISS); Chairperson, Centre for Global Environmental Culture (CGEC) and Program on Education for Sustainability at IUBAT University. Professor Rahman is a Plantation Crop, Forestry and Agriculture Specialist and also Coordinator of WWOOF Bangladesh. He teaches Ecology and Crop Climatology, Environmental Science, Sustainable and Urban Agriculture and Plant Physiology. Professor Rahman worked as Contributing Author with Late Professor M Alimullah Miyan, Former Vice Chancellor and Lead Author IPCC Working Group II and Asian Disaster Preparedness Centre (ADPC) and Disaster Management. Moreover, he is the Executive Director and Coordinator of RCE Greater Dhaka acknowledged by UNU-IAS. He obtained specialized training on Sustainability from the Association of Advancement of Sustainability in Higher Education (aashe), USA and participated in the UNESCO-UNFCCC webinar series on Climate Change Education in 2023. Professor Rahman is also working as Chief Consultant of Forestry, Environment, Plantation Crops and Permaculture Consultancy and Research (FEPPCAR), Dhaka, Bangladesh. Previously, He worked as a Senior Manager, Forestry and Plantation Crops, with 13 and a half years of experience in the Finlays' Plantation Industry of Bangladesh working with a number of the World's leading brands including Finlay Tea, Timber, Bamboo, Rubber, Canes and Black pepper, Betel Leaf etc. He possess specialised knowledge of environmental and habitat of Tropical Forest Plants, especially, Tea and Rubber and Non-Timber Forest crops. He is also experienced with plantation establishment and management of plantation crops viz. Tea, Timber, Coconut, Bamboo, Fruits and Spices, etc. and forest resource exploitation and with specialisation in wood preservation techniques. He is actively associated with multidisciplinary activities like ecoagriculture, organic farming, permaculture, multiple and multi-strata cropping culture for sustainable rural livelihood. He worked for Sustainable Agriculture and Development with the World Bank's International Assessment of Agricultural Science and Technology (IAAST). Before joining Finlays, he served Bangladesh Forest Industries Development Corporation (BFIDC) and Rural Electrification Board (REB) of Bangladesh for twelve years with significant contribution in development of forest industries and timber products. He is an Author of 12 books, booklets and book chapters. Scientific Articles are being published regularly in the English and Bangla News Papers like The Daily Star, Bangladesh Observer, News Today, Financial Express and Inquilab, Destiny etc.

Sakik Rahman is a student in the Department of Agricultural Science at Daffodil International University, Bangladesh. A dedicated plant lover, his academic and research interests revolve around plant health, crop improvement, and sustainable farming practices that balance productivity with environmental stewardship. He is particularly curious about the interactions between plants and their surrounding ecosystems, and how innovative agricultural approaches can enhance biodiversity while ensuring food security. Alongside his coursework, he is developing skills in field observation, laboratory analysis, and academic writing, allowing him to connect scientific learning with real-world agricultural challenges. In the long run, he aspires to contribute to research and technologies that support greener farming systems and empower smallholder farmers in South Asia.

Sakibul Islam Ratul is a researcher in the Department of Agricultural Science of Daffodil International University, Bangladesh. He is highly academic and committed towards research-driven innovation. His scholarly and research work is on sustainable agrifood solutions, especially in post-harvest technology, plant pathology, soil health, and climate-resilient agrifood systems. He is also pursuing more advanced research on solar-biomass hybrid drying systems, soil microplastic pollution and its effects on water retention, carbon sequestration measures, plant extract to combat fungal pathogens, UV light treatment of farm soils, and microbial interactions of soil ecologies. In addition to coursework, he has acquired hands-on experience in fieldwork, laboratory work, and academic writing, which has allowed him to relate theoretical learning to practical agricultural problems. His interests lie in supporting a science-based approach to solving the problems of smallholder and off-grid farm communities in South Asia. He would like to advance to a new stage of education and to participate in the research work of the whole world, and in the long run, he hopes to develop sustainable technolog

Tasnim Taranni Shuchi is a student of the Department of Agricultural Science at Daffodil International University, Dhaka, Bangladesh. Her academic interests focus on sustainable agriculture, environmental management, and resource-efficient technologies. She is a nature enthusiast and actively involved in scouting, reflecting her commitment to environmental stewardship, teamwork, and community service.

Index

A

B

C

R

S

W

Z